Cambridge Astrophysics Series

Accretion-driven stellar X-ray sources

Accretion-driven stellar X-ray sources

Edited by
WALTER H.G. LEWIN
Professor of Physics, Massachusetts Institute of Technology

EDWARD P.J. VAN DEN HEUVEL
Professor of Astronomy, University of Amsterdam

CAMBRIDGE UNIVERSITY PRESS
Cambridge
London   New York   New Rochelle
Melbourne   Sydney

Published by the Press Syndicate of the University of Cambridge
The Pitt Building, Trumpington Street, Cambridge CB2 1RP
32 East 57th Street, New York, NY 10022, USA
296 Beaconsfield Parade, Middle Part, Melbourne 3206, Australia

© Cambridge University Press 1983

First published 1983

Printed in Great Britain at the University Press, Cambridge

Library of Congress catalogue card number: 83-1858

British Library cataloguing in publication data

Accretion driven stellar x-ray sources. – (Cambridge astrophysics series)

1. X-ray astronomy
I. Lewin, W. H. G. II. Heuvel, E. P. J. van den
III. Series
523.01'97'222   QB472
ISBN 0 521 24521 4

DJ

CONTENTS

Foreword   ix
Bruno B. Rossi

Introduction   xv
Walter H. G. Lewin and Edward P. J. van den Heuvel

1   X-ray pulsars in massive binary systems
Saul A. Rappaport and Paul C. Joss

1.1 Introduction	1
1.2 Pulse profiles and spectra	2
1.3 Studies of pulse period variations	8
1.4 Orbital determinations and the evaluation of binary system parameters	13
1.5 Summary	33

2   X-ray bursters and the X-ray sources of the galactic bulge
Walter H. G. Lewin and Paul C. Joss

2.1 Introduction and brief summary	41
2.2 Lack of pulsations and eclipses	45
2.3 Globular cluster X-ray sources	46
2.4 X-ray burst sources	49
2.5 Fast transients (bursts?)	84
2.6 Theory	88
2.7 Concluding remarks	106

3   X-ray emission from normal galaxies
Knox S. Long and Leon P. Van Speybroeck

3.1 Introduction	117
3.2 Detailed observations of sources in nearby galaxies	119
3.3 The survey of normal galaxies	136
3.4 Very bright non-nuclear sources	141
3.5 Summary	142

4 Accreting degenerate dwarfs in close binary systems
France A. Córdova and Keith O. Mason

4.1 Introduction	147
4.2 Classification of cataclysmic binaries	148
4.3 Dynamical properties	150
4.4 X-ray emission	154
4.5 The ultraviolet and optical spectrum	161
4.6 Magnetic fields in cataclysmic variables	169
4.7 Short timescale variability	171
4.8 The outburst	176
4.9 Conclusions	179

5 Optical observations of compact galactic X-ray sources
Jan van Paradijs

5.1 Introduction	189
5.2 Class II sources: low-mass X-ray binaries	192
5.3 Class I sources: massive X-ray binaries	231

6 Accreting magnetic neutron stars
John G. Kirk and Joachim E. Trümper

6.1 Introduction	261
6.2 Pulsating X-ray sources	262
6.3 Gamma-ray bursts	278

7 SS 433
Bruce Margon

7.1 Spectroscopy and the basic kinematic model	287
7.2 Photometry and polarimetry	293
7.3 X-ray and radio observations	294
7.4 Outstanding problems	297

8 Formation and evolution of X-ray binaries
Edward P. J. van den Heuvel

8.1 Introduction and summary	303
8.2 Observational data on compact objects in binary systems	303
8.3 Evolution of close binaries and the formation of compact objects	308
8.4 (Quasi-) conservative evolution: formation of the massive X-ray binaries	316
8.5 The formation of young low-mass X-ray binaries and young runaway pulsars	332
8.6 Possible formation mechanisms of Type II sources	333

9 Evolution and mass transfer in X-ray binaries
Gert Jan Savonije

9.1 Introduction	343

9.2	The main principles of mass transfer in binary systems	344
9.3	Application to X-ray binaries	351

10 Accretion disks in close-binary systems
Jacobus A. Petterson

10.1	The occurrence of accretion disks	367
10.2	The structure of a disk model	373
10.3	Basic stationary axisymmetric models	380
10.4	Additional features of disk models	386

11 Spinup and spindown of accreting neutron stars
Huib F. Henrichs

11.1	Introduction and overview	393
11.2	Observed characteristics of X-ray pulsars	394
11.3	Accretion torques on a neutron star in a binary system	396
11.4	The rotational history of a neutron star in a binary system	419
	Author index	431
	Object index	438
	Subject index	443

FOREWORD

Bruno B. Rossi
Institute Professor, Emeritus, Massachusetts Institute of Technology

Whenever technical advances opened a new window in the electromagnetic spectrum, astronomers, looking through this window, saw a surprisingly different picture of the heavens. New celestial bodies came into sight and new phenomena were discovered.

As long as only the narrow optical window was available, the most striking features of the sky were the myriads of bright stars, sparsely distributed over vast regions of seemingly empty space. Eventually, thermonuclear reactions were recognized as the process powering their radiation.

Then came the radio telescope. Now the stars disappeared, while the stellar space itself came to life. Working in radio bands around 1-m wavelength, the telescope saw mainly streams of relativistic electrons circulating in magnetic fields and radiating via the synchrotron process. Sharply tuned to the 21-cm wavelength, the telescope saw clouds of neutral hydrogen, the radiation process being here a flipping of the electron spin from a parallel direction to one anti-parallel to that of the proton spin.

The most recent development was space technology, which made it possible to observe the sky in spectral regions for which the Earth's atmosphere is an opaque screen. So far, of the various branches of space astronomy, the one that has produced the most abundant crop of unexpected results is X-ray astronomy. The X-ray sky was found to be much more active in X-rays than anyone had anticipated. Surprisingly strong galactic and extragalactic X-ray sources were discovered. Once more, astronomical observations in a new spectral band led to the discovery of a new celestial phenomenon. It was found that most bright galactic X-ray sources are powered by accretion of matter upon collapsed objects, most frequently neutron stars. This is indeed an exceptionally efficient mechanism for the production of X-rays. The gravitational potential energy set free by matter reaching the vicinity of a neutron star is truly enormous, about one tenth of its rest energy. Most importantly, this energy is released in a very

Foreword x

small volume. Therefore the matter contained in this volume is heated to an exceedingly high temperature so that its thermal radiation falls in the spectral range of X-rays. Contrast this situation with that typical of ordinary stars. Here very large amounts of energy are released by nuclear reactions; but the energy is spread over very large volumes and consequently generates much lower temperatures, so that the star radiates in the optical rather than in the X-ray spectral band.

Extrasolar X-ray astronomy was born in 1962 with the discovery, during a rocket flight, of an exceedingly strong X-ray source, later to be named Sco X-1. When an estimate of its distance became available, it turned out that its intrinsic X-ray luminosity was about ten thousand times the total luminosity of the Sun at all wavelengths.

In the following years, several additional celestial X-ray sources were observed. However, only in one case — that of the atypical source in the Crab Nebula — could the nature of the physical process responsible for the X-ray emission be discovered (it was the same synchrotron process responsible for the optical and radio emission). The nature of all other sources remained a mystery.

The first mention of accretion as a possible mechanism for X-ray emission came in the late 1960s. It was a purely hypothetical model, describing an X-ray star as a close-binary system, formed by a collapsed object and an 'ordinary' star, whose function was that of providing a sufficiently abundant supply of accreting material. At that time, however, no evidence of a binary nature had been detected in any of the known X-ray stars. Therefore the accretion model was shelved along with other proposed models. Observational evidence definitely supporting the accretion model came in the early 1970s. It was obtained from a detailed study of two X-ray sources, Cyg X-1 and Cen X-3.

In the case of Cyg X-1, the first step was the identification of the X-ray source with a weak radio source which, in turn, was identified with a supergiant. The second step was the discovery that the spectral lines of this star displayed the periodic Doppler shifts characteristic of a spectroscopic binary. It was concluded that Cyg X-1 was a binary system, with the supergiant as one of the components. Its X-ray activity was regarded as evidence that the second component was a collapsed object, accreting matter from the supergiant. From the optical data it was found that the orbital period was 5.6 days. These data also made it possible to estimate the mass of the collapsed component, which appeared to be greater than the upper limit for the mass of a white dwarf or a neutron star. It was thus concluded that, in all likelihood, the collapsed object was a black hole.

In the case of Cen X-3 the first step was the discovery that its X-ray emission showed regular pulsations with a period of about 4.8 seconds, one of the early

Foreword

important results obtained by means of the first X-ray satellite, Uhuru. It was then found that the frequency of these pulsations underwent small periodical changes, clearly suggesting a Doppler effect due to an orbital motion of the X-ray source in a binary system. The observation of periodic eclipses provided further evidence for the binary nature of Cen X-3. It was then practically unavoidable to identify the X-ray source with a collapsed object accreting matter from a yet unseen non-degenerate companion. Observation of the eclipses, in agreement with observations of the Doppler curve showed that the period of the orbital motion was 2.09 days. It was also clear that the collapsed object was a neutron star, for it was believed that black holes cannot produce pulsations, and white dwarfs could be ruled out on several grounds, one of them being that, as an energy source, accretion on a white dwarf is about one thousand times less efficient than accretion on a neutron star. The binary companion of the neutron star was detected some time later, and found to be a supergiant as was the case for Cyg X-1.

In the following years, a number of pulsing, often eclipsing, X-ray sources were discovered. Like Cen X-3, each of them could be shown to consist of a neutron star and a massive non-degenerate star in orbit around one another. The neutron star was believed to be strongly magnetized. The magnetic field would cause the accreting hot X-ray emitting material to become distributed unevenly over the surface of the neutron star. Rotation of this object would then produce the observed pulsations.

To this day (January 1983), Cyg X-1 is the only celestial object which has provided strong evidence for the existence of stellar-mass black holes. It is rather strange that such a rare object as an X-ray binary containing perhaps a black hole instead of a neutron star should have been the first to be identified (the discovery of the nature of Cyg X-1 was announced a few weeks before that of Cen X-3).

While the detailed study of the X-ray binaries with heavy optical partners – often called high-mass binaries – was proceeding, the existence of a different class of non-pulsing X-ray sources became apparent. These objects, in general, did not eclipse and did not pulse. Also, in none of them had a massive bright star been detected. In fact, until the middle 1970s, only one of them (Sco X-1) had been convincingly identified optically; the visible object was found to be much less luminous than the optical counterparts of the high-mass binaries.

Then, after the launching of SAS-3 (1975) and of other satellites capable of accurate positional determinations, the optical counterparts of many sources belonging to this class were discovered. They were all remarkably similar, very faint bluish objects, with spectra unlike those of any other known star.

Although the observational evidence for the structure of the X-ray sources of this group is not as compelling as the evidence for the structure of the high-mass binaries, it appears practically certain that they, too, are binary systems. They are supposed to be formed by a collapsed object, most likely a neutron star, and a non-degenerate low-mass dwarf. Therefore they are often called low-mass binaries; sometimes they are also referred to as bulge sources because their density is greatest in the galactic bulge.

It soon became clear that the faint luminescence did not originate from the non-degenerate partner, but was a secondary effect, due to heating of some sort of cloud or disk of the accreting material by the X-ray flux. The much fainter partner could be seen only on the rare occasions when the X-ray flux, and therefore the secondary optical luminescence, became extinct. This happened in three cases, and the optical objects then seen were found to have spectra characteristic of K-dwarfs.

Besides the strong binary X-ray sources powered by accretion upon neutron stars (and black holes?) there exist sources of a similar structure in which, however, the accreting object is a white dwarf. These sources are very weak, and their radiation lies mostly in the range of soft X-rays. Very few of them were known before the launching, in 1978, of the Einstein Observatory, whose grazing incidence telescope was much more sensitive than the X-ray detectors flown on previous satellites.

It is hardly necessary to point out that not all celestial X-ray sources are powered by accretion. Several strong galactic X-ray sources have been identified with supernova remnants (I have already mentioned the Crab Nebula). In these, as far as is known, the physical processes responsible for the X-ray emission do not include accretion. Accretion is also not the agent responsible for the heating of the solar and stellar coronae, which are weak sources of soft X rays.

A most important result of X-ray astronomy has been the discovery that many extragalactic objects generate strong fluxes of X-rays; but, as yet, little is known about the physical processes by which they are produced.

Finally, there exists a diffuse X-ray radiation, known to be partly of galactic, partly of extragalactic origin, which is produced presumably by processes other than accretion.

Despite the discovery of many kinds of celestial X-ray sources different from accreting binaries, these objects still remain at the center of the interest of X-ray astronomers. After the basic question concerning their structure was answered, many challenging problems remained. One of them has been how best to use such observational data as the orbital period, the Doppler effects of the X-ray pulsations and of the optical spectral lines, the duration of the eclipses, in order to compute the parameters of the binary system of which the mass of the neutron star is physically the most significant. Another important but very difficult problem is a detailed understanding of the complex accretion process.

Foreword

Any realistic theoretical prediction about this phenomenon must take into account a number of factors, including the structure of the gravitational field, the presence or absence of 'wind' from the non-degenerate partner, the magnetic field, if any, around the accreting collapsed object, the distance between the two components of the binary system, etc. Other problems concern the interpretation of certain peculiar phenomena, such as the transient behavior of several high-mass and low-mass binaries, and the very short X-ray 'bursts' which are produced only by low-mass binaries. Finally, there is the problem of the apparently different origin and evolution of high-mass and low-mass binaries.

This book contains an up-to-date and detailed presentation of the observational and theoretical results concerning celestial X-ray sources that are powered by accretion. Its eleven chapters were written by distinguished scientists who, because of their personal contributions, are particularly well qualified to discuss this matter. I wish to add that the book appears at a very appropriate time, when many isolated, scattered, seemingly unrelated results gathered during the last decade have coalesced into a coherent picture. In its broad traits this picture is not likely to change in the future, although much still needs to be done to fill in important details.

The book is a most useful source of information for anyone who wishes to be brought up to date on an important branch of X-ray astronomy. It will also provide a sound basis for specialists who wish to do further work on accreting stellar X-ray sources.

INTRODUCTION

The idea for this book was born in 1981 when Saul A. Rappaport and one of us (WHGL) spent a large part of their sabbatical year at the Astronomical Institute in Amsterdam. We organized a series of ten colloquia on X-ray binary systems. Rather than publishing the proceedings of these colloquia, we decided on a book with complete up to date review articles on the entire family of accretion driven stellar X-ray sources including the cataclysmic variables.

We believe that there is a need for this book and that the timing is just right. It is impressive how much new astrophysics was uncovered in only one decade. Starting with the revolutionary discoveries of the Uhuru satellite and culminating in the recent results of the Einstein Observatory, the seventies will probably always be remembered as the golden decade of X-ray astronomy. We consider ourselves fortunate that we had the opportunity to work in this rich field.

The accretion driven stellar X-ray sources can be crudely divided into three groups: (1) the young massive X-ray binaries, (2) the old low-mass X-ray binaries (including the burst sources and globular cluster X-ray sources), and (3) the cataclysmic variables. In the first two groups the X-ray emitting compact object is almost always a neutron star (in a few cases perhaps a black hole), whereas in the last group the compact X-ray emitter is a white dwarf. This book deals with these three groups separately. However, where possible (e.g., in the case of the behavior of accretion disks) they are combined. Much attention is paid not only to the X-ray behavior but also to the optical data which have crucially contributed to our present understanding.

We expect that this book will not 'age' fast due to the unfortunate near absence in the eighties of fresh data from X-ray observatories with sufficient sensitivity and accuracy to substantially alter our present views. This book is written for astronomers, physicists, and for students in astronomy and astrophysics; it can certainly be used for a specialized graduate course. The first seven chapters of the book cover the observational data, their interpretation

Introduction

and some theory. The last four chapters are largely theoretical; they deal with the evolution of the binary systems, with accretion disks, and with the changing spin rates of neutron stars.

We thank our co-authors who spent countless hours on their manuscripts and the Netherlands Organization for Pure Research (ZWO) who made our collaboration possible; without their generous support (from grants B78-178 and B78-183) this book would not have been written.

> Walter H. G. Lewin
> *Cambridge, Massachusetts*
> Edward P. J. van den Heuvel
> *Amsterdam*
> April 1983

1

X-RAY PULSARS IN MASSIVE BINARY SYSTEMS†

Saul A. Rappaport and Paul C. Joss
*Department of Physics and Center for Space Research,
Massachusetts Institute of Technology, Cambridge, MA 02139*

1.1 Introduction

There are approximately 100 known 'bright' galactic X-ray sources with fluxes above $\sim 10^{-10}$ erg cm^{-2} s^{-1} in the energy range 1–10 keV. Many fainter sources have recently been discovered with the Einstein X-ray Astronomy Observatory (Giacconi 1981 and references therein), but most of these are too faint for the type of detailed study described in this review. About 20 of the bright galactic X-ray sources have been identified with binary systems containing early-type companion stars (Bradt, Doxsey & Jernigan 1979). A substantial fraction of these X-ray sources are known to exhibit regular X-ray pulsations and are called 'X-ray pulsars' (see Note 1). The X-ray pulsars are widely accepted to be neutron stars (see Note 2) that are undergoing accretion from binary stellar companions. About 14 of the ~ 19 known X-ray pulsars are identified with early-type companions (Bradt, Doxsey & Jernigan 1979), while the only X-ray pulsars definitely associated with other types of systems are 4U 1626−67 (Joss, Avni & Rappaport 1978; Middleditch *et al*. 1981), which has been found to be a low-mass binary, and GX 1 + 4, which is associated with an M6 giant (Glass & Feast 1973; Davidsen, Malina & Bowyer 1976). Much of our detailed knowledge of X-ray binaries has been derived over the past decade from extensive studies of the X-ray pulsars in massive binaries. Our understanding of these systems is often applied to lower-mass binaries where there is considerably less detailed information available, as well as to sources where there is no direct evidence of binary membership.

In this article we examine how information on X-ray pulsars in massive binaries is extracted from the observations. Specifically, in Section 1.2 the phenomenology of the X-ray pulsations and pulsar energy spectra is briefly

† Since this article was written, three new binary X-ray pulsars have been discovered: A 0538−66 ($P = 0.069$ s; Skinner *et al*. 1982), 1E 2259+586 ($P = 3.49$ s; Gregory & Fahlman 1980), and 2S 1553−54 (9.26 s; Kelley, Ayasli & Rappaport 1982).

described; in Section 1.3 pulse timing studies and their implications for the accretion process and the properties of neutron stars are discussed; and in Section 1.4 measurements of orbits and the determination of binary system parameters, including masses, radii, and internal structure of the companion stars, as well as the masses of the neutron stars, are presented.

1.2   Pulse profiles and spectra

The pulse periods of the known binary X-ray pulsars range over three decades in period, from 0.7 s to 835 s (see Figure 1.1). The apparent 'gap' in the

Fig. 1.1. The known X-ray pulsars (see Note 1) as of September 1981.

SOURCE	PULSE PERIOD(s)	DISCOVERY OF PULSATION
SMC X-1	0.714	Lucke et al. (1976)
Her X-1	1.24	Tananbaum et al. (1972)
4U0115+63	3.61	Cominsky et al. (1978)
Cen X-3	4.84	Giacconi et al. (1971)
4U1626-67	7.68	Rappaport et al. (1977)
LMC X-4	13.5	Kelley et al. (1981c)
2S1417-62	17.6	Kelley et al. (1981a)
OAO1653-40	38.2	White and Pravdo (1979)
AO535+26	104	Rosenberg et al. (1975)
GX1+4	122	Lewin, Ricker and McClintock (1971) White et al. (1976)
A1239-59 4U1230-61 }?	191	Huckle et al. (1977)
GX304-1	272	McClintock et al. (1977); Huckle et al. (1977)
4U0900-40	283	McClintock et al. (1976)
4U1145-61	292	White et al. (1978) ,
1E1145.1-6141	297	White et al. (1978); Lamb et al. (1980)
A1118-61	405	Ives, Sanford and Bell-Burnell (1975)
4U1538-52	529	Davison (1977); Becker et al. (1977a)
GX301-2	696	White et al. (1976)
4U0352+30	835	White, Mason and Sanford (1976)

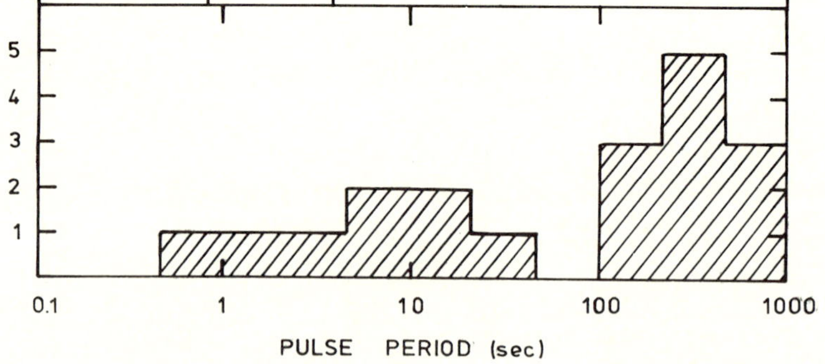

distribution of pulsar periods between 10 and 100 s has recently been closed with the discoveries of three pulsars with periods of 13, 17, and 38 s (Kelley *et al.* 1981*c*; Kelley *et al.* 1981*a*; White & Pravdo 1979). Sample pulse profiles for 14 of the ~19 known pulsars are presented in Figure 1.2. The pulse profiles are characterized by (i) large 'duty cycles' of ~50 per cent (compared to the ~3 per

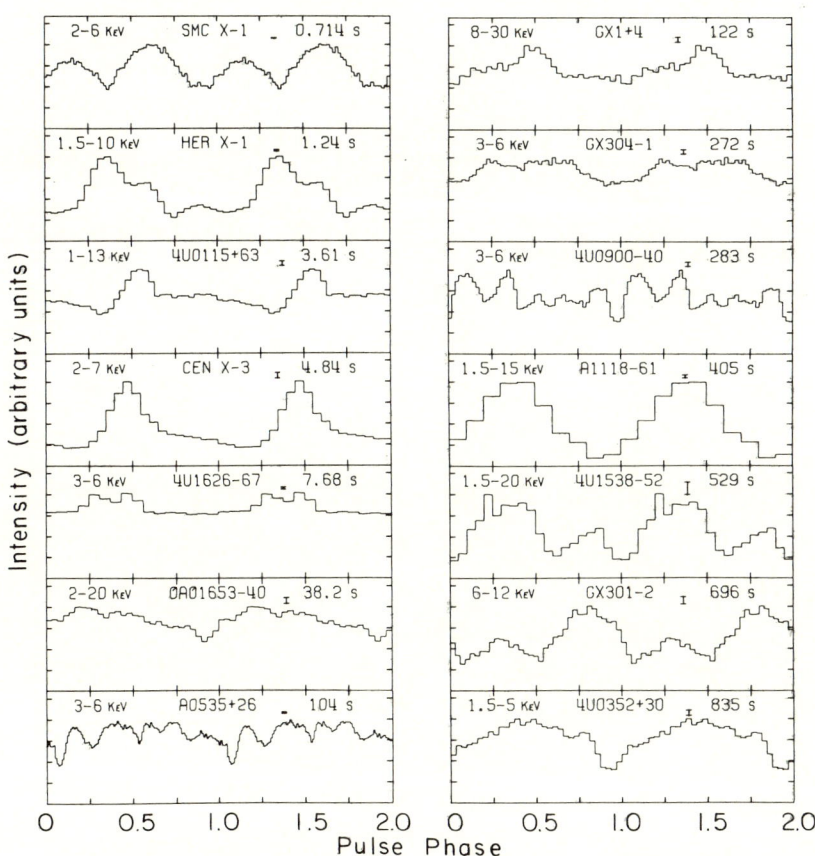

Fig. 1.2. Sample pulse profiles for fourteen X-ray pulsars (from Rappaport & Joss 1981). The profiles for Her X-1, 4U 0115 + 63, Cen X-3, OAO 1653−40, A 1118−61, and 4U 1538−52 are from Joss *et al.* (1978); Johnston *et al.* (1978); Ulmer (1976); White & Pravdo (1979); Ives, Sanford & Bell Burnell (1975); and Becker *et al.* (1977*a*) respectively; the remaining profiles are from SAS-3 data. (For other references see Rappaport & Joss 1977*a*.) In each case, the data are folded modulo the pulse period and plotted against pulse phase for two complete cycles. The approximate pulse periods and X-ray energy intervals are indicated for each pulsar. Non-source background counting rates have been subtracted. A typical ±1σ error bar, derived from photon counting statistics, is indicated for each pulse profile.

cent duty cycles typical of radio pulsars; Manchester & Taylor 1977), (ii) modulation factors that range from ~25 to 90 per cent, (iii) a range from symmetric to highly asymmetric shapes, (iv) *no* obvious trend in pulse morphology as a function of pulse period, (v) *no* obvious correlation of pulse morphology with X-ray luminosity, L_x, over a range of nearly 10^5 in L_x, and (vi) a slight but significant positive correlation between pulse period and binary orbital period (van den Heuvel & Rappaport 1981). For several of the sources (e.g., A 0535+26 and 4U 0900−40) the pulse profiles vary radically with X-ray energy (see Figure 1.3), while for others (e.g., Cen X-3 and 4U 0352+30) the basic pulse profile is retained over a wide range of energies.

Representative X-ray pulsar spectra from six sources are shown in Figure 1.4. In five cases the spectrum has been averaged over pulse phase; all spectra are plotted to the same scale for comparison. Though these spectra differ in detail, the spectra of X-ray pulsars have the following general features: (i) broad-band emission not dominated by sharp spectral features (see Figure 1.4 and references

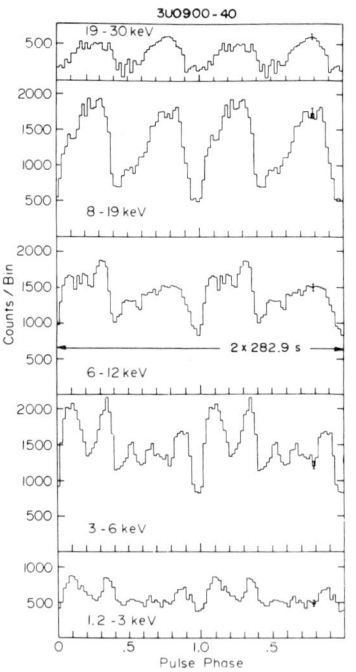
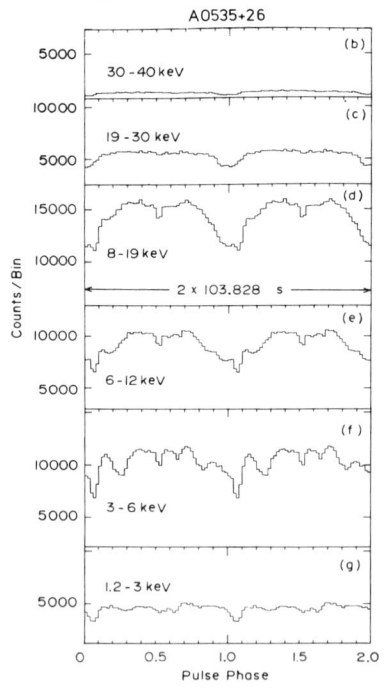

Fig. 1.3. Sample pulse profiles as a function of X-ray energy for two X-ray pulsars (from McClintock *et al.* 1976; Bradt *et al.* 1976). For each energy interval, the data are folded modulo the pulse period and plotted against pulse phase for two complete cycles. Non-source background counting rates have been subtracted.

Fig. 1.4. Sample X-ray spectra for six X-ray pulsars. The spectra are replotted from the work of White et al. (1980; 4U 1145−61); White & Pravdo (1979; OAO 1653−40); Pravdo et al. (1979; 4U 1626−67); Rose et al. (1979; 4U 0115+63); Becker et al. (1978; 4U 0900−40); and Becker et al. (1977b; Her X-1). For five of the sources the X-ray spectra are averaged over all pulse phases, while for 4U 0115+63 the spectrum is from a relatively narrow (but representative) interval in pulse phase. In all cases we have suppressed the error bars on the individual data points. The solid curves are schematic renditions of the continuum spectra. The ordinates for 4U 1145−61 and OAO 1653−40 have been multiplied by factors of 10 and 3 respectively.

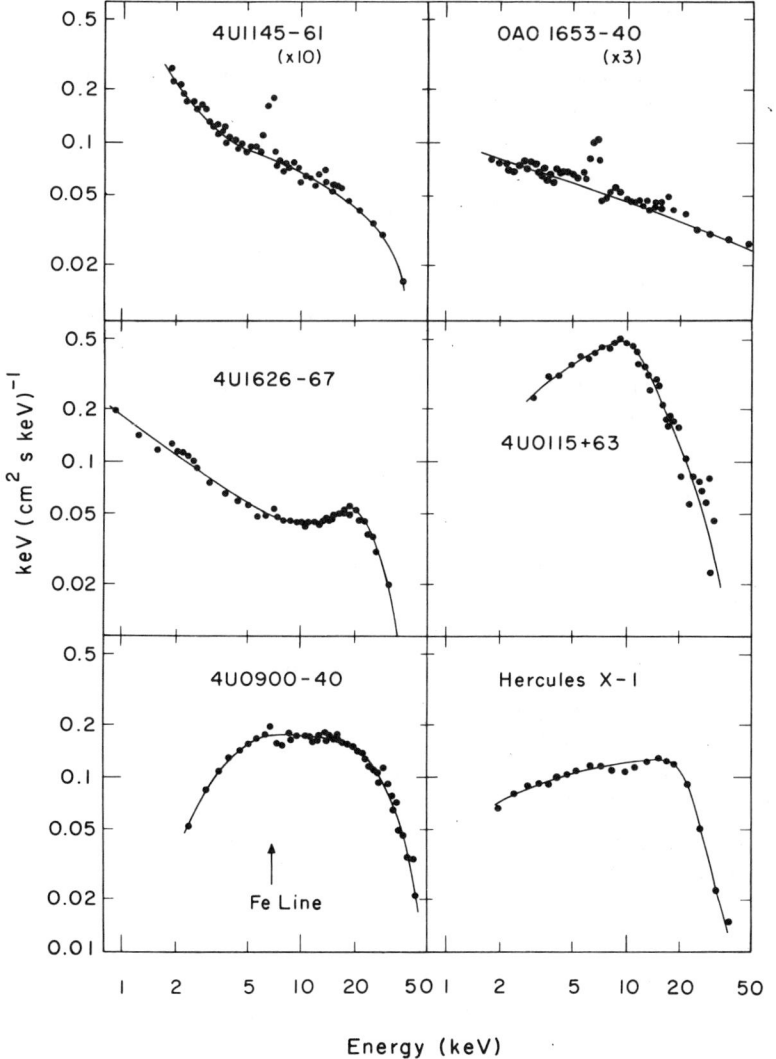

in the figure caption), (ii) power mostly emitted in the range 2-20 keV, (iii) rapid falloff in flux for energies above $\sim$20 keV, (iv) no simple description in terms of a blackbody, thermal bremsstrahlung, or power-law shape, (v) an iron K-shell emission feature near 6.7 keV in many, but not all, cases (see, e.g., Becker et al. 1978; White & Pravdo 1979; White et al. 1980), and (vi) evidence for a cyclotron line feature in a few sources (see, e.g., Trümper et al. 1977; Wheaton et al. 1979; Chapter 6, this book).

At present, we are far short of a complete theoretical understanding of the complex pulse shapes (Figure 1.2) or their dependence on X-ray energy (Figures 1.3 and 1.4). It is widely accepted, however, that the pulsations result from an X-ray beam pattern that is misaligned with the rotation axis of an accreting, magnetized neutron star and viewed from different directions as the star rotates (Pringle & Rees 1972; Davidson & Ostriker 1973; Lamb, Pethick & Pines 1973; Baan & Treves 1973). The X-ray luminosity is generated by the release of the gravitational potential energy of the accreting matter, which is, in turn, supplied by a close-binary stellar companion. The X-ray beam pattern is determined by the external magnetic field of the neutron star, the resultant funneling of the accretion flow, and the complex radiative transfer processes for X-rays propagating from the vicinity of the neutron-star surface through regions of accreting, magnetized plasma.

The infalling matter may be decelerated and its energy thermalized in one or more of the following ways:

(i) Deceleration by radiation pressure (a 'photon cushion'). In this picture (Davidson 1973; Inoue 1975; Basko & Sunyaev 1976; Wang & Frank 1981), an essentially blackbody radiation field, with a temperature corresponding to the soft X-ray range, is present within the accretion column due to its high optical depth. The incoming electrons are decelerated by interactions with the photons, and this deceleration is then transmitted electrostatically to the infalling ions, whose kinetic energy ($\sim$100 MeV per nucleon) is thereby thermalized.

(ii) Collisions beneath the neutron-star surface. Here, the infalling matter is confined to beams that strike the magnetic polar caps of the neutron star at nearly the free-fall velocity (Lamb, Pethick & Pines 1973; Basko & Sunyaev 1975; Pavlov & Yakovlev 1976; Kirk & Galloway 1981). The beams consist essentially of free particles, and their kinetic energy is not thermalized until after they reach the surface of the neutron star. The infalling ions are then stopped by binary collisions (nuclear or Coulomb) within a short distance beneath the neutron-star surface. A difficulty with this picture is that the scale height for a gas at a temperature of $\sim 10^8$ K (the characteristic temperature of the emitted X-radiation) at the surface of a neutron star is only $\sim$1 m. The X-radiation cannot then emerge in directions transverse to the accretion flow through the small area comprising the sides of the raised, heated polar cap without violating the blackbody limit, since this area is only ~ 3 km $\times$ 1 m. If, on the other hand, the radiation emerges in the direction of the accretion flow, the radiation pressure resulting from even relatively small accretion rates (corresponding to $L_x \gtrsim 10^{35}$ erg s^{-1}) should disrupt the flow.

(iii) Passage through a shock followed by settling. In this picture, the incoming matter is assumed to undergo a strong shock above the surface of the neutron star (Inoue 1975; Shapiro & Salpeter 1975; Basko & Sunyaev 1976; Wang & Frank 1981; Langer & Rappaport 1981). At the shock a large fraction of the kinetic energy of the incoming matter is thermalized. The matter in the post-shock region then radiates and cools as it settles onto the neutron-star surface. Only recently have the effects of the strong magnetic field upon the emission processes and the heat exchange between electrons and ions been incorporated into such a model (Langer & Rappaport 1981).

The success of any such ideas depends ultimately on their ability to describe the emergent X-ray spectra and beam patterns. At present only crude X-ray spectra have been calculated for several simplified models (Davidson 1973; Inoue 1975; Pavlov & Shibanov 1979 and references therein; Yahel 1980 and references therein; Kanno 1980; Langer & Rappaport 1981; Mészàros, Nagel & Ventura 1980 and references therein). We do not yet even know how the energy characterizing the observed X-ray spectra ($\sim$10 keV) arises. There are at least two characteristic energies associated with the accretion problem that are about the correct magnitude. One is the characteristic cyclotron line energy, E_c, at the surface of the neutron star (see also Chapter 6, this book):

$$E_c \approx \hbar \omega_B \approx 12 \left(\frac{B}{10^{12} \text{ gauss}} \right) \text{keV}, \qquad (1.1)$$

where B is the surface magnetic field strength. The other is the thermalization energy, E_T, characteristic of a hot polar cap of area A, radiating luminosity L_x at temperature T:

$$E_T \approx kT \approx k \left(\frac{L_x}{\sigma A} \right)^{1/4} \approx 10 \left(\frac{L_x}{10^{38} \text{ erg s}^{-1}} \right)^{1/4} \left(\frac{A}{10^{10} \text{ cm}^2} \right)^{-1/4} \text{keV}. \qquad (1.2)$$

In any case, the complex frequency and directional dependence of photon scattering, spatial variations in the magnetic field, and thermal Doppler broadening effects may contribute to distortion of the spectra into the shapes that we observe (see, e.g., Langer & Rappaport 1981).

Early attempts to derive theoretical pulse profiles (see, e.g., Bisnovatyi-Kogan 1973; Lamb, Pethick & Pines 1973; Gnedin & Sunyaev 1973; Basko & Sunyaev 1975; Tsuruta 1975; Wang 1975) utilized highly simplifying approximations for the actual physical processes. These efforts suggest emission patterns ranging from pencil beams directed along the magnetic axes of the neutron star to fan beams perpendicular to these axes. Recently, a few workers (Yahel 1980 and references therein; Pravdo & Bussard 1981) have used Monte Carlo techniques to simulate the generation of the pulse profiles and X-ray spectra. These calculations have taken into account many of the important scattering processes for X-rays propagating through the accretion column. However, such calculations have not yet been carried out in a self-consistent manner (i.e., the effects of the radiation upon the plasma and the emission of radiation by the plasma are not fully considered).

Studies of the general accretion flow problem, in which the emission and transfer of radiation in the intense magnetic field are coupled to the mass flow, are now needed. Some progress is currently being made in this area (Langer, Castor & Rappaport 1981; Klein, Arons & Lea 1981). The similarities among the pulse profiles and energy spectra for pulsars with a very wide range of pulse periods and luminosities provide hope that a basic understanding of the accretion processes may in fact be attainable in the near future.

1.3 Studies of pulse period variations

The pulse periods for a number of X-ray pulsars have been sufficiently well measured over the past decade to provide important information regarding the torques exerted on the neutron stars by the accreting material (see also Chapter 11, this book). Nearly all of the available pulse period measurements (as of June 1981) for the eight best-measured X-ray pulsars are shown in Figure 1.5. When we examine the pulse period histories of these eight pulsars, it is apparent that the 'spin-up' trend first noted in Her X-1 and Cen X-3 (Giacconi 1974; Gursky & Schreier 1975; Schreier & Fabbiano 1976) is very prominent in at least five of them.

The trend in most of the X-ray pulsars toward a secular decrease in pulse period can be understood in terms of torques exerted by the matter accreting onto the neutron star. These torques can be readily calculated for the case where the matter has roughly circular Keplerian velocities at the magnetopause of the neutron star, as would be the case if the accretion were mediated by a disk. Such calculations (Pringle & Rees 1972; Lamb, Pethick & Pines 1973; Ghosh & Lamb 1979 and references therein; a thorough review is given in Chapter 11 of this book and see also Chapter 6, Section 6.2.1) show that the rate of change, $\dot{P}$, of the intrinsic pulse period, P, is related to the X-ray luminosity and the physical properties of the neutron star:

$$\frac{\dot{P}}{P} \approx -3 \times 10^{-5} \left(\frac{\xi v_r}{v_{ff}}\right)^{1/7} \left(\frac{M_x}{M_\odot}\right)^{-10/7} \left(\frac{R}{10 \text{ km}}\right)^{6/7} \left(\frac{R_g}{10 \text{ km}}\right)^{-2}$$
$$\times \left(\frac{\mu}{10^{30} \text{ gauss cm}^3}\right)^{2/7} \left(\frac{L_x}{10^{37} \text{ erg s}^{-1}}\right)^{6/7} \left(\frac{P}{1 \text{ s}}\right) \text{ yr}^{-1}. \quad (1.3)$$

Here, ξ is the fractional solid angle subtended at the neutron star by the infalling matter at the magnetopause; v_r/v_{ff} is the ratio of the average radial infall velocity of a particle to its free-fall velocity just outside the magnetopause; M_x, R, R_g, and μ are the mass, radius, radius of gyration, and magnetic dipole moment of the neutron star, respectively; and L_x is the accretion-driven luminosity. The quantity $(\xi v_r/v_{ff})^{1/7}$ is not expected to differ greatly from unity (see, e.g., Lamb, Pethick & Pines 1973).

The overall minus sign in Equation (1.3) is explicitly for the case where the sense of the orbital angular momentum in the accreting matter is the same as that of the rotation of the neutron star. This is, in fact, to be expected if (i) the rotation of the progenitor star was in the same sense as the binary orbital motion, (ii) the neutron star retained this sense of rotation during its formation, and (iii) the sense of the orbital angular momentum of the matter leaving the companion star is retained as it enters the accretion disk that surrounds the neutron star.

For simplicity, we can rewrite Equation (1.3) as

$$\frac{\dot{P}}{P} = -3 \times 10^{-5} f\left(\frac{P}{1 \text{ s}}\right) \left(\frac{L_x}{10^{37} \text{ erg s}^{-1}}\right)^{6/7} \text{ yr}^{-1} \quad (1.4)$$

(Rappaport & Joss 1977b), where the dimensionless function f is expected to be of order unity for a neutron star and contains parameters that are not yet measurable for most or all of the X-ray pulsars.

Following Rappaport & Joss (1977b), we have estimated the spin-up rate for each of ten X-ray pulsars in the following way. For the five sources in Figure 1.5 that show a clear secular trend toward decreasing pulse period (where we exclude 4U 0900−40 on the basis of recent observations by Nagase et al. 1981), we have adopted the average value of $\dot{P}$ (i.e., the dashed lines in Figure 1.5). In the cases of the remaining three sources shown in Figure 1.5 we have adopted the average value of $\dot{P}$ during the longest interval with generally decreasing pulse period as a representative measure of the accretion torques during times when Equations (1.3) and (1.4) are most likely to be valid (see Elsner & Lamb 1976; Ghosh & Lamb 1979). For two other pulsars (4U 0015+63 and 4U 1626−67), not indicated in Figure 1.5, a reliable value of $\dot{P}$ is available (Rappaport et al.

Fig. 1.5. Pulse period histories for eight X-ray pulsars. The heavy dots are individual measurements of pulse period; the vertical bars represent the 1σ uncertainties in the period determinations. The data are from the *Uhuru*, Copernicus, Ariel 5, SAS-3, OSO-8, HEAO-1 and Einstein satellites, the Apollo-Soyuz Test Project, and a balloon observation and sounding rocket observation. In general, the original references are quoted elsewhere in the text (for additional references see Rappaport & Joss 1977a, b; White, Mason & Sanford 1977; Lamb, Pines & Shaham 1978; Nagase 1981). The dashed lines are minimum chi-squared fits of a straight line to the data points. (In some cases, where the pulse period was measured with very high accuracy, a larger error bar, of fixed size, was assigned in the fitting procedure to take into account the known systematic deviations of the pulse period variations from a simple linear time dependence.)

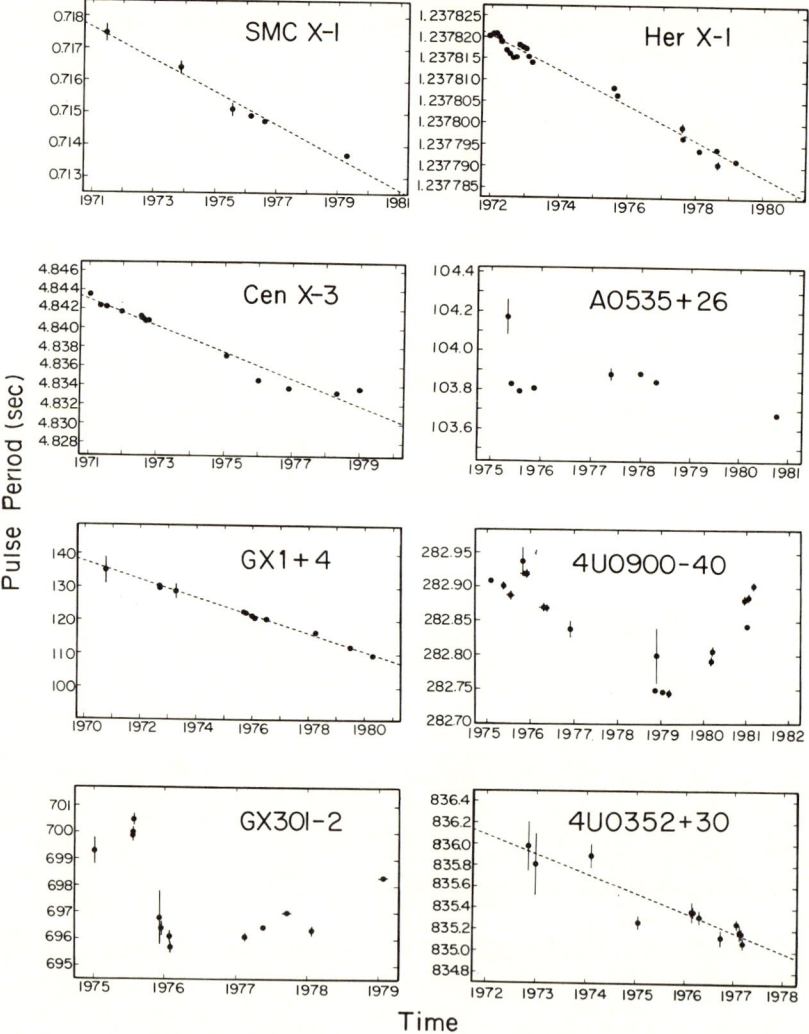

1977, 1978), but the distance to the source, and hence the X-ray luminosity, is not known.

The empirical data relating the spin-up rate and the quantity $PL_x^{6/7}$ (see Equation (1.4)) for these ten X-ray pulsars are shown in Figure 1.6. The solid line is the relation expected from Equation (1.4) if the X-ray stars are neutron stars with commonly accepted values of mass, radius, and magnetic moment as indicated in Equation (1.3) (i.e., if $f = 1$). The dashed line, lying ~2 orders of magnitude below the data points, is the expected relation for accreting degenerate dwarfs. This result (see also Mason 1977) provides a compelling quantitative argument that the X-ray pulsars are, in fact, accreting neutron stars.

Fig. 1.6. The empirical relation between the fractional rate of change of pulse period, $\dot{P}/P$, and the parameter (pulse period × luminosity$^{6/7}$) for ten binary X-ray pulsars (adapted from Rappaport & Joss 1977b and Rappaport & Joss 1981). The units of $\dot{P}/P$, P, and L are yr^{-1}, s, and 10^{37} erg s^{-1} respectively. The solid line is the best fit to a straight line with logarithmic slope 1.0 (see Equation (1.4)); the data point for Her X-1 and the limits for 4U 1626−67 and 4U 0115+63 were excluded from the fit (see Elsner & Lamb 1976 and Ghosh & Lamb 1979 for discussions of the rather special case of Her X-1). The intercept at $\log(\dot{P}/P) = -4.6$ is in good agreement with the expected value if the X-ray emitting objects are accreting neutron stars (see text and Equation (1.4)). The dashed line is the expected relation for ~1 M$_\odot$ degenerate dwarfs.

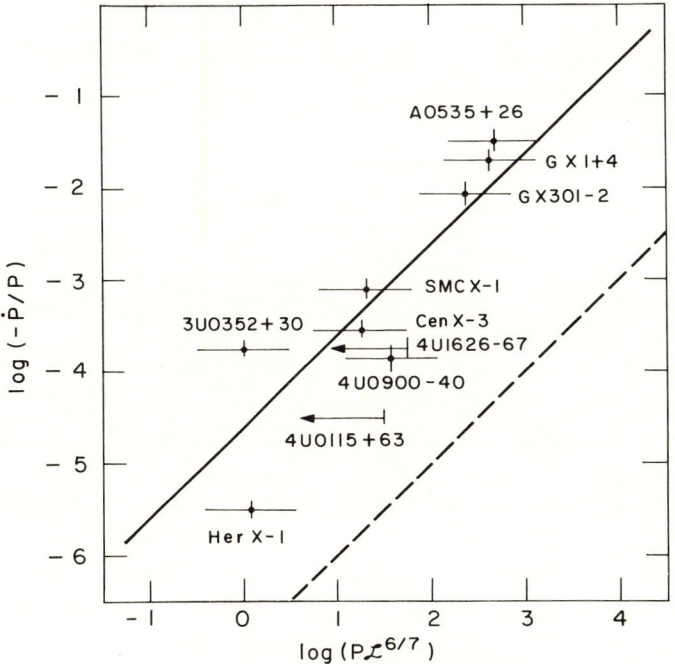

The foregoing discussion is based on the premise that the accreting matter has sufficient angular momentum to enter a circular Keplerian orbit outside the magnetosphere of the neutron star. This situation is expected if the accreting matter is lost from the companion star at relatively low velocity through the inner saddle point of the critical potential lobe (see Section 1.4.1); in fact, if the specific angular momentum of accreting matter is sufficiently large, the accretion will be mediated by a large disk that surrounds the neutron star. If, on the other hand, matter is accreted from a high-velocity stellar wind via the Bondi–Hoyle (1944) mechanism (Davidson & Ostriker 1973), then the net specific angular momentum of the captured matter is likely to be too small to result in the formation of an accretion disk (Shapiro & Lightman 1976; Wang 1981). In this case, the accretion torques exerted on the neutron star will be determined by the properties of the stellar wind and binary orbit and will be nearly independent of the properties of the neutron-star magnetosphere (Ghosh & Lamb 1979; Arons & Lea 1980; Wang 1981). In fact, the sense of the accreted angular momentum may even be *reversed* over that given in Equation (1.4) (see Wang 1981). The dependence of the spin-up (or spin-down) timescale on pulse period and X-ray luminosity for the case of wind capture will still be qualitatively similar to that resulting from disk accretion, but the dependence on the binary system parameters will differ (Ghosh & Lamb 1979; Arons & Lea 1980). At present, the observed pulse period variations do not permit a conclusive comparison between these two accretion modes (see Arons & Lea 1980). However, even if wind accretion prevails in some of the X-ray pulsars that we have considered, the conclusion that the underlying object is a neutron star (see Figure 1.6) would be unaffected.

An alternative method of displaying the pulsar spin-up data, for the case of disk accretion, is to plot $\log \dot{P}$ against $\log(PL_x^{3/7})$ (Ghosh & Lamb 1979). This has the advantage of sorting the pulsars according to their 'fastness' (i.e., the ratio of the rotational angular velocity of the neutron star to that of the orbiting matter in an accretion disk just outside the magnetopause), and may better reveal possible systematic deviations from a simple linear relation. However, in view of the fact that severe complications, such as those discussed in the preceding paragraph, may affect the pulse period behavior of even the persistently luminous 'slow' X-ray pulsars (e.g., GX 301−2 and 4U 0900−40), it may be premature to expect a simple correlation of 'fastness' with average spin-up rate.

We note that even for the five sources in Figure 1.5 that exhibit a clear 'spin-up' trend, the change in pulse period is not always monotonic on short timescales (see, e.g., Giacconi 1974; Fabbiano & Schreier 1977; Ögelman *et al.* 1977; Darbro *et al.* 1981 and Chapter 11, this book). As noted above, for three of the sources in Figure 1.5 (GX 301−2, 4U 0900−40, and A 0535+26; Kelley,

Rappaport & Petre 1980; Nagase *et al.* 1981; Li *et al.* 1979) there are long intervals where the pulse period is nearly constant or even increasing with time. Recent observations of 4U 0900−40 with the Hakucho satellite (Nagase 1981) indicate that after its pulse period had decreased for at least four years (1975.0-1979.0) the period has been consistently increasing for the past two years; as a result, the pulse period as of March 1981 is substantially the same as it was six years earlier.

It is possible that A 0535 + 26 becomes a 'fast pulsar' (Elsner & Lamb 1976) when in the low part of its transient cycle and thereby undergoes periods of spin-down (Li *et al.* 1979; Ziolkowski 1980; Elsner, Ghosh & Lamb 1980). The pulse period behavior of GX 301−2 and 4U 0900−40 may possibly be explained by occasional reversals in the sense of rotation of the accreted matter at the magnetopause (see Wang 1981). (For further discussions of such effects, see Ghosh, Lamb & Pethick 1977; Ghosh & Lamb 1978, 1979; Davies, Fabian & Pringle 1979; and references therein.) It is suggestive in this regard that the three sources in Figure 1.5 which do not exhibit a consistent spin-up trend are most likely to be undergoing accretion from a stellar wind whose velocity is large compared to the orbital velocity (van den Heuvel & Henrichs 1981), rather than being fed by Roche lobe overflow (in which case the angular momentum of the accreting matter should always retain the sense of the orbital angular momentum, as described above).

Finally, we note that important additional information concerning the internal structure and dynamics of neutron stars may eventually be gleaned from studies of the observed smaller-scale fluctuations in pulse period (see, e.g., Lamb, Pines & Shaham 1978; Boynton & Deeter 1979; Lamb & Boynton 1980; Boynton 1981).

For a more detailed review of pulse period variations and accretion torques, see Chapter 11, this book.

1.4 Orbital determinations and the evaluation of binary system parameters

1.4.1 Assumptions and techniques

Measurements of the pulse arrival times from a number of binary X-ray pulsars have been used very successfully to determine the orbits of these systems. As will be discussed below, the measured binary orbital parameters constitute the key ingredients that allow a determination of many of the physically interesting properties of these systems. In the early 1970s the relatively tight orbits of the luminous X-ray pulsars Cen X-3 and Her X-1 were measured with the *Uhuru* satellite (Schreier *et al.* 1972a; Tananbaum *et al.* 1972); sources such as 4U 0115 + 63 and GX 301−2 have lower average X-ray luminosities and relatively wide orbits, and the orbital parameters of these systems were measured only recently (Rappaport et al. 1978; White, Mason & Sanford 1978; Kelley, Rappaport & Petre 1980; Watson, Warwick & Corbet 1981).

The method of determining orbits from the Doppler delays of X-ray pulse arrival times (Schreier et al. 1972a), which is analogous to classical optical measurements of Doppler shifts in spectral lines, is illustrated schematically in Figure 1.7. For a perfect clock emitting pulses at uniform intervals and moving with constant velocity, a plot of pulse arrival time as a function of pulse number will reveal a simple linear relation. If the intrinsic rate of the clock increases with time (as is the case for a neutron star that is spinning up; see Section 1.3), the same type of plot yields a curved line such as the one shown in Figure 1.7. If, in addition, the clock is in a Keplerian orbit, then periodic Doppler delays in the arrival times, due to the time-of-flight of the pulses across the orbit, will be superposed. For the case where the curvature due to orbital motion is much greater than that due to changes in the intrinsic pulse period (or where the measurement interval contains a number of orbital cycles), the orbit can be determined by subtracting a simple polynomial (in time) from the arrival-time plot. Complications arise when the reverse of the above condition obtains, and in such circumstances the orbits are difficult to measure. This is the case for slow X-ray pulsars (with their concomitantly large values of $|\dot{P}|$) in long-period orbits (see, e.g., Li et al. 1979).

An example of an orbital determination from a measurement of the Doppler delays of X-ray pulses is shown in Figure 1.8. The Doppler delays in the arrival

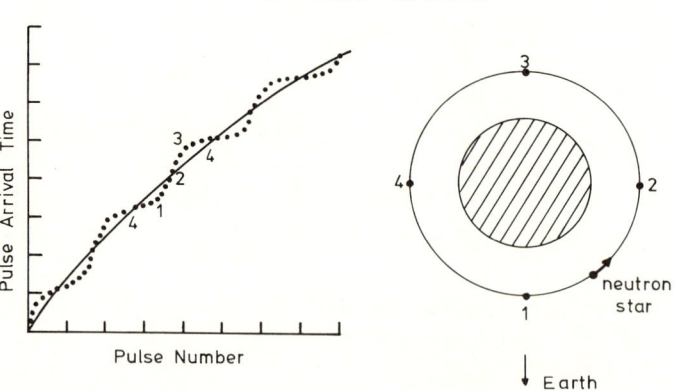

Fig. 1.7. Schematic sketch of the Doppler delays in X-ray pulse arrival times. The numbered points on the Doppler delay curve (left) correspond to the indicated points on the orbit (right). Such measurements are used to determine the orbital elements in binary X-ray pulsar systems (see text).

times of pulsations from SMC X-1 (Primini, Rappaport & Joss 1977), from which the expected arrival times for a constant pulse period have been subtracted, are shown for one orbital cycle. The resultant data points can be fitted extremely well by a simple sinusoidal function, which indicates that the orbit is very nearly circular (eccentricity $<7 \times 10^{-4}$). The projected semi-major axis of the X-ray star can be read directly from the figure ($a_x \sin i = 53.5$ lt s, where a_x is the semi-major axis and i is the inclination of the orbit to the plane of the sky). At present, the main limitation on the accuracy of this orbital determination is the statistical precision with which the pulse arrival times can be measured. Even now, however, the orbit of SMC X-1 is the most precisely determined binary stellar orbit in an extragalactic system.

The measured Doppler delay curve for the spectrally 'hard' transient source 4U 0115+63 is shown in Figure 1.9 (Rappaport *et al.* 1978). These data were obtained with the SAS-3 X-ray astronomy satellite in observations that spanned an interval of 26 d. Analysis of these data revealed a short pulse period ($P \approx 3.6$ s; Cominsky *et al.* 1978), a long orbital period ($P_{orb} \approx 24$ d), a large projected semi-major axis ($a_x \sin i \approx 140$ lt s), and a moderately high eccentricity ($e \approx 0.34$). The

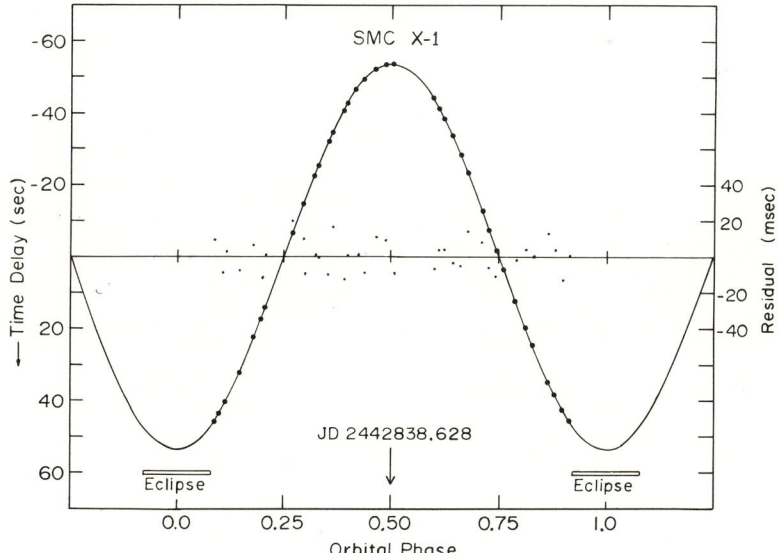

Fig. 1.8. Doppler delay data for SMC X-1 (from Primini, Rappaport & Joss 1977). The large dots are the measured Doppler delays in the pulse arrival times from SMC X-1 as a function of the 3.892 d orbital phase, with the scale indicated on the left-hand side of the figure. The solid curve represents the best-fit circular orbit. Small dots indicate the residual differences between the measured delays and the best-fit orbit, with the scale indicated on the right-hand side of the figure. The rms scatter in the residuals is ~10 ms.

companion is thought to be a B emission-line star that does not fill its critical potential lobe; the episodes of X-ray flaring are probably attributable to periods of enhanced spontaneous mass loss from the companion (Rappaport et al. 1978; Rappaport & van den Heuvel 1981). This measurement ended speculation about the basic nature of 4U 0115+63, and, by inference, the other hard X-ray transients as well.

All the measured orbits of binary X-ray pulsars (with the exception of the recently measured orbit of LMC X-4) are shown to scale in Figure 1.10, in order of increasing size of the semi-major axis (see also Table 1.1). Nominal values for the masses and radii of the companion stars are indicated in the figure; the derivation of these parameters will be discussed below. A few other X-ray binaries that lack detectable X-ray pulsations (e.g., 4U 1700−37 and Cyg X-2) have been sufficiently well studied in optical light (Hutchings 1976; van Paradijs, Hammerschlag-Hensberge & Zuiderwijk 1978; Cowley, Crampton & Hutchings 1979) to yield important information about the binary system parameters; however, the information is much less complete than for the systems where the orbit of the X-ray star can be measured by pulse timing observations.

There are six binary X-ray pulsars for which sufficient information is now available to allow a determination of many of the system parameters. In particular, for five of these systems, the following four key quantities have been

Fig. 1.9. Doppler delay data for 4U 0115+63 (from Rappaport et al. 1978). The vertical bars are the measured delays in pulse arrival time (left-hand scale); the length of each bar is considerably greater than the uncertainty in the measurement. The solid curve represents the expected delays for a Keplerian orbit with the best-fit orbital parameters (see Table 1.1). Small circles (right-hand scale) indicate the residual differences between the measured delays and the best-fit curve.

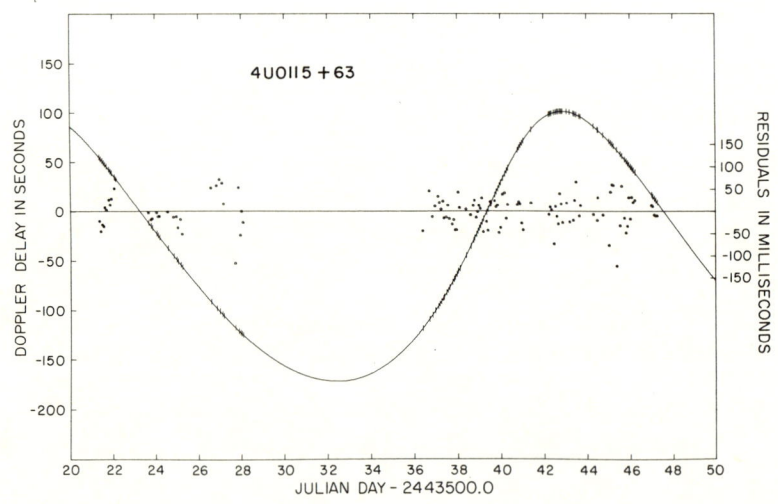

measured: (i) the orbital period; (ii) the projected semi-major axis of the X-ray star; (iii) the amplitude of the Doppler velocity curve for the optical companion; and (iv) the duration of the X-ray eclipse. These systems are SMC X-1, Cen X-3, 4U 0900−40, 4U 1538−52, and LMC X-4 (see Tables 1.1 and 1.2 for references).

The orbital period, P_{orb}, and the projected semi-major axis of the X-ray star yield the mass function:

$$f(M) = \frac{4\pi^2 (a_x \sin i)^3}{GP_{orb}^2} = \frac{M_c \sin^3 i}{(1+q)^2}, \qquad (1.5)$$

where M_c is the mass of the companion star, $q \equiv M_x/M_c$ is the mass ratio, and G is the universal gravitational constant. The system geometry is shown schematically in Figure 1.11. The curve that encircles both the companion star and the neutron star is a contour of constant effective (gravitational plus centrifugal)

Fig. 1.10. Schematic sketch, to scale, of the orbits and companion stars of seven binary X-ray pulsars (from Rappaport & Joss 1981). The approximate mass of the companion star is indicated below each orbit. See text (Section 1.4) for a discussion of the derivation of the orbital parameters, stellar masses, and stellar radii. Another, slightly larger, orbital solution for GX 301−2 is admissible by the available data (see note to Table 1.1).

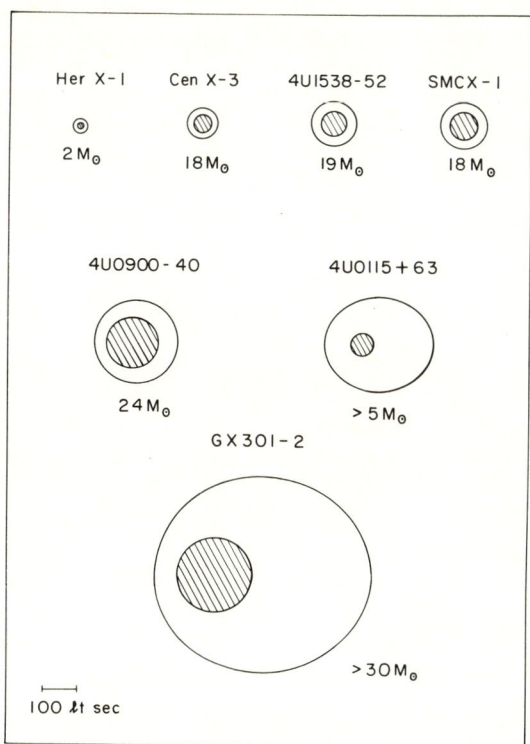

potential in the frame of reference that is centered on, and rotates with, the companion star. The particular contour shown is called the critical potential lobe because it represents the largest volume that a dynamically stable companion star can occupy before mass is transferred through the inner saddle point of the effective potential.

The mass ratio q can be determined directly from the ratio of the velocity of the companion star to that of the X-ray star:

$$q = \frac{a_c \sin i}{a_x \sin i} = \frac{K_c P_{orb} \sqrt{(1-e^2)}}{2\pi a_x \sin i}, \tag{1.6}$$

where $a_c \sin i$ is the projected semi-major axis of the orbit of the companion, K_c is the semi-amplitude of the optical Doppler velocity curve, and e is the orbital eccentricity.

The final ingredient for determining the system parameters is the orbital-inclination angle. This can be computed approximately, with the aid of a simple model. If we replace the companion star by a sphere of radius R_c whose volume equals the actual volume of the star, we obtain the relation

$$R_c \approx a \, [\cos^2 i + \sin^2 i \sin^2 \theta_e]^{1/2}, \tag{1.7}$$

Table 1.1. *Orbital parameters for eight binary X-ray pulsars*

Source	$a_x \sin i$[a] (lt s)	P_{orb} (d)	$f(M)$ ($M_\odot$)	e[a]	References
Her X-1	13.1831 ± 0.0003	1.700	0.85	<0.0003	[1]
LMC X-4	30.0 ± 5.0	1.408	14.6	–	[2]
Cen X-3	39.792 ± 0.005	2.087	15.5	0.0008 ± 0.0001	[3]
SMC X-1	53.46 ± 0.03	3.892	10.8	<0.0007	[4]
4U 1538−52	55.2 ± 3.7	3.730	13	–	[5]
4U 0900−40	113.0 ± 0.8	8.965	19.3	0.092 ± 0.005	[6]
4U 0115+63	140.13 ± 0.1	24.31	5.00	0.3402 ± 0.0002	[7]
GX 301−2[b]	304 ± 3	35.0	25	0.44 ± 0.01	[8]
	367 ± 3	41.4	31	0.47 ± 0.01	

[a] Wherever possible, we have given 1σ confidence limits for the error bars. The quoted upper limits on e are 95 per cent confidence limits.
[b] Two values for the orbital period are admissible by the available data. Recent evidence suggests that the longer period may be preferred (Watson, Warwick & Corbet 1981).

[1] Tananbaum *et al.* (1972); Schreier & Fabbiano (1976); Fechner & Joss (1977); Joss *et al.* (1980); Deeter, Boynton & Pravdo (1981).
[2] Li, Rappaport & Epstein (1978); White (1978); Kelley *et al.* (1981c).
[3] Schreier *et al.* (1972a); Fabbiano & Schreier (1977).
[4] Primini, Rappaport & Joss (1977).
[5] Becker *et al.* (1977a); Davison, Watson & Pye (1977).
[6] Rappaport, Joss & McClintock (1976); Rappaport, Joss & Stothers (1980).
[7] Rappaport *et al.* (1978).
[8] White, Mason & Sanford (1978); Kelley, Rappaport & Petre (1980); Watson, Warwick & Corbet (1981).

Table 1.2. *Parameters used in the Monte Carlo analysis*

Source	$a_x \sin i$ [a] (lt sec)	K_c [a] (km s^{-1})	θ_e [b] (degrees)	T_e [d] (K)	References
Her X-1	13.1831 ± 0.0003	20.2 ± 3.5 [c]	24.4–24.7	–	[1]
LMC X-4	30.0 ± 5.0	37.9 ± 5	25.5–33	37 000	[2]
Cen X-3	39.79 ± 0.01	24 ± 6	35–40	39 000	[3]
SMC X-1	53.46 ± 0.03	19 ± 2	26.5–29	28 000	[4]
4U 1538–52	55.2 ± 3.7	33 ± 7	25–33	30 000	[5]
4U 0900–40	113.0 ± 0.8	21.8 ± 1.2	31–37	25 000	[6]

[a] The quoted uncertainties are 1σ confidence limits.
[b] The extreme range of eclipse half-angles allowed by the observations.
[c] Semi-amplitude of the Doppler velocity curve for the optically reprocessed X-ray pulsations.
[d] The values used for T_e are chosen to be representative of those that have appeared in the observational literature for each source (see references in the text and elsewhere in this table).

[1] Middleditch & Nelson (1976); Deeter, Boynton & Pravdo (1981).
[2] Chevalier & Ilovaisky (1977); Hutchings, Crampton & Cowley (1978); Li, Rappaport & Epstein (1978); White (1978); Kelley *et al.* (1981c).
[3] Pounds *et al.* (1975); Pounds (1976); Fabbiano & Schreier (1977); Hutchings *et al.* (1979).
[4] Schreier *et al.* (1972b); Primini *et al.* (1976); Primini, Rappaport & Joss (1977); Hutchings *et al.* (1977).
[5] Davison, Watson & Pye (1977); Becker *et al.* (1977a); Crampton, Hutchings & Cowley (1978).
[6] Forman *et al.* (1973); van Paradijs *et al.* (1977); Watson & Griffiths (1977); Rappaport, Joss & Stothers (1980).

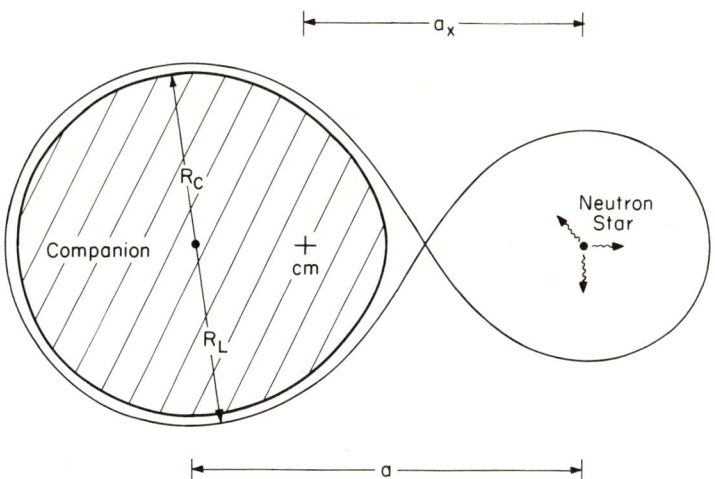

Fig. 1.11. Schematic sketch of a binary X-ray system viewed from above the orbital plane. The companion star (hatched region) and neutron star are shown encircled by their critical potential lobes, which join at the inner saddle point (see text).

where $a \equiv a_x + a_c$ is the separation of the centers of mass of the two stars and the eclipse half-angle, θ_e, is π times the fraction of the orbital period that the X-ray star is in eclipse behind its companion. This approximation is surprisingly accurate, even for stars that suffer appreciable tidal or rotational distortion (see, e.g., Avni 1976; Rappaport 1979). Moreover, the size and shape of the critical potential lobe can be computed from the mass ratio and the rotation rate of the companion star. Its effective radius, R_L, has been found (Plavec 1968; Avni 1976) to be reasonably well fitted by the expression

$$R_L \approx a[A + B \log q + C \log^2 q], \tag{1.8}$$

where A, B, and C are constants that depend on the ratio, Ω, of the rotational frequency of the companion star to the orbital frequency. We have calculated R_L/a as a function of Ω and q and devised the following fitting formulae:

$$A \approx 0.398 - 0.026\Omega^2 + 0.004\Omega^3,$$
$$B \approx -0.264 + 0.052\Omega^2 - 0.015\Omega^3, \tag{1.9}$$
$$C \approx -0.023 - 0.005\Omega^2.$$

These expressions give R_L to within an accuracy of ~ 2 per cent over the range $0 \leqslant \Omega \leqslant 2$ and $0.02 \leqslant q \leqslant 1$ (see also Plavec 1968; Avni 1976; Bahcall 1978b; Rappaport 1979). If we define the radius of the companion as some fraction, β, of the radius of the critical potential lobe, then Equations (1.7) and (1.8) may be combined to yield an expression for the inclination angle:

$$\sin i \approx [1 - \beta^2 (R_L/a)^2]^{1/2}/\cos \theta_e. \tag{1.10}$$

We have ascertained that the use of Equations (1.7)–(1.10) over a wide range of values of q, θ_e, and Ω appropriate to X-ray binaries yields inclination angles with typical errors of only $\sim 1°$–$2°$.

For the case of 4U 0900–40 Equations (1.7), (1.8), and (1.10) have been modified to take into account the finite eccentricity of the orbit. Specifically, $\sin i$ (in Equation (1.10)) can be written as an analytic function of β, R_L/a, and the eccentric anomaly at ingress to eclipse. The quantities β and R_L are then redefined as the fraction of the critical potential lobe filled by the companion star and the size of the critical potential lobe, respectively, both evaluated at the time of periastron passage. A modified version of Equations (1.8) and (1.9), describing the size of the critical potential lobe at periastron passage, was calculated by use of the prescription given by Avni (1976).

From the above relations, we see that if K_c, $a_x \sin i$, and θ_e are measured, and if we make reasonable assumptions about the rotation rate of the companion star and fraction of its critical potential lobe that it occupies, then all of the system parameters mentioned above may be determined. The amplitude of the observed ellipsoidal light variations in several of these systems indicates that the companion stars nearly fill their critical potential lobes (i.e., $\beta \gtrsim 0.9$; see, e.g., Avni & Bahcall 1975a, b). It can then be argued that tidal dissipation should force these systems into approximately synchronous rotation (i.e., $\Omega \approx 1$; see Zahn 1975, 1977; Lecar, Wheeler & McKee 1976). However, in view of the ongoing evolution of such binaries under the influence of mass exchange within the system and mass and angular momentum losses from the system, it is uncertain whether synchronous rotation can always be enforced. Measurements of the rotational velocities of the companion stars in several of the X-ray binaries (see, e.g., Conti 1978; Crampton, Hutchings & Cowley 1978; Hutchings, Crampton & Cowley 1978), while lacking high precision, do yield values consistent with $0 \lesssim \Omega \lesssim 1.5$. (For other discussions of these issues, see van den Heuvel 1975; Avni & Bahcall 1976; Rappaport & Joss 1977c; Avni 1978; Bahcall 1978a, b; Conti 1978; and Petterson 1978.)

The best-fit values for the system parameters are easily found from Equations (1.5), (1.6), (1.7), and (1.10) by inserting the most probable values for K_c, $a_x \sin i$, and θ_e, and a best guess for Ω and β. The values that we used for K_c, $a_x \sin i$, and θ_e in our analysis are given in Table 1.2. A difficulty arises in propagating both the experimental errors in these quantities and the theoretical and experimental uncertainties in Ω and β. We have therefore evaluated the system parameters and their uncertainties by means of a Monte Carlo error propagation technique (Rappaport, Joss & Stothers 1980). In 2×10^4 trial evaluations, $a_x \sin i$ and K_c were chosen randomly with respect to Gaussian distributions of the appropriate widths, while the values of θ_e were chosen randomly and uniformly in the appropriate range, to reflect the experimental uncertainties. To simulate the theoretical and observational uncertainties in the values of β and Ω, we chose values of β randomly and uniformly over the range 0.9–1.0 (see Note 3) and Ω randomly and uniformly between 0 and 1.5.

1.4.2 Properties of the companion stars

The ranges of companion-star masses and radii for SMC X-1, Cen X-3, 4U 0900−40, and 4U 1538−52, as determined by the Monte Carlo error propagation technique, are shown in Figure 1.12 as contours of constant probability, and are also listed in Table 1.3. The most probable parameter values (see Table 1.3) are in good agreement with previously reported values (see Tables 1.1 and 1.2 for references). The outer contour, which contains at least 95 per cent of the Monte Carlo events, represents reasonably secure error limits for the masses and radii.

If the effective surface temperature of the companion star, T_e, is measured, the same Monte Carlo error propagation technique also yields its absolute luminosity and hence its position in the Hertzsprung-Russell (H-R) diagram.

The most probable values of T_e used in our analysis of these companion stars are given in Table 1.2. Two independent uncertainties in the temperature were taken into account (Underhill *et al.* 1979; Lamers 1981a; Remie & Lamers 1981): the uncertainty in spectral classification was represented by a flat probability distribution of ±8 per cent around the most probable value of T_e, while the uncertainty in the effective temperature calibration scale for early-type stars was taken to be a Gaussian distribution with a width (1σ) of 0.05 T_e.

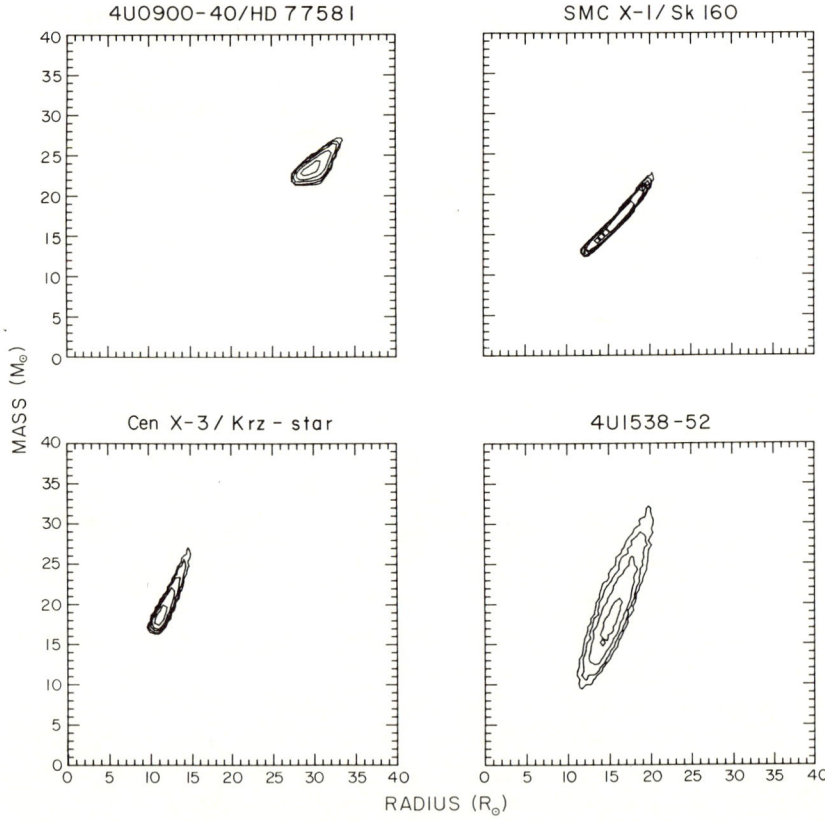

Fig. 1.12. Computer-generated error contours of the mass and radius of four companion stars in binary X-ray pulsar systems (updated from Rappaport & Joss 1981). The input data are the orbital period, the projected semi-major axis of the neutron-star orbit, the X-ray eclipse duration, and the amplitude of the optical Doppler velocity curve (see text). The error contours are derived from a Monte Carlo analysis (Rappaport, Joss & Stothers 1980; see text) and are generated directly from the raw output of the Monte Carlo analysis. The various contours represent arbitrary confidence levels; however, the outer contour contains ~95 per cent of the Monte Carlo events.

The resultant locations and corresponding uncertainties of the above four companion stars in the H–R diagram are shown in Figure 1.13. Also included in Figure 1.13 is the optical counterpart to LMC X-4. This eclipsing X-ray binary system, with $P_{orb} = 1.4$ d (Chevalier & Ilovaisky 1977; Li, Rappaport & Epstein 1978; White 1978; Hutchings, Crampton & Cowley 1978), has recently been discovered to be an X-ray pulsar with a pulse period of 13.5 s (Kelley et al. 1981c). These observations have also yielded a semi-major axis for the orbit of the neutron star of 30 ± 5 lt s. Thus, all the required ingredients for the type of analysis we have been discussing are now available for the LMC X-4 system.

The uncertainties in M_c, R_c, and position in the H–R diagram (see Figures 1.12 and 1.13) are important to bear in mind when trying to fit the companion stars into an evolutionary scenario involving mass-losing supergiant companions (see, e.g., van den Heuvel 1976, 1977; Conti 1978; Chapter 8, this book). In particular, the claim that the companion stars in X-ray binaries are undermassive for their luminosity or overluminous for their mass (see, e.g., Conti 1978; Hutchings et al. 1979) is apparently borne out, though we caution that in several cases the uncertainties displayed in Figure 1.13 are of the same order as the effect in question.

In the case of the Her X-1-HZ Her system, a reliable optical Doppler velocity curve is not yet available; however, extensive studies of optical pulsations

Table 1.3. *Derived binary system parameters*[a]

	Companion mass (M_c)	Companion radius (R_c)	Neutron-star mass (M_x)	Inclination angle (i)
SMX X-1	$17.0 \pm ^{4.5}_{4.0}$	16.5 ± 4	$1.05 \pm ^{0.40}_{0.30}$	$57°$–$77°$
Cen X-3	$19.0 \pm ^{6.0}_{2.5}$	$12.2 \pm ^{2.8}_{2.2}$	$1.07 \pm ^{0.63}_{0.57}$	$>63°$
4U 0900−40	$23.0 \pm ^{3.5}_{1.5}$	$31.0 \pm ^{4}_{3}$	$1.85 \pm ^{0.35}_{0.30}$	$>73°$
4U 1538−52	$18.5 \pm ^{12}_{6.5}$	$16.0 \pm ^{5}_{4}$	$1.87 \pm ^{1.33}_{0.87}$	$>60°$
LMC X-4	$19.0 \pm ^{35}_{13}$	$9.0 \pm ^{4.5}_{3.5}$	$1.70 \pm ^{1.90}_{1.00}$	$>58°$
Her X-1	$2.35 \pm ^{0.20}_{0.40}$	$4.05 \pm ^{0.25}_{0.40}$	$1.45 \pm ^{0.35}_{0.40}$	$>80°$
PSR 1913+16	–	–	1.35 ± 0.15 [b]	$40°$–$76°$

[a] All dimensioned quantities are in solar units; the quoted uncertainties are 95 per cent confidence limits.
[b] From Taylor et al. (1979); the mass of the companion star is also measured, but not included in the table because its nature is very different from the other companion stars (it is probably another neutron star).

(Middleditch & Nelson 1976), which are apparently produced by the reprocessing of X-ray pulsations that strike the surface of the companion star, have provided additional information needed to estimate the system parameters (Bahcall & Chester 1977; Chester 1978). We have determined the most probable Her X-1 system parameters and their uncertainties by use of a modified version of our Monte Carlo analysis program. This analysis incorporates the measurements of

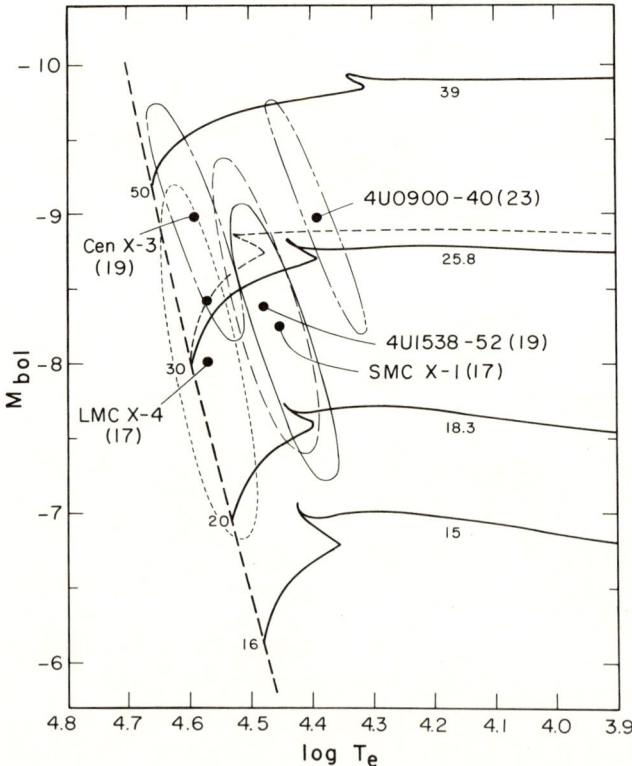

Fig. 1.13. Location in the H–R diagram of the companion stars in five massive binary X-ray sources. The region of uncertainty (~90 per cent confidence) in the H–R diagram for each source was computed with the Monte Carlo error propagation technique discussed in the text. We note that the absolute bolometric magnitude derived in this manner effectively determines the distance to the source and is independent of any other measurements of the source distance. Numbers in parentheses are the most probable masses, in solar units, of the companion stars. The solid curves are evolutionary tracks computed by van der Linden (1981); these include the effects of stellar-wind mass loss according to an empirical fitting formula derived by Lamers (1981b). The smaller numbers near each curve denote the mass, in solar units, of the companion when it has $\log T_e \sim 4.2$. The dashed curve, shown for comparison, is an evolutionary track for a constant mass of 30 $M_\odot$ (van der Linden 1981).

X-ray pulsars in massive binary systems

P_{orb}, $a_x \sin i$, θ_e, and the Doppler velocity of the optical pulsations, and utilizes a geometrical model for the reprocessing of X-ray pulsations into optical pulsations developed by Middleditch & Nelson (1976) and Bahcall & Chester (1977; in particular their equation (3)). For Her X-1 only we restricted β to the range of $0.95 < \beta < 1.0$ (see, e.g., Bahcall & Chester 1977). The results of this analysis are again given in Table 1.3.

1.4.3 Neutron-star masses

In Figure 1.14, we show the probability distributions for additional system parameters of SMC X-1, Cen X-3, 4U 0900−40, 4U 1538−52, LMC X-4, and Her X-1, as determined by the Monte Carlo error propagation technique described above. Among these system parameters, probably the ones of greatest physical and astrophysical interest are the masses of the neutron stars that such systems contain.

The measured neutron-star masses for these six systems, derived from the probability distributions displayed in Figure 1.14, are shown in Figure 1.15 (cf.

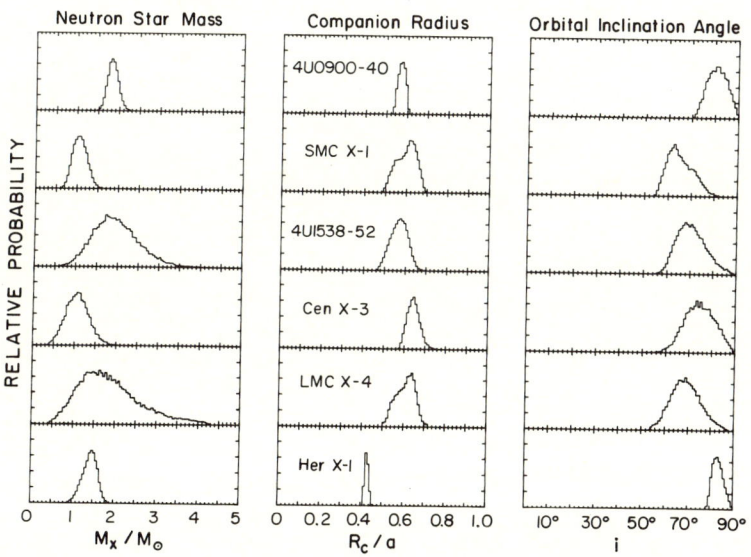

Fig. 1.14. Several of the binary system parameters and their uncertainties, for the SMC X-1, 4U 0900−40, Cen X-3, 4U 1538−52, LMC X-4, and Her X-1 systems, as determined by the Monte Carlo analysis technique described in the text. The quantities M_x, R_c/a, and i are the neutron-star mass, the ratio of the radius of the companion star to the orbital separation, and the orbital-inclination angle respectively. Note that the inclination-angle distribution plotted here will differ from the distribution of $\sin i$ by a factor of $\cos i$ (so that, for example, the latter distribution may peak at $\sin i = 1$ even though the distribution shown here goes to zero as i approaches $90°$).

Joss & Rappaport 1976; Rappaport & Joss 1977c; Avni 1977; Bahcall 1978b; Rappaport & Joss 1981). The most probable masses (indicated by the filled circles) and the corresponding uncertainties are obtained directly from the distributions in Figure 1.14. The shorter error bars shown in Figure 1.15 represent the uncertainties in the neutron-star masses for the less conservative assumptions that $\beta = \Omega = 1$ (Roche geometry). Also included in the figure is the mass of the binary radio pulsar PSR 1913+16, which is taken from Taylor, Fowler & McCulloch (1979) and is added to our sample for completeness. These seven cases represent the only reliable mass measurements for objects known to be neutron stars. Not included are the masses of the X-ray stars in such systems as 4U 1700−37 (Hutchings 1976) and Cyg X-2 (Cowley, Crampton & Hutchings

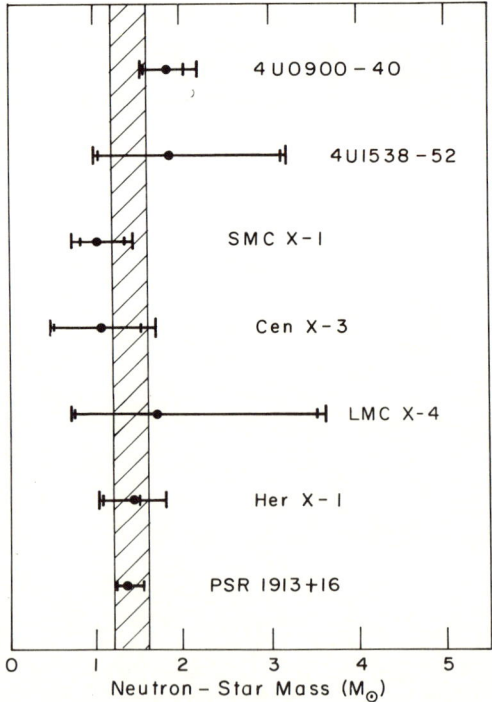

Fig. 1.15. Empirical knowledge of neutron-star masses. Six of the neutron-star masses are derived from observations of binary X-ray pulsars (see text). PSR 1913+16 is a binary radio pulsar (Taylor, Fowler & McCulloch 1979) and is added for completeness. The most probable value for the mass of each neutron star is indicated by the filled circle. For the X-ray binary systems, an inner set of error limits is also shown, corresponding to the less conservative assumptions that $\beta = 1$ and $\Omega = 1$ (see text). The hatched region represents the range of neutron-star masses (1.2–1.6 $M_\odot$) that might be expected on the basis of current theoretical scenarios for neutron-star formation (see text).

Fig. 1.16. Theoretical equations of state (pressure, P, plotted against gravitational mass density, ρ) for extremely dense matter. At the indicated densities, thermal contributions to the pressure are negligible unless the temperature is extremely high (note that $\rho_N \approx 2.5 \times 10^{14}\,\text{g cm}^{-3}$ is typical of the density within an atomic nucleus). (1) The mean-field (MF) equation of state for nuclear matter (Pandharipande & Smith 1975a), which is relatively stiff (i.e., $dP/d\rho$ is relatively large). (2) The tensor-interaction (TI3) equation of state for nuclear matter, obtained by Pandharipande & Smith (1975b) from their nuclear-interaction potential 3, which is of intermediate stiffness. Note the presence of a first-order phase transition (dashed line) between a neutron liquid and a neutron solid between $\rho \approx 7 \times 10^{14}\,\text{g cm}^{-3}$ and $\rho \approx 9 \times 10^{14}\,\text{g cm}^{-3}$. (3) The BJVH equation of state for nuclear matter (Malone, Johnson & Bethe 1975), which is relatively soft (i.e., $dP/d\rho$ is relatively small). (4) The TI3 equation of state joined to an equation of state for quark matter, under the assumption that there is a first-order phase transition (dashed line) from nuclear matter to quark matter between $\rho \approx 5 \times 10^{14}\,\text{g cm}^{-3}$ and $\rho \approx 9 \times 10^{14}\,\text{g cm}^{-3}$ (Freedman & McLerran 1978; Fechner & Joss 1978). (5) The BJVH equation of state joined to an equation of state for quark matter, under the assumption that there is a second-order phase transition from nuclear matter to quark matter at $\rho \approx 6 \times 10^{14}\,\text{g cm}^{-3}$. Somewhat softer equations of state than those shown here are possible if, for example, a pion condensate forms at densities in excess of ρ_N (see Baym & Pethick 1979 and references therein).

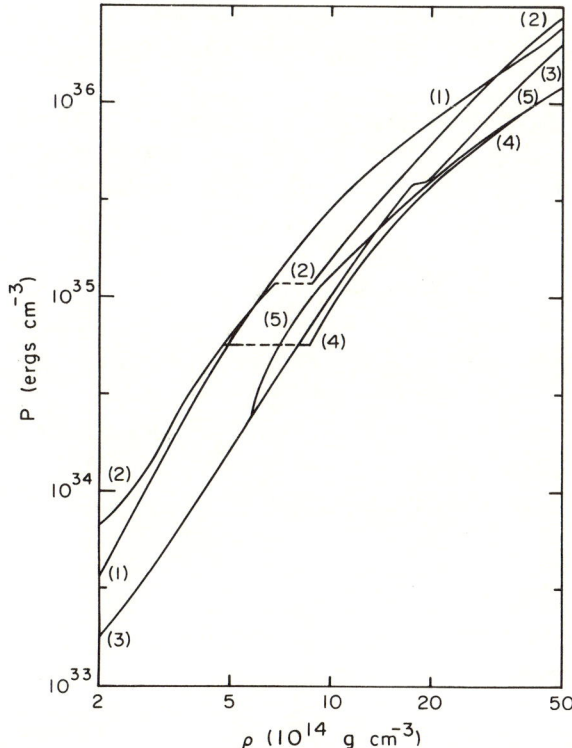

1979), where, in view of the lack of detectable pulsations, we do not know whether these objects are neutron stars.

The masses given in Figure 1.15 are all consistent, within the uncertainties, with a mass of $1.4 \pm 0.2\,M_\odot$ (the shaded region on Figure 1.15). This is the range of masses that might be expected for neutron stars if they are formed in the collapse of the degenerate cores of highly evolved stars (Arnett & Schramm 1973; Iben 1974 and references therein; see Chapter 8, this book, for an extensive discussion of binary evolution) or the collapse of accreting degenerate dwarfs in close-binary stellar systems (see, e.g., Canal & Schatzman 1976; Nomoto 1981 and references therein).

Furthermore, all of the measured neutron-star masses are consistent with neutron-star and quark-star models based on conventional many-body nuclear and high-energy physics (Arnett & Bowers 1977; Fechner & Joss 1978; Baym & Pethick 1979; and references therein). Some of the equations of state for the deep interiors of such objects that have been put forth in the literature are illustrated in Figure 1.16. For a given equation of state and choice of central density, ρ_c, and under the assumption of spherical symmetry, the general relativistic equation of hydrostatic equilibrium (Oppenheimer & Volkoff 1939) can be simply integrated to find the neutron-star mass, radius, moment of inertia, and so forth. In Figure 1.17 the gravitational mass derived from this type of calculation is shown as a function of ρ_c for the equations of state given in Figure 1.16. Maximum allowed masses for non-extreme equations of state lie between ~ 1.4 and $\sim 2.7\,M_\odot$. Thus, the currently available observational mass estimates for neutron stars are marginally sufficient to constrain the equation of state of matter at very high densities.

1.4.4 The apsidal motion test in X-ray binaries

Almost all of our knowledge of stellar interiors is derived from theoretical studies. At present, very few observations can be performed to test any theoretical predictions. Two direct measurements are, however, possible: the measurement of the neutrino flux from the sun (see Davis 1978 for a review) and the apsidal motion test for binary stellar systems (Russell 1928). In the former case, one attempts to gain information about the thermal and chemical structure of the solar interior, while in the latter, a one-parameter measure of the mass distribution within a star may be obtained. Reliable measurements of apsidal motion, which results from the tidally and rotationally induced gravitational quadrupole moment of a star, have been made for about 16 close-binary systems (Kopal 1965; see Stothers 1974 for more recent references), wherein both stars are non-degenerate and are usually on the main sequence. The rate at which the longitude of periastron, ω, of an eccentric orbit advances can then be related to the 'apsidal motion constant', k, which is calculated for various stellar model mass distributions under standard theoretical assumptions (see Schwarzschild 1958). In general, the measured values of k are smaller than the standard calculated values by factors of ~ 2. In all cases, however, the interpretation is complicated by the fact that both stars of the binary system contribute to the apsidal motion.

Binary X-ray pulsars, on the other hand, provide a potentially less ambiguous means of carrying out measurements of this type. In such systems, the neutron star acts essentially as a point mass and has a completely negligible apsidal motion constant compared to that of the companion star. Of the well studied X-ray binaries, only GX 301−2 (White, Mason & Sanford 1978; Kelley, Rappaport & Petre 1980; Watson, Warwick & Corbet 1981), 4U 0900−40

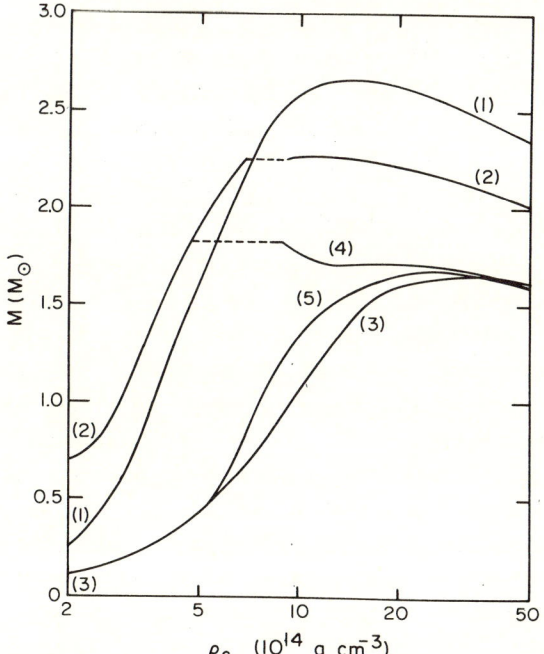

Fig. 1.17. Gravitational mass, M, plotted against central density, ρ_c, for hydrostatic neutron-star and quark-star models based on the five equations of state shown in Figure 1.16. Regions where $dM/d\rho_c > 0$ are dynamically stable, while regions where $dM/d\rho_c < 0$ are unstable. Note the region where $dM/d\rho_c = 0$ (dashed line) for the TI3 equation of state, which corresponds to a first-order phase transition from a neutron liquid to a neutron solid, and a similar region corresponding to the first-order phase transition between the TI3 and quark-matter equations of state. The maximum stable masses for these neutron-star and quark-star models range from ~ 1.7 to $\sim 2.7\, M_\odot$.

(Rappaport, Joss & McClintock 1976), and 4U 0115+63 (Rappaport *et al.* 1978) have measurably eccentric orbits. The orbit of GX 301−2 has only recently been measured, and the observational baseline is too short to allow a determination of the rate of apsidal motion, $\dot\omega$. The 4U 0900−40 and 4U 0115+63 systems are therefore the best candidate systems in which to attempt measurements of $\dot\omega$.

The orbit of 4U 0900−40 was first measured in July 1975 (Rappaport, Joss & McClintock 1976) and found to have a small but detectable eccentricity of $e \approx 0.1$. The pulse phase of 4U 0900−40 was again tracked for one orbital cycle with SAS-3 during November 1978 (Rappaport, Joss & Stothers 1980). In order to obtain constraints on the rate of apsidal motion in the 4U 0900−40 system, we performed an orbital fit to the joint 1975-1978 data set (Rappaport, Joss & Stothers 1980). We obtained only an upper limit to the apsidal motion: $\dot\omega < 3.8°$ yr^{-1} (97 per cent confidence). More recent measurements of ω by Nagase *et al.* (1981) confirm this limit and marginally improve on it.

For a binary system where the orbital and stellar parameters are known, limits on $\dot\omega$ can be used to constrain the allowed values of k through the relation

$$\dot\omega \approx \left(\frac{2\pi k}{P_{\rm orb}}\right)\left(\frac{R_{\rm c}}{a}\right)^5 [15qg(e) + \Omega^2(1+q)h(e)] \qquad (1.11)$$

(Cowling 1938; Sterne 1939). The functions $g(e)$ and $h(e)$ in this expression are slightly greater than unity and incorporate the contributions to $\dot\omega$ from the non-infinitesimal eccentricity of the orbit. The contribution to $\dot\omega$ from general relativistic effects (see, e.g., Landau & Lifshitz 1962) is negligible in this system. (Equation (1.11) is derived from considerations of equilibrium tides only; see Zahn (1977) and Papaloizou & Pringle (1980) for a discussion of the possible effects of dynamical tides.) We have evaluated the quantity $(R_{\rm c}/a)^5 \times [15qg + \Omega^2(1+q)h]$ and its uncertainties for the 4U 0900−40 system by means of the Monte Carlo technique described in Section 1.4.2. Our limit on apsidal motion then yields a limit on $\log k$ ($\lesssim -2.5$) that is a direct measure of the structure of the companion star HD 77581.

In collaboration with R. Stothers, we have calculated values of k for a wide range of stellar models (Rappaport, Joss & Stothers 1980). The results of some of these calculations are shown in Figure 1.18 as contours of constant $\log k$ in the $M_{\rm c}$-$R_{\rm c}$ plane. We have also evaluated $M_{\rm c}$, $R_{\rm c}$, and the corresponding uncertainties for HD 77581 by the methods discussed in Section 1.4.2 (Rappaport, Joss & Stothers 1980). The results are shown superposed on the contours of $\log k$ in Figure 1.18.

The heavy curve in each of Figures 1.18(*a*) and (*b*) separates the allowed from the excluded regions, based on our limit for k. The experimental limit on k and the theoretical calculations of k are inconsistent for $T_{\rm e} \lesssim 18\,000$ K (not shown in Figure 1.18). For $T_e = 20\,000$ K the lack of observed apsidal motion is just consistent with theoretical expectations, while for $T_e \gtrsim 23\,000$ K apsidal motion should not have been observed. Thus, these results in principle constrain the effective temperature of HD 77581.

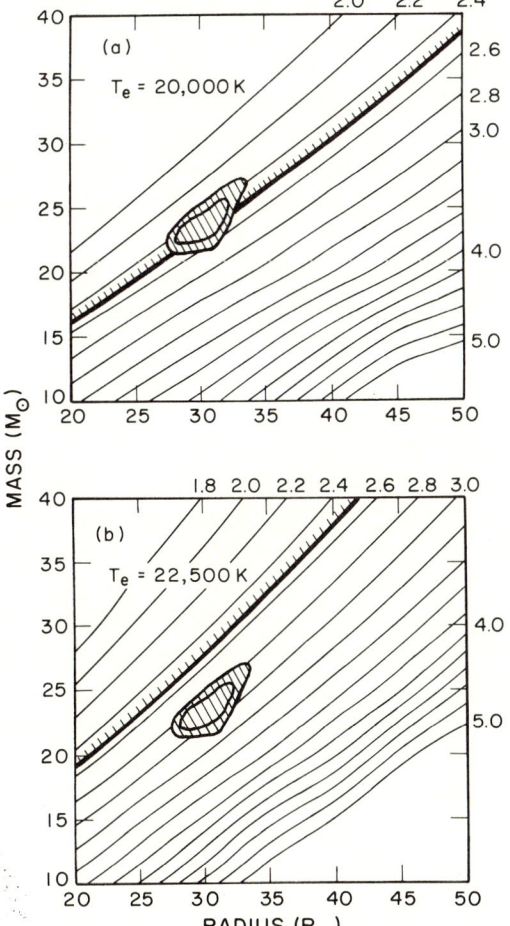

Fig. 1.18. Contours of apsidal motion constant, k, in the mass–radius plane for HD 77581, calculated for two assumed values of the effective temperature, T_e. (a) $T_e = 20\,000$ K; (b) $T_e = 22\,500$ K (from Rappaport, Joss & Stothers 1980). Numbers to the top and to the right of each plot denote the value of $-\log k$ for the various contours. The shaded error region for M_c and R_c and the region enclosed by the inner curve contain 95 per cent and 70 per cent of the Monte Carlo events respectively. The most probable parameter values are $M_c = 23\,M_\odot$ and $R_c = 31\,R_\odot$ (see Table 1.3). The heavy curve is the limit on k as measured with SAS-3 and separates the allowed (lower) from the excluded (upper) region.

We anticipate that another accurate measurement of $\dot\omega$, a few years in the future, should be able to detect the apsidal motion of 4U 0900−40. This will provide a unique measure of the internal mass distribution of an evolved supergiant. Such a measurement will, in turn, yield a direct check on theoretical stellar models and on our understanding of the evolution of close-binary stellar systems.

The transient source 4U 0115+63 has been detected in X-rays on only three occasions during the past decade (Cominsky et al. 1978; Ricketts et al. 1981). The orbit of this system was measured with high accuracy (see Section 1.4.1) during the outburst in 1978 (Rappaport et al. 1978); the longitude of periastron at this time was $\omega = 47.7° \pm 0.1°$. Significant pulse timing data are also available (Kelley et al. 1981b) from the original *Uhuru* discovery observation (Giacconi et al. 1972) and from a recent 15 d observation with the UK-6 satellite (Ricketts et al. 1981).

An analysis (Kelley et al. 1981b) of the 1971 *Uhuru* data on 4U 0115+63 has yielded a value for the longitude of periastron seven years prior to the SAS-3 observation. The best-fit orbital solution gave $\omega = 51° \pm 19°$ (95 per cent confidence), which implies an upper limit to a positive apsidal motion in the 4U 0115+63 system of $\dot\omega \lesssim 2°$ yr^{-1} (95 per cent confidence). Observations of 4U 0115+63 with the UK-6 satellite (Ricketts et al. 1981) have yielded an even more stringent upper limit on apsidal motion of $\dot\omega \lesssim 10$ arcmin yr^{-1}. The measured Doppler delays from these latter observations are shown in Figure 1.19.

There have been very few optical observations of the companion star in this system (see, e.g., Johns et al. 1978; Hutchings & Crampton 1981); however, the available evidence indicates that it is probably a B emission-line star of moderate mass and radius (Hutchings & Crampton 1981). A main-sequence B star is expected to have $\log k \approx -2.0$ (see, e.g., Schwarzschild 1958; Kopal 1965; Stothers 1974). The value of $\dot\omega$ for the 4U 0115+63 system, as estimated from Equation (1.11), is then only ~ 0.3 arcmin yr^{-1} for a non-rotating companion star with a nominal radius of ~ 10 R$_\odot$. For rapid rotation rates the apsidal motion increases as Ω^2 (see Equation (1.11)); a companion star rotating at a substantial fraction of the breakup speed (as is often observed in Be stars [Slettebak 1979]), could produce a value of $\dot\omega$ as large as 10 arcmin yr^{-1} (see Kelley et al. 1981b). Thus, the present observational upper limit is still marginally consistent with the largest value of $\dot\omega$ that could reasonably be expected for this system. A rate of apsidal motion as small as 1 arcmin yr^{-1} in 4U 0115+63 will be detectable if this system can be observed for a complete orbital cycle during an outburst at least ~ 5 yr in the future. Such a measurement would comprise an accurate determination of k for a seemingly unevolved companion star in an X-ray binary system.

1.5 Summary

During the past decade, the study of X-ray pulsars in massive binary systems has provided a great stimulus to many areas of astrophysics. Of particular importance are the measurements of the masses of neutron stars (Section 1.4.2) and their relation to stellar evolution, nuclear and high-energy physics, and gravitation theory. Detailed information has been obtained on the companion stars (Section 1.4.2), and this should in turn assist the development of a coherent picture of binary stellar evolution in these systems (see Chapter 8, this book). Binary X-ray pulsars have also provided a significant new probe into the mass transfer processes in binary stellar systems and an impetus to theoretical investigations of accretion flows, accretion disks (see Chapter 10, this book), and accretion torques (see Section 1.3 and Chapter 11, this book). Preliminary efforts have also been made to utilize the X-ray pulsations as probes of the internal structure of both the emitting neutron stars (see references at the end of Section 1.3) and their companion stars (Section 1.4.4).

At the present time, the data banks of the Ariel 5, OSO-8, SAS-3, HEAO-1, Hakucho, UK-6, and Einstein X-ray astronomy satellites may well contain

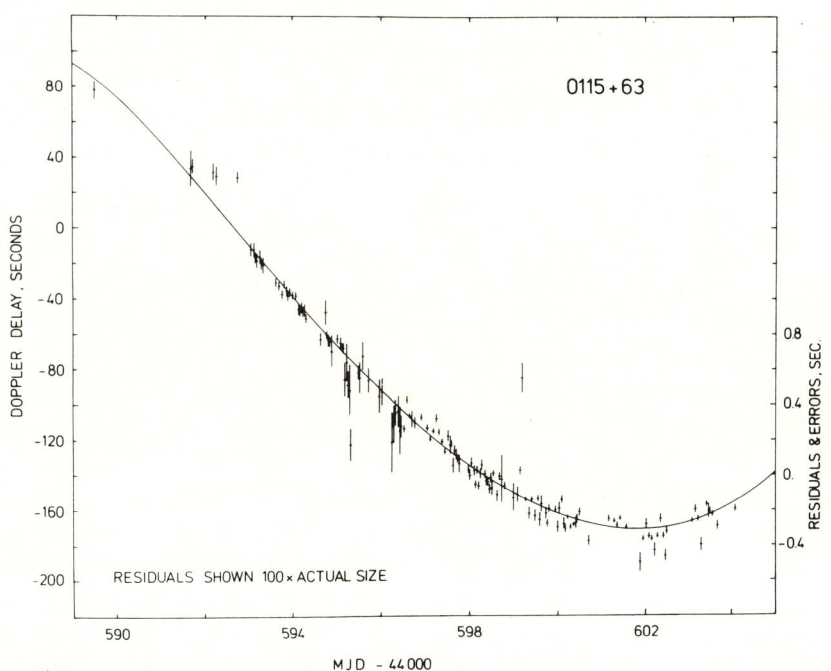

Fig. 1.19. Doppler delay data from 4U 0115+63 measured with the UK-6 satellite (Ricketts *et al.* 1981; Ricketts 1981) during the 1981 X-ray outburst. The solid curve is an eccentric-orbit fit to the data.

significant new data on binary X-ray pulsars. These data will require several more years for a complete analysis. We are hopeful that the European satellite EXOSAT (due for launch in 1983) will be used to pursue new X-ray pulsar observations over the next few years.

In the future, a dedicated X-ray timing satellite (XTE) will continue and substantially advance the study of X-ray pulsars and other compact X-ray sources that was carried out so successfully during the 1970s.

Acknowledgements

The authors acknowledge numerous stimulating discussions with H. Henrichs, R. Kelley, S. Langer, R. McCray, G. Savonije, R. Takens, E. P. J. van den Heuvel, F. Verbunt, and J. van Paradijs. We thank R. Kelley and W. Fechner for carrying out a number of the calculations utilized in this work, T. van der Linden for providing us with his stellar evolution calculations, and M. Ricketts for providing the Doppler delay data on 4U 0115+63. We also thank T. Dobson for her able assistance in the preparation of the manuscript. We are grateful to J. Sullivan for his valuable advice concerning the development of the computer programs used in the generation of some of the figures. S.R. acknowledges the hospitality of E. P. J. van den Heuvel and the Astronomical Institute of the University of Amsterdam during his visit. This work was supported in part by the National Aeronautics and Space Administration under contract NAS5-24441 and grants NSG-7643 and NGL-22-009-638, by the National Science Foundation under grant AST7 8-21993, and by the Netherlands Organization for Pure Research, ZWO, under contract Nr. B78-183.

Notes

[1] In this review, we shall restrict our attention to sources that are widely believed to be neutron stars undergoing accretion in binary stellar systems, and we shall use the term 'X-ray pulsars' to refer exclusively to these sources. We shall not discuss other sources that have been observed to emit periodic X-ray pulsations. In particular, we exclude from further consideration such sources as the Crab pulsar NP 0532 (which is thought to be an isolated neutron star whose radiated energy is derived from its rotational kinetic energy; Manchester & Taylor 1977), the 5 March 1979 γ-ray transient (whose underlying character is unknown, but which shows no evidence of binary membership; Mazets et al. 1979), and sources, including H 2252−035 (White & Marshall 1981), which are thought to be accreting degenerate dwarfs rather than neutron stars.

[2] Recent theoretical efforts (see Baym & Pethick 1979 for references) have pointed increasingly to the possibility that neutron-rich matter undergoes a transition to a fluid composed of 'free' (but strongly interacting) quarks at densities not far in excess of nuclear-matter densities ($\rho \approx 2.5 \times 10^{14}$ g cm^{-3}). If this is the case, then much of the matter within a 'neutron' star may actually be composed of quark matter, and such objects might be more accurately called 'quark stars'. However, many of the observable macroscopic properties of quark stars (masses, radii,

moments of inertia, and so forth) need not be different from those of neutron stars (Fechner & Joss 1978). For simplicity, in this review we shall refer to all such objects as 'neutron stars'.

[3] Note that our definition of β uses the effective radii of the critical potential lobe and the companion star, and does *not* refer to distances along the line joining the two stars.

References

Arnett, W. D. & Bowers, R. L. 1977, *Ap. J. Suppl.*, **33**, 415.
Arnett, W. D. & Schramm, D. N. 1973, *Ap. J.*, **184**, L47.
Arons, J. & Lea, S. M. 1980, *Ap. J.*, **235**, 1016.
Avni, Y. 1976, *Ap. J.*, **209**, 574.
Avni, Y. 1977, talk presented at the 16th General Assembly of the IAU. In: *Highlights of Astronomy*, ed. E. Müller. Vol. 4, Part 1, p. 137.
Avni, Y. 1978. In: *Physics and Astrophysics of Neutron Stars and Black Holes*, eds. R. Giacconi & R. Ruffini (Amsterdam: North Holland), p. 43.
Avni, Y. & Bahcall, J. N. 1975a, *Ap. J.*, **197**, 675.
Avni, Y. & Bahcall, J. N. 1975b, *Ap. J. (Lett.)*, **202**, L131.
Avni, Y. & Bahcall, J. N. 1976, *X-Ray Binaries* (NASA SP-389), p. 615.
Baan, W. A. & Treves, A. 1973, *Astron. Ap.*, **22**, 421.
Bahcall, J. 1978a, In: *Physics and Astrophysics of Neutron Stars and Black Holes*, eds. R. Giacconi & R. Ruffini (Amsterdam: North Holland), p. 63.
Bahcall, J. 1978b, *Ann. Rev. Astron. Ap.*, **16**, 241
Bahcall, J. N. & Chester, T. J. 1977, *Ap. J. (Lett.)*, **215**, L21.
Basko, M. M. & Sunyaev, R. A. 1975, *Astron. and Ap.*, **42**, 311.
Basko, M. M. & Sunyaev, R. A. 1976, *MNRAS*, **175**, 395.
Baym, G. & Pethick, C. 1979, *Ann. Rev. Astron. and Ap.*, **17**, 415.
Becker, R. H., Swank, J. H., Boldt, E. A., Holt, S. S., Pravdo, S. H., Saba, J. R. & Serlemitsos, P. J. 1977a, *Ap. J. (Lett.)*, **216**, L11.
Becker, R. H., Boldt, E. A., Holt, S. S., Pravdo, S. H., Rothschild, R. E., Serlemitsos, P. J., Smith, B. W. & Swank, J. H. 1977b, *Ap. J.*, **214**, 879.
Becker, R. H., Rothschild, R. E., Boldt, E. A., Holt, S. S., Pravdo, S. H., Serlemitsos, P. J. & Swank, J. H. 1978, *Ap. J.*, **221**, 912.
Bisnovatyi-Kogan, G. S. 1973, *Astron. Zh.*, **50**, 902.
Bondi, H. & Hoyle, F. 1944, *MNRAS*, **104**, 273.
Boynton, P. E. 1981, In: *IAU Symposium No. 95, Pulsars*, eds. W. Sieber & R. Wielebinski (Dordrecht: Reidel), p. 279.
Boynton, P. & Deeter, J. 1979, In: *Compact Galactic X-Ray Sources*, eds. F. Lamb & D. Pines (Urbana, Ill., Physics Dept., Univ. of Illinois), p. 168.
Bradt, H. V., Mayer, W., Buff, J., Clark, G. W., Doxsey, P. et al. 1976, *Ap. J. (Lett.)*, **204**, L67.
Bradt, H., Doxsey, R. & Jernigan, J. 1979, In: *X-Ray Astronomy*, eds. W. A. Baity & L. E. Peterson (Oxford: Pergamon), p. 3.
Canal, R. & Schatzman, E. 1976, *Astron. Ap.*, **71**, 217.
Chester, T. J., 1978, *Ap. J.*, **222**, 652.
Chevalier, C. & Ilovaisky, S. A. 1977, *Astron. Ap.*, **59**, L9.
Cominsky, L., Clark, G. W., Li, F., Mayer, W. & Rappaport, S. 1978, *Nature*, **273**, 367.
Conti, P. S. 1978, *Astron. and Ap.*, **63**, 225.
Cowley, A. P., Crampton, D. & Hutchings, J. B. 1979, *Ap. J.*, **231**, 539.
Cowling, T. G. 1938, *MNRAS*, **98**, 734.
Crampton, D., Hutchings, J. B. & Cowley, A. P. 1978, *Ap. J. (Lett.)*, **225**, L63.
Darbro, W., Ghosh, P., Elsner, R. F., Weisskopf, M. C., Sutherland, P. G. & Grindlay, J. E. 1981, *Ap. J.*, **246**, 231.
Davidsen, A., Malina, R. & Bowyer, S. 1976, In: *X-Ray Binaries* (NASA SP-389), p. 691.

Davidson, K. 1973, *Nature Phys. Sci.*, **246**, 1.
Davidson, K. & Ostriker, J. P. 1973, *Ap. J.*, **179**, 585.
Davies, R. E., Fabian, A. C. & Pringle, J. E. 1979, *MNRAS*, **186**, 779.
Davis, R. Jr. 1978, *In: Proceedings of the Brookhaven Solar Neutrino Conference* (Brookhaven National Laboratory Report No. BNL 50878), Vol. 1, p. 1.
Davison, P. J. N. 1977, *MNRAS*, **179**, 35P.
Davison, P. J. N., Watson, M. G. & Pye, J. P. 1977, *MNRAS*, **181**, 73P.
Deeter, J. E., Boynton, P. E. & Pravdo, S. H. 1981, *Ap. J.*, **247**, 1003.
Elsner, R. F. & Lamb, F. K. 1976, *Nature*, **262**, 356.
Elsner, R. F., Ghosh, P. & Lamb, F. K. 1980, *Ap. J. (Lett.)*, **241**, L155.
Fabbiano, G. & Schreier, E. J. 1977, *Ap. J.*, **214**, 235.
Fahlman, G. G. & Gregory, P. C. 1981, *Nature*, **293**, 202.
Fechner, W. B. & Joss, P. C. 1977, *Ap. J. (Lett.)*, **213**, L57.
Fechner, W. B. & Joss, P. C. 1978, *Nature*, **274**, 347.
Forman, W., Jones, C., Tananbaum, H., Gursky, H., Kellogg, E. & Giacconi, R. 1973, *Ap. J. (Lett.)*, **182**, L103.
Freedman, B. & McLerran, L. 1978, *Phys. Rev.*, **D17**, 1109.
Ghosh, P., Lamb, F. K. & Pethick, C. J. 1977, *Ap. J.*, **217**, 578.
Ghosh, P. & Lamb, F. K. 1978, *Ap. J. (Lett.)*, **223**, L83.
Ghosh, P. & Lamb, F. K. 1979, *Ap. J.*, **234**, 296.
Giacconi, R. 1974, *In: Astrophysics and Gravitation. Proceedings of the 16th International Solvay Congress*, p. 27 (Brussels: Universite de Bruxelles), p. 27.
Giacconi, R., Gursky, H., Kellogg, E., Schreier, E. & Tananbaum, H. 1971, *Ap. J. (Lett.)*, **167**, L67.
Giacconi, R., Murray, S., Gursky, H., Kellogg, E., Schreier, E. & Tananbaum, H. 1972, *Ap. J.*, **178**, 281.
Giacconi, R. (ed.) 1981, *X-ray Astronomy with the Einstein Satellite* (Dordrecht: Reidel).
Glass, I. S. & Feast, M. W. 1973, *Nature Phys. Sci.*, **245**, 39.
Gnedin, Yu. N. & Sunyaev, R. A. 1973, *Astron. and Ap.*, **25**, 233.
Gregory, P. C. & Fahlman, G. G. 1980, *Nature*, **887**, 805.
Gursky, H. & Schreier, E. 1975, *In: Neutron Stars, Black Holes and Binary X-Ray Sources*, eds. H. Gursky & R. Ruffini (Dordrecht: Reidel), p. 175.
Huckle, H. E., Mason, K. O., White, N. E., Sanford, P. W., Maraschi, L., Tarenghi, M. & Tapia, G. 1977, *MNRAS*, **180**, 21P.
Hutchings, J. B. 1976. *In: X-Ray Binaries* (NASA SP-389), p. 531.
Hutchings, J. B. & Crampton, D. 1981, *Ap. J.*, **247**, 222.
Hutchings, J. B., Crampton, D., Cowley, A. P. & Osmer, P. S. 1977, *Ap. J.*, **217**, 186.
Hutchings, J. B., Crampton, D. & Cowley, A. P. 1978, *Ap. J.*, **225**, 548.
Hutchings, J. B., Cowley, A. P., Crampton, D., van Paradijs, J. & White, N. E. 1979, *Ap. J.*, **229**, 1079.
Iben, I. 1974, *Ann. Rev. Astron. Ap.*, **12**, 215.
Inoue, H. 1975, *Pub. Astron. Soc. Japan*, **27**, 311.
Ives, J. C., Sanford, P. W. & Bell Burnell, S. J. 1975, *Nature*, **254**, 578.
Johns, M., Koski, A., Canizares, C. & McClintock, J. 1978, *IAU Circ., No. 3171*.
Johnston, M., Bradt, H., Doxsey, R., Gursky, H., Schwartz, D. & Schwarz, J. 1978, *Ap. J. (Lett.)*, **223**, L71.
Joss, P. C. & Rappaport, S. 1976, *Nature*, **264**, 219.
Joss, P. C., Avni, Y. & Rappaport, S. 1978, *Ap. J.*, **221**, 645.
Joss, P. C., Fechner, W. B., Forman, W. & Jones, C. 1978, *Ap. J.*, **225**, 994.
Joss, P. C., Li, F., Nelson, J. & Middleditch, J. 1980, *Ap. J.*, **235**, 592.
Kanno, S. 1980, *Pub. Astron. Soc. Japan*, **32**, 105.
Kelley, R., Rappaport, S. & Petre, R. 1980, *Ap. J.*, **238**, 699.
Kelley, R., Apparao, K., Doxsey, R., Jernigan, G., Naranan, S. & Rappaport, S. 1981*a*, *Ap. J.*, **243**, 251.
Kelley, R., Rappaport, S., Brodheim, M., Cominsky, L. & Stothers, R. 1981*b*, *Ap. J.*, **251**, 630.

Kelley, R., Jernigan, J., Levine, A., Petro, L. & Rappaport, S. 1981c, *Ap. J. (Lett.)*, in press.
Kelley, R., Ayasli, S. & Rappaport, S. 1982, *IAU Circ.*, No. 3667.
Klein, R. I., Arons, J. & Lea, S. M., 1981, in preparation.
Kopal, Z. 1965, *Adv. Astron. Astroph.*, 3, 89.
Kirk, J. R. & Galloway, J. J. 1981, *MNRAS*, 195, 45P.
Lamb, F. K., Pethick, C. J. & Pines, D. 1973, *Ap. J.*, 184, 271.
Lamb, F. K., Pines, D. & Shaham, J. 1978, *Ap. J.*, 224, 969.
Lamb, F. K. & Boynton, P. 1980, private communication.
Lamb, R. C., Markert, T. H., Hartman, R. C., Thompson, D. J. & Bignami, G. F. 1980, *Ap. J.*, 239, 651.
Lamers, H. J. G. L. M. 1981a, private communication.
Lamers, H. J. G. L. M. 1981b, *Ap. J.*, 245, 593.
Landau, L. D. & Lifshitz, E. M. 1962, *The Classical Theory of Fields* (Oxford: Pergamon), p. 374.
Langer, S. & Rappaport, S. 1981, *Ap. J.*, 257, 733.
Langer, S., Castor, J. & Rappaport, S. 1981, in preparation.
Lecar, M., Wheeler, J. C. & McKee, C. F. 1976, *Ap. J.*, 205, 556.
Lewin, W. H. G., Ricker, G. R. & McClintock, J. E. 1971, *Ap. J. (Lett.)*, 169, L17.
Li, F., Rappaport, S. & Epstein, A. 1978, *Nature*, 271, 37.
Li, F., Rappaport, S., Clark, G. W. & Jernigan, J. G. 1979, *Ap. J.*, 228, 893.
Lucke, R., Yentis, D., Friedman, H., Fritz, G. & Shulman, S. 1976, *Ap. J. (Lett.)*, 206, L25.
Malone, R. C., Johnson, M. B. & Bethe, H. A. 1975, *Ap. J.*, 199, 741.
Manchester, R. N. & Taylor, J. H. 1977, *Pulsars* (San Francisco: Freeman), p. 18.
Mason, K. O. 1977, *MNRAS*, 178, 81P.
Mazets, E. P., Goleneˇtskii, S. V., Il'inskii, Y. N., Aptekar, R. L. & Guryan, Yu. A. 1979, *Nature*, 282, 587.
McClintock, J. E. et al. 1976, *Ap. J. (Lett.)*, 206, L99.
McClintock, J. E., Rappaport, S., Nugent, J. J. & Li, F. K. 1977, *Ap. J. (Lett.)*, 216, L15.
Mészàros, P., Nagel, W. & Ventura, J. 1980, *Ap. J.*, 238, 1066.
Middleditch, J. & Nelson, J. 1976, *Ap. J.*, 208, 567.
Middleditch, J., Mason, K. O., Nelson, J. & White, N. 1981, *Ap. J.*, 244, 1001.
Nagase, F. 1981, In: *X-Ray Astronomy, Proceedings of the 15th ESLAB Symposium*, ed. R. D. Andersen (Dordrecht: Reidel), p. 395.
Nagase, F. et al. 1981, *Nature*, 290, 572.
Nomoto, K. 1981, In: *IAU Symposium 93. Fundamental Problems in the Theory of Stellar Evolution*, eds. D. Sugimoto, D. Q. Lamb & D. N. Schramm (Dordrecht: Reidel), p. 295.
Ögelman, H., Beuermann, K. P., Kanbach, G., Mayer-Hasselwander, H. A., Capozzi, D., Fiordilino, E. & Molteni, D. 1977, *Astron. Ap.*, 58, 385.
Oppenheimer, J. R. & Volkoff, G. 1939, *Phys. Rev.*, 55, 374.
Pandharipande, Y. R. & Smith, R. A. 1975a, *Phys. Lett.*, 59B, 15.
Pandharipande, Y. R. & Smith, R. A. 1975b, *Nuclear Physics*, A237, 507.
Papaloizou, J. & Pringle, J. E. 1980, *MNRAS*, 193, 603.
Pavlov, G. G. & Shibanov, Yu. A. 1979, *JETP*, 49, 741.
Pavlov, G. G. & Yakovlev, Yu. A. 1976, *JETP*, 43, 389.
Petterson, J. A. 1978, *Ap. J.*, 224, 625.
Plavec, M. 1968, *Adv. in Astron. and Ap.*, 6, 201.
Pounds, K. A., Cooke, B. A., Ricketts, M. J., Turner, M. J. & Elvis, M. 1975, *MNRAS*, 172, 473.
Pounds, K. A. 1976, Talk presented to the Meeting of the American Astronomical Society (HEAD), Cambridge, MA, January 1976.
Pravdo, S. H., White, N. E., Boldt, E. A., Holt, S. S., Serlemitsos, P. J., Swank, J. H. & Szymkowiak, A. E. 1979, *Ap. J.*, 231, 912.
Pravdo, S. H. & Bussard, R. W. 1981, *Ap. J. (Lett.)*, 246, L115.

Primini, F., Rappaport, S., Joss, P. C., Clark, G. W., Lewin, W., Li, F., Mayer, W. & McClintock, J. 1976, *Ap. J. (Lett.)*, **210**, L71.
Primini, F., Rappaport, S. & Joss, P. C. 1977, *Ap. J.*, **217**, 543.
Pringle, J. E. & Rees, M. J. 1972, *Astron. Ap.*, **21**, 1.
Rappaport, S. 1982, In: *Galactic X-Ray Sources*, eds. P. W. Sanford, P. Laskarideo & J. Salton (Chichester: Wiley), p. 159.
Rappaport, S., Joss, P. C. & McClintock, J. E. 1976, *Ap. J. (Lett.)*, **206**, L103.
Rappaport, S. & Joss, P. C. 1977a, *Nature*, **266**, 123.
Rappaport, S. & Joss, P. C. 1977b, *Nature*, **266**, 683.
Rappaport, S. & Joss, P. C. 1977c, *Ann. N.Y. Acad. Sci.*, **302**, 460.
Rappaport, S., Markert, T., Li, F. K., Clark, G. W., Jernigan, J. G. & McClintock, J. E. 1977, *Ap. J. (Lett.)*, **217**, L29.
Rappaport, S., Clark, G. W., Cominsky, L., Joss, P. C. & Li, F. K. 1978, *Ap. J. (Lett.)*, **224**, L1.
Rappaport, S., Joss, P. C. & Stothers, R. 1980, *Ap. J.*, **235**, 570.
Rappaport, S. & Joss, P. C. 1981, In: *X-Ray Astronomy with the Einstein Satellite*, eds. R. Giacconi (Dordrecht: Reidel), p. 123.
Rappaport, S. & van den Heuvel, E. P. J. 1982, In: *IAU Symposium No. 98 on Be Stars*, eds. M. Jaschek & H.-G. Groth (Dordrecht: Reidel) p. 327.
Remie, H. & Lamers, H. J. G. L. M. 1982, *Astron. Ap.*, **105**, 85.
Ricketts, M. 1981, private communication.
Ricketts, M., Hall, R., Page, C. G. & Pounds, K. A. 1981, In: *X-Ray Astronomy, Proceedings of the 15th ESLAB Symposium*, ed. R. D. Andersen (Dordrecht: Reidel), p. 395.
Rose, L. A., Pravdo, S. H., Kaluzienski, L. J., Marshall, F. E., Holt, S. S., Boldt, E. A., Rothschild, R. E. & Serlemitsos, P. J. 1979, *Ap. J.*, **231**, 919.
Rosenberg, F. D., Eyles, C. J., Skinner, G. K. & Willmore, A. P. 1975, *Nature*, **256**, 628.
Russell, H. N. 1928, *MNRAS*, **88**, 641.
Schreier, E. J., Levinson, R., Gursky, H., Kellogg, E., Tananbaum, H. & Giacconi, R. 1972a, *Ap. J. (Lett.)*, **172**, L79.
Schreier, E. J., Giacconi, R., Gursky, H., Kellogg, E. & Tananbaum, H. 1972b, *Ap. J. (Lett.)*, **178**, L71.
Schreier, E. J. & Fabbiano, G. 1976, In: *X-Ray binaries* (NASA Sp-389), p. 197.
Schwarzschild, M. 1958, *Structure and Evolution of the Stars* (Princeton: Princeton University Press).
Shapiro, S. L. & Salpeter, E. E. 1975, *Ap. J.*, **198**, 671.
Shapiro, S. L. & Lightman, A. P. 1976, *Ap. J.*, **204**, 555.
Skinner, G. K., Bedford, D. K., Elsner, R. F., Leaky, D., Weisskopf, M. C. & Grindlay, J. 1982, *Nature*, **297**, 568.
Slettebak, A. 1979, *Sp. Sci. Rev.*, **23**, 541.
Sterne, T. E. 1939, *MNRAS*, **99**, 451.
Stothers, R. 1974, *Ap. J.*, **194**, 651.
Tananbaum, H., Gursky, H., Kellogg, E. M., Levinson, R., Schreier, E. & Giacconi, R. 1972, *Ap. J. (Lett.)*, **174**, L143.
Taylor, J., Fowler, L. A. & McCulloch, P. M. 1979, *Nature*, **277**, 437.
Trümper, J., Pietsch, W., Reppin, C., Sacco, B., Kendziorra, E. & Staubert, R. 1977, *Ann. N.Y. Acad. Sci.*, **302**, 538.
Tsuruta, S. 1975, *Ann. N.Y. Acad. Sci.*, **262**, 391.
Ulmer, M. P. 1976, *Ap. J.*, **204**, 548.
Underhill, A. B., Divan, L., Prevot-Burnichon, M.-L. & Doazan, V. 1979, *MNRAS*, **189**, 601.
van den Heuvel, E. P. J. 1975, *Ap. J. (Lett.)*, **198**, L109.
van den Heuvel, E. P. J. 1976, In: *Structure and Evolution of Close Binary Systems*, ed. P. Eggleton (Dordrecht: Reidel), p. 35.
van den Heuvel, E. P. J. 1977, *Ann. N.Y. Acad. Sci.*, **302**, 14.

van den Heuvel, E. P. J. & Henrichs, H. 1981, private communication.
van den Heuvel, E. P. J. & Rappaport, S. 1981, *In: Proceedings of the 2nd Asian Pacific Regional Meeting of the IAU*, Bandung, Indonesia, ed. B. Hidayat, in press.
van der Linden, T. 1981, private communication.
van Paradijs, J., Zuiderwijk, E. J., Takens, R. J., Hammerschlag-Hensberge, G., van den Heuvel, E. P. J. & de Loore, C. 1977, *Astron. Ap. Suppl.*, **30**, 195.
van Paradijs, J. A., Hammerschlag-Hensberge, G. & Zuiderwijk, E. J. 1978, *Astron. Ap. Suppl.*, **31**, 189.
Wang, Y.-M. 1975, *Nature*, **253**, 249.
Wang, Y.-M. 1981, *Astron. Ap.*, **102**, 36.
Wang, Y.-M. & Frank, J. 1981, *Astron. Ap.*, **93**, 255.
Watson, M. G. & Griffiths, R. E. 1977, *MNRAS*, **178**, 513.
Watson, M. G., Warwick, R. S. & Corbet, R. H. D. 1982, *MNRAS*, **199**, 915.
White, N. E. 1978, *Nature*, **271**, 38.
White, N. E., Mason, K. O. & Sanford, P. W. 1976, *MNRAS*, **176**, 201.
White, N. E., Mason, K. O., Huckle, H. E., Charles, P. A. & Sanford, P. W. 1976, *Ap. J. (Lett)*, **209**, L119.
White, N. E., Mason, K. O. & Sanford, P. W. 1977, *Nature*, **267**, 229.
White, N. E., Mason, K. O. & Sanford, P. W. 1978, *MNRAS*, **184**, 67.
White, N. E., Parkes, G. E., Sanford, P. W., Mason, K. O. & Murdin, P. G. 1978, *Nature*, **274**, 665.
White, N. E. & Pravdo, S. H. 1979, *Ap. J. (Lett.)*, **233**, L121.
White, N. E., Pravdo, S. H., Becker, R. H., Boldt, E. A., Holt, S. S. & Serlemitsos, P. J. 1980, *Ap. J.*, **239**, 655.
White, N. E. & Marshall, F. E. 1981, *Ap. J. (Lett.)*, **249**, L25.
Wheaton, W. A. *et al.* 1979, *Nature*, **282**, 240.
Yahel, R. Z. 1980, *Astron. Ap.*, **90**, 26.
Zahn, J.-P. 1975, *Astron. Ap.*, **41**, 329.
Zahn, J.-P. 1977, *Astron. Ap.*, **57**, 383; erratum in **67**, 162 (1978).
Ziolkowski, J. 1980, *In: Close Binary Stars: Observations and Interpretation*, eds. M. J. Plavec, D. M. Popper & R. K. Ulrich (Dordrecht: Reidel), p. 335.

2

X-RAY BURSTERS AND THE X-RAY SOURCES OF THE GALACTIC BULGE†

Walter H. G. Lewin and Paul C. Joss

Center for Space Research, Center for Theoretical Physics and Department of Physics, Massachusetts Institute of Technology, Cambridge, MA 02139

2.1 Introduction and brief summary

There is a class of bright ($\gtrsim 10^{34}$ erg s^{-1}) X-ray sources which distinguish themselves from the massive X-ray binaries by the following characteristics:

Their star-like optical counterparts are faint ($M_v \approx +2$) in contrast to the luminous massive binary systems (for which $M_v \approx -6$).

Their spectra are generally devoid of normal stellar absorption features (see Figure 2.1).

The ratio of their X-ray to optical luminosities, L_x/L_{opt}, ranges from $\sim 10^2$ to $\sim 10^4$. For the massive systems this ratio ranges from $\sim 10^{-3}$ to $\sim 10^1$ (see Figure 2.1).

With one exception (4U 1626−67), their X-ray spectra are softer than the spectra of the massive X-ray binary systems.

With one exception (4U 1626−67), they show no periodic pulsations such as are often observed from highly magnetized, rotating neutron stars in the massive X-ray binary systems.

They show no X-ray eclipses as are commonly observed in the massive binary systems.

Many of them produce X-ray bursts. In contrast, no X-ray bursts have definitely been observed from any of the massive binary systems (see Figure 2.1).

In our galaxy, about 11 bright X-ray sources are located in globular clusters. Their X-ray properties are similar to those listed above. About nine of them produce X-ray bursts (see Table 2.1). None of the globular cluster X-ray sources have been optically identified with stellar objects. Such identifications would be

† An earlier version of this review was published in *Space Science Reviews*, **28**, 3 (1981). It is complete up to September 1980; however, most of the relevant information that reached the authors before 15 July 1981 is included.

very difficult, since most of these globular clusters are centrally condensed (see Figure 2.2) and several are highly reddened due to interstellar extinction. The similarities between the globular cluster X-ray sources and the above class of X-ray sources lead one to consider them as a single class of objects. They are often referred to as the 'galactic bulge sources' since they are concentrated toward the galactic center. Their apparent location outside regions of active star formation identify them as being probable members of an older stellar population.

Figure 2.1 shows the ratios of the X-ray luminosities, L_x, to the optical luminosities, L_{opt}, of all optically identified compact galactic X-ray sources with $L_x > 10^{34}$ erg s^{-1}. The figure also shows which of them (i) are associated with an O or a B star (i.e., contained in massive X-ray binary systems), (ii) show no normal stellar absorption features at maximum light, and (iii) are X-ray burst sources. The sources under (ii) and (iii) have values for $L_x/L_{opt} > 10$, whereas the sources under (i), with one exception, all have values for $L_x/L_{opt} < 10$. This strongly suggests that we are dealing here with two different classes: the massive

Table 2.1. *Globular cluster X-ray sources*

		R_c (arcsec)	Burst source?	References[b]
47 Tuc	1E 002151−7221.5	29	No bursts observed	[100]
NGC 1851	MX 0513−40	7	Yes	[14, 48, 78, 134, 196, 211]
Terzan 2	4U 1722−30 1E 172420−3045.6	7	Yes	[92, 103, 268]
Liller 1	MXB 1730−335	7	Yes	[67, 180, 206, 218]
Terzan 1	1732−303	−	Yes (only bursts observed)	[311, 312, 313]
Terzan 5	1745−248	−	Yes (only bursts observed)	[311, 312, 313]
NGC 6441	4U 1746−37	9	Probably	[50, 134, 196]
NGC 6624	4U 1820−30	6	Yes	[35, 46, 79, 98, 134, 136]
NGC 6712	A/4U 1850−08	49	Almost certainly	[51, 56, 66, 134, 197, 272]
NGC 7078	4U 2131+11	10	No bursts observed	[79, 134, 136]
Grindlay–Hertz 1	MXB 1728−34	−	Yes	[122, 124, 168, 331, 338]
[a]Kron 3	4U 0026−73	−	No bursts observed	[79, 92]
[a]NGC 6440	MX 1746−20	−	No bursts observed	[79, 136, 211]

[a] These two associations are uncertain. According to J. Grindlay (private communication) his earlier tentative identification of 4U 0026−73 with Kron 3 [92] is almost certainly incorrect.
[b] For more detailed references and for accurate source locations, see Table 2.4 and [28].

X-ray bursters and the X-ray sources of the galactic bulge

binaries, on the left in Figure 2.1, and the galactic bulge sources, on the right in Figure 2.1. (Her X-1 is special, as it does not fit neatly into either class.)

In this article we shall discuss mainly the observed X-ray properties of the galactic bulge sources (called Type II or Class II sources by some authors, e.g. in Chapters 1, 5, 8 and 9, this book), with an emphasis on those that produce type I X-ray bursts. The 'type I/type II' burst classification (not to be confused with Type I and Type II sources) will be discussed in Section 2.4. We shall also discuss some of their optical, infrared and radio properties (Section 2.4.2.7). However, the optical properties are discussed in much greater detail in Chapter 5,

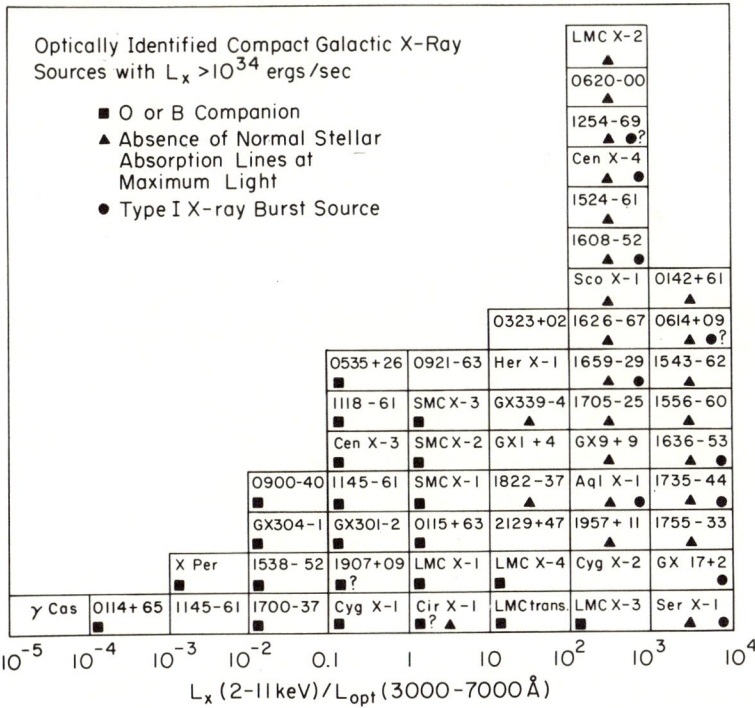

Fig. 2.1. Histogram of the optically identified compact galactic X-ray sources (with $L_x > 10^{34}$ erg s^{-1}) as a function of L_x/L_{opt}. There seem to be two classes: (i) the massive binaries (at left) with an O or B stellar companion, indicated by ■; (ii) the galactic bulge sources (on the right) which, in general, are optically faint objects that show no normal stellar absorption lines in their spectra (at maximum light), indicated by ▲. Several objects in this class are type I X-ray burst sources, indicated by ●. Many of the sources on the right are known low-mass close-binary systems (see Sections 2.4.2.2 and 2.4.2.7). It is very likely that the galactic bulge sources, as a class, are low-mass close-binary systems. This figure was provided by J. E. McClintock, who used [28], [38], [138], [284], and [363]; it is a revised version of the one previously published by us [368] (see also [363]).

this book. There is persuasive evidence that these burst sources and many other galactic bulge sources are neutron stars in low-mass, close-binary stellar systems. As noted above, several burst sources are found in globular clusters with high central densities (see Figure 2.2); they were probably formed by capture during close encounters between neutron stars and nuclear-burning stars.

The commonly observed X-ray bursts (of type I) are almost certainly due to thermonuclear flashes within freshly accreted material on the surfaces of neutron stars. Optical bursts, associated with the type I X-ray bursts, have been observed from three sources. The optical flux, which arrives a few seconds after the X-ray flux, is probably due to X-ray heating of an accretion disk surrounding the neutron star and the delay in emission is predominantly caused by travel time differences.

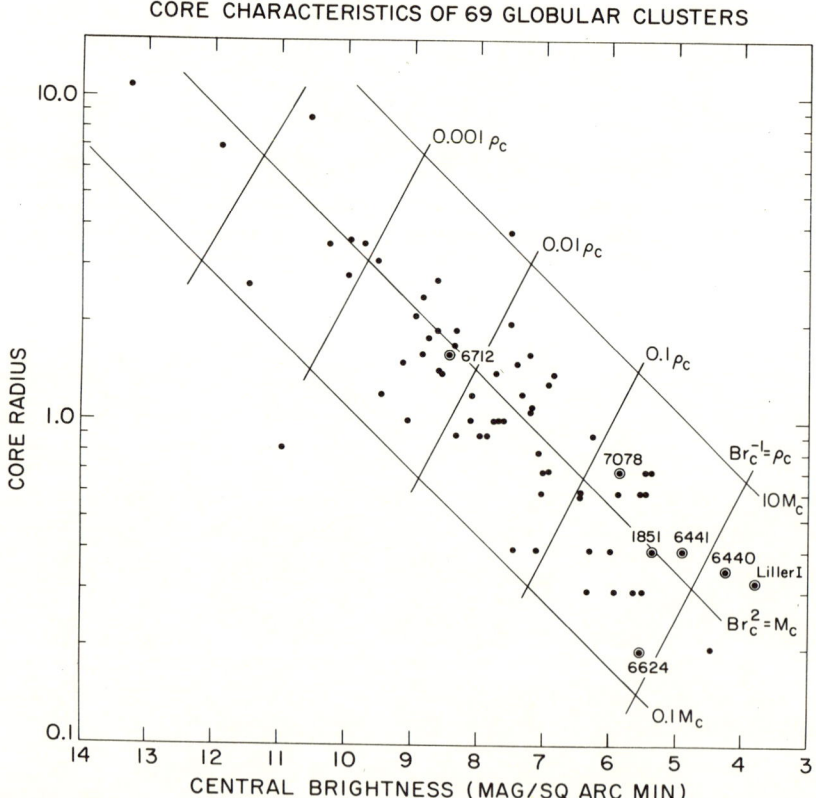

Fig. 2.2. Log-log plot of the core radius r_c (in parsec), versus the central core brightness (B) for 69 globular clusters. The grid lines are loci of constant core mass ($M_c = Br_c^2$) and core density ($\rho_c = M_c/r_c^3$). The core densities have a range of 10^5. The globular clusters that contain X-ray sources (except NGC 6712) have relatively high core densities (see text). Not all X-ray clusters are indicated (see Table 2.1). This figure is from F. K. Li, Ph.D. thesis, MIT.

The X-ray bursts of type II, as observed from the Rapid Burster (MXB 1730−335), are almost certainly due to an accretion instability which converts gravitational potential energy into heat and radiation, and are thus of a fundamentally different nature than the type I bursts.

2.2 Lack of pulsations and eclipses
2.2.1 Pulsations

More than twenty galactic bulge sources have been observed intensively in search of periodic pulsations. Except for 4U 1626−67, no pulsations were found. Cominsky *et al.* [52] analyzed the persistent flux of seven burst sources and found upper limits on the pulsed fractions of 1-10 per cent for periods from ~1.6 to 10^3 s (Table 2.2). Spada, Rappaport & Li (private communication) searched for periodicities in seven of the brightest (non-bursting) sources near the galactic center. They also found none with typical upper limits of 3 per cent in the range ~2 ms-2 s. Upper limits to pulsed fractions of 10 per cent (periods in the range 3-100 s) and 30 per cent (periods in the range 2 min-2 d) were obtained by Parsignault & Grindlay [248]. Upper limits to pulsations from the general region of the galactic center have been reported by Córdova, Garmire & Lewin [57]. (See also [374].)

Pulsations are expected from accreting neutron stars if they have magnetic fields misaligned with their rotation axes and sufficiently strong to channel the accretion flow, as is presumably the case among the binary X-ray pulsars of

Table 2.2. *Upper limits to periodicities in the persistent emission from type* I *burst sources*[a]

Source	Upper limits (90 per cent confidence) to pulsed fraction, in per cent, for periods in the ranges listed		
	2 ms-2 s	1.6 s-66 s	66 s-10^3 s
1636−53	−	1.9	3.5
1659−29	−	3.5	9.5
1728−34	−	3.6	9.2
1735−44	8	1.9	13.0
1820−30	1.5[b]	1.5	1.5[c]
1837+05	−	1.1	6.1
1916−05	−	1.3	15

[a] Most of the data in this table are from Cominsky *et al.* [52]. Upper limits of ~3 per cent in the range ~2 ms-2 s were found for seven bright sources (not burst sources) near the galactic center (see text).
[b] F. K. Li, private communication.
[c] Period range 66-200 s.

stellar Population I. The critical surface magnetic field strength is uncertain but is most likely of the order of 10^{12} gauss [8]. The surface magnetic fields of older neutron stars may have largely decayed away, or their rotation and magnetic dipole axes may have become coaligned (see [77] and [258] and references therein). It is, perhaps, also possible that differences in the formation may have prevented the development of strong surface magnetic fields in neutron stars that now comprise the galactic bulge sources [144]. The strong magnetic fields of young neutron stars funnel the material towards the magnetic poles and may suppress thermonuclear flashes in the accreted material. This could explain why type I X-ray burst sources do not pulse and why X-ray pulsars do not burst [142, 148, 275] (see Section 2.6.3.4).

Thus, the absence of detected pulsations does not constitute a strong argument against the neutron-star nature of the collapsed objects in these systems.

2.2.2 Eclipses

No X-ray eclipses have been found among the galactic bulge sources (but see also Section 2.6.4). Joss & Rappaport [149] have shown that an X-ray source powered by accretion through Roche-lobe overflow from a low-mass companion (of ~ 0.3 solar masses) has a low probability (~ 25 per cent) of being eclipsed by its companion. The eclipse probability varies as the cube root of the mass of the companion and would therefore still be about 10 per cent for a companion of only 0.02 solar masses. The available estimates of the masses of companion stars of galactic bulge sources (see Section 2.4.2.7.3 and the Chapter 5, this book) indicate that the very low-mass model of Joss & Rappaport is probably insufficient by itself to account for the absence of X-ray eclipses in all ~ 20 sources for which a careful search for eclipses was made (see Section 2.6.4). Milgrom [232] proposed a model in which an accretion disk casts an X-ray shadow in the orbital plane of a low-mass binary, thereby excluding from our detection those systems which would otherwise show eclipses (for causes of disk thickening see Chapter 10, Section 10.2.4, this book).

Thus, the absence of eclipses does not argue compellingly against the binary nature of these systems. In Section 2.4.2.7 we shall discuss the observational evidence that these systems are highly-compact binary systems containing low-mass (<1 solar mass) or degenerate companion stars (see also Chapter 5, this book). Theoretical models and the evolution of systems of this type will be discussed in Section 2.6.

2.3 Globular cluster X-ray sources

It was first pointed out by Gursky [330] and later by Katz [156] that X-ray sources occur with unusually high frequency in globular clusters. In terms of their typical intrinsic properties, these sources seem to be indistinguishable from other galactic bulge sources [33]. An examination of the circumstances surrounding the frequent occurrence of such sources in globular clusters may throw light on the nature of the entire class of galactic bulge X-ray sources [175, 176, 177].

2.3.1 Centrally condensed globular clusters

The structure of a globular cluster can be broadly characterized by three observables: the core radius r_c, the tidal radius, and the central brightness B [159]. Peterson & King [252] summarized the data as of 1975, and several studies updating and extending their work have been carried out since then, with special attention to the clusters containing X-ray sources (see, e.g., [13, 14, 35, 160, 251]). The data show a striking correlation between the presence of an X-ray source and the quantity B/r_c, which is a measure of the central density of luminous stars (Figure 2.2).

In recognition of the correlation with high central density, it was widely expected that the cluster sources (listed in Table 2.1) would be found to lie close to the cluster centers. This expectation was confirmed by position measurements, carried out with the rotating modulation collimator detectors on the SAS-3 X-ray Observatory, which showed that in five cases the positional error circles with radii of 20–30 arcsec (90 per cent confidence) included the optical centers of the clusters [134]. Work with the Einstein X-ray Observatory has further refined the positional measurements of X-ray sources in globular clusters [93] (see Section 2.3.4).

2.3.2 Formation of the X-ray sources in globular clusters

The evidence clearly points away from any explanation of the cluster sources based on the evolution of primordial binary stellar systems that might be found in all clusters, and toward an explanation based on the formation of the X-ray sources under the circumstances peculiar to the dense cores of the most centrally condensed clusters. (It is conceivable, however, that the central density of a cluster is somehow related to the number of primordial binaries that were able to form within it.) One plausible circumstance is a high concentration of collapsed remnants of the short-lived massive stars that were presumably present at the birth of the cluster (i.e., degenerate dwarfs, neutron stars and/or black holes that were not ejected from the clusters by the processes of their formation and that subsequently settled toward the cluster centers). The frequency of close encounters between such remnants and other stars certainly increases with increasing central density of the cluster.

Clark [42] suggested that the cluster sources are massive compact remnants (neutron stars or black holes) that have acquired close stellar companions by capture. Fabian, Pringle & Rees [75] estimated that captures could occur with sufficient frequency in dense cluster cores through tidal dissipation of orbital energy in two-body encounters, and Press & Teukolsky [254] confirmed the viability of this mechanism by detailed calculations. Hills [114] examined the formation of binary systems through star-exchange interactions between neutron stars and primordial low-mass binaries. This can be important if the unknown frequency of binaries in globular clusters is sufficiently high. Capture of collapsed remnants by other stars through three-body encounters in a dense subcluster near the cluster center may also be significant [76]. Hut & Verbunt [370] have recently demonstrated that in star-exchange interactions neutron-star capture is more efficient than white dwarf capture. For more details see Chapter 8, Section 6 and for a discussion on globular clusters in M 31, see Chapter 3, this book.

2.3.3 Massive black holes?

It has been proposed that the deep potential well of a centrally condensed cluster might retain gas lost by stars and that a black hole of sufficient mass, formed perhaps in a collapse of the dense core, could accrete this gas at the rate required to produce the observed X-ray luminosities [11, 266]. Evidence has been reported [101] for the existence of gas in the cores of four X-ray globular clusters in amounts sufficient to supply a black hole if its mass exceeds 10^2 solar masses. Shortly after the discovery of X-ray bursts (Section 2.4.1), Grindlay & Gursky [95] argued that the X-ray burst source located in the globular cluster NGC 6624 is a very massive (larger than a few hundred solar masses) black hole. Their argument was largely based on the spectral hardening that was observed in two bursts [98]. However, Canizares [34] showed that their calculations were erroneous and that the observed effect does not constitute a strong constraint on the properties of the X-ray star. In addition, the discovery of more burst sources (Section 2.4.1) showed that as a rule the spectra soften during burst decay, and this further weakened the argument for massive black holes. It was also pointed out by Lewin [172] and Lewin, Hoffman & Doty [186] that the galactic distribution of the observed burst sources argues against an association with massive black holes.

2.3.4 Mass determinations of X-ray sources in globular clusters

A new approach to determine the mass of X-ray sources in globular clusters has recently become available by use of precise position measurements. From the work of Bahcall & Wolf [12], it follows that the expectation value of the projected distance of an object of mass M from the center of a dynamically relaxed cluster of stars, each of mass m, is $0.7 R_c (m/M)^{1/2}$ arcsec. In typical

X-ray clusters R_c is about 7 arcsec (see Table 2.1). Thus an X-ray source with four times the average mass of cluster stars [246] (~0.5 solar mass) would have a projected expectation distance of about 3 arcsec from the center, while a black hole with more than 100 times the average mass would rarely be found outside of 1 arcsec. The position measurements with SAS-3 permitted the conclusion that the cluster X-ray sources are, on average, more massive than the visible stars in the clusters [134].

Grindlay and co-workers [93], using the Einstein Observatory, determined the locations of eight X-ray sources located in globular clusters to an accuracy of several arcsec. They found that at a 90 per cent confidence level the mass of each X-ray source lies between 1 and 5 solar masses under the assumption that the masses of all eight sources are about equal, that the cluster cores are isothermal and that the masses of the cluster stars are approximately constant. This important result is an independent confirmation of the low-mass character of these systems (see [175, 367], and Section 2.4.2.7).

Another point to keep in mind regarding the low-mass binary picture for these objects is that it implies a fair chance of finding two bright X-ray sources within the same cluster. No such pairs have been found so far by the Einstein Observatory [93, 375]. Of the ~25 known, centrally condensed clusters, seven contain a bright X-ray source (see Figure 2.2 and Table 2.1). Thus, it is easily found that the *a priori* probability of finding more than one X-ray source in any one of the ~25 centrally condensed clusters is ~0.6 if the low-mass binary model is correct.

2.3.5 NGC 6712

Among globular clusters that contain X-ray sources, NGC 6712 has an exceptionally low central density (see Figure 2.2). This anomaly is intriguing in the light of calculations by Shapiro [264], which suggest that core collapse may be followed by distension and dissolution of the remaining cluster through the injection of kinetic energy by interactions of stars with a massive central body. Perhaps X-ray binaries as well as a massive black hole are produced in the processes of core collapse, and in NGC 6712 we are witnessing the aftermath as the cluster proceeds to self-destruct. Some consequences of this scenario of the evolution and death of globular clusters have been explored by Lightman, Press & Odenwald [202].

An alternative, and more mundane, possibility is that NGC 6712, despite its low central density, has nonetheless managed to produce a binary X-ray source.

2.4 X-ray burst sources

In this chapter we shall review the observational data on X-ray burst sources. A large amount of observational material has been obtained in the past six years. The Rapid Burster, with all its idiosyncrasies, will be treated in Section

2.4.3 separately from the other burst sources. We begin with a short historical review of the important issues and contributions as we perceived them at MIT, where much of the observational work was carried out.

2.4.1 Brief history and some highlights

X-ray bursts were discovered in December 1975 independently by Grindlay & Heise [98, 99] and by Belian, Conner & Evans [21, 22]. Grindlay *et al.* [98, 99] were able to associate the two bursts they observed with a known X-ray source in the globular cluster NGC 6624, while the Los Alamos group [21-3] could not make an association with a known X-ray source, since their positional accuracy for the many burst events they observed was insufficient. Subsequently, using existing SAS-3 data from May 1975, Clark *et al.* [46] found a series of ten bursts from the source in NGC 6624.

Within two months, in February 1976, two additional burst sources (possibly transients) were discovered by Lewin, Hoffman & co-workers [166, 167, 187], within a few degrees of the galactic center. The presence of a third burst source was suspected and searched for in early March 1976. It was found [168] (MXB 1728−34; MXB stands for MIT X-Ray Burst Source), and in the process another source with extraordinary bursting behavior was discovered [168, 180]. Bursts from this object were observed in rapid succession on time scales of seconds to minutes. This behavior gave the source its name: the Rapid Burster (the official designation is MXB 1730−335). The bursts from the Rapid Burster varied by about a factor of ∼100 in duration, but bursts with durations in excess of ∼15 s had nearly equal peak fluxes (Figure 2.11). The energy in a burst was approximately proportional to the waiting time to the next burst; the system clearly behaved like a relaxation oscillator [180].

The activity of this source stopped by mid-April 1976. About a year later, White & Burnell [304], using the Ariel 5 satellite, observed a recurrence of bursts from the Rapid Burster. This established that the burst activity was a recurrent phenomenon. We now know that the periods of burst activity occur at intervals of approximately 6 mth and persist for ∼2-6 wk (see Section 2.4.3.6).

A previously unknown, highly reddened globular cluster [160] was found in March 1976 by Liller [204] in the error circle (radius ≈4 arcmin) for the Rapid Burster [111]. The globular cluster was called Liller 1 with the expectation that more such globular clusters would be found. Some workers indeed believed that perhaps all burst sources would be associated with globular clusters (see [78] and the discussion following Lewin's talk at the Texas Symposium [172] in December 1976). Within a year, over 20 burst sources were found, largely due to observations made with SAS-3 (by W. Lewin and his co-workers) and OSO-8 (by J. Swank and her co-workers). A first catalogue of burst sources was compiled in September 1976 and contained 15 sources [171]. By December 1976 the number of burst sources was about 22 [172], and it became clear from the distribution of these sources that, as a class, they were not associated with globular clusters [172, 186].

Whether or not burst sources, as a class, are associated with globular clusters was at the time an important issue, since it had immediate implications for the nature of the burst sources, the globular cluster X-ray sources and the whole class of galactic bulge sources. As described in Section 2.3.3, it was proposed by Grindlay & Gursky [95] that burst sources are associated with massive (greater than a few hundred solar masses) black holes in the cores of globular clusters, and it was thought by some that most, perhaps all, burst sources would be found in globular clusters [78]. However, by December 1976 it became clear that this was not the case [172], and the massive black hole idea evaporated (see Section 2.3.3 for other problems with the massive black hole picture).

In 1977, Swank et al. [268], using the OSO-8 satellite, were the first to show that the spectral evolution of a particular long burst (of ~ 600 s duration) was consistent with the cooling of a collapsed object whose surface behaved like a black body (Figure 2.8). Hoffman, Lewin & Doty [120, 121] confirmed the consistency with black-body cooling for a series of bursts from two burst sources (Figure 2.4), and they showed that throughout the cooling the effective black-body radii of these two objects did not change. They found values of ~ 10 km for both sources if source distances of ~ 10 kpc were assumed. In 1978, van Paradijs [291] showed evidence that bursts of average intensity are 'standard candles' at burst maximum and that their radii are all near 7 km if the average values of their peak burst luminosities are near the Eddington limit of a ~ 1.4 solar-mass object with a hydrogen-rich surface. (When examined in detail, bursts are not precise standard candles; in a series of bursts from one source, peak fluxes can differ substantially [194].) The measured black-body radii strongly indicated that the collapsed objects are neutron stars.

Several models were proposed in 1976 and 1977 to explain the bursts. Svestka [267], Henriksen [113], Joss & Rappaport [319], Baan [10], and Lamb *et al.* [163] considered various possible instabilities in the accretion flow at or near the magnetopause of a neutron star or white dwarf, while Wheeler [302] and Liang [201] investigated potential instabilities in an accretion disk surrounding a collapsed object, and Grindlay [91] suggested a thermal instability in accretion onto a massive black hole.

A very different mechanism was proposed by Woosley & Taam [308] and Maraschi & Cavaliere [208], who suggested thermonuclear flashes in the surface layers of an accreting neutron star. However, after the discovery of the Rapid Burster [168], it became apparent that the rapidly repetitive bursts could not be due to thermonuclear flashes; if they were, a very high flux of persistent X-ray emission due to the release of gravitational potential energy by the accretion of the matter onto the neutron-star surface should be present, and this was not observed. Thus, the nuclear-flash theory could not explain all burst phenomena, and that was not in its favor.

A break came in September 1977 when Hoffman, Marshall & Lewin discovered that the Rapid Burster emits two very different kinds of bursts [126, 127] (Figures 2.12 and 2.13). They introduced the classification of type I and type II bursts (Table 2.3) and suggested that the rapidly repetitive type II bursts are due to accretion instabilities, thus deriving their energy from (and accounting for the release of) gravitational potential energy, but that the type I bursts are due to thermonuclear flashes. This is still the most plausible explanation. Other burst sources, which emit only type I bursts, evidently release the gravitational potential energy (due to accretion) as a persistent X-ray flux (see Section 2.4.2.2.). Detailed calculations performed since 1977 have strengthened the idea that thermonuclear flashes on the surfaces of neutron stars produce the type I bursts [142] and other references given in Section 2.6).

In the summer of 1977, the burst source MXB 1735−44 was the first to be firmly optically identified by J. McClintock; its spectrum was very similar to that of Sco X-1 [223, 225]. This identification was made possible by the precise (∼20 arcsec) X-ray source position obtained by the SAS-3 rotating modulation

Table 2.3. *Classification of X-ray bursts*[a]

Type I (∼30 sources)
- Burst intervals of hours − days (longer?)
- Distinct spectral 'softening' during burst decay
  - cooling of a 'black body' during decay[b]
  - thermonuclear flashes on a neutron star
  - energy source is nuclear energy

Type II (Rapid Burster + others?)
- Burst intervals of seconds−minutes
- No distinct spectral softening during burst decay
  - 'black-body' radiation with no significant cooling during decay
  - instability in accretion flow onto a neutron star
  - energy source is gravitational potential energy

[a] This classification was introduced by Hoffman, Marshall & Lewin [127]. It was based on the two phenomenological differences listed under the solid dots. The characteristics listed under the open circles are interpretations (see e.g., [127, 142, 144, 218]).
[b] Gray bodies with an emissivity of >0.1 are allowed. This would increase the size of the emitting regions by a factor <3 [120, 121, 268, 273, 291, 292].

collimator (RMC) observations. (The RMC observations were directed by
H. Bradt, G. Jernigan & R. Doxsey.) MXB 1735—44 was also the first burst
source from which, in the summer of 1978, a simultaneous optical and X-ray
burst was observed. This observation was the result of a combined Harvard–MIT
effort [104, 105, 228], and it came after an extensive world-wide coordinated
burst watch in 1977 failed to detect any radio, infrared or optical burst
coincident with an observed X-ray burst [1, 24, 137, 277, 281, 286].

Infrared bursts from the Rapid Burster were first reported by Kulkarni *et al.*
[162] in 1979, and radio bursts from this source were reported by Calla *et al.*
[30, 31]. The connection between these bursts (if they are real) and the X-ray
bursts is still unclear to date.

It was long suspected that, as a class, burst sources are low-mass close-binary
stellar systems (see Sections 2.4.2.2, 2.4.2.7 and 2.6.4). In early 1980,
Kaluzienski, Holt & Swank [155] obtained the first direct indication that the
transient [19, 55, 153] burst source Cen X-4 is a close-binary system with
an orbital period of ~ 8 h. There are now several burst sources for which an
orbital period may have been found (see Section 2.4.2.7.3).

We shall now discuss in detail the observational characteristics of X-ray burst
sources.

2.4.2 Type I burst sources
2.4.2.1 Angular distribution

Type I bursts show distinct spectral softening during burst decay (see
Figure 2.3) and, in general, recur on time scales of hours to days [127] (Table
2.4). A sky map of type I burst sources is shown in Figure 2.3. The sources are
listed in Table 2.4. It is immediately clear from this map that the burst sources
are associated with the Galaxy, and their apparent locations outside regions of
current or recent active star formation identify them as probable members of an
old stellar population. Nine of the 32 sources shown in Figure 2.3 are located in
globular clusters (see Table 2.1). Six (perhaps seven) have been optically
identified with faint stellar objects (see Section 2.4.2.7). The positions of
many more burst sources have been reported in the literature [20-3, 73], but we
omit them because it is presently uncertain that all of these are type I burst
sources. However, it is very likely that many of the events observed by the Los
Alamos group [20-3, 73] were of type I and, in retrospect, there seems to be
little doubt that they discovered type I X-ray bursts from the Norma region
[21-3].

The type I burst sources are a subset of the known galactic bulge X-ray
sources [172-86], which have the general characteristics listed in Section 2.1. It
is believed that many, and perhaps all of the galactic bulge sources are accreting

Table 2.4

August 15, 1981

Burst Source				Assoc. Persistent Source				4U Catalog Intensity		Comments	References
Name	α (1950)/δ (degrees)	l^{II}, b^{II} (degrees)	Pos Acc (deg²)	Names	(1950) h m s ° ' "	error circle radius (arcsec)		Average or Maximum Uhuru	Max/Min		
MXB 0512-40 (in NGC 1851)				2S0512-400 2A0512-399 4U0513-40	05 12 28.7 -40 05 53	20		18	3		48, 50, 56, 78, 79, 101, 134, 140, 174, 175, 177, 186, 190, 198, 296
XB 0611-??		200	-5	~8 × 8	4U0614+09? or 4U0621+11? star?	06 21 38 11 46 48 06 14 22.3 09 09 25	3	120 2.1	5	1 burst observed	62, 64, 79, 174, 175, 186, 190, 236, 269, 306, see Fig. 3
MXB 0???-??		263	-14	~24 × 24	?					Ten 4U sources in error box 1 Type I burst observed	63, 171, 172, 174, 175, 186, 190, see Fig. 3
MXB 14??-6?		315	-4	~24 × 24	?					Many 4U sources in error box 1 Type I burst observed	7, 177, 220, see Fig. 3
XB 1455-31 (blue star)				~1	Cen x-4 star	14 55 19.5 -31 28 07	5			Transient	19, 36, 38, 55, 115, 153, 155, 207, 222, 239, 262, 293, 295, 309
1535-29?					4U1535-29?	15 35 53 -29 13 12		200	>10	Only 1 burst-like event. May not be a Type I burst source	79
XB 1608-52 (star)					MXB1608-52 4U 1608-52 2S1608-523 star	16 08 51 -52 18 02 16 08 52.2 -52 17 43	20 ±0.5	10^3	~25	Transient	4, 21, 22, 23, 47, 68, 70, 79, 89, 94, 96, 102, 152, 154, 171, 172, 174, 175, 177, 181, 195, 214, 234, 235, 238, 278
MXB 1636-53 (blue star)					4U1636-53 2S1636-536	16 36 56.2 -53 39 15	3	250	2	40 optical bursts observed, many simultaneously with X-Ray bursts	37, 79, 107, 119, 120, 132, 174, 175, 177, 186, 190, 224, 226, 249, 250, 271, 281, 291, 296, 306, 307, 333, 337, 342, 345
MXB 1659-29 (blue star)					4U1704-30? H1658-298 star	16 58 55.5 -29 52 26	5	3.1 80	>15	Transient. Star visible in summer 1978 (X-Ray high state) not in summer 1979 (X-Rays undetectable). Bursts not observed when persistent source in high state	37, 54, 69, 88, 90, 172, 173, 174, 175, 177, 183, 186, 189, 190, 265, 291, 296, 297
XB1702-42					4U1702-42 2S1702-424	17 02 40.4 -42 57 58	30	30	3		79, 133, 172, 174, 175, 177, 217, 241, 272
MXB 1715-32					MX 1716-31 2S1715-321	17 15 31.9 -32 07 15	60	~2000 ~15			123, 212, 213, 241
XB1724-30 (in Terzan 2)					4U1722-30 1E 172420 -3045.6	17 24 20 -30 45 36	5	7			79, 92, 93, 103, 174, 175, 177, 268, 291
MXB 1728-34 (in Grindlay-Hertz)					4U1728-33 1E 172839 -3347.8	17 28 39.46 -33 47 49.2	5	150	5	Some bursts have two peaks at high-energy X-Rays	25, 41, 81, 84, 118, 121, 122, 124, 125, 132, 168, 171, 172, 174, 175, 177, 181, 190, 206, 296, 307, 331, 338
MXB1730-335 Rapid Burster (in Liller 1)					H1730-333	17 30 07.4 -33 20 34		~50 (Type II bursts)	>10	Type I and Type II bursts. Recurrent transient with period of ~6 months. IR and radio bursts reported	6, 16, 25, 30, 31, 41, 67, 97, 111, 112, 117, 124, 126, 127, 128, 131, 135, 137, 139, 150, 160, 168, 171, 172, 174, 175, 177, 178, 180, 181, 182, 186, 188, 190, 192, 204, 205, 206, 215, 216, 218, 219, 240, 241, 253, 259, 280, 287, 294, 296, 297, 304, 305, 307
											311, 312, 313

Name						R.A.	Dec.			Notes	References
(blue star)									1.7	2 distinct bursts observed, one of which simultaneously with X-ray burst. Burst intervals very erratic.	27, 37, 79, 104, 105, 107, 132, 174, 175, 177, 223, 225, 228, 291, 296, 297, 307, 336
MXB1742-29	265.7	-29.6	359.5		0.34	2S1735-444 star	17 35 19.0		>2	Association with transient "persistent" source is possible but uncertain in view of high source density in Gal. Center region.	26, 44, 57, 59, 79, 133, 158, 167, 171, 172, 174, 175, 184, 186, 187, 255
MXB1743-28	265.9	-28.5	0.5	+0.0	0.28	A1742-294? 2S1742-294? in GCX (4U1743-29)	17 42 53.6	-29 29 50	30	90	5	
						in GCX (4U1743-29)				40		
MXB1743-29	265.75, either one of two positions 265.58	-29.02, -29.29	359.68, 359.67	-0.16, -0.17	0.09, 0.09	A1742-289? in GCX (4U1743-29)	17 42 26.4	-28 59 55	25	40	5	Three bursts observed in quick succession (~4 min and 17 min apart). Association with transient "persistent" source is possible but uncertain in view of high source density in Gal. Center region. Double-peaked bursts.
											57, 59, 116, 118, 158, 171, 172, 184, 186, 187, 255, 291
										2×10^3	>200	44, 57, 59, 158, 166, 167, 171, 172, 174, 184, 186, 187, 190, 255, 291
XB1744-26					$\sim 2\times 10^{-3}$	4U1744-26 GX 3+1 2S1744-265	17 44 49.1	-26 32 50	4	600	3		28, 79, 311, 340, 351
XB1745-24 (in Terzan 5)						Terzan 5	17 45 0.1	-24 45 52.1				Only bursts observed, no persistent emission.	311, 312, 313
MXB1746-37 (in NGC 6441)					~1	4U1746-37 2S1746-370	17 46 48.8	-37 02 25	30	40	1.5	Association with persistent source probable, not certain (see Table 2)	50, 79, 98, 134, 174, 175, 177, 196
XB1907-27			4	-1	~4	?						No catalogued persistent source in error box. Globular cluster NGC 6553 is in error box.	172, 174, 175, 177, 186, 190, 272, see Fig. 3
XB1813-14						GX17+2 4U1813-14 Radio Source	18 13 10.8	-14 03 13.0	3.0	950	4	It is uncertain whether the radio source is the X-ray source, but that is very likely	349, 350, 351, 352, 353
MXB1820-30 (in NGC 6624)			11	-8	~6 x 8	4U1820-30 2S1820-303	18 20 27.7	-30 23 11	20	320	3, >10	Bursts only observed when persistent source in low state.	34, 35, 39, 43, 45, 46, 49, 50, 79, 80, 98, 99, 101, 134, 136, 137, 171, 172, 174, 175, 177, 181, 186, 190, 203, 296, 297, 307
XB 1837-27						4U1831-237	18 31 47	-23 12 18		6		4U1831-23 is the only known persistent source in the large error box of burst source. Association uncertain. 4 Globular clusters in error box.	17, 171, 172, 174, 175, 177, 186, 190, see Fig. 3
MXB1837+05 (blue star)						4U1837+04 Ser X-1 2S1837+049 star	18 37 29.6	04 59 21	~3	260	2	Blue star ($m_B=19.2$) within 7" of 18.5 magn. star. Burst intervals very erratic. Optical burst observed simultaneously with X-ray burst.	1, 24, 61, 65, 66, 79, 80, 108, 109, 137, 171, 172, 174, 175, 177, 179, 181, 186, 190, 199, 200, 210, 270, 277, 282, 286, 290, 291, 314, 354
MXB1850-08 (in NGC 6712)					<0.3	A1850-08 4U1850-08 2S1850-087	18 50 21.9	-08 45 54	~30	9		Association almost certain (see Table 2)	51, 56, 66, 79, 118, 134, 172, 174, 175, 177, 190, 197, 272
MXB1906+00						4U1857+01 A1905+00 2S1905+005	19 05 54.9	00 05 37	35	4.1±1			56, 66, 79, 137, 171, 172, 174, 175, 177, 181, 186, 190, 191, 199, 277, 291, 296
Aql MXB XB 1908+00 (blue star)						4U 1908+00 Aql X-1 2S1905+005 star	19 08 42.9	00 30 05	2	20	10^3	Transient. Reanalysis of SAS-3 data shows that Aql MXB is almost certainly Aql X-1. Optical burst observed.	28, 65, 171, 172, 174, 175, 177, 181, 186, 190, 191, 242, 283, 344
MXB1916-05						4U1915-05 2S1916-053	19 16 08.5	-05 19 51	20	20	2		1, 18, 66, 79, 122, 174, 175, 177, 185, 186, 190, 272, 277, 285, 291, 296, 304, 334, 348, 351
XB 2??? +37			79	-9	~50	?						4U 2058+32, 2104+31, and 2130+32 (all very weak) in large error region of burst source.	174, 175, 177, 186, 190, see Fig. 3

55

neutron stars that produce type I bursts under some circumstances (e.g., if the accretion rate is neither too large nor too small) (see Sections 2.4.2.2, 2.4.2.4, 2.6.1.2 and 2.6.3.1).

2.4.2.2 Persistent X-ray emission, transients and binaries

Almost all burst sources emit a detectable, persistent (though variable) flux of X-rays. In Table 2.4 we list the names of burst sources, the names of the associated sources of persistent X-ray emission, and their persistent fluxes in Uhuru flux units (also for those burst sources which are not listed in the 4U-catalogue [79]). We also indicate the variability of the persistent flux by showing the ratio of the maximum to minimum observed flux.

The suggestion of a connection between burst sources and transient X-ray sources [3, 53, 151] was first made by Lewin *et al.* [187] for the sources MXB 1742−29 and MXB 1743−29 near the galactic center. MXB 1659−29 is a burst source [183] and a transient [69, 173, 174, 189, 265]. When bursts from this source were first detected in October 1976, the persistent flux was below the SAS-3 detectability threshold. In March 1978 the source flared up, but X-ray bursts were not detected [173, 189]. The Rapid Burster is a recurrent transient with a recurrence period of about six months (see Section 2.4.3.6). Cen X-4, 4U 1608−52 and Aql X-1 are transient sources of the spectrally 'soft'

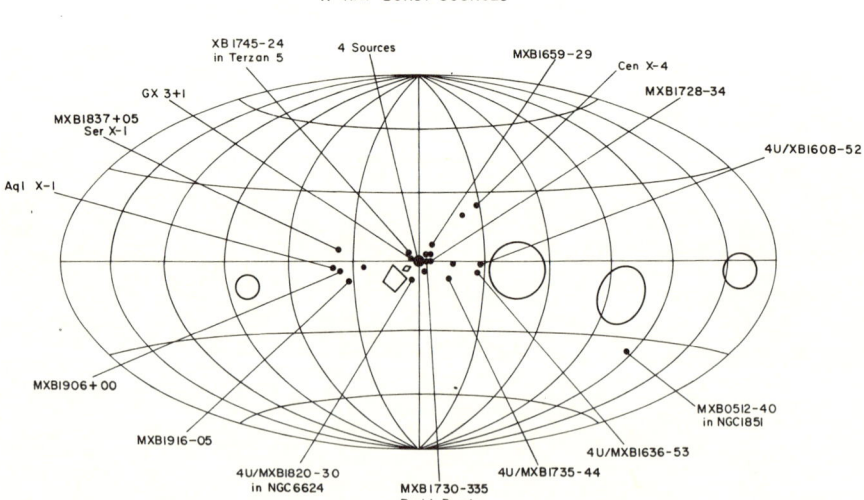

Fig. 2.3. Sky map (galactic coordinates) of the 32 type I X-ray burst sources listed in Table 2.4 (based on data available as of 15 August 1981). The many burst events observed by the Los Alamos group [21, 22] are not shown here (see text). At least 27 but possibly all 32 burst sources shown here emit type I bursts.

type, and they all produce type I X-ray bursts [222, 234, 238, 239, 371]. The bursts are probably only produced when the persistent X-ray flux is within certain limits (for references see Table 2.4). The association of burst sources with soft transients has provided circumstantial evidence supporting the binary nature of these systems [74, 174].

Kaluzienski, Holt & Swank [155] reported an indication of a 8.2 ± 0.2 h period in the persistent X-ray flux from the burst/transient source Cen X-4. This period, if real, most likely results from binary orbital motion, and such a short period strongly supports earlier suggestions that these systems are low-mass close-binary stellar systems [149]. Orbital periods have since been reported for several other galactic bulge sources (including a few burst sources). (See Sections 2.3.2, 2.4.2.7 and 2.6.4.)

2.4.2.3 Burst profiles

Bursts come in large varieties, with rise times in the range of ~ 1 s–~ 10 s and decay times in the range of seconds to minutes. The decay times, in general, are much shorter at high energies than at low energies. This can be seen clearly in Figure 2.4, which shows a superposition of five bursts from MXB 1728−34. In the range 1.2–3.0 keV, the burst decay takes several minutes; in contrast, in the range 19–27 keV, the decay time is only a few seconds (see Figures 2.5-7). This is the result of the cooling of the neutron-star photosphere (see Sections 2.4–6). It has been suggested that the observed lengths and X-ray energy dependences of the burst 'tails' might be explained by scattering of X-rays due to large (~ 3 μm) interstellar grains [2]. However, this idea is strongly contradicted by the observational data [297].

Figure 2.5 shows bursts from five different burst sources. Figure 2.6 shows a series of bursts from 4U/MXB 1636−53. Bursts from a given source often look similar, but this tendency is by no means a universal phenomenon. Some bursts have a distinct double-peaked structure (see Section 2.4.2.6).

Statistically significant intensity changes on time scales of tens of milliseconds were detected in a burst from MXB 1728−34 [125]; the presence of ~ 12.5 ms quasi-periodic fluctuations was reported in the same burst [81].

Figure 2.7 shows double-peaked burst profiles from three different burst sources; they look very similar. The depth of the dips and the separation between the peaks increase with energy. Hoffman, Cominsky & Lewin [118] suggest that perhaps many type I burst sources, on occasion, exhibit this type of structure (see also [103]). For a possible explanation see [376].

A relation between burst profiles and persistent emission was first found in 1976 by Clark *et al.* [49] for 3U/MXB 1820−30. They observed a gradual shortening in decay time (i.e., in the range 1-6 keV the decay time changed

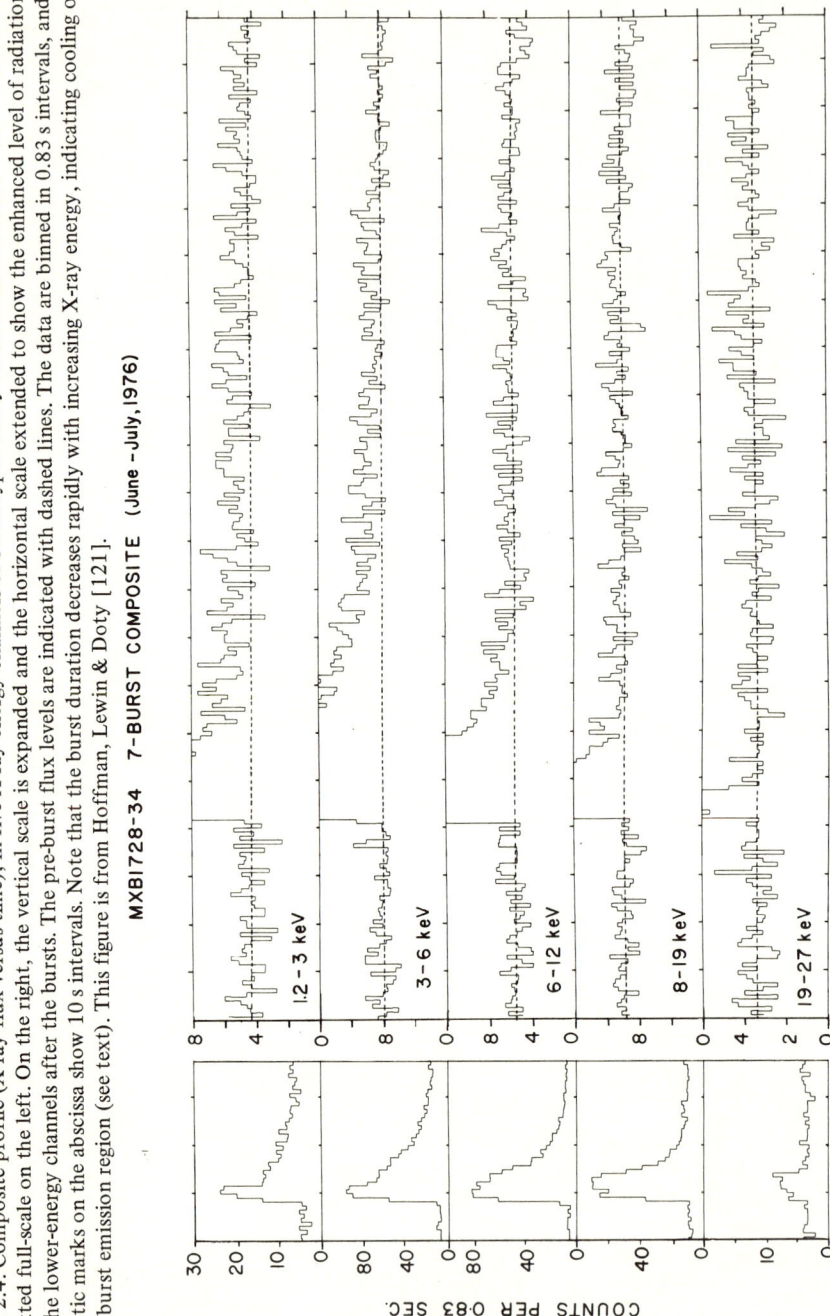

Fig. 2.4. Composite profile (X-ray flux versus time), in five X-ray energy channels of seven type I X-ray bursts from MXB1728−34. The bursts are plotted full-scale on the left. On the right, the vertical scale is expanded and the horizontal scale extended to show the enhanced level of radiation in the lower-energy channels after the bursts. The pre-burst flux levels are indicated with dashed lines. The data are binned in 0.83 s intervals, and the tic marks on the abscissa show 10 s intervals. Note that the burst duration decreases rapidly with increasing X-ray energy, indicating cooling of the burst emission region (see text). This figure is from Hoffman, Lewin & Doty [121].

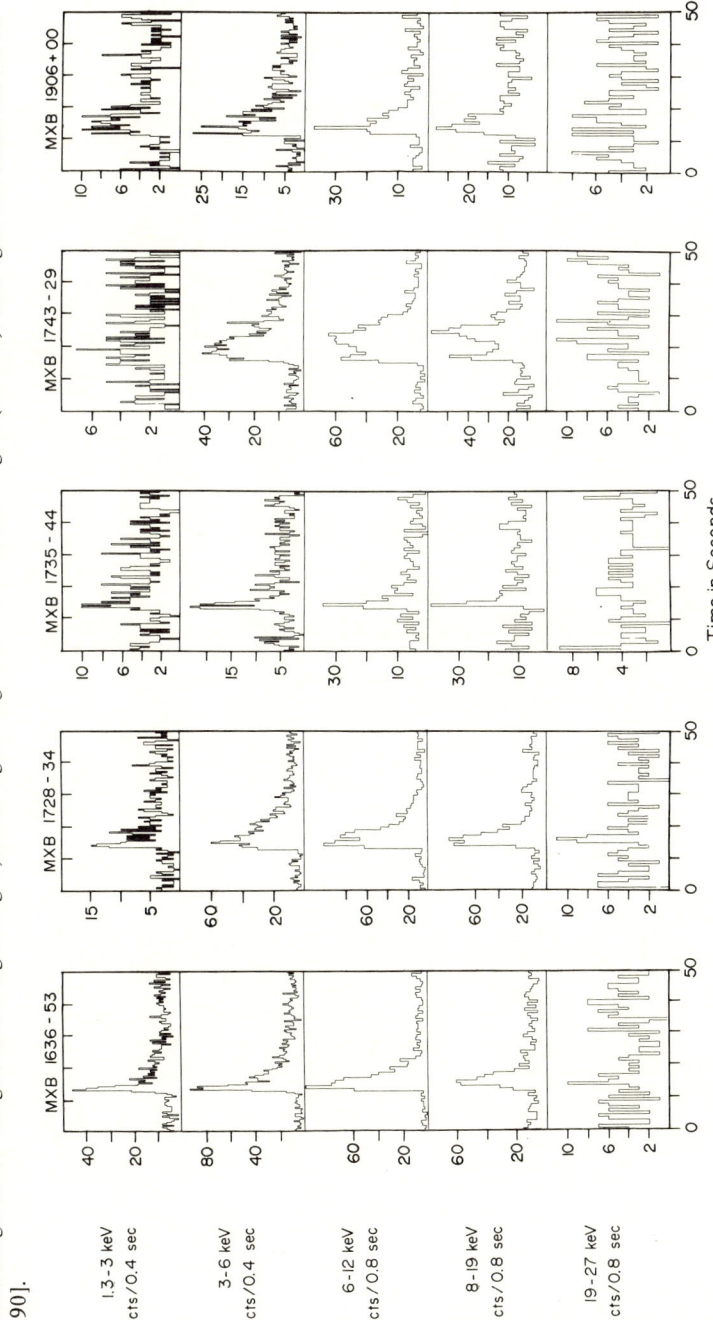

Fig. 2.5. Profiles of type I X-ray bursts from five different sources (SAS-3 data). Note again (as in Figure 2.4) that the gradual decay (burst 'tail') persists longer at low energies than at high energies, indicating cooling of the burst emission region (see text). This figure is from Lewin & Joss [190].

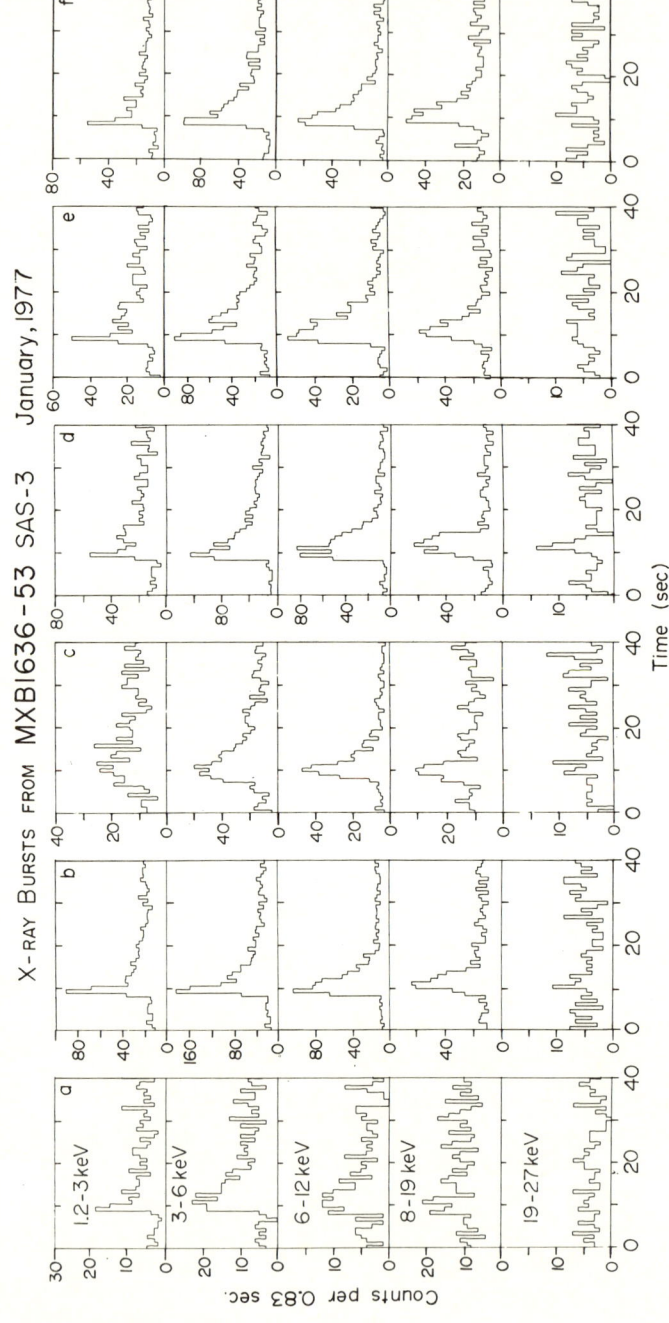

Fig. 2.6. Profiles of six type I X-ray bursts from 4U/MXB1636−53. The bursts have similar shapes except for the third one (c), which does not show the narrow feature in the 1.2–3.0 keV channel that is evident in the other bursts. This burst occurred unusually soon (2.7 h) after the preceding burst. Intervals between the other bursts range from 9.5 h to 12.2 h. This figure is from Hoffman, Lewin & Doty [120].

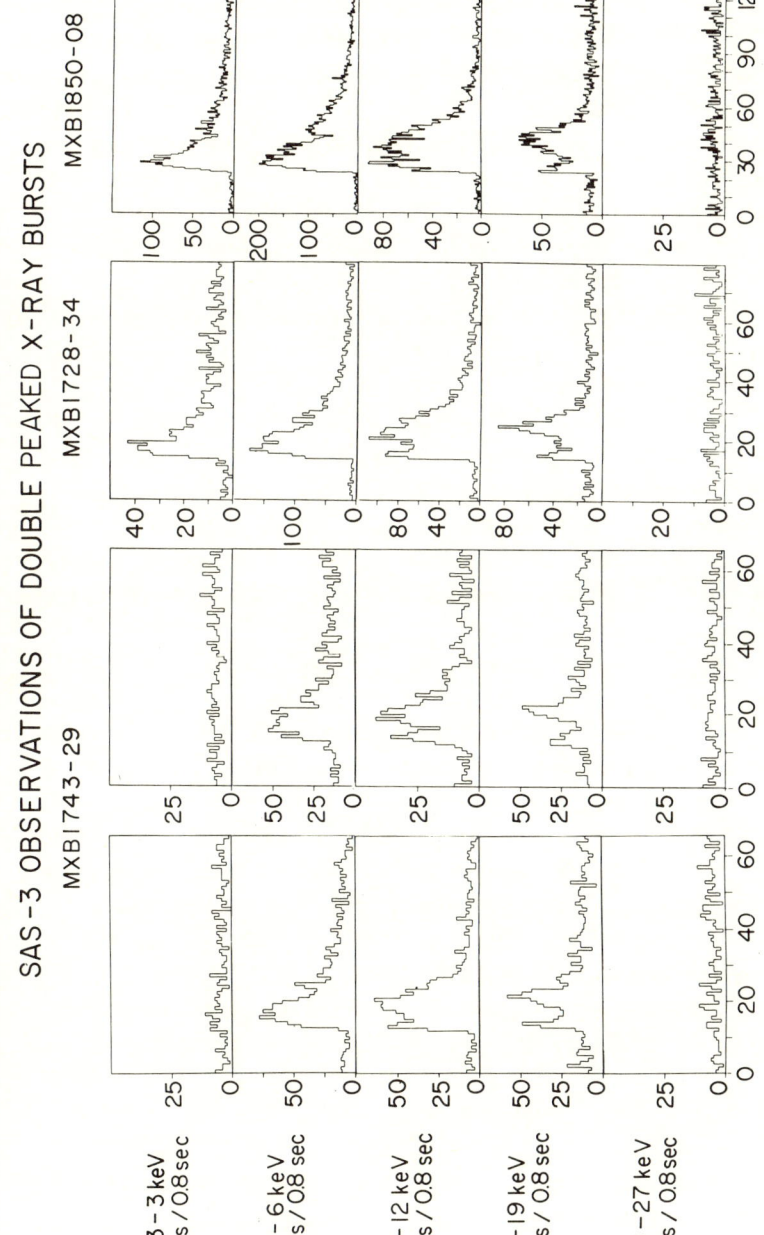

Fig. 2.7. Profiles of type I X-ray bursts from three different sources, all showing a characteristic multiple-peaked structure (see text). The spectrum from MXB 1743−29 (this source is located near the galactic center) is highly cut-off below 3 keV due to interstellar absorption. The burst source MXB 1850−08 is located in the globular cluster NGC 6712. This figure is from Hoffman, Cominsky & Lewin [118].

from ~4.7 s to <3 s) while the level of persistent emission increased by a factor of ~5. Another change in burst profile correlated with the level of persistent emission was observed by the Hakucho team. Murakami *et al.* [234] observed 4U 1608−52 during an active phase in April 1979. The majority of the bursts had a fast rise ($\lesssim$2 s) and a fast decay ($\lesssim$15 s). Two months later, when the persistent flux had decreased substantially (by a factor of ~5) the bursts had a slower rise and a slower decay. In addition, the maximum observed burst flux was lower at that time (by a factor of ~3) than it was during the early part of the active phase (see Section 2.6.1.2). The possible relation of such effects to the thermonuclear-flash model for type I X-ray bursts has been discussed by various authors [82, 234] (see Section 2.6.1.2). No general correlation exists between the rise times of bursts and the associated persistent flux [314, 338].

2.4.2.4 Burst intervals

Burst intervals can be regular or irregular on time scales of hours to days. The burst occurrence rate is sometimes (but not always) related to the level of persistent X-ray emission.

In May 1976, the persistent flux of 3U/MXB 1820−30 increased by a factor of ~5 while the burst intervals gradually decreased by 50 per cent. The persistent flux continued to increase, and the bursts stopped completely [49]. Bursts have only been detected from this source when its persistent flux was relatively low (for references see Table 2.4).

In October 1976, MXB 1659−29 produced bursts at fairly regular intervals of ~2.5 h; no persistent flux could be detected (i.e., its flux was less than 2 per cent of that of the Crab Nebula). Bursts were also detected in June 1977. In March 1978, the persistent flux was much higher (~8 per cent of that of the Crab Nebula), and no bursts were observed (see Table 2.4 for references). Bursts from the other transient sources Cen X-4, 4U 1608−52 and Aql X-1 were only observed after the persistent X-ray flux had flared up substantially (see Section 2.4.2.2).

In several other cases (MXB 1837+05, 4U/MXB 1735−44, and 4U/MXB 1636−53), large fluctuations in the observed burst intervals were not noticeably correlated with variations in the persistent X-ray flux [120, 194, 200, 314, 338].

According to the thermonuclear-flash model in its present, still rather simple, form (see Section 2.6 and references therein), one would expect no type I bursts at very low [165] or very high [142, 144] values of the accretion rate onto the neutron star. Moreover, the burst intervals are expected to decrease if the accretion rate and the resulting persistent X-ray flux increase. This is sometimes observed [49] (see above); however, more theoretical work is required to explain why, in several cases, the burst intervals can change by an order of magnitude or more without an appreciable change in the associated persistent X-ray fluxes.

A possibly interesting connection between the irregular burst behavior of 4U/MXB 1735−44 and MXB 1837+05 (= Ser X-1 = 4U 1837+04) and the thermonuclear-flash model is suggested by the following considerations [194, 296]. One prediction of these calculations is that thermonuclear flashes will not occur above a certain accretion rate, which corresponds to an accretion-driven luminosity that is a substantial fraction of the Eddington limit for a solar-mass object [142]. If the maximum burst luminosity is of the order of the Eddington limit, as indicated by theoretical considerations [141, 142] and by the phenomenological work of van Paradijs et al. [296], the ratio γ of the persistent X-ray flux to the maximum burst flux is an approximate measure of the X-ray luminosity in units of the Eddington limit. For 4U/MXB 1735−44 and MXB 1837+05 this ratio equals 0.2 and 0.3 respectively. These values are the two highest among 12 burst sources for which γ was reasonably well determined [296]. This result suggests that the irregular burst behavior of 4U/MXB 1735−44 and MXB 1837+05 is related to their large persistent luminosities, which may be near the critical values above which X-ray bursts do not occur.

2.4.2.5 Energy ratio (α) between persistent emission and bursts

Let α be the ratio of time-averaged energy in the persistent flux to that emitted in bursts. In most cases $\alpha \gtrsim 10^2$ [171, 296]. This is consistent with the idea that type I bursts are produced by thermonuclear flashes of helium and/or heavier elements. The gravitational potential energy liberated at the surface of a neutron star and released as persistent X-ray emission should be ~ 100 MeV per accreted nucleon, while the energy released by nuclear reactions and released as burst emission should be ~ 1 MeV per accreted nucleon.

For MXB 1659−29, in October 1976, α was less than 25 [190]. Values of α moderately below 100 could be the result of the entrainment of hydrogen in the helium flashes. There are four reported cases where type I burst intervals were very short. In 1976, three type I bursts in quick succession were observed with SAS-3 from MXB 1743−28, with intervals of only 17 min and 4 min [187]. In 1979, the Hakucho Observatory detected two type I bursts from 4U 1608−52 with an interval of only ~ 10 min [235]; this leads to a value for α of at most 2.5. Two optical bursts were observed from 4U/MXB 1636−53 with an interval of ~ 6 min ([333]; see also Section 2.4.2.7.4), and two type I X-ray bursts were observed from XB 1745−24 (in Terzan 5) with an interval of ~ 8 min [340]. A reconciliation of the thermonuclear flash model with these very short burst intervals has not yet been achieved (Section 2.6.3.1).

2.4.2.6 Burst spectra, 'standard candles', Eddington limit, black-body radii and neutron stars

Why do we believe that burst sources are neutron stars? Of course, the ability of the thermonuclear-flash model to explain many of the burst properties is very persuasive. However, substantial evidence in favor of neutron stars preceded the development of the thermonuclear-flash model. It was first pointed out by Swank *et al.* [268] that the burst spectra fit that of a black body reasonably well (Figure 2.8). For one particular burst of ~600 s duration, they derived radii of the success of the thermonuclear-flash model [142]. It was first pointed out by a distance of 10 kpc) and found values of ~100 km during the first 20 s and ~15 km later on in the burst. Hoffman, Lewin & Doty [120, 121] found for MXB 1728−34 and 4U/MXB 1636−53 that near burst maximum, before the decay begins, the black-body temperature is ~3×10^7 K. The object cooled

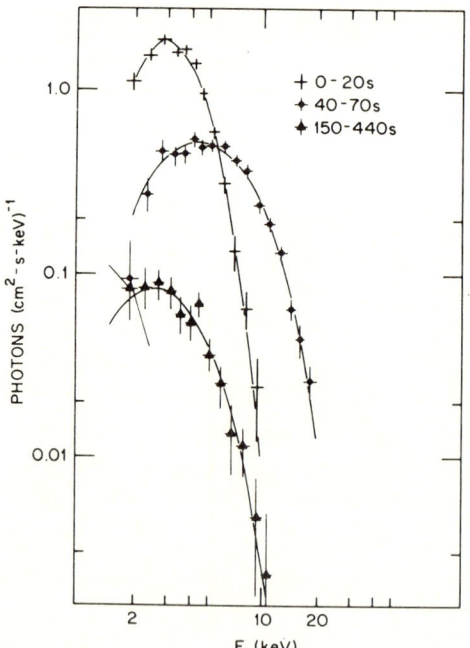

Fig. 2.8. Average spectra, in three time intervals, of a very long (~600 s) type I X-ray burst from XB 1724−30 which is probably the burst source located in Terzan 2 [103]. Time zero is near the burst onset. The solid curves show the best fits to black-body spectra. The values for kT are ~0.9 keV (0–20 s), ~2.3 keV (40–70 s) and ~1.2 keV (150–440 s). Under the assumption of a spherical emitting surface and a source distance of 10 kpc the best-fit black-body radii were ~100 km during the first 20 s of the burst and ~15 km during the remainder of the burst. This figure is from Swank *et al.* [268].

rapidly during the decay portion of the burst (note the spectral softening in Figures 2.4–2.7), but the black-body radii remained constant (at $\sim$10 km) throughout the burst decay. However, no satisfactory fits to black-body spectra could be found during the early part of the bursts [120, 121] (see also [125]). The above dimensions suggest that the burst emission regions are on or near the surfaces of neutron stars.

Van Paradijs [291] carried this analysis a step further in a study of the average characteristics of bursts from each of ten sources. He showed that all ten sources, interpreted as black bodies emitting isotropically, have nearly the same ratio of radiating surface area to average peak-burst luminosity. If the peak luminosity is a 'standard candle', the effective black-body radius is approximately the same for all burst sources. If, moreover, the peak luminosities are assumed to be equal to the Eddington limit for a 1.4 solar-mass object with a hydrogen envelope, the radii are found to have an average value of $\sim$7 km with a scatter of only $\sim$20 per cent [291]. Van Paradijs [292] later adjusted the maximum burst luminosity so that the ten sources would be distributed evenly about the galactic center (at an assumed distance of 9 kpc). He obtained an average black-body radius of 8.5 km. However, this result is complicated by relativistic corrections ([86, 273, 292]; see Section 2.6.1.3).

Lewin et al. [194] first showed that, when examined in detail, individual burst maxima are not standard candles. They showed that the standard deviation in the maximum flux of 53 bursts from MXB 1735−44 was 37 per cent of the mean value. The highest observed flux at burst maximum was about seven times higher than the lowest. They found that the energy in a burst is approximately proportional to the maximum flux in that burst [199]. Similar results were subsequently found by Murakami et al. [234] for bursts from 4U 1608−52, by Ohashi et al. [342] for bursts from 4U/MXB 1636−53, and by Sztajno et al. [314] for bursts from Ser X-1. Basinska et al. [338] showed that, in the case of MXB 1728−34, the maximum burst flux does not exceed a certain value regardless of the amount of energy in that burst. It is unclear at this time whether this maximum level is a true 'standard candle'. The observed spreads may still be small enough to justify the use of the average values in the analysis of global properties of X-ray burst sources [291].

There is evidence that in many cases the Eddington limit is exceeded near the peak of bursts ([86, 103, 121, 338, 339, 341, 361, 362]; see Section 2.6.1.4).

If the double-peaked bursts, shown in Figure 2.7, are fitted to black-body spectra [118], it is found that temperature changes are anticorrelated with radii changes. The fitted radii remain constant (typically $\sim$10 km) during the smooth portion of burst decay as the temperature decreases. Similar fluctuations in the fitted radii and temperatures during a burst from XB1724−30 (in Terzan 2) have been reported by Grindlay *et al.* [103]. It is likely that the photosphere expands during intense bursts [103, 118, 376].

On the basis of (i) the derived black-body radii (~10 km) of the emitted bursts and (ii) the success of the thermonuclear-flash model, it seems safe to conclude that type I X-ray burst sources are neutron stars. The reported values of black-body radii are not reliable enough to come to any meaningful conclusion regarding the equation of state of nuclear matter in neutron stars. The uncertainty in the measurements was recently discussed by Lewin [362].

2.4.2.7 Identifications, optical properties and optical-X-ray bursts

We shall here review some of the optical properties of type I X-ray burst sources. For more details see Chapter 5, this book.

2.4.2.7.1 Optical identifications

Of the 31 burst sources listed in Table 2.4, six (and perhaps a seventh) have been optically identified (for references see Table 2.4, [28] and references therein). The optical counterparts are intrinsically faint blue objects. With the exception of the nearby transient burst source Cen X-4 (see Section 2.4.2.2. and Figure 2.9), their apparent visual magnitudes are all greater than 17 mag. Studies of these optical counterparts and of the X-ray properties of burst sources (and other galactic bulge sources) have provided persuasive evidence that they are low-mass close-binary stellar systems. Recent identifications by Grindlay [351] of X-ray bursters with ordinary G stars are bound to be wrong.

2.4.2.7.2 Optical spectra of burst sources

Canizares, McClintock & Grindlay [37] made spectroscopic studies of the optical counterparts of the X-ray burst sources 4U/MXB 1735−44, 4U/MXB 1636−53 and MXB 1659−29 [90, 223, 224, 226]. The spectra are remarkably similar in overall shape and in the locations and strengths of spectral features. They lack normal stellar absorption lines and are dominated by emission features. They lack strong Balmer emission lines but occasionally show weak Balmer absorption and/or emission at Hα and Hβ. The spectra show evidence for Doppler shifts due to velocity differences of $\gtrsim 10^3$ km s^{-1} and possibly $\sim 10^4$ km s^{-1}. The high-velocity features, several of which have symmetrical redshifted and blueshifted components, suggest that the objects contain gas streams and accretion disks [37]. The ratio of X-ray to optical luminosity is $\gtrsim 500$–1000 (see Figure 2.1), and the optical emission is probably largely the result of X-ray heating of the accretion disk. The spectra of 4U/

MXB 1735−44 and MXB 1659−29 are variable on a time scale of hours, and it seems likely that this is also the case for 4U/MXB 1636−53 [37]. The absence of normal stellar features in the spectra and the faintness of the objects require that if the suspected binary companions are main-sequence stars, their spectral types are later than F0 [37]. See also Section 2.4.2.7.3 for a discussion of the spectra of burst-transient sources.

2.4.2.7.3 *Orbital periods, spectral types and masses of companion stars*

There is direct evidence for the existence of low-mass companion stars in four X-ray sources: the burst-transient X-ray sources Cen X-4 and Aql X-1 (see Section 2.4.2.2.), the transient source A 0620−00, and the X-ray pulsar 4U 1626−67. During transient flareups, the optical counterparts of A 0620−00, Cen X-4 (see Figure 2.9) and Aql X-1 brighten by several magnitudes and then fade ([28, 36, 38, 153, 209, 242, 243, 262, 283, 303, 310, 332, 337] and references therein). The optical counterparts were very blue at maximum light, and their spectra showed emission lines but none of the absorption lines normally observed from stellar atmospheres [38, 209, 233, 283, 310, 332]. As the X-ray flux decreased, the optical counterpart faded and became redder and stellar absorption lines appeared [243, 283, 293, 299]. The spectra then indicated the presence of main-sequence stars with spectral types between K5 V and K7 V for A 0620−00 [243 and 343], between K3 V and K7 V for Cen X-4 [293, 299] and between G3 V and K3 V for Aql X-1 [283]. The transient source A 1742−28 may have a K3 V companion [343], but this is very uncertain. The observed spectral

Fig. 2.9. The optical brightening of the transient X-ray source Cen X-4. (*a*) The Palomar Observatory sky survey blue print showing Cen X-4 in its quiescent state. (*b*) The discovery plate obtained by M. Liller at the prime focus of the CTIO 4 m telescope on 19 May 1979. This figure is from Canizares, McClintock & Grindlay [38].

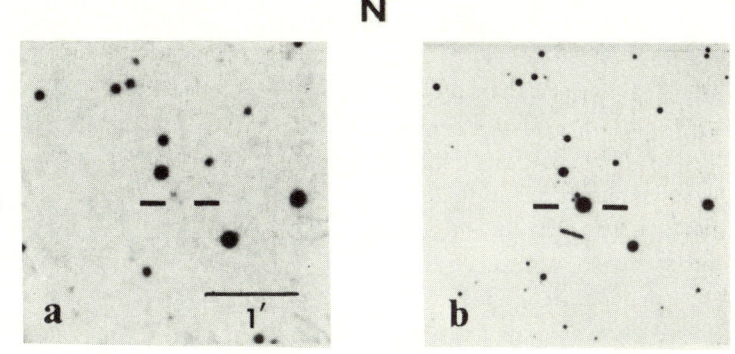

types suggest that the masses of the companion stars are in the range of ~0.5-~1.0 solar mass (see also Chapter 5, Section 5.2.4.4., this book).

In these three cases, the companion stars were only found after the X-ray flux, and thus the optical flux, due to X-ray heating, had greatly decreased. Unless the X-ray flux is very low, the intrinsic optical flux of a low-mass companion star is negligible compared to that due to X-ray heating.

The burst-transient X-ray sources 4U 1608−52 and MXB 1659−29 were optically identified [90, 92] during X-ray flareups. However, during X-ray quiescence, these objects are extremely faint. A year after the flareup of MXB 1659−29 (see Section 2.4.2.2), its optical counterpart was so faint (V > 23) that it could no longer be detected (H. Pedersen, private communication). Such optical faintness during X-ray quiescence greatly inhibits the measurement of the spectral type of the companion stars.

4U 1626−67 is a 7.7 s X-ray pulsar; it is not a burst source. Its orbital period is ~42 min [231], and the mass of the companion star is less than ~0.1 solar mass [198]. It required an extremely arduous observational effort to measure the optical pulsations from the companion star in this system [231]. These pulsations apparently result from the reprocessing of a relatively small amount of pulsed X-rays which reach the companion star despite the shadowing by an accretion disk that surrounds the pulsar, and the existence of these pulsations revealed the binary character of this system [231]. The great difficulty of demonstrating the binary nature of 4U 1626−67, even in the presence of X-ray pulsations, suggests that it may be very difficult to obtain direct evidence for the binary character of many other galactic bulge sources. In Figure 2.10 we show the dimensions of the 4U 1626−67 system.

Of all sources listed in Figure 2.1 with $L_x/L_{opt} > 10$, the following have well established orbital periods: 4U 1626−67 (~42 min); the burst source MXB 1916−05 (~50 min); 4U 2129+47 (~5.3 h), 4U 1822−37 (~5.57 h), Sco X-1 (~0.78 d), 4U 0921−63 (8.99 d) and Cyg X-2 (~9.8 d) ([28, 230, 231, 263, 284, 320, 321, 326, 334, 348, 369, 372] and references therein). There are several other burst sources for which the orbital periods are less certain: Cen X-4 (~8.2 h), Aql X-1 (~1.3 d), 4U/MXB 1735−44 (~4.3 h) and 4U/MXB 1636−53 (~4 h) [155, 301, 336, 344, 345]. In the case of A 0620−00, there is a suggestion [221, 243] of an orbital period of ~7.8 d. A complete list of orbital periods is given in Table 5.2.4 of Chapter 5, this book (see also Chapter 5, Section 5.2.4).

If the companion star fills its Roche lobe (which is likely in view of the rate of mass transfer needed to power the persistent X-ray emission of such sources), and if the companion star is on the main sequence, then the mass of the companion, M_c, depends only on the orbital period, P_{orb} [198, 346, 347]: $P_{orb} \approx 8.7 (M_c/M_\odot)$ h. Orbital periods in the range ~1.0-~10 h may therefore be expected for M_c in the range of $\sim 0.1 M_\odot$-$\sim 1 M_\odot$ (see Figure 2.10 and [256]). We note that the long orbital periods reported for 4U 0921−63 and Cyg X-2 are consistent with a low-mass system; the mass-losing stars appear to be evolved but low-mass ($<1 M_\odot$) subgiants in relatively wide binary systems [320, 321, 369].

2.4.2.7.4 Simultaneous optical-X-ray observations and optical bursts

Simultaneous optical and X-ray bursts have been observed from three burst sources: 4U/MXB 1735−44, Ser X-1 (= MXB 1837+05 = 4U 1837+04) and 4U/MXB 1636−53 ([105, 109, 228, 250]; see also Chapter 5, Section 5.2.4.2, this book). The observations were made during the 1978, 1979 and 1980 worldwide coordinated burst watches [327, 328, 329]. In these observations, the X-ray emission can be used as a probe that illuminates the surroundings of the X-ray star; some of the X-radiation is reprocessed into visible and ultraviolet light. A careful comparison of the X-ray flux (i.e., the probing signal) and the responding optical flux allows one to deduce important information on the character and geometry of the optically emitting regions. In all cases the optical burst was delayed by a few seconds: ~ 2.8 s for one burst from 4U/MXB 1735−44 [228], ~ 1.4 s for one burst from Ser X-1 and ~ 3 s for each of three events from 4U/MXB 1636−53 observed in 1979 [249]. This delay is apparently due to travel time differences (the reprocessing times have been shown to be negligible here

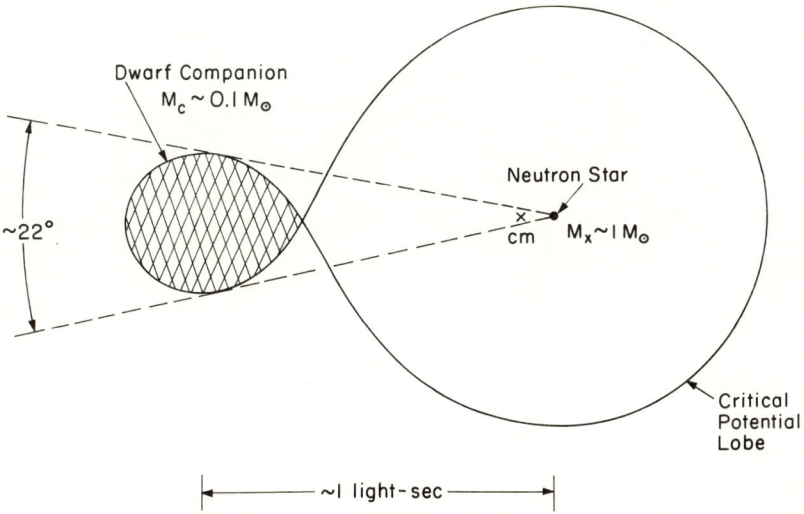

Fig. 2.10. Highly-compact binary model for the X-ray pulsar 4U 1626−67. The figure is drawn to scale for a companion star of mass $\sim 0.1 M_\odot$, a neutron star of mass $\sim 1 M_\odot$ and an orbital separation of ~ 1 lt s (which is appropriate to a main-sequence dwarf companion). This figure is from Li *et al.* [198] (see also [149] and [231]).

[249]) between the X-rays and the optical photons, which are most likely produced in an accretion disk that is a few light seconds across, and on the side of the companion star facing the neutron star [249]. A total of ~40 optical bursts have been observed to date by H. Pedersen [333] from 4U/MXB 1636−53; many of them were simultaneously observed with the Hakucho X-ray Observatory. Two of the optical bursts occurred ~6 min apart [333] (see Section 2.4.2.5). A few of these optical bursts were simultaneously observed in the U, B and V bands and in the R and I bands [364]. The time delay between the optical and X-ray bursts from MXB 1636−53 seems to vary (Hakucho team-MIT group, private communication). This would indicate that the position of the reprocessor, relative to the neutron star, changes. It is possible that this is caused by a bulge in the accretion disk at the location where the accreting matter flows into the disk [345]. If the reprocessing were to take place largely on the surface of the companion star, as seems less likely, a variable time delay is also expected. The presence of a bulge in the accretion disk of MXB 1636−53 was suggested by Pedersen, van Paradijs & Lewin [345] based on an observed ~4 h modulation in their optical data which they interpret as the orbital period. They predict that both the time delays (between optical and X-ray bursts) and the ratio of the integrated optical burst flux to integrated X-ray burst flux will be correlated with the orbital period.

Table 2.5 lists the ratios of optical (B band) to X-ray (~2-~20 keV) energy fluxes observed for both the persistent emission and the bursts from three sources. (Data for only one particular burst from MXB 1636−53 are shown here; for other bursts from this source see [333].) Corrections for extinction

Table 2.5. *Ratios of optical (B band) to X-ray energy for persistent emission and for bursts* (values in parentheses have been corrected for interstellar extinction)

	E_{opt}/E_x		X-ray band	References
	Persistent	Burst		
4U/MXB 1735−44	$1.3 \pm 0.1 \times 10^{-4}$	$2.0 \pm 0.6 \times 10^{-5}$ [a]	2.5−10 keV	[105, 226, 228]
	($<4 \times 10^{-4}$)	($<5 \times 10^{-5}$)[a]		[249]
MXB 1837+05 [c] = Ser X-1	$2.2 \pm 0.4 \times 10^{-5}$	$2.8 \pm 0.9 \times 10^{-6}$ [a]	1.2−12 keV	[109]
4U/MXB 1636−53 [d]	~5.0×10^{-4}	~1.4×10^{-4} [b]	1−25 keV	[249]
	($>2 \times 10^{-3}$)	($>5 \times 10^{-4}$)[b]		[249]

[a] ~10 s integration.
[b] ~10 s integration at maximum flux.
[c] The interstellar extinction in the B band for Ser X-1 is very uncertain (see [179]).
[d] See also [333].

could only be made for two of the three sources; the corrected values are listed in parentheses.

An optical burst was observed from the transient burst source Aql X-1 (= Aql MXB = 4U 1908 + 00) when the source was an outburst [337]. Mason et al. [220] observed an optical burst from the counterpart of 2S 1254−690 [87]. The burst had the gross temporal characteristics of a type I X-ray burst, and it is likely that it was associated with such a burst (though there were no simultaneous X-ray observations). The source lies in the very large error region of the burst source which we designate 14??-6? in Table 2.4 [7]. There was an indication that the optical burst was modulated with a frequency of 36.4 Hz [220]. We note that if the optical burst was modulated at such a high frequency, it is unlikely that it was associated with a type I X-ray burst; if the burst comes from a region with dimensions of a few light seconds [249], such a high-frequency modulation should be washed out by light–travel–time effects.

McClintock et al. [227] found that in the case of 4U 1626−67 (the 7.7 s X-ray pulsar), a significant portion ($\gtrsim 8$ per cent) and possibly the bulk of the optical emission is produced within ~ 0.5 lt s of the neutron star. This indicates that relatively little optical emission comes from the X-ray heated surface of the dwarf companion; most of it probably comes from an accretion disk [227] (see also Section 2.4.2.7.3 and [231]). Although 4U 1626−67 is not a burst source, these results lend support to the emerging model of galactic bulge sources as highly-compact binary systems with low-mass companion stars, wherein most of the optical light results from the reprocessing of X-radiation in an accretion disk [149, 198, 232, 256] (see also Sections 2.4.2.7.3 and 2.6.4).

2.4.2.7.5 The need for a companion star

From the above considerations, it seems very likely that galactic bulge sources, as a class, are low-mass close-binary stellar systems. There is an additional argument in favor of binary systems, which comes from the X-ray observations alone and which historically predates those based on the optical and the combined optical-X-ray observations. In principle, a collapsed object of roughly solar mass could be adequately powered by accretion from the interstellar medium to produce the observed persistent high X-ray luminosity if it were in a dense ($>10^6$ cm^{-3}) interstellar cloud [245]. This seems very unlikely, however, since no intrinsic low-energy absorption is observed in the spectra of the persistent emission. Moreover, the galactic latitude of some of these sources argues strongly against the presence of such dense clouds. Hence, another supply of accreting matter must be present. In some cases, the sources might be isolated neutron stars with massive fossil accretion disks [247]; however, if this were so in all cases, it is difficult to understand why they occur preferentially in the cores of centrally condensed globular clusters.

2.4.2.8 Infrared bursts from type I burst sources?

It is unclear at this time whether the reported infrared bursts [5, 139, 162] are real and associated with type I X-ray bursts [179] (see also Section 2.4.3.10).

2.4.2.9 Radio properties of type I burst sources

During the 1977 coordinated burst observations, five different type I X-ray burst sources were observed at radio frequencies. No radio bursts were detected [137, 281, 286].

2.4.3 The Rapid Burster (MXB 1730−335)

The Rapid Burster (MXB 1730−335) was discovered by Lewin *et al.* in March 1976 [168, 180]. Its behavior is unlike any other known source. When the source is burst active, it can produce bursts in quick succession with intervals as short as ~ 10 s (see Figure 2.11). As discovered by Hoffman, Marshall & Lewin [127], two distinctly different types of bursts are produced by this source, which led to the classification of type I and type II X-ray bursts [127] (see Figures 2.12 and 2.13).

2.4.3.1 Type I and type II bursts

In contrast to the type I bursts, the type II bursts do not show a significant amount of spectral softening during burst decay, and, in general, they recur on time scales about one to two orders of magnitude shorter than those of type I bursts [127] (see Table 2.3). The Rapid Burster is the only object known to produce both type I and type II bursts [127] (see Table 2.3, Figures 2.12 and 2.13). Type I bursts were detected at intervals of several (~ 3-4) hours [127] and are only (but not always) observed when there is also type II activity [127, 218]. The type I bursts sometimes have a noticeable effect on the type II bursts [127].

The average power in type II bursts is about 120 times that in type I bursts (when the latter are present) [127]. This is consistent with the idea that the type I bursts result from the release of nuclear energy (~ 1 MeV per nucleon for helium-burning) and the type II bursts result from the release of gravitational potential energy (~ 100 MeV per nucleon) ([141, 142, 144, 165, 208, 319]; see also Section 2.6.3.2).

2.4.3.2 Type II burst profiles and saturation

The type II bursts from the Rapid Burster have rise times of ~ 1 s, and they last from a few seconds up to about ten minutes [16, 131, 172, 180, 215, 218, 287, 305]. Structure on time scales down to 50 ms has been observed [287], and a quasi-period of ~ 0.5 s was found in a few long bursts [373]. When the

X-ray bursters and the X-ray sources of the galactic bulge 73

burst duration is in excess of about 15 s, they saturate in intensity and have 'flat' maxima, as shown in Figures 2.11 and 2.14. Some authors introduce a distinction between long bursts and short bursts. They call the long bursts 'trapezoidal' (e.g., [131]) and the short bursts 'exponential' (e.g., [131 and 365]). This is unfortunate as it implies that different physical processes are involved, which is not the

Fig. 2.11. Type II X-ray bursts from the Rapid Burster (MXB 1730−335). These are ~24 min stretches of data from eight different orbits of the SAS-3 Observatory on 2-3 March 1976, when this unique object was discovered [180]. Bursts with durations in excess of ~15 s show 'flux saturation' (i.e., flat tops). This is also apparent in the very long bursts shown in Figure 2.14. Note the strong correlation between the integrated counts in a burst and the duration of the quiescent period that follows (see Figure 2.15). The arrow indicates a type I burst from MXB 1728−34. This figure is from Lewin [172].

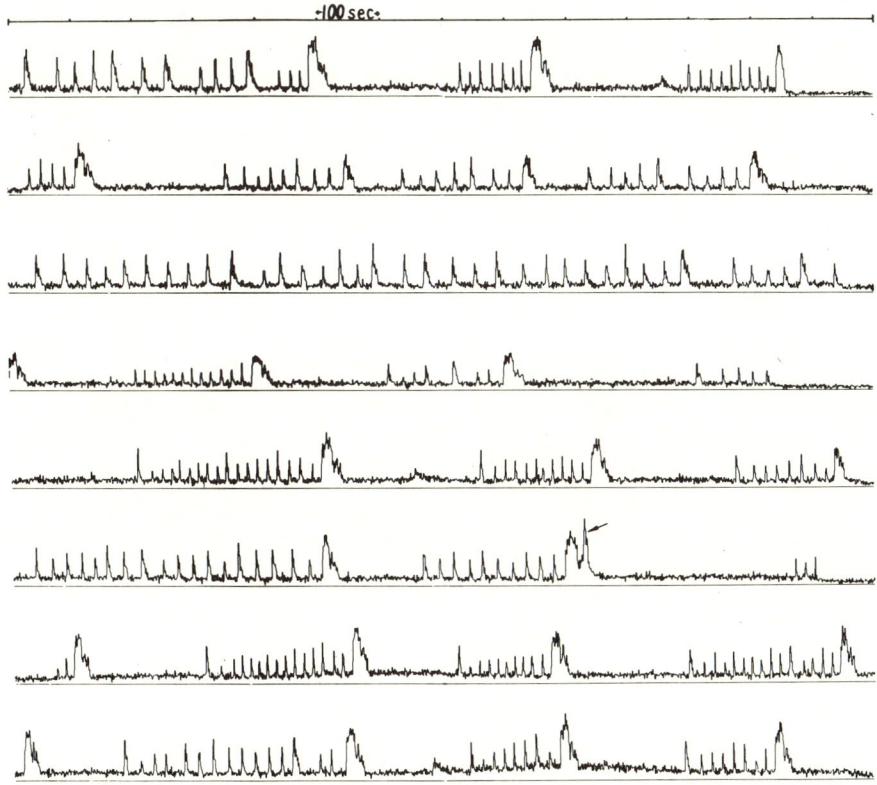

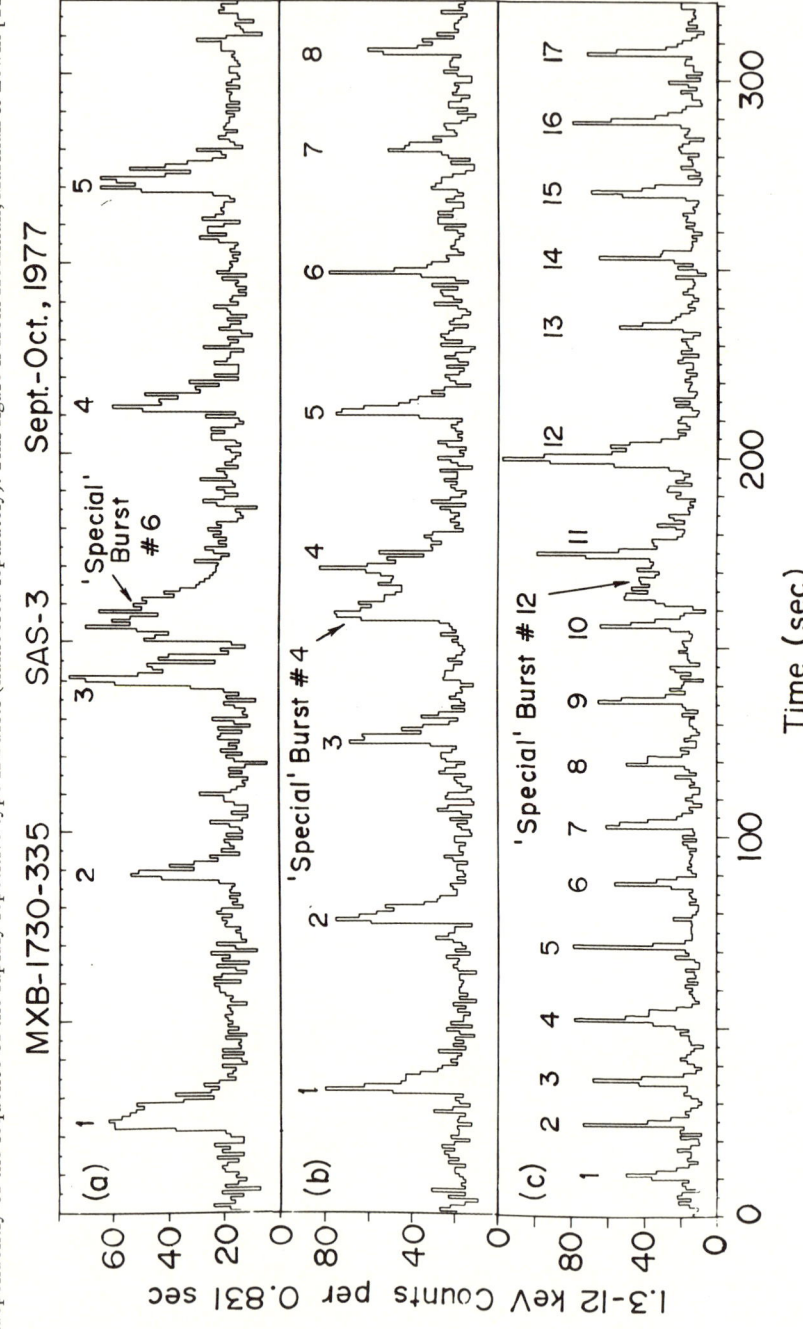

Fig. 2.12. Discovery of type I X-ray bursts from the Rapid Burster (MXB 1730−335) (see Table 2.3). The type I bursts (marked as 'special') occur independently of the sequence of the rapidly repetitive type II bursts (numbered separately). This figure is from Hoffman, Marshall & Lewin [127].

case. The type II bursts are most certainly due to accretion instabilities regardless of their duration. As first observed by Lewin et al. [180], for an increasing energy release the burst fluxes 'saturate' and reach a maximum; the bursts get longer but not higher (see Figure 2.11). To draw a line between the long, flat-topped bursts and the short ones, without flat tops, is somewhat artificial.

Inoue et al. [131] reported that in observations with Hakucho in August 1979, the 'saturated' peak flux varied by a factor of 4. Basinska et al. [16] subsequently found that in March 1976 the saturated peak flux (among a total of ~220 type II bursts observed with SAS-3) varied by a factor of ~3; during ~3-5 March 1979, they varied by a factor of 1.7. Both sets of data suggest that the variability of the peak flux (in saturated bursts) is highest when

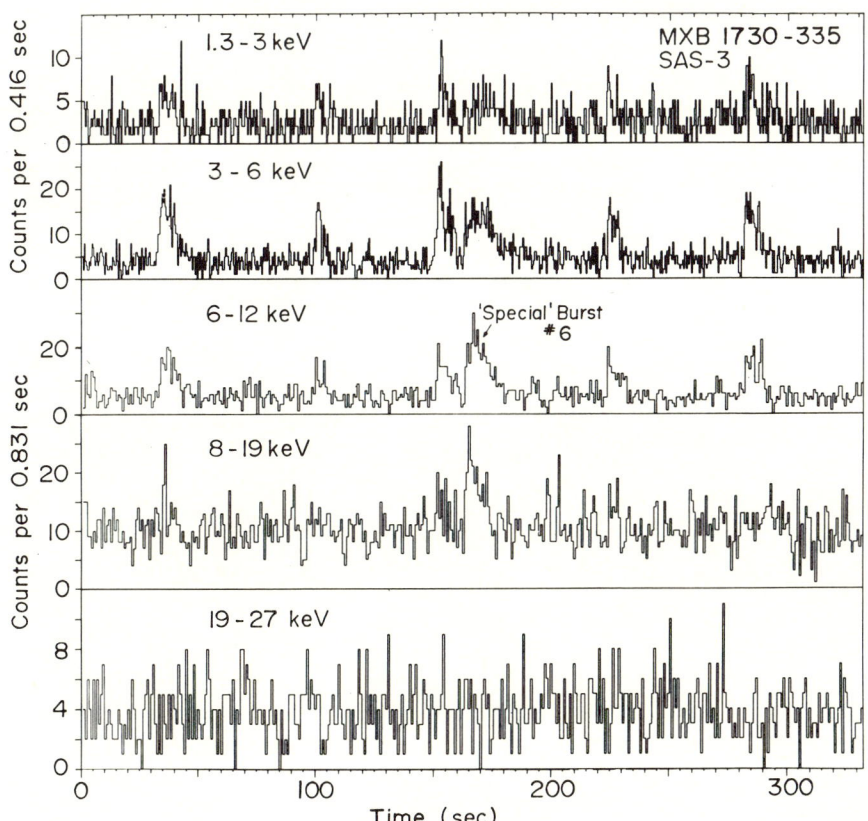

Fig. 2.13. Spectral information for the data shown in Figure 2.12(a). Note that the type I ('special') burst spectra soften during burst decay; this softening is absent in the type II bursts. This figure is from Hoffman, Marshall & Lewin [127].

Fig. 2.14. Type II X-ray bursts from the Rapid Burster (MXB 1730–335), as observed with SAS-3 between 3 and 5 March 1979. The bursts labelled I and III last ~400 s and ~250 s, respectively. About five months later, many more of these very long type II bursts were observed with Hakucho [131]. The type I burst indicated with an arrow is from MXB 1728–34. This figure is from Basinska et al. [16] (see also [215]).

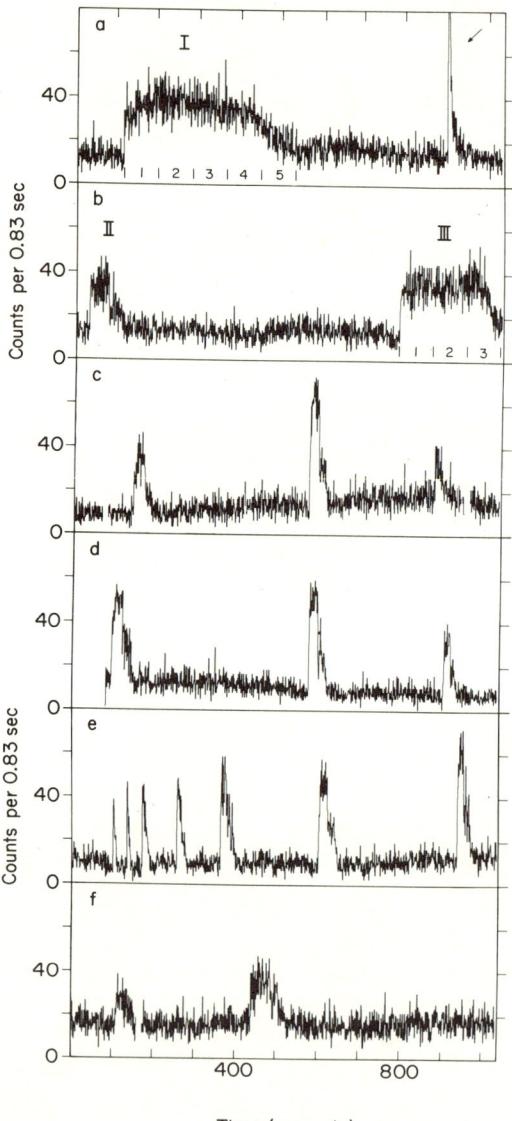

the time-averaged X-ray luminosity is high, which seems to occur early in each burst-active period [218].

2.4.3.3 E-Δt relation for type II bursts

For moderately energetic type II bursts from the Rapid Burster (total emitted energies in excess of $\sim 4 \times 10^{38}$ erg for isotropic emission and an assumed distance of 10 kpc), the integrated burst energy, E, is approximately proportional to the 'waiting time', Δt, to the next burst [131, 172, 180, 218, 305]. For weak bursts, this approximately linear relation breaks down [172, 180, 305] (see Figure 2.15).

2.4.3.4 Type II bursting patterns (modes)

There seem to be preferred type II burst recurrence patterns in the Rapid Burster. H. Marshall et al. [218] distinguished two patterns (modes), as illustrated in Figures 2.16 and 2.17 (see also [172]). They suggested that mode I occurs near turn-ons of the burst-active periods and mode II later on. In mode I, there is a large spread in burst energies (by a factor of ~ 100), and histograms of the distribution of the burst energies show a distinct double peak (Figure 2.17). In mode II, the bursts do not vary as much in energy as in mode I, and the energy distribution is not double-peaked.

At times, the type II bursts from the Rapid Burster can be of almost equal energy and spaced at nearly equal time intervals. On 10.7 March 1976, UT the bursts were nearly constant in energy and spaced almost regularly at ~ 16 s intervals. On 4.2 April 1976, UT the burst energies were again nearly constant, and the bursts were then spaced at intervals of ~ 30 s (see Figure 2.16 in [172]). On the latter occasion one could select trains of bursts of ~ 10 min duration whose recurrence intervals would fit nearly constant periods [219].

There seems to be at least one other stable bursting mode [131] whereby the bursts are very energetic (they last ~ 1-10 min). These very long bursts were first observed with SAS-3 by F. Marshall in March 1979, probably during the early part of the turn-on of a burst-active phase [16, 215] (Figure 2.14). The very long bursts observed ~ 5 months later by Inoue et al. [131] with Hakucho [161] were detected for more than a week, and they too were observed early on in a burst-active period.

2.4.3.5 Anomalous bursts

'Anomalous' bursts from the Rapid Burster were reported by Ulmer et al. [287]. These bursts have rise times of about 3 s and low peak intensities (no more than 20 per cent of the intensity of saturated type II bursts), and they violate the E-Δt relation for type II bursts. They appeared only after the detec-

Fig. 2.15. Log–log plots of the integrated type II X-ray burst energy, E, versus the waiting time, Δt, to the next burst, for some of the data indicated with arrows in the left part of Figure 2.16. The data points would follow a straight line with slope 45° if the relation between E and Δt were strictly linear. See Section 2.4.3.4 for the definitions of modes I and II. This figure is from Marshall et al. [218].

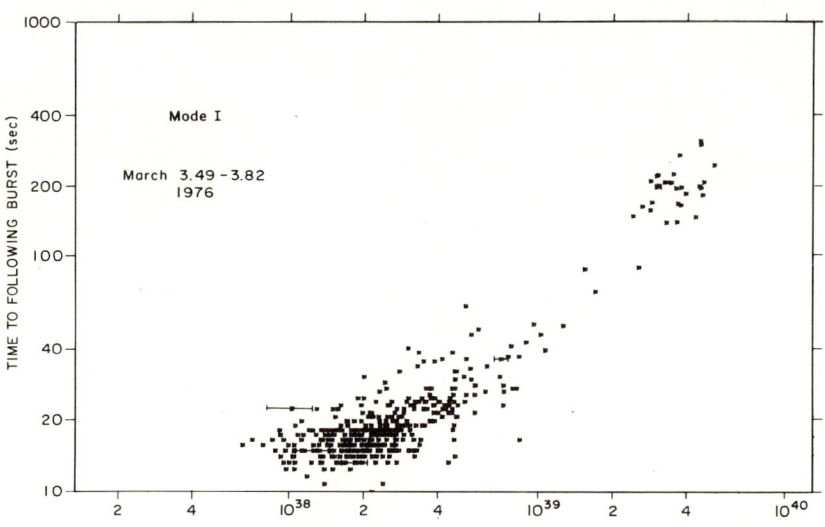

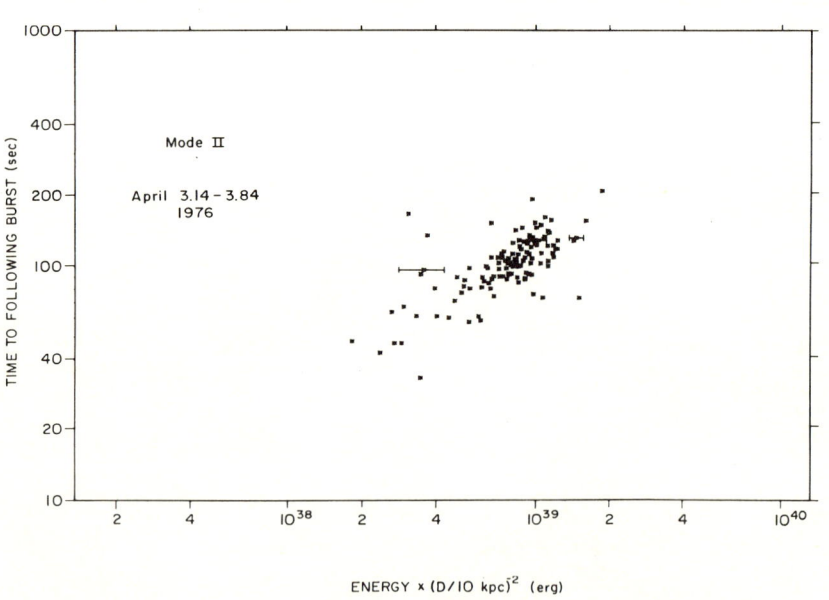

ENERGY x $(D/10\ kpc)^2$ (erg)
(~1.2 –12 keV)

tion of very energetic type II bursts. They probably do not have a distinct spectral softening during burst decay (as is also the case for the type II bursts), but this is somewhat uncertain [127]. It is unclear whether these 'anomalous' bursts are more closely related to type I or type II bursts.

2.4.3.6 Recurrent transient behavior

The Rapid Burster is a recurrent transient X-ray source with a recurrence period of about 6 months [97, 170, 218, 304]. The duration of the periods of burst activity are not well known since X-ray observations of the Rapid Burster are scant (see Figure 2.18). One period of burst activity (in 1976) lasted at least 5 wk, another (in 1978) could not have lasted longer than 3 wk. The time-averaged type II burst flux can vary by a factor of ~ 2 in a few days [218] and tends to be higher during the early part of a burst-active phase than later on [16, 218]. The 'light curve' of the time-averaged type II burst flux [218] is not as smooth as that of some of the classical 'soft' transients. The intervals between transient outbursts are ~ 6 months [3, 53, 151]. There is some doubt whether the transient outbursts occurred 'on schedule' in 1981 and 1982 [340]. The recurrent outbursts are most likely due to periodic

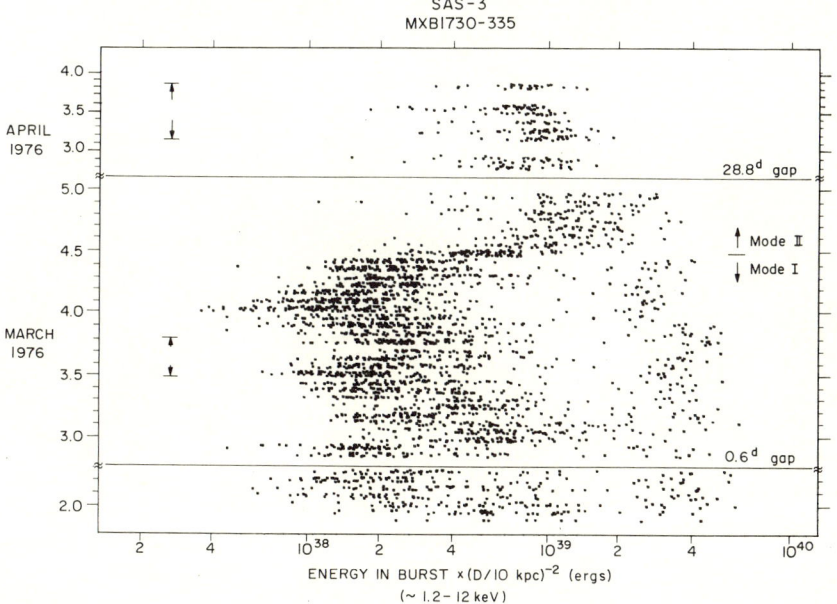

Fig. 2.16. Integrated energy in each type II X-ray burst (abscissa) versus the time in UT of its occurrence (ordinate). Each dot represents one type II burst from the Rapid Burster (MXB 1730−335). D is the distance to the source in kpc. Two different burst patterns are distinguishable (see also Figure 2.17). This figure is from Marshall *et al.* [218].

2.4.3.7 X-ray emission between type II bursts

An upper limit to the X-ray emission between type II bursts from the Rapid Burster was set at less than 4×10^{37} erg s^{-1} (for an assumed source distance of 10 kpc) in early March 1976, when the time-averaged burst luminosity was about 2×10^{37} erg s^{-1} [180]. White et al. [305] set an upper limit for interburst emission during April 1977 of 3×10^{36} erg s^{-1}, or 20 per cent of the time-averaged burst luminosity that was then observed.

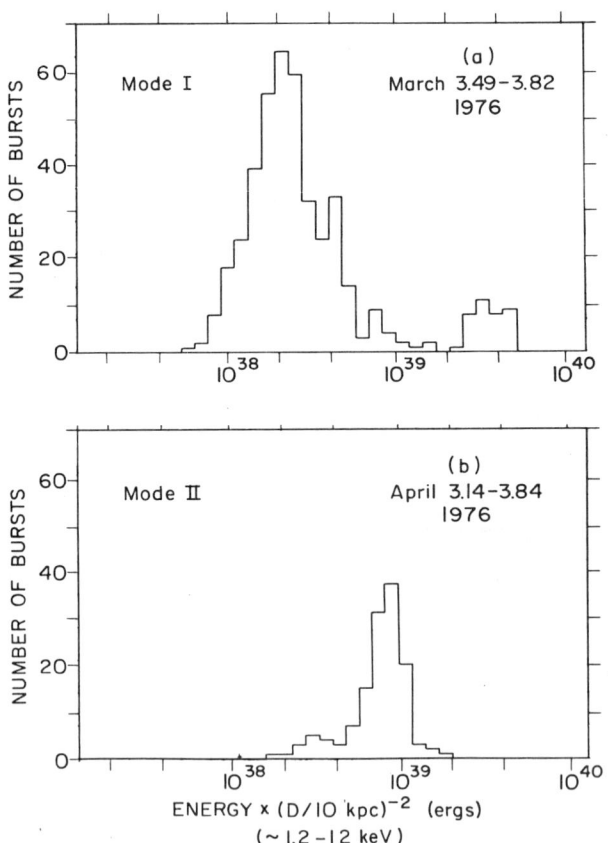

Fig. 2.17. Distribution of integrated energies in type II X-ray bursts from the Rapid Burster (MXB 1730−335). (a) Data from an ∼8 h period in March 1976 (indicated in Figure 2.16). Note the double-peaked distribution. (b) Data from a ∼17 h period in April 1976 (indicated in Figure 2.16). This figure is from Marshall et al. [218].

Fig. 2.18. Complete record of observations of the Rapid Burster (MXB 1730−335). The scale of the horizontal axis is in fractions of a year; the months of the year are also indicated. Periods of X-ray coverage where bursts were not observed are indicated by open rectangles, while burst-active periods are marked by filled rectangles. Only a few (one or two) bursts were observed during periods marked by open rectangles with one or two diagonal lines. These bursts were probably type I bursts from the nearby source MXB 1728−34. Data are from Uhuru, Copernicus, ANS, SAS-3, Ariel 5, HEAO-1, Einstein and Hakucho (see Table 2.4 for references). The eight vertical dashed lines indicate observing periods of less than one day during which no bursts were observed. The line marked 'a' represents a ~7 h period during which the source may have been burst-active, however, this is very uncertain [305]. The source may have been burst-active during the period marked 'b', however, this is somewhat uncertain [305]. During the period marked 'c' (from about mid-August to 26 September 1980), observations of the Rapid Burster were made intermittently with ~3 d intervals; no burst activity was observed (M. Oda, private communication). The date of the discovery of the Rapid Burster (March, 1976) is marked. This figure was prepared by Ewa Basinska.

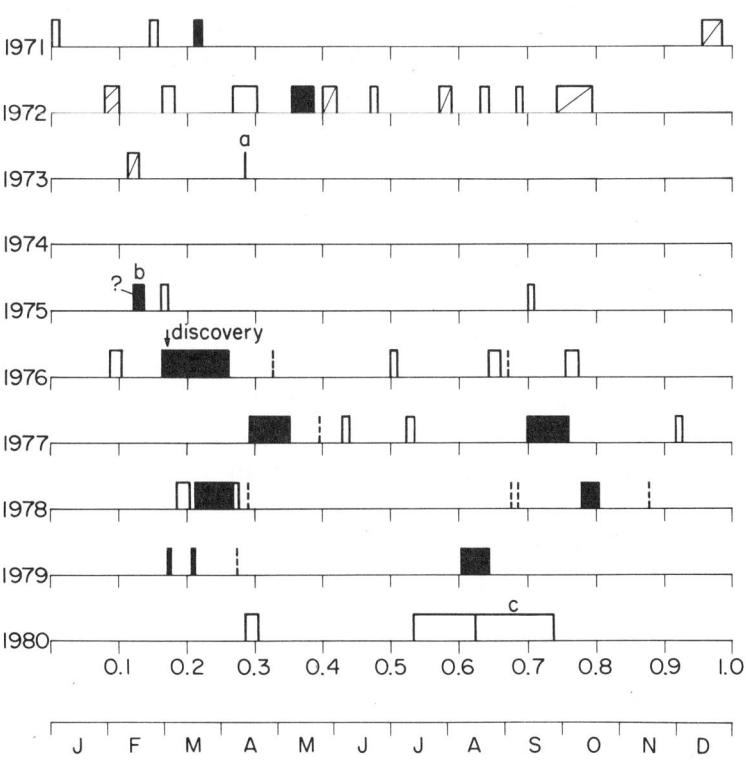

Further detailed analysis of the March 1976 SAS-3 data established the presence of X-ray emission between type II bursts. The enhanced emission which can be seen by careful inspection of Figure 2.11 was observed after the occurrence of very energetic type II bursts [218, 295] ($\sim 3 \times 10^{39}$ erg for an assumed source distance of 10 kpc). This emission can last for $\sim$1-5 min; it rises in $\lesssim$20 s and decays in $\lesssim$30 s shortly before the next type II burst occurs. In March 1976 this emission (at maximum intensity) was about half the time-averaged type II burst flux, and the mean flux contained in this emission equalled about 5 per cent of the time-averaged type II burst flux.

Emission between type II bursts was also observed with HEAO-1 in March 1978 [128] and with Hakucho in August 1979 [280]. During the latter observation, between August 8 and 10, the observed luminosity between bursts was $\sim 2 \times 10^{37}$ erg s^{-1} (for an assumed source distance [160] of 10 kpc). This value was comparable to or somewhat smaller than the time-averaged type II burst luminosity.

2.4.3.8 Persistent emission in burst-off state?

Several attempts have been made to find out whether the Rapid Burster emits X-rays when there are no bursts. The best upper limit was reported by Grindlay [93], based on observations with the Einstein Observatory. On 10 April 1979 the mean X-ray luminosity was less than 10^{34} erg s^{-1} (for an assumed source distance [160] of 10 kpc).

2.4.3.9 Black-body radii of the Rapid Burster

The spectral data for both type I and type II bursts from the Rapid Burster can be fitted reasonably well by black-body spectra [218]. For the type I bursts, a radius of about 9 ± 2 km is found for the emitting region (under the assumptions of a spherical emitting surface and a source distance of 10 kpc), and the radius remains approximately constant throughout each burst. During the type I burst decay the temperature decreases rapidly from a maximum temperature of $\sim 3 \times 10^7$ K. For type II bursts a constant temperature of $\sim 1.8 \times 10^7$ K is found and the radius of the emitting region decreases from $\sim$16 km to $\lesssim$10 km. For short type II bursts this can happen in only a few seconds [218]. These scale sizes indicate that the Rapid Burster is probably a neutron star as earlier suggested by Brecher, Morrison & Sadum [318]. (See also Sections 2.4.2.6 and 2.6.1.3.)

2.4.3.10 Infrared bursts from the Rapid Burster?

Infrared bursts have been detected from the direction of the Rapid Burster by two groups, first during 4-5 April 1979 by Apparao et al. [6] and Kulkarni et al. [162] (at 1.6 μm) and then during 5-12 September 1979 by

Jones et al. [139] (at 2.2 μm). No simultaneous X-ray observations were made during these observations. It is not clear whether the Rapid Burster was in a burst-active phase during either of these observation periods (see Figure 2.18). It was probably not in a burst-active phase during 5-12 September 1979, since by 23 August the preceding burst-active phase had already almost ceased (M. Oda, private communication). A total of eight bursts were detected in ~ 8 h. Sometimes the bursts came in pairs ~ 2 min apart [6, 139, 162].

During 9-12 August 1979, Sato et al. [259] made infrared observations of the Rapid Burster. The Rapid Burster was definitely in a burst-active phase [131, 259] (see Figure 2.18). No infrared bursts were detected in a total of ~ 6 h of observing. For one particular type II X-ray burst (of >126 s duration) the upper limits to the integrated burst fluxes at 1.6 μm and at 2.2 μm are more than one order of magnitude lower than those reported by Kulkarni et al. [162] and Jones et al. [139]. A few days later, during 16-23 August 1979, Glass [85] made infrared observations (at 2.2 μm) of the Rapid Burster for a total of ~ 3.5 h while type II X-ray bursts were again detected simultaneously with Hakucho [85]. Again, no infrared bursts were observed [85]. The upper limits (at a 3σ level of confidence) for 15 s of integration are an order of magnitude below the integrated infrared flux values reported earlier [139, 162].

If the infrared bursts [139, 162] are real and are emitted isotropically, and for an assumed source distance of 10 kpc, the emitted energy in the infrared bursts is in the range of $\sim 6 \times 10^{37}$ erg to $\sim 3 \times 10^{38}$ erg.

It would be very valuable to obtain more simultaneous infrared and X-ray observations of the Rapid Burster. To date, infrared bursts have only been detected when no simultaneous X-ray observations were made [139, 162]; they were not detected when simultaneous X-ray observations were made and type II X-ray bursts were observed [85, 259]. A detailed review of this is given by Lawrence et al. [366].

In view of the null results described above, it is presently uncertain whether the reported infrared bursts are real and, if they are, whether they are associated with the X-ray bursts.

2.4.3.11 Radio bursts from the Rapid Burster?

Calla et al. [30, 31], have reported microwave bursts (at 4.1 GHz) from the direction of the Rapid Burster. The rise times of the bursts ranged from a few seconds to ~ 90 s. Peak amplitudes ranged from ~ 0.2 db to ~ 3 db [30 and 31], where a peak amplitude of 0.25 db is estimated [253] to correspond to ~ 270 Jy.

Simultaneous X-ray and radio (at 2695 MHz and 8085 MHz) observations of the Rapid Burster were made by Johnson et al. [137] in April 1977. The radio observations were made for a total of ~ 10 h with the NRAO Green Bank interferometer and the X-ray observations with Ariel 5 and Copernicus. At this time the Rapid Burster was in an active phase (see Figure 2.18); the two X-ray observatories detected a total of 64 X-ray bursts during the simultaneous radio observations. No radio bursts were detected. One-sigma upper limits at both 2695 MHz and 8085 MHz were ~20 mJy, which is ~4 orders of magnitude smaller than the peak fluxes of the radio bursts reported by Calla et al. [30, 31].

Simultaneous X-ray and radio (at 327 MHz) observations were also made during 13–14 August 1979 [253]. The Rapid Burster was definitely in a burst-active phase [131] (see Figure 2.18); at least two type II X-ray bursts were observed [131] during the ~10.5 h of observation with the Ooty radio telescope. However, no radio bursts were observed. Upper limits to the peak radio flux densities were ~1 Jy. A detailed review is given by Lawrence et al. [366].

In view of these null results it is presently very uncertain whether the reported radio bursts are real and, if so, whether they are associated with X-ray bursts. It would be useful to obtain further simultaneous radio and X-ray observations of the Rapid Burster.

2.4.4 Type II bursts from other sources?

Type II bursts distinguish themselves from type I bursts in two ways: (i) the time scales of recurrence are two to three orders of magnitude shorter; and (ii) there is no strong spectral 'softening' during burst decay (see Table 2.3). The 'flare-burst-like' events observed from the X-ray pulsars GX 304-1 [229] and GX 301-2 [29] and from the X-ray binaries Cyg X-1 [40] and LMC X-4 [71] meet the phenomenological definition for type II bursts and, as suggested by Hoffman, Marshall & Lewin [127], they may be caused by the same mechanism as the type II bursts in the Rapid Burster. In any case, it seems likely that the 'burst-like' events from these four sources and the type II bursts from the Rapid Burster are all due to instabilities in the accretion flow.

2.5 Fast transients (bursts?)

Many fast X-ray transient events of duration from several minutes up to a few hours have been reported [123, 268, 324, 325]. Intensity profiles and spectral evolution are not known for any of the ~20 high galactic latitude events observed with Ariel 5 (K. Pounds, private communication and [325]), nor for the high galactic latitude event observed by Rappaport et al. [324] with SAS-3. It is unclear at this time whether these events are related to type I X-ray bursts. However, an event of ~600 s duration observed by Swank et al. [268] exhibited spectral characteristics and spectral evolution that are indistinguishable from an ordinary type I X-ray burst (see Section 2.4.2.6), and it seems highly likely that this event is generically related to type I bursts.

Hoffman et al. [123] have reported two events which may have been type I
X-ray bursts [123] (Figures 2.19-2.21). An event on 7 February 1977 lasted
~1500 s, and an event on 28 June 1977 lasted ~150 s. Both events were preceded by precursor pulses which lasted only a few seconds and which rose and
fell in less than 0.4 s. The precursors were separated from the 'main' events by
several seconds, during which no X-rays (other than those due to background
radiation) were detected. The spectra of the main events started out much softer
than the spectra of the precursors. There are similarities between the two main
events and ordinary type I X-ray bursts in both their temporal and spectral
evolution. The spectra of the main events became harder as they approached
maximum intensity and softened substantially as they decayed. In the main
event of 7 February 1977, X-rays with energies greater than 10 keV were delayed
by about 80 s relative to the 1.5-6.0 keV X-rays. A black-body fit to the spectral
data of the main event of 7 February 1977 gives a maximum temperature of
2.9×10^7 K (Figure 2.21) and a radius of the emitting region of $\gtrsim 9$ km (under

Fig. 2.19. Profiles of the early portion of the 7 February 1977 event, as observed
with the SAS-3 Center Slat detectors. A precursor is clearly visible but shows no
obvious spectral evolution. After the precursor the signal drops back to its original
level before the 'main' event begins. The delay between the precursor and the rise
of the 'main' event increases with energy and has a duration of ~80 s in the
8-35 keV channel. Thus, in this energy channel the 'main' event rises near
t ~115 s (off-scale); the rise is shown in Figure 2.20. The 'main' event displays
significant spectral softening during its decay (see Figure 2.20 and 2.21). This
figure is a modified version of one shown by Hoffman et al. [123].

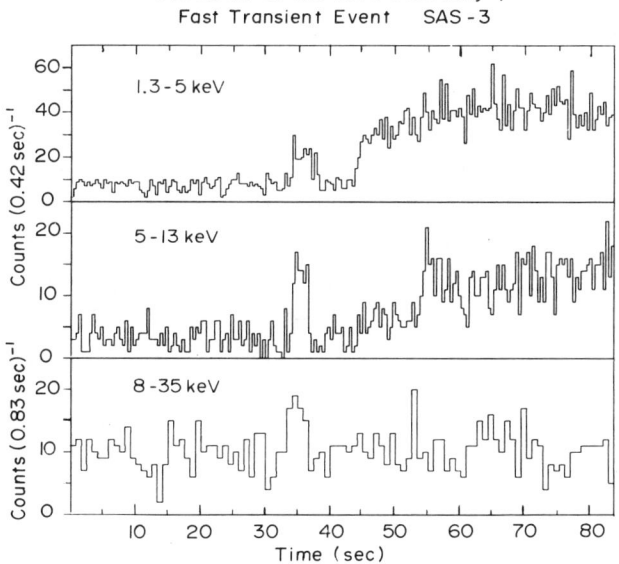

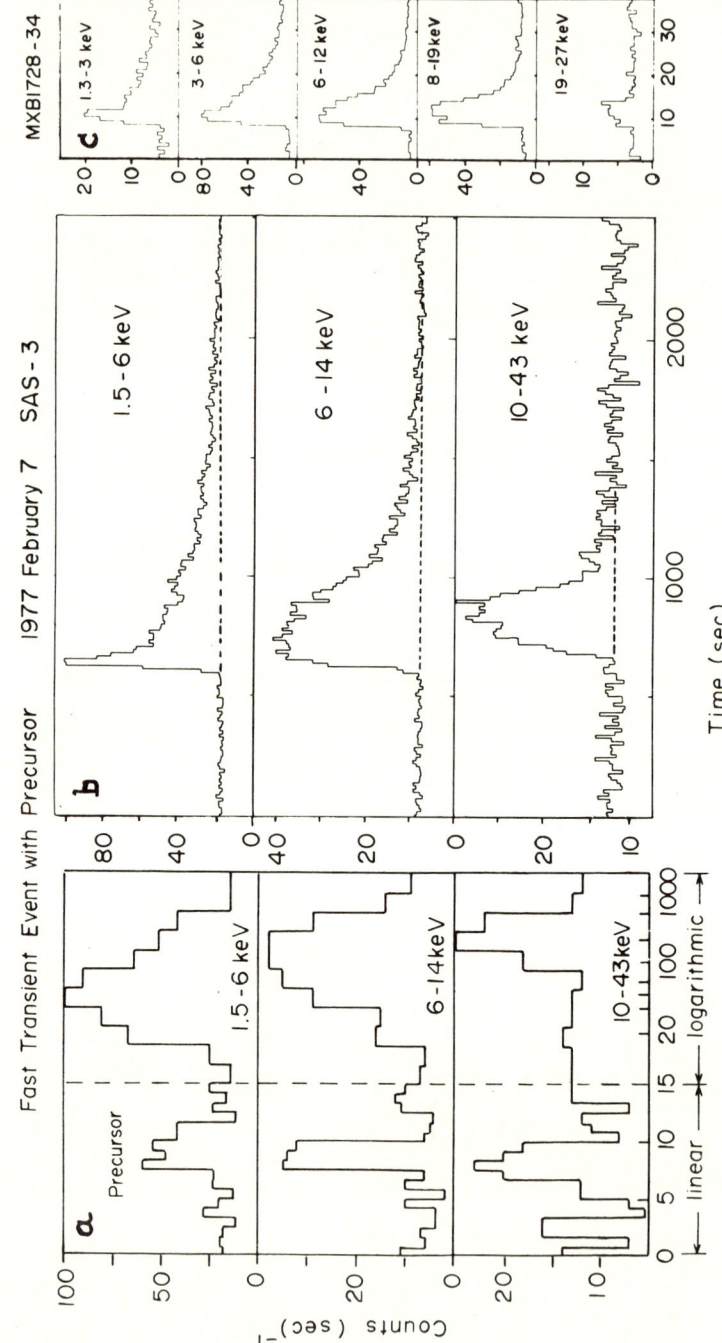

Fig. 2.20. (*a*) and (*b*) Profiles of the complete 7 February 1977 event, which lasted ~1500 s. Note the distinct spectral softening during the decay of the 'main' event. The precursor is not visible in (*b*) because of the lumping of the data in coarse time bins. (*c*) Composite profile of type I X-ray bursts from MXB 1728−34 (see also Figure 2.4). The resemblance between the event shown in (*b*) and the type I X-ray bursts is striking. This figure is from Hoffman *et al.* [123].

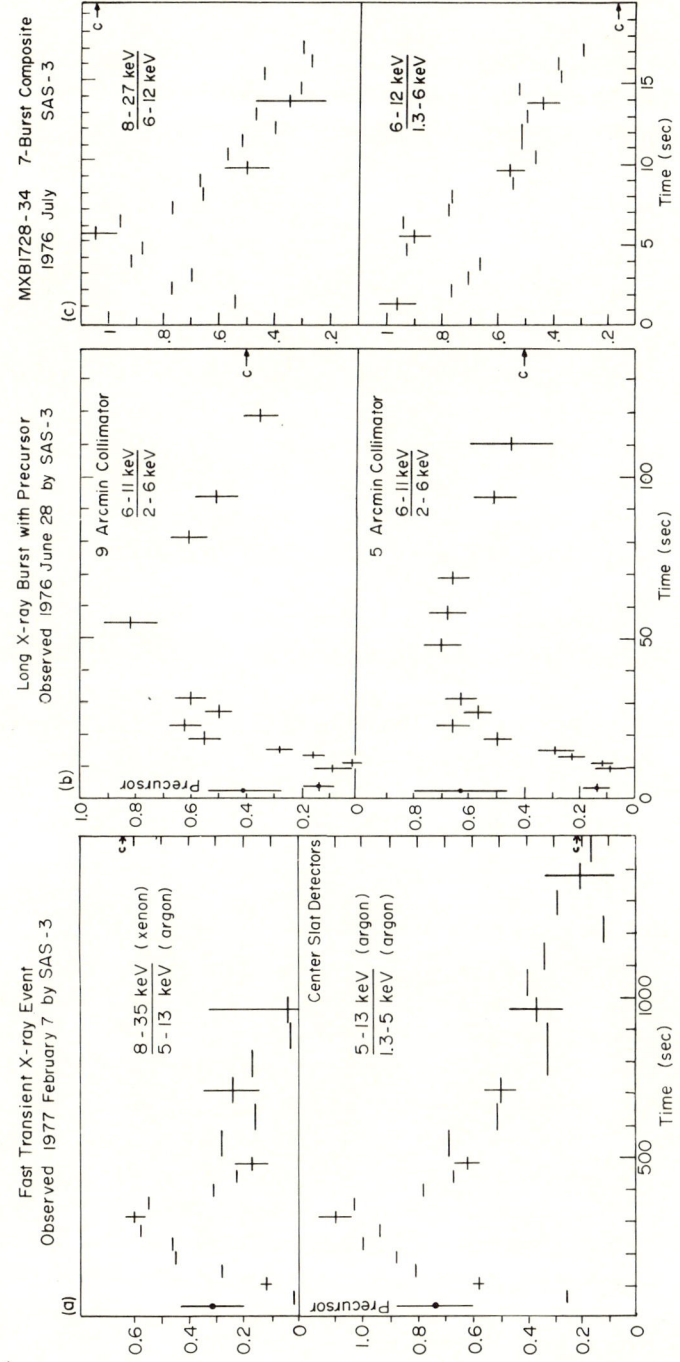

Fig. 2.21. The evolution of the ratios of X-ray fluxes in different spectral channels with time for (a) the 7 February 1977 event, (b) the 28 June 1976 event and (c) a composite X-ray burst from MXB 1728–34. All cases display initial spectral hardening and subsequent softening. The 'c' with an arrow (right) in each case represents the corresponding ratio for the Crab Nebula. The spectral evolution during the decay of these three profiles is strikingly similar. This suggests that the events of June 1976 and February 1977 are generically related to type I X-ray bursts. This figure is from Hoffman et al. [123].

the assumption of a spherical surface and a source distance of 10 kpc). This is similar to the properties of many type I X-ray bursts (see Sections 2.4.2.6 and 2.4.3.9). If the 7 February 1977 event came from a catalogued X-ray source, it was probably 4U 1708−23 (see Figure 2 of [123]).

As pointed out by Hoffman et al. [123], if the 28 June 1976 event came from a catalogued X-ray source it must have been MX 1716−31 [213]. This source was also the probable origin of a ~ 10 min X-ray flare on another occasion [212], and ordinary type I X-ray bursts have also been detected from this source [241]. If MX 1716−31 is the origin of the 28 June 1976 event, then the integrated energy flux was $\sim 6 \times 10^{-8}$ erg cm^{-2} in the precursor and $\sim 3 \times 10^{-6}$ erg cm^{-2} in the main event.

Extended SAS-3 observations, lasting for weeks and covering the large regions of the sky from which these events were seen, detected only these two. It thus seems that if these events are repetitive, the intervals are quite long. It might be possible to account for the durations, recurrence intervals and precursors in these events in terms of thermonuclear flashes that occur relatively deeply beneath the surfaces of slowly accreting neutron stars (see Section 2.6.3.3).

2.6 Theory

Nearly all mechanisms that have been proposed to account for the X-ray burst phenomenon utilize accretion of matter onto a collapsed object (degenerate dwarf, neutron star or black hole). In every instance, the collapsed object serves at least one of two functions: its deep gravitational potential well allows the release of large amounts of gravitational energy in the form of X-radiation by the accreting matter, and its small dimensions permit the X-ray emission to be released on the short time scale of an X-ray burst.

The proposed models can be broken down into two broad classes: (i) those utilizing instabilities in the accretion flow onto a collapsed object; and (ii) those that invoke thermonuclear flashes in the surface layers of an accreting neutron star. During the last few years, the greatest amount of progress has been achieved in developing the thermonuclear-flash model, which has proved to be amenable to detailed numerical computations. As will be documented below, these calculations have been remarkably successful in accounting for the general properties of type I X-ray bursts. Moreover, the theoretical work to date strongly suggests that the characteristics of type I bursts should be capable of imparting substantial constraints upon the properties (e.g., masses, radii and internal temperatures) of the underlying neutron stars. However, as will be discussed in Section 2.6.2, it presently seems virtually certain that an accretion instability is responsible for type II bursts (see also Section 2.4.3.1). Hence, more theoretical work on such instabilities is presently called for.

2.6.1 Thermonuclear-flash models for type I bursts
2.6.1.1 The overall physical picture

Consider a neutron star undergoing accretion from a binary stellar companion. The freshly accreted matter will be rich in hydrogen and/or helium. However, at depths greater than $\sim 10^4$ cm beneath the surface of the neutron star, the density is sufficiently high that nuclear statistical equilibrium will be swiftly achieved; the predominant nuclei will have maximal binding energies, with atomic weights of ~ 60. (Still deeper in the star, these nuclei dissolve into a fluid in which neutrons are the primary constituent.) Hence, the accreting matter must pass through a series of nuclear-burning shells as it is gradually compressed by the accretion of still more material. If the core of the neutron star is sufficiently hot or the accretion rate is sufficiently high, the temperature in the surface layers will be high enough that the burning will proceed via thermonuclear reactions, rather than electron capture or pycnonuclear reactions (which are driven by high densities rather than high temperatures). A sketch of the resultant structure of the neutron-star surface layers is given in Figure 2.22.

It was first realized by Hansen & van Horn [110] that these burning shells will tend to be unstable to thermal runaway. The instability, known as the 'thin-shell instability', was first discovered in a different context by Schwarzschild & Härm [261]. The existence and strength of the instability are a direct result of the strong temperature-dependence of the thermonuclear reaction rates. In the case of neutron-star envelopes, the instability is further enhanced by the partial degeneracy of the burning material. A cogent and thorough technical discussion of this type of instability has been given by Giannone & Wiegert [83] (see also [15 and 82]).

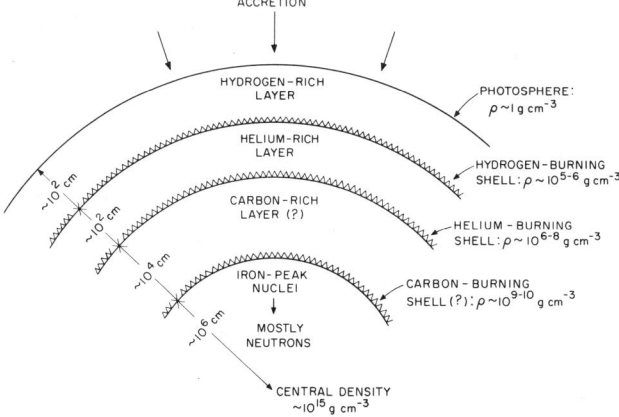

Fig. 2.22. Schematic sketch of the surface layers of an accreting neutron star. (This figure is from Joss [143].)

The pp chains are insufficiently temperature-sensitive to produce a thermal runaway in the hydrogen-burning shell of a neutron star. The instability of this shell is thus largely quenched by the saturation of the CNO cycle at very high reaction rates [141, 165]. The saturation results from the appreciable lifetimes ($\sim 10^2$–10^3 s) of the beta-unstable nuclei ^{13}N, ^{14}O, ^{15}O and ^{17}F that participate in the cycle. For low neutron-star core temperatures ($\lesssim 1 \times 10^8$ K) the shell can be unstable, but any runaways will be halted by the saturation effect before the release of a substantial amount of energy (see, however, the discussion of interacting hydrogen–helium shells in Section 2.6.1.2 below). The next shell inward is the helium-burning shell, which should be unstable over a wide range of conditions. It is uncertain whether there will be any other significant burning shells, as the matter might already burn to quite heavy elements in the helium shell [142, 275]. However, if a carbon shell exists, it is very likely to be unstable also [275, 308].

Dimensional anlysis [141, 165] indicates that the helium-burning flashes should have the following properties: (i) They should occur after the accumulation of $\lesssim 10^{21}$ g of fuel and release total energies of $\lesssim 10^{39}$ erg per flash. (ii) For accretion rates comparable with those observed in X-ray pulsars ($\lesssim 10^{17}$ g s^{-1}), the time interval between flashes should be $\sim 10^4$ s, very roughly. (iii) The transport of energy through the surface layers should result in the emission of bursts of electromagnetic radiation from the neutron-star photosphere with rise times of ~ 0.1 s, peak luminosities of $\sim 10^{38}$ erg s^{-1}, decay time scales of ~ 10 s, and peak black-body temperatures of $\sim 3 \times 10^7$ K (if a full 10^{39} erg of energy is indeed released in a single flash).

Carbon-burning flashes, if they exist, would occur much deeper beneath the neutron-star surface ($\sim 10^4$ cm, compared to $\sim 3 \times 10^2$–10^3 cm for the helium shell) and would result in the release of substantially more energy. Hence, the rise time and duration of a 'burst' resulting from a carbon flash should be much longer than those of a burst following a helium flash [142], unless dynamical effects are generated in the outermost surface layers.

2.6.1.2 Numerical models

The above estimates, though very crude, suggest that thermonuclear flashes on accreting neutron stars could account for the observed properties of type I burst sources. With this encouragement, detailed numerical computations of the evolution of the surface layers of an accreting neutron star have been carried out.

Joss [142] explored the evolution of the helium-burning shell. A simplified nuclear reaction network was used, incorporating the dominant reactions linking the nuclei from ^{4}He to ^{28}Si and allowing the release of most of the available nuclear energy. The accretion was assumed to be spherical and the star was taken to be non-rotating and unmagnetized, so that spherical symmetry could be assumed throughout the calculations. Moreover, the effects of hydrogen burning upon the structure of the surface layers was neglected. The importance of these assumptions and approximations will be discussed below.

Joss' models [142] contain two free parameters: the mass accretion rate, $\dot{m}$, and the core temperature of the neutron star, T_c. If the core of the neutron star is in thermal equilibrium (i.e., if the heat flow into the core from the surface layers during thermonuclear flashes is just balanced by the heat lost from the core between flashes), then there is a unique relationship between $\dot{m}$ and T_c [165]; three of the four models calculated by Joss [142] were chosen to lie along the estimated locus of these equilibrium values. Of these four models, three displayed thermonuclear flashes in the helium-burning shell (see Figures 2.23 and 2.24). The properties of these flashes were in good agreement with those expected from dimensional analysis [141, 165]. More importantly, these calculations indicated that (i) a full $\sim 10^{21}$ g of matter accumulates on the neutron-star surface before each helium flash, (ii) a flash consumes virtually all the available nuclear fuel and probably synthesizes mostly iron-peak elements, and (iii) most of the energy of a flash is transported to the photosphere and lost as X-radiation, rather than carried inward to heat the interior of the star. These properties of the flashes had not been discerned prior to the performance of detailed evolutionary computations, at least partly because they depend upon the highly non-linear characteristics of the flash growth and decay. The behavior of the helium-burning shell was further explored by Joss & Li [148] (see also Hoshi [130]), who investigated the sensitivity of the flash properties to the assumed mass and radius of the neutron star.

Joss & Li [148] also investigated the evolution of the helium-burning shell in models wherein the effects of an intense magnetic field upon the surface layers were taken into account. They argued that if the magnetic field is sufficiently strong to funnel the accretion onto the magnetic polar caps of the neutron star, then the effective accretion rate in the polar cap regions is enhanced by a factor of $\sim 10^3$ (for a fixed total accretion rate $\dot{m}$) and the instability of the nuclear-burning shells should be reduced (see also [141 and 275]). The surface magnetic field strength, B, required to funnel the accretion is not well determined, but available estimates (see, e.g., Arons & Lea [8]) yield $B \approx 10^{12}$ G. Joss & Li [148] also incorporated into their model calculations the significant reductions in the radiative [322] and conductive [323] opacities that would be produced by magnetic fields in excess of $\sim 10^{12}$ G, and found that these effects further reduce the instability of the nuclear-burning shells.

It is now clear that hydrogen-burning has a major influence on the behavior of the helium-burning shell [9, 60, 72, 275, 300, 315, 355]. The saturation of

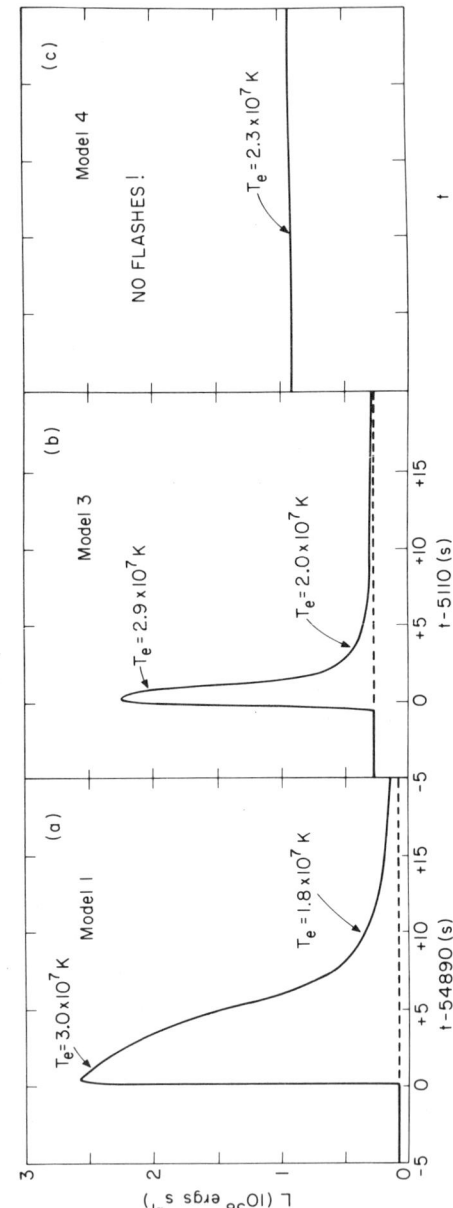

Fig. 2.23. (a) The behavior of the surface luminosity L following a thermonuclear flash in model 1 by Joss [142]. (b) Same for model 3. In each case, time $t = 0$ is at the start of accretion onto the neutron-star surface, the dashed line denotes the level of persistent accretion-driven luminosity, and the effective black-body temperature (T_e) is indicated at a few points. The properties of these luminosity variations are in remarkably good agreement with the typical properties of observed X-ray bursts (see text). (c) The surface luminosity behavior for model 4. No flashes occur at the high core temperature and accretion rate of this model, so that the nuclear energy generation rate does not vary greatly and never produces more than a small perturbation on the accretion-driven luminosity. In all cases, the neutron star was assumed to have a mass of 1.4 $M_\odot$ and a radius of 6.6 km. This figure is from Joss [143].

Fig. 2.24. Structure of the surface layers of model 1 by Joss [142] prior to and during the first helium-burning flash, which begins near time t_0. M is the total mass of the neutron star and $m(r)$ is the mass enclosed within a sphere of radius r, with $r = 0$ at the stellar center; thus $(M-m)$ is the total mass of the surface layers above level r. T is the temperature (left-hand scale), ρ the density (right-hand scale), and Y the fractional abundance of helium by mass (left-hand scale). The hatched regions indicate the extent of the convection zone generated by the flash. (a) Just prior to the flash; (b) near the start of the flash; (c) at the time when ~50 per cent of the available fuel has been consumed; and (d) near the time of peak shell-burning temperature and peak surface luminosity. This figure is from Joss [142].

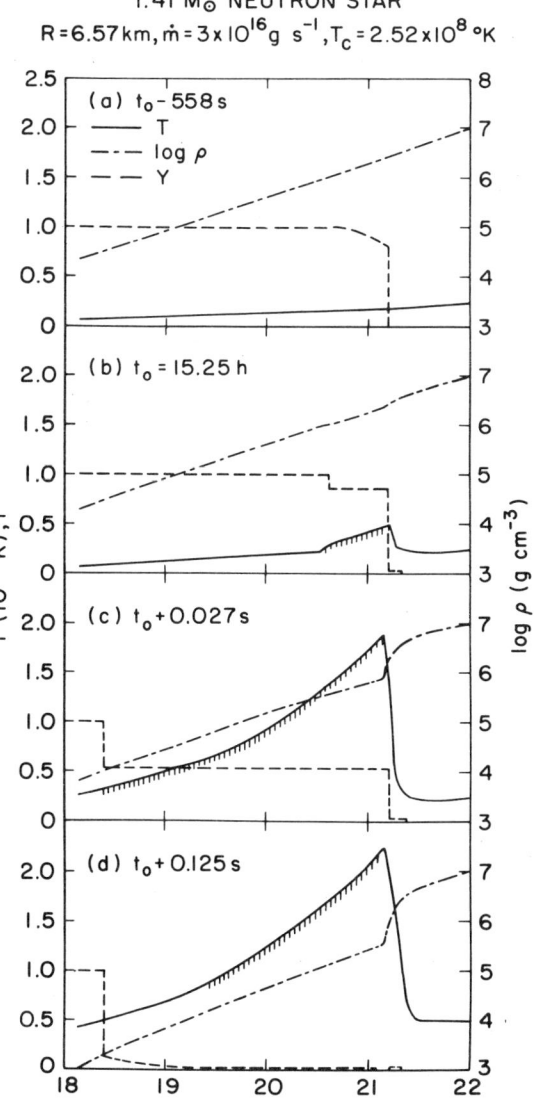

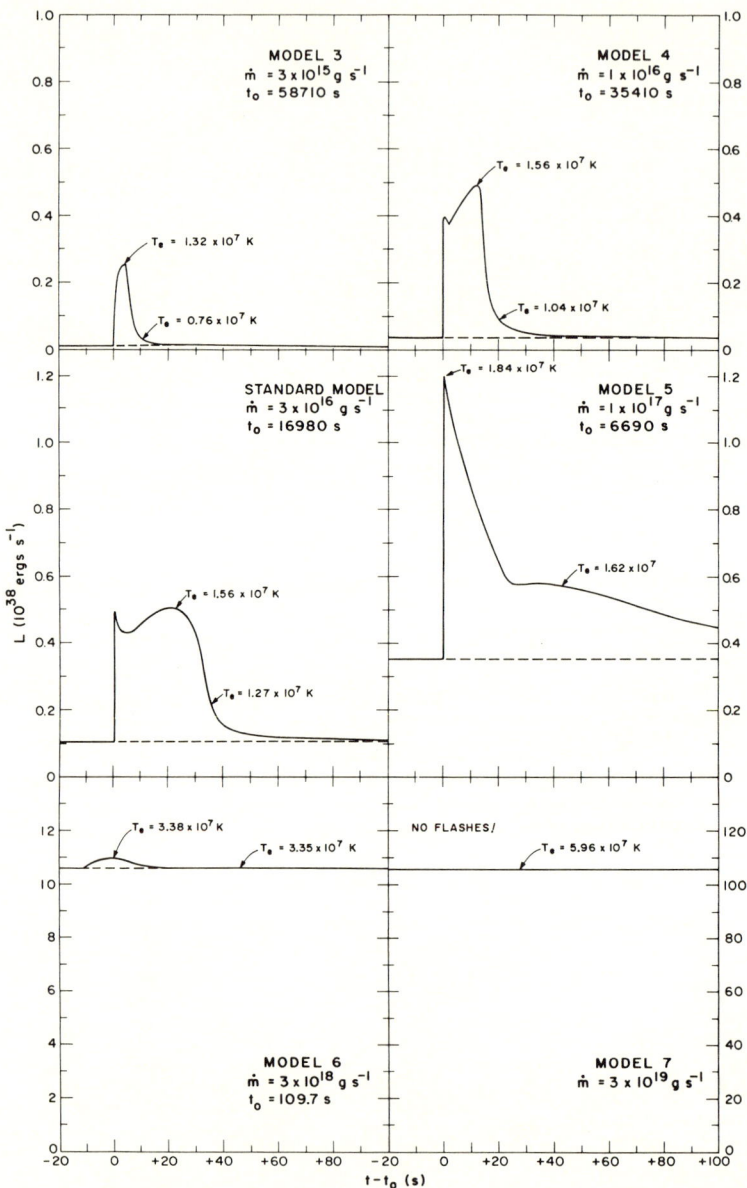

Fig. 2.25. The behavior of the surface luminosity, L, following a thermonuclear flash in the standard model and five additional models calculated by Ayasli and Joss [9]. All significant general relativistic corrections to the equations of stellar structure and evolution were taken into account in these calculations, and the numerical values of all quantities are those that would be measured by a distant observer. In each case the neutron star is assumed to be non-rotating and unmagnetized and to have a mass of 1.41 $M_\odot$, a radius of 6.57 km and a core

the CNO cycle by the non-negligible lifetimes of the beta-unstable nuclei that participate in the cycle limits the hydrogen-burning rates to such an extent that, at sufficiently high accretion rates ($\gtrsim 1 \times 10^{16}\,\text{g s}^{-1}$), the hydrogen-burning shell is forced inward until it overlaps the helium-burning shell. Taam & Picklum [276] and Taam [274, 355] carried out the first fully time-dependent computations of the evolution of the surface layers of a neutron star with the hydrogen-burning shell included and found that the hydrogen- and helium-burning shells can, indeed, interact in a complex way.

Fujimoto, Hanawa & Miyaji [82] have also studied, by semi-analytical methods, the interaction between the hydrogen- and helium-burning shells. They have suggested that different 'modes' of observed bursting behavior in 4U 1608−52 (see Section 2.4.2.3 and [234]) could be identified with changes in the character of the interaction between the hydrogen- and helium-burning shells as the accretion rates vary. However, in view of the great complexity of the observational data (see Section 2.4) and the remaining theoretical problems, we believe that such suggestions are presently very speculative. Our caution in this regard seems to gain support from 1980 observations of 4U 1608−52 with the Hakucho satellite (M. Oda, private communication), in which the bursting behavior does not fit into either of the 'modes' identified by Murakami *et al.* [234].

Ayasli & Joss [9] have calculated a new series of models with both hydrogen-burning and helium-burning shells included. In these models, the conditions

> Caption for Fig. 2.25 (*cont.*).
> temperature, as measured by an observer on the neutron-star surface, of $1.5 \times 10^8\,\text{K}$; the accreting matter is assumed to be hydrogen-rich and to accrete spherically onto the neutron-star surface. Time t_0 is the interval from the onset of accretion (at time $t = 0$) to the start of the flash, the effective temperature, T_e, is indicated at a few points, and the dashed lines indicate the level of persistent accretion-driven luminosity in each case. The accretion rate, $\dot{m}$, is successively higher in models 3 and 4, the standard model, and models 5, 6, and 7, as indicated. In model 3, $\dot{m}$ is sufficiently low that the hydrogen at the base of the freshly accreted matter is consumed by steady hydrogen-burning prior to any flash, and a pure helium flash results. In both model 4 and the standard model, hydrogen is still present at the base of the freshly accreted matter when helium ignites; the resultant X-ray bursts are double-peaked, with the first luminosity maximum reflecting the helium flash and the second maximum reflecting the subsequent rapid burning of hydrogen due to proton-capture reactions involving the moderately heavy nuclei that are produced by the helium flash. In model 5, the helium flash is very strong and generates sufficiently high temperatures to swiftly process the moderately heavy elements into iron-peak elements, so that the hydrogen is unable to burn very rapidly and the second luminosity maximum is replaced by an extended plateau. As $\dot{m}$ continues to be increased, in models 6 and 7, the thermonuclear flashes become weaker and disappear, since the surface layers become sufficiently hot to consume the helium fuel virtually as fast as it accumulates. This figure is from Ayasli & Joss [9].

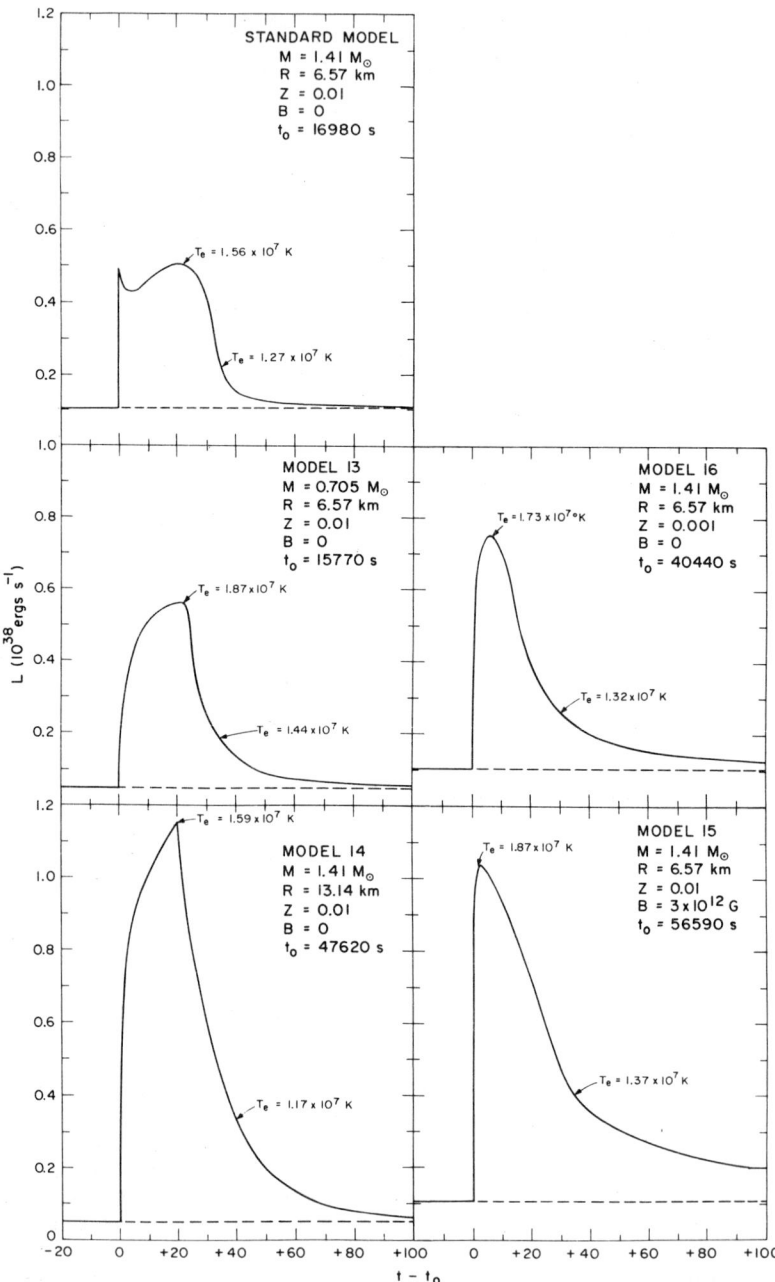

Fig. 2.26. Same as Figure 2.25, for the standard model and four additional models calculated by Ayasli & Joss [9]. The notation is the same as in Figure 2.25. The parameter values assumed in each value are identical to those of the standard

required for thermal equilibrium of the neutron-star core were determined directly by tracking the heat flow into and out of the core throughout a complete thermonuclear-flash cycle (cf. [142, 165]), and all significant general relativistic corrections to the equations of stellar structure and evolution were taken into account. The models, which are the most physically realistic that have yet been calculated, again displayed great complexity in the interactions between the hydrogen- and helium-burning shells. Samples of some of the burst profiles obtained from these calculations are given in Figures 2.25 and 2.26.

2.6.1.3 General relativistic effects

The work by Swank et al. [268] and Hoffman, Lewin & Doty [120, 121] showed that the spectra of X-ray bursts following peak luminosity could be well represented by those of black bodies. Utilizing the assumption of black-body spectra, van Paradijs [291] demonstrated that in many cases the scale size of the X-ray emitting region is nearly constant; if the emitting region is a spherical surface, then its radius is ~7 km (see Section 2.4.2.6). This result lends support to the idea that type I X-ray bursts are thermal emission from the photospheres of neutron stars.

However, it has become apparent that this argument is complicated by general relativistic corrections, such as the effect of gravitational redshift upon the X-radiation emitted by the neutron-star photosphere [86, 273, 292]. The potential importance of general relativity can be seen by inspection of the parameter

$$\frac{2GM}{Rc^2} \approx 0.60 \left(\frac{M}{1.4\,\mathrm{M}_\odot}\right)\left(\frac{R}{7\,\mathrm{km}}\right)^{-1}. \tag{6.1}$$

The left-hand side of this expression is just the ratio of the Schwarzschild radius of the neutron star ($2GM/c^2$) to its actual radius. For the indicated values of M and R, which have been used in many of the actual model calculations of thermonuclear flashes, it is evident that this parameter is not very much smaller than unity, so that general relativistic effects should be substantial.

Caption for Fig. 2.26 (cont.).
model (see Figure 2.25), except that in model 13 the neutron-star mass is half that of the standard model, in model 14 the neutron-star radius is twice that of the standard model, in model 15 a surface magnetic field of strength $B = 3 \times 10^{12}$ gauss is assumed to be present, and in model 16 the metallicity of the accreting matter ($Z = 10^{-3}$) is one-tenth that of the standard model. Model 16 is of special interest, since the low metallicity, in combination with relatively high temperatures achieved in the helium flash, reduces the hydrogen-burning rate to the extent that more than half of the original hydrogen is still unburned after the X-ray burst; hence, a second burst in rapid succession is energetically feasible (see text). This figure is from Ayasli & Joss [9].

It was shown by Goldman [86] and van Paradijs [292] that when general relativistic corrections are included in determinations of the luminosities and effective black-body temperatures of X-ray bursts, one obtains, at least in principle, strong constraints on the masses and radii of the underlying neutron stars. If taken at face value, these constraints would, in turn, severely constrain the equation of state of matter at densities in excess of nuclear-matter densities ($\rho > 2 \times 10^{14}$ g cm^{-3}). However, such constraints cannot yet be taken seriously, as other complications, including possible violations of spherical symmetry (see Section 2.6.4) and deviations of the emitted spectrum from a simple black-body spectrum [273, 298], may turn out to be important. Once these additional complexities have been untangled, the observed properties of type I bursts may prove to be a powerful probe of the basic properties of neutron stars.

General relativistic corrections to the equations of stellar structure and evolution also play a significant role in determining the behavior of the thermonuclear flashes themselves. Some of these corrections have been included in some calculations (e.g., those by Taam & Picklum [275, 276], Ergma & Tutukov [72], Czerny & Jaroszynski [60] and Taam [274, 355]). The recent calculations by Ayasli & Joss [9] were the first fully time-dependent computations to incorporate all of the relevant general relativistic corrections. See Figure 2.2.7.

2.6.1.4 Future work

We have just begun to grasp all of the intricacies of the thermonuclear-flash model for type I bursts. The theoretical problems are fascinating, not only for their probable applications to X-ray burst sources and other observational phenomena, but also as investigations in fundamental physics and as a potentially powerful tool for probing the basic properties of neutron stars.

The only complete evolutionary computations of neutron-star thermonuclear flashes that have been carried out to date [9, 142, 148, 274, 276, 355] relied on a number of simplifying assumptions and approximations, such as the assumption of spherical symmetry and the neglect of possible dynamical effects. These approximations will have to be relaxed in future studies before this phenomenon and its observational implications can be more fully understood.

Small but significant violations of spherical symmetry might result from the effects of a relatively weak ($\lesssim 10^{11}$ G) surface magnetic field or the residual angular momentum of the accreting matter. If thermonuclear flashes result in X-ray bursts, such violations could be the key to the observed complexities in burst structure and recurrence patterns (see Sections 2.4.2.3 and 2.4.2.4). For example, a thermonuclear flash that ignites on one portion of the neutron-star surface may propagate around the star, in a pattern that varies from flash to flash and from one star to another [142]. A thorough investigation of such possibilities will require two- or three-dimensional numerical computations, which will be much more difficult than the computations of spherically symmetric models that have been attempted to date.

Complexities in radiative transfer in the outer surface layers of the neutron star and possible mass ejection from the photosphere near the peak of a burst may also substantially complicate the observational properties of X-ray bursts and render their physical interpretation much more difficult. There are strong indications that the Eddington limit is exceeded near the peak of some type I bursts ([86, 103, 121, 338, 339, 341, 361, 362]; see Section 2.4.2.6); Ayasli & Joss [9] have suggested that super-Eddington luminosities may result from the suppression of radiative opacities near the neutron-star photosphere by a moderately strong magnetic field, but the nature of such luminosities is not yet well understood. Some preliminary results on deviations from black-body emission by a hot neutron-star atmosphere have been reported [273, 298], but much work in this area also remains to be done.

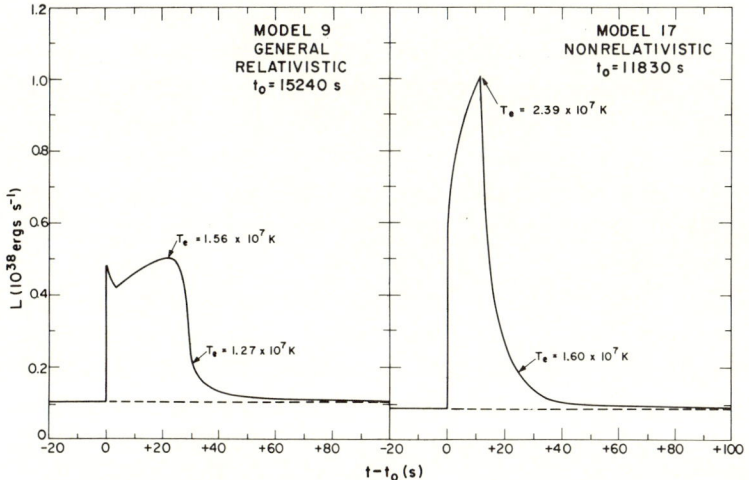

Fig. 2.27. Same as Figure 2.25 for two further models calculated by Ayasli & Joss [9]. The notation is the same as in Figure 2.25. The parameter values assumed for model 9 are identical to those of the standard model, except that the neutron-star core temperature is taken to be 2×10^8 K. Model 17 assumes identical parameter values to those of model 9, but all general relativistic corrections to the equations of stellar structure and evolution have been suppressed and all parameters have Newtonian definitions. The longer X-ray burst duration and recurrence time (t_0), the lower peak X-ray luminosity and effective temperature, the higher level of persistent accretion-driven luminosity, and the more complex burst shape in model 17 compared to model 9 are all the result of gravitational redshift and other general relativistic effects. This figure is from Ayasli & Joss [9].

2.6.2 Accretion-instability models for type II bursts

Since the discovery of X-ray bursts, a number of theoretical mechanisms have been proposed that relied on instabilities associated with accretion onto a collapsed object [10, 91, 113, 129, 163, 201, 237, 267, 302, 317, 319, 358]. Much of the early work in this area was thoroughly reviewed by Lamb & Lamb [164]. For the reasons discussed in Sections 2.6.1 and 2.6.3.1, it now seems likely that type I bursts are caused by thermonuclear flashes rather than an accretion instability. However, it is also evident from the properties of the Rapid Burster that its type II bursts are, in fact, the result of an accretion instability [127] (see Sections 2.4.3 and 2.6.3.2), and more theoretical work on such instabilities is sorely needed.

Some of the earlier proposals [91, 163, 267] for accretion instabilities involved spherical accretion onto a collapsed object. Grindlay's model [91] was developed following an early suggestion by Grindlay & Gursky [95] that 3U/MXB 1820−30 was associated with a massive black hole in the globular cluster NGC 6624 (see Sections 2.3.3 and 2.4.1). This model entails a purely thermal instability that does not require the presence of a magnetic field, so that the central collapsed object could be a black hole; moreover, the requisite mass of the collapsed object was well in excess of the probable maximum stable mass of a white dwarf or neutron star. However, as discussed in Sections 2.3.3 and 2.4.1, the phenomenological reasons for favoring the idea of a massive black hole have disappeared. Moreover, Cowie, Ostriker & Stark [58] have placed rather severe physical constraints on models of this general type, and it now seems unlikely that any such model will prove to be viable.

Several models that involve the angular momentum of the accreting matter, as well as an intense magnetic field from the collapsed object, have also been advanced [10, 113, 129, 267, 317, 319]. In principle, such models appear to be more promising to develop into a completely satisfactory model for type II bursts. However, the presence of non-spherical accretion substantially complicates the relevant physics, and it seems clear that none of the models advanced to date will be entirely adequate. Other models [201, 302], which invoke instabilities in an accretion disk surrounding a collapsed object, are presently both rather primitive and rather far removed from observational tests. However, Wheeler's model [302], which requires the presence of a collapsed object with a mass of $\sim 10 M_\odot$ (almost certainly greater than the maximum stable mass of a neutron star), might be eliminated if the neutron-star nature of the Rapid Burster is accepted (see Sections 2.4.3.1, 2.4.3.9 and 2.6.3.2).

2.6.3 Theory versus observation
2.6.3.1 Type I X-ray bursts

The results of the numerical calculations of thermonuclear flashes (see Figures 2.23–2.27) strongly support the conjecture that type I bursts result from such flashes. In particular, the typical burst rise times, decay time scales, peak luminosities, total emitted energies, spectral properties, low-energy 'tails', and recurrence intervals (see Section 2.4.2) are reproduced remarkably well by such calculations.

However, there are some difficulties with this model. In particular, the apparent violations of the Eddington limit near the peak of some bursts (see Section 2.4.2.6 and [103, 331, 359, 361, 362]), the corresponding violation of the maximum effective temperature, $\sim 2.1 \times 10^7$K, attainable by a neutron star radiating at or below the Eddington limit (see Section 2.4.2.6 and [86, 292, 360]), and the substantial variations in apparent black-body radius during the course of some bursts (see Section 2.4.2.6 and [103, 118, 268]) are all problematic for the model. (The high effective temperatures may constitute a more serious problem than the apparent super-Eddington luminosities, since the measured temperatures are independent of the imprecisely known source distances [9, 362].) Another problem is the narrow range of persistent X-ray luminosities (and, presumably, a correspondingly narrow range of accretion rates) for which bursting behavior has been observed in some sources (see Sections 2.4.2.2 and 2.4.2.4). It is quite possible that these and other difficulties will be better understood when some of the approximations of the present model calculations are relaxed (see Section 2.6.1.4 and the models recently presented in [9]).

A particularly interesting problem for the nuclear-flash model is the ratio, α, of time-averaged persistent X-ray luminosity to time-averaged burst luminosity from the observed burst sources (see Section 2.4.2.5). In this model, α should just be the ratio of the gravitational energy released by accretion (more or less continuously) to the nuclear energy released in the flashes. For helium-burning flashes, the numerical value of α should thus be [141, 165]

$$\alpha \approx 100 \left(\frac{M}{M_\odot}\right)\left(\frac{R}{10\,\text{km}}\right)^{-1}. \tag{6.3}$$

However, some type I burst sources have reported values of α significantly less than 100; in several cases two bursts were separated by an interval of only $\sim$5–10 min, with a correspondingly small upper limit of $\alpha < 2.5$ during that interval [235] (see Section 2.4.2.5). Moreover, as discussed in Section 2.4.2.4, in some sources the recurrence intervals have been observed to increase (decrease) when the persistent luminosity increased (decreased), which is opposite to the trend expected from this model.

If the nuclear-flash model for type I bursts is correct, it is possible that the observed values of α are sometimes reduced by the storage of nuclear fuel during phases of low burst activity [165]. Such a 'battery' mechanism may be provided by large-scale violations of spherical symmetry (see Section 2.6.1.4), so that only a fraction of the surface of the neutron star participates in each flash, or from the incomplete consumption of the available nuclear fuel during a flash. Moreover, the entrainment of hydrogen into a helium-burning flash may reduce the value of α by up to a factor of ~ 5, due to proton captures onto the heavier nuclei being synthesized and a concomitant increase in the energy yield per unit mass. The model calculations recently presented in [9] demonstrate that both incomplete fuel consumption and hydrogen entrainment can be important in this regard, so that even the shortest observed intervals between X-ray bursts may not prove to be an insurmountable problem for the nuclear-flash model.

In some cases, α-values considerably larger than 10^2 have been observed during burst-inactive periods in some sources. However, in the context of the nuclear-flash model, it will probably not be difficult to account for large α-values. For example, such values might result from episodes when the nuclear-burning shells become relatively stable and the bursting phenomenon is intermittently suppressed, and/or from one or more inefficiencies in the nuclear-flash process, such as loss of some flash energy to the core of the neutron star and partial burning of the relevant nuclear fuel between flashes (see, e.g., [9]).

2.6.3.2 Type II bursts and the Rapid Burster

It is important to realize that the type II bursts from the Rapid Burster almost certainly cannot be the result of thermonuclear flashes (see Section 2.4.3.1). Hoffman, Marshall & Lewin [127] were the first to propose that the type I bursts from the Rapid Burster are the result of thermonuclear flashes on an accreting neutron star, while the type II bursts are the result of an unstable accretion flow onto the same object (see Section 2.6.2 for a discussion of the various proposed instabilities that may produce the type II bursts).

2.6.3.3 The fast X-ray transients

The morphology of some of the fast X-ray transients is suggestively similar to that of ordinary type I X-ray bursts [123] (see Section 2.5). Joss [145] suggested that the fast transients may be the result of helium-burning flashes relatively deep within the surface layers of very slowly accreting neutron stars $(\dot{m} \lesssim 10^{15}\,\mathrm{g\,s^{-1}})$. If this picture is correct, one would expect outbursts from the fast transients to recur, but only on time scales of weeks or longer. Moreover, the low accretion rate should result in a relatively large temperature contrast between the nuclear-flashing shell and the outermost surface layers of the neutron star, so that the precursors observed in some fast transients (see Section 2.5) might be the result of shock heating of the outer surface layers in response to the heating and subsonic expansion of the nuclear-flashing shell. Recent detailed numerical computations of a deep helium-burning flash by Wallace, Woosley & Weaver [356] seem to confirm the viability of the above scenario, although the shock heating of the outer surface layers in their calculations failed to generate a precursor similar to the precursors in observed fast transients.

2.6.3.4 *Relation to binary X-ray pulsars and the ages of burst sources*

Let us accept the reasonable inference that type I bursts can be understood as thermonuclear flashes on accreting neutron stars. Since X-ray pulsars are also widely believed to be accreting neutron stars, it is then puzzling, at first sight, that these objects do not also display bursting behavior. However, the strong magnetic field ($\gtrsim 10^{12}$G; see [8]) that funnels the accretion onto the magnetic polar caps of an X-ray pulsar will also enhance the efficiency of radiative and conductive heat transport within and above its nuclear-burning shells. Moreover, the heat released by accretion will have a much greater influence upon the inner burning shells if the freshly accreted matter is confined to the polar caps, rather than spread uniformly over the neutron-star surface (see Section 2.6.1.2). These effects should reduce the instability of the nuclear-burning shells of an X-ray pulsar against thermonuclear flashes [142, 148, 275]. Evolutionary models of the helium-burning shell in the presence of an intense magnetic field confirm the assertion that such fields reduce the instability of the shell [9, 148].

With this picture, we can also understand why the persistent X-ray flux from type I burst sources is unpulsed: those neutron stars whose magnetic fields are too weak to funnel the accreting matter may be precisely those that can undergo thermonuclear flashes [142, 148, 275]. If the magnetic field was originally as strong as in an X-ray pulsar but has since decayed, then the neutron star must be fairly old (probably older than 10^7yr; see Section 2.2.1 and [77, 258]). The lack of X-ray eclipses in burst sources may also reflect membership in relatively old binary systems [149] and may result from X-ray beaming effects that set in after the neutron-star magnetic field has decayed (see Sections 2.2.2 and 2.6.4 and [232]). The concentration of X-ray burst sources in the direction of the galactic center (see Figure 2.3) and the identification of several of them with globular clusters (see Section 2.3) may well be other manifestations of membership in an older galactic population than the X-ray pulsars, which are distributed through the disk of the galaxy and whose binary companion stars are often of early spectral type.

2.6.4 The highly-compact binary model for galactic bulge sources

If the X-ray burst sources are a physical subset of the larger class of galactic bulge sources, the conclusion seems inescapable that all of these sources derive their X-ray luminosities, directly or indirectly, by accretion of matter onto a collapsed object which, in the case of the burst sources, is evidently a neutron star. If this is the case, then the observed X-ray luminosities (up to $\sim 10^{38}$ erg s^{-1}) require accretion rates as high as $\sim 10^{18}$ g s^{-1}. Such accretion rates, in turn, seem to require the presence of close-binary companion stars to supply the accreting matter (provided that very massive black holes are excluded; see Sections 2.3.3 and 2.4.1). Many of the ideas related to the basic concept of galactic bulge sources as binary stellar systems have been mentioned in earlier sections, but for the sake of clarity we shall collect and elaborate upon these ideas in this section.

The faintness of the optical counterparts of most of these sources (see Sections 2.4.2.7) rules out giant, supergiant, or early-type main-sequence stellar companions. It therefore seems likely that many of the companion stars, with perhaps only a few exceptions, are low-mass ($\lesssim 0.5$ M$_\odot$), main-sequence dwarfs or degenerate dwarfs [149] (see Section 2.4.2.7.3). It is, moreover, probable that such a star could only transfer sufficient mass to the collapsed star if it fills its critical potential lobe, in which case the orbital separation would be $\lesssim 10^6$ km and the orbital period $\lesssim 0.3$ d (see Figure 2.10). For such highly-compact binary systems, the mass transfer could be effected through the decay of the orbit due to gravitational radiation, possibly augmented by a self-excited stellar wind and/or the evolution of the companion (which could be the evolved remnant of a more massive star, see [149, 256] and references therein).

The properties of such systems can be reconciled with the observational characteristics of most of the galactic bulge sources [149]. In particular, the faintness of the optical counterparts is a natural consequence of the intrinsic faintness of the companion star; in fact, most of the very blue light that is seen results from reprocessing of X-radiation within the system, rather than from the intrinsic luminosity of the companion (see Section 2.4.2.7).

The general lack of observed X-ray eclipses (see Section 2.2.2) is a bit harder to understand. The recent discovery [334, 348, 378] of 50 min periodic absorption events in the persistent X-ray flux from the burst source MXB 1916−05 is extremely important, in that these events very probably reflect the orbital periodicity of an underlying low-mass binary. However, these events cannot be true eclipses by the companion star, since they do not recur on every orbital cycle and have other properties that are incommensurate with true eclipses.

Due to the small size of the companion compared to the dimenions of the system, the probability of observing true eclipses in any one system is only

$$P_{ecl} \approx 0.2 \left(\frac{1+q}{10}\right)^{-1/3} \quad (q>1), \tag{6.3}$$

where q is the ratio of the mass of the X-ray star to that of its companion and it is assumed that the companion star fills its Roche lobe [149]. However, the *a priori* probability is quite high ($\gtrsim$99 per cent; cf. [149]) for having detected true eclipses in at least one of the twenty or so galactic bulge sources for which adequate observations are available (see Section 2.2.2). As first suggested by Milgrom [232], the companion may be largely shielded from the X-radiation by an accretion disk surrounding the collapsed star, which may account for the lack of eclipsing behavior. Moreover, the shielding of the companion from most of the X-radiation may account for the general lack of optical photometric variability at the orbital period (see Section 2.4.2.7).

As was pointed out in Section 2.4.2.7, the observed properties of the X-ray pulsar 4U 1626−67 also provide indirect but persuasive evidence in favor of the highly-compact binary model for the galactic bulge sources. There is evidence that some of the spectrally 'soft' X-ray transients have somewhat more massive companions ($\gtrsim$0.5 $M_\odot$; see Section 2.4.2.7.5), but it is possible that their transient nature results from an instability in the mass transfer process that is, in turn, causally related to these relatively high masses [256].

If the surface magnetic field of a neutron star has largely decayed away, as is probable for neutron stars in systems with ages of $\gtrsim 10^9$ yr [77, 258], then the accretion disk may extend downward all the way to the surface of the neutron star [232]. When this occurs, up to $\sim\frac{1}{2}$ the gravitational potential energy may be released in the disk rather than on the surface of the neutron star. The relatively large X-radiating surface area of the inner disk plus neutron star may then account for the 'soft' X-ray spectra of the galactic bulge sources compared to those of the X-ray pulsars, whose magnetic fields are evidently strong enough to funnel the accretion flow onto the magnetic polar caps of the neutron star.

A highly-compact binary stellar system of the type described above may evolve from a cataclysmic-variable system, wherein a degenerate dwarf accretes matter until it exceeds the Chandrasekhar limiting mass ($\sim$1.4 $M_\odot$), at which time it undergoes a dynamical collapse to form a neutron star [106, 149, 288]. It is conceivable that the formation of a neutron star under such circumstances could be relatively 'quiet', resulting in the ejection of sufficiently little mass to leave the system gravitationally bound (see, e.g., [32, 357]). This scenario and the subsequent evolution of the binary system have most recently been discussed by Li *et al.* [198] and Rappaport, Joss & Webbink [256]. For those sources with X-ray luminosities up to $\sim 10^{36}$ erg s^{-1}, it is plausible that a sufficient rate of mass transfer can be driven by the decay of the orbit due to gravitational radiation. However, detailed numerical calculations [256] show that even when gravitational radiation controls the mass transfer rate, the evolutionary history of the system is substantially influenced by the thermal evolution of the companion star. Much more work on the evolution of such systems remains to be done.

We note that the evolution of the X-ray sources in globular clusters is likely to be somewhat different from those of other galactic bulge sources. It is possible that many or all of the galactic bulge sources outside globular clusters are primordial binaries, but the observational statistics suggest that the globular cluster X-ray sources are binary systems formed by capture rather than primordial binaries (see Section 2.3). We also note that some of the non-bursting galactic bulge sources might conceivably contain black holes with masses of the order of a few solar masses, formed when a neutron star accretes sufficient matter to exceed its maximum stable mass [146]. (Most current theoretical estimates place the maximum stable mass of a neutron star at $\sim 2 M_\odot$.) However, there is presently no compelling theoretical or observational evidence for the existence of black holes of any mass in any of the galactic bulge sources.

2.7 Concluding remarks

The past several years have witnessed an enormous increase in observational and theoretical information on the galactic bulge X-ray sources in general, and on the X-ray burst sources in particular. In the preceding sections, we have attempted to distill from this information those aspects which seem to have the greatest relevance to the continued development of our understanding of the nature of these sources. We have undoubtedly failed in some respects. However, we hope that, at a minimum, the material we have compiled will serve to supply the interested reader with both an overview and a fairly comprehensive bibliography from which more detailed information might be obtained.

The basic character of the galactic bulge sources now seems clear. They are collapsed objects of roughly solar mass, probably neutron stars in most cases, which are accreting matter from low-mass stellar companions. Type I bursts very likely result from thermonuclear flashes in the surface layers of some of these neutron stars, while the type II bursts from the Rapid Burster almost certainly result from an instability in the accretion flow onto a neutron star.

There are, however, a number of outstanding phenomenological and theoretical problems. What is the prior evolutionary history of the binary stellar systems that become galactic bulge sources? How do the globular cluster X-ray sources differ in character and genealogy from other galactic bulge sources? Are the fast X-ray transients the result of thermonuclear flashes on neutron stars? What mechanisms are responsible for the long-term transient outbursts in the Rapid Burster and several other galactic bulge sources? What is the nature of the accretion instability that is operative in the Rapid Burster? Why is this instability not comparably active in any other known X-ray burst source or X-ray pulsar? Has the same instability ever, in fact, been observed in any other X-ray source? This list of problems is hardly more than illustrative; a complete list of the unanswered questions raised in the preceding sections would be very much longer.

Nonetheless, we have come a remarkably long way. Six years prior to the time of this writing, the galactic bulge sources were a complete enigma and X-ray bursts were unknown. Now, we seem not only to be on our way to satisfying understanding of these phenomena, but also to have gained a new and powerful observational handle on the fundamental properties of neutron stars and of the interacting binary systems in which they are often contained.

Acknowledgements

An earlier version of this review was published in *Space Science Reviews* [368].

We are grateful to Ewa Basinska, Claude Canizares, George Clark, Lynn Cominsky, Josh Grindlay, Jeff McClintock, Bruno Rossi, Scott Tremaine, and Jan van Paradijs for their valuable comments on this manuscript, and we thank Susan Black for her assistance and patience in preparing this manuscript. One of us (WHGL) wants to thank Ed van den Heuvel and the Astronomical Institute of the University of Amsterdam for their generous hospitality during his sabbatical visit and the Netherlands Association for Fundamental Research (ZWO) for support. This work was also supported by the National Aeronautics and Space Administration under contract NAS5-24441 and grants NSG-7643 and NGL-22-009-638 and by the National Science Foundation under grant AST78-21993.

References

[1] Abramenko, A. N., Gershberg, R. E., Pavlenko, E. P. *et al.* 1978, *MNRAS*, **184**, 27P.
[2] Alcock, C. & Hatchett, S. 1978, *Ap. J.*, **222**, 456.
[3] Amnuel, P. R. & Guseinov, O. H. 1979, *Ap. Space Sci.*, **63**, 131.
[4] Apparao, K. M. V., Bradt, H. V., Dower, R. G. *et al.* 1978, *Nature*, **271**, 225.
[5] Apparao, K. M. V. & Chitre, S. M. 1980, submitted to *Ap. Space Sci.*
[6] Apparao, K. M. V., Chitre, S. M., Ashok, N. M. & Kulkarni, P. V., 1979, *IAU Circ.*, No. 3344.
[7] Apparao, K. & Narranan, S. 1978, private communication, SAS-3 data.
[8] Arons, J. & Lea, S. M. 1980, *Ap. J.*, **235**, 1016.
[9] Ayasli, S. & Joss, P. C. 1982, *Ap. J.*, **256**, 637.
[10] Baan, W. 1977, *Ap. J.*, **214**, 245; and 1979, *Ap. J.*, **227**, 987.

[11] Bahcall, J. N. & Ostriker, J. P. 1975, *Nature*, **256**, 23.
[12] Bahcall, J. N. & Wolf, R. A. 1976, *Ap. J.*, **209**, 214.
[13] Bahcall, N. A. & Hausman, M. A. 1977, *Ap. J.*, **213**, 93.
[14] Bahcall, N. A., Lasker, B. M. & Wamsteker, W. 1977, *Ap. J.*, **213**, L105.
[15] Barranco, M., Buchler, J. R. & Livio, M. 1980, *Ap. J.*, **242**, 1226.
[16] Basinska, E., Lewin, W. H. G., Cominsky, L. *et al.* 1980, *Ap. J.*, **241**, 787.
[17] Becker, R. H., Pravdo, S. H., Serlemitsos, P. J. & Swank, J. H. 1976, *IAU Circ.*, No. 2953.
[18] Becker, R. H., Smith, B. W., Swank, J. H. *et al.* 1977, *Ap. J.*, **216**, L101.
[19] Belian, R. D., Conner, J. P. & Evans, W. D. 1972, *Ap. J.*, **171**, L87.
[20] Belian, R. D., Conner, J. P. & Evans, W. D. 1976, *IAU Circ.*, No. 2669.
[21] Belian, R. D., Conner, J. P. & Evans, W. D. 1976, *Bull. Amer. Astron. Soc.*, **8**, 396.
[22] Belian, R. D., Conner, J. P. & Evans, W. D. 1976, *Ap. J.*, **206**, L135.
[23] Belian, R. D., Conner, J. P. & Evans, W. D. 1976, *Ap. J.*, **207**, L33.
[24] Bernacca, P. L., Bianchini, A., Walker, A. *et al.* 1979, *MNRAS*, **186**, 287.
[25] Birmingham Group (Ariel V) 1976, *IAU Circ.*, No. 2929.
[26] Birmingham Group (Ariel V) 1976, *IAU Circ.*, No. 2934.
[27] Bond, H. E. 1977, *IAU Circ.*, No. 3085.
[28] Bradt, H., Doxsey, R. E. & Jernigan, J. G. 1979. *In:* 'Advances in space exploration', (*Proceedings of IAU/COSPAR Symposium on X-Ray Astronomy*, Innsbruck, Austria, May 1978), eds. W. A. Baity & L. E. Peterson (Oxford: Pergamon), Vol. 3, p. 3.
[29] Bradt, H., Kelley, R. & Petro, L. 1982, *In: Galactic X-Ray Sources*, eds. P. Sanford, P. Laskarides & J. Salton. John Wiley and Sons, New York.
[30] Calla, O. P. N., Barathy, S., Snagal, A. K. *et al.* 1980, *IAU Circ.*, Nos. 3458 and 3467.
[31] Calla, O. P. N., Bhandari, S. M., Deshpande, M. R. & Vats Hari, O. M. 1979, *IAU Circ.*, No. 3347.
[32] Canal, R. & Schatzman, E. 1976, *Astron. Ap.*, **46**, 229.
[33] Canizares, C. R. 1975, *Ap. J.*, **201**, 589.
[34] Canizares, C. R. 1976, *Ap. J.*, **207**, L101.
[35] Canizares, C. R., Grindlay, J. E., Hiltner, W. A. *et al.* 1978, *Ap. J.*, **224**, 39.
[36] Canizares, C. R., McClintock, J. E. & Grindlay, J. E. 1979, *IAU Circ.*, No. 3362.
[37] Canizares, C. R., McClintock, J. E. & Grindlay, J. E. 1979, *Ap. J.*, **234**, 556.
[38] Canizares, C. R., McClintock, J. E. & Grindlay, J. E. 1980, *Ap. J.*, **236**, L55.
[39] Canizares, C. R. & Neighbours, J. E. 1975, *Ap. J.*, **199**, L97.
[40] Canizares, C. R. & Oda, M. 1977, *Ap. J.*, **214**, L119.
[41] Carpenter, G. F., Skinner, G. K., Wilson, A. M. & Willmore, A. P. 1976, *Nature*, **262**, 473.
[42] Clark, G. W. 1975, *Ap. J.*, **199**, L143.
[43] Clark, G. W. 1976, *IAU Circ.*, No. 2907.
[44] Clark, G. W. 1976, *IAU Circ.*, No. 2922.
[45] Clark, G. W. 1976, *IAU Circ.*, No. 2932.
[46] Clark, G. W., Jernigan, G., Bradt, H. *et al.* 1976, *Ap. J.*, **207**, L105.
[47] Clark, G. W. & Li, F. K. 1977, *IAU Circ.*, No. 3090.
[48] Clark, G. W. & Li, F. K. 1977, *IAU Circ.*, No. 3092.
[49] Clark, G. W., Li, F. K., Canizares, C. R. *et al.* 1977, *MNRAS*, **179**, 651.
[50] Clark, G. W., Markert, T. H. & Li, F. K. 1975, *Ap. J.*, **199**, L93.
[51] Cominsky, L., Forman, W., Jones, C. & Tananbaum, H. 1977, *Ap. J.*, **211**, L9.
[52] Cominsky, L., Jernigan, J. G., Ossmann, W. *et al.* 1980, *Ap. J.*, **242**, 1102.
[53] Cominsky, L., Jones, C., Forman, W. & Tananbaum, H. 1978, *Ap. J.*, **224**, 46.
[54] Cominsky, L., Lewin, W. H. G., Ossmann, W. *et al.* 1983, *Ap. J.* (in press).
[55] Conner, J. P., Evans, W. D. & Belian, R. D. 1969, *Ap. J.*, **157**, L157.
[56] Cooke, B. A., Ricketts, M. J., Maccacaro, T. *et al.* 1978, *MNRAS*, **182**, 489 (2A Catalogue).
[57] Cordova, F. A., Garmire, G. P. & Lewin, W. H. G. 1979, *Nature*, **278**, 529.
[58] Cowie, L. L., Ostriker, J. P. & Stark, A. A. 1978, *Ap. J.*, **226**, 1041.

[59] Cruddace, R. G., Fritz, G., Shulman, S. et al. 1978, Ap. J., **222**, L95.
[60] Czerny, M. & Jaroszynski, M. 1979, Acta Astronomica, **30**, 157.
[61] Davidsen, A. 1975, IAU Circ., No. 2824.
[62] Davidsen, A., Malina, R., Smith, H. et al. 1974, Ap. J., **193**, L25.
[63] Doty, J. 1976, IAU Circ., No. 2922.
[64] Dower, R. G., Apparao, K. M. V., Bradt, H. V. et al. 1978, Nature, **273**, 364.
[65] Doxsey, R. E. 1975, IAU Circ., No. 2820.
[66] Doxsey, R. E., Apparao, K. M. V., Bradt, H. et al. 1977, Nature, **269**, 112.
[67] Doxsey, R. E., Bradt, H., Gursky, H. et al. 1978, Ap. J., **221**, L53.
[68] Doxsey, R., Clark, G. W. & Li, F. 1977, IAU Circ., No. 3094.
[69] Doxsey, R. E., Grindlay, J., Griffiths, R. et al. 1978, Ap. J., **228**, L67.
[70] Duldig, M., Greenhill, J., Thomas, R. et al. 1977, IAU Circ., No. 3108.
[71] Epstein, A., Delvaille, J., Helmken, H. et al. 1977, Ap. J., **216**, 103.
[72] Ergma, E. V., & Tutukov, A. V. 1980, Astron. Ap., **84**, 123.
[73] Evans, W. D., Belian, R. D. & Conner, J. P. 1976, Ap. J., **207**, L91.
[74] Fabbiano, G. & Branduardi, G. 1979, Ap. J., **227**, 294.
[75] Fabian, A. C., Pringle, J. E. & Rees, M. J. 1975, MNRAS, **172**, 15P.
[76] Fall, S. M. & Malkan, M. A. 1978, MNRAS, **185**, 899.
[77] Flowers, E. & Ruderman, M. 1977, Ap. J., **215**, 302.
[78] Forman, W. & Jones, C. 1976, Ap. J., **207**, L177.
[79] Forman, W., Jones, C., Cominsky, L. et al. 1978, Ap. J. Suppl., **38**, 357 (4U Catalogue).
[80] Forman, W., Jones, C. & Tananbaum, H. 1976, Ap. J., **208**, 849.
[81] Sadeh, D. et al. 1982, Ap. J., **257**, 214.
[82] Fujimoto, M., Hanawa, T. & Miyaji, S. 1981, Ap. J., **246**, 267.
[83] Giannone, P. & Weigert, A. 1967, Zs. Ap., **67**, 41.
[84] Glass, I. S. 1978, Nature, **273**, 35.
[85] Glass, I. S. & Oda, M. 1980, private communication.
[86] Goldman, Y. 1979, Astron. Ap., **78**, L15.
[87] Griffiths, R. E., Gursky, H., Schwartz, D. A. et al. 1978, Nature, **276**, 247.
[88] Griffiths, R., Johnston, M., Bradt, H. et al. 1978, IAU Circ., No. 3190.
[89] Grindlay, J. 1977, IAU Circ., No. 3101.
[90] Grindlay, J. 1978, IAU Circ., No. 3229.
[91] Grindlay, J. E. 1978, Ap. J., **221**, 234.
[92] Grindlay, J. E. 1978, Ap. J., **224**, L107.
[93] Grindlay, J. E. 1981, In: 'X-ray astronomy with the Einstein satellite' (Proceedings of the HEAD/AAS Meeting, Cambridge, Massachusetts, January 1980), ed. R. Giacconi (Dodrecht: Reidel), p. 79.
[94] Grindlay, J. & Gursky, H. 1976, IAU Circ., No. 2932.
[95] Grindlay, J. & Gursky, H. 1976, Ap. J., **205**, L131.
[96] Grindlay, J. E. & Gursky, H. 1976, Ap. J., **209**, L61.
[97] Grindlay, J. E. & Gursky, H. 1977, Ap. J., **218**, L117.
[98] Grindlay, J., Gursky, H., Schnopper, H. et al. 1976, Ap. J., **205**, L127.
[99] Grindlay, J. & Heise, J. 1976, IAU Circ., No. 2879.
[100] Grindlay, J., Marshall, H. L. et al. 1980, Ap. J., **240**, L121.
[101] Grindlay, J. E. & Liller, W. 1977, Ap. J., **216**, L105.
[102] Grindlay, J. E. & Liller, W. 1978, Ap. J., **220**, L127.
[103] Grindlay, J., Marshall, H., Hertz, P. et al. 1980, Ap. J., **240**, L121.
[104] Grindlay, J., McClintock, J., Canizares, C. & van Paradijs, J. 1978, IAU Circ., No. 3230.
[105] Grindlay, J. E., McClintock, J. E., Canizares, C. R. et al. 1978, Nature, **274**, 567.
[106] Gursky, H. 1976, In: 'Structure and evolution of close binary systems', IAU Symposium, No. 73, eds. P. Eggleton, S. Mitton & J. Whelan (Dordrecht: Reidel), p. 19.
[107] Gursky, H., Bradt, H., Schwartz, D. A. et al. 1978, Ap. J., **223**, 973.
[108] Hackwell, J. A., Gehrz, R. D., Grasdalen, G. L. et al. 1979. IAU Circ., No. 3331.
[109] Hackwell, J. A., Gehrz, R. D., Grasdalen, G. L. et al. 1979, Ap. J., **233**, L115.
[110] Hansen, C. J. & van Horn, H. M. 1975, Ap. J., **195**, 735.

[111] Hearn, D. 1976, *IAU Circ.*, No. 2925.
[112] Heise, J. & Grindlay, J. 1976, *IAU Circ.*, No. 2929.
[113] Henriksen, R. N. 1976, *Ap. J.*, **210**, L19.
[114] Hills, J. G. 1975, *Astron. J.*, **80**, 1075.
[115] Hjellming, R. M. 1979, *IAU Circ.*, No. 3369.
[116] Hoffman, J. 1976, *IAU Circ.*, No. 2946.
[117] Hoffman, J. 1976, *IAU Circ.*, No. 2953.
[118] Hoffman, J. A., Cominsky, L. & Lewin, W. H. G. 1980, *Ap. J.*, **240**, L27.
[119] Hoffman, J., Doty, J. & Lewin, W. H. G. 1977, *IAU Circ.*, No. 3025.
[120] Hoffman, J. A., Lewin, W. H. G. & Doty, J. 1977, *Ap. J.*, **217**, L23.
[121] Hoffman, J. A., Lewin, W. H. G. & Doty, J. 1977, *MNRAS*, **179**, 57P.
[122] Hoffman, J., Lewin, W. H. G., Doty, J. *et al.* 1976, *Ap. J.*, **210**, L13.
[123] Hoffman, J. A., Lewin, W. H. G., Doty, J. *et al.* 1978, *Ap. J.*, **221**, L57.
[124] Hoffman, J., Lewin, W. H. G., Marshall, H. *et al.* 1978, *IAU Circ.*, No. 3190.
[125] Hoffman, J. A., Lewin, W. H. G., Primini, F. A. *et al.* 1979, *Ap. J.*, **233**, L51.
[126] Hoffman, J. A., Marshall, H. & Lewin, W. H. G. 1977, *IAU Circ.*, No. 3117.
[127] Hoffman, J. A., Marshall, H. L. & Lewin, W. H. G. 1978, *Nature*, **271**, 630.
[128] Hoffman, J. A., Wheaton, W. A., Primini, F. A. *et al.* 1978, *Nature*, **276**, 587.
[129] Horiuchi, R., Kadonaga, T. & Tomimatsu, A. 1981, *Prog. Theor. Phys.*, **66**, 172.
[130] Hoshi, R. 1980, *Prog. Theor. Phys.*, **64**, 820.
[131] Inoue, H., Koyama, K., Makishima, K. *et al.* 1980, *Nature*, **283**, 358.
[132] Jernigan, J. G., Apparao, K. M. V., Bradt, H. V. *et al.* 1977, *Nature*, **270**, 321.
[133] Jernigan, J. G., Apparao, K. M. V., Bradt, H. V. *et al.* 1977, *Nature*, **272**, 701.
[134] Jernigan, J. G. & Clark, G. W. 1979, *Ap. J.*, **231**, L125.
[135] Jernigan, J. G., McClintock, J., Marshall, H. *et al.* 1978, *IAU Circ.*, No. 3204.
[136] Johnson, H. M. 1976, *Ap. J.*, **208**, 706.
[137] Johnson, H. M., Catura, R. C., Lamb, R. A. *et al.* 1978, *Ap. J.*, **222**, 664.
[138] Johnston, M. D., Bradt, H. V. & Doxsey, R. E. 1979, *Ap. J.*, **233**, 514.
[139] Jones, A. W., Selby, M. J., Mountain, C. M. *et al.* 1980, *Nature*, **283**, 550.
[140] Jones, C. & Forman, W. 1976, *IAU Circ.*, No. 2913.
[141] Joss, P. C. 1977, *Nature*, **270**, 310.
[142] Joss, P. C. 1978, *Ap. J.*, **225**, L123.
[143] Joss, P. C. 1979, *Comments on Ap.*, **8**, 109.
[144] Joss, P. C. 1980, *Annals N. Y. Acad. Sci.*, **336**, 479. (*Proceedings of the 9th Texas Symposium on Relativistic Astrophysics*, Munich, West Germany, December 1978.)
[145] Joss, P. C. 1979, *In: Compact Galactic X-Ray Sources*, eds. D. Pines & F. Lamb (Urbana, Illinois: Physics Dept., Univ. of Illinois), p. 89.
[146] Joss, P. C. 1980, *In:* 'X-ray astronomy with the Einstein satellite' (*Proceedings of the HEAD/AAS Meeting*, Cambridge, Massachusetts, January 1980), ed. R. Giacconi (Dordrecht: Reidel), p. 111.
[147] Joss, P. C., Avni, Y. & Rappaport, S. 1978, *Ap. J.*, **221**, 645.
[148] Joss, P. C. & Li, F. K. 1980, *Ap. J.*, **238**, 287.
[149] Joss, P. C. & Rappaport, S. A. 1979, *Astron. Ap.*, **71**, 217.
[150] Joss, P. C., Ricker, G. R., Mayer, W. & Hoffman, J. 1977, *IAU Circ.*, No. 3108.
[151] Kaluzienski, L. 1977, Ph.D. Thesis, University of Maryland.
[152] Kaluzienski, L. & Holt, S. 1977, *IAU Circs.*, Nos. 3099, 3108, 3129 and 3349.
[153] Kaluzienski, L. & Holt, S. 1979, *IAU Circs.*, Nos. 3360 and 3362.
[154] Kaluzienski, L. J., Holt, S., Boldt, E. A. & Serlemitsos, P. J. 1975, *IAU Circ.*, No. 2859.
[155] Kaluzienski, L. J., Holt, S. S. & Swank, J. H. 1980, *Ap. J.*, **241**, 779.
[156] Katz, J. I. 1975, *Nature*, **253**, 698.
[157] Kelley, R., Rappaport, S. & Petre, R. 1980, *Ap. J.*, **238**, 699.
[158] Kellogg, E., Gursky, H., Murray, S. *et al.* 1971, *Ap. J.*, **169**, L99.
[159] King, I. R. 1966, *Astron, J.*, **71**, 64.
[160] Kleinmann, D. E., Kleinmann, S. G. & Wright, E. L. 1976, *Ap. J.*, **210**, L83.
[161] Kondo, I., Inoue, H., Koyama, K. *et al.* 1981, *Space Sci. Instr.*, **5**, 211.

[162] Kulkarni, P. V., Ashok, N. M., Apparao, K. M. V. & Chitre, S. M. 1979, *Nature*, **280**, 819.
[163] Lamb, F. K., Fabian, A. C., Pringle, J. E. & Lamb, D. Q. 1977, *Ap. J.*, **217**, 197.
[164] Lamb, D. Q. & Lamb, F. K. 1977, *Annals N.Y. Acad. Sci.*, **302**, 261. (*Proceedings of the 8th Texas Symposium on Relativistic Astrophysics*, Boston, Massachusetts, December 1976.)
[165] Lamb, D. Q. & Lamb, F. K. 1978, *Ap. J.*, **220**, 291.
[166] Lewin, W. H. G. 1976, *IAU Circ.*, No. 2911.
[167] Lewin, W. H. G. 1976, *IAU Circ.*, No. 2918.
[168] Lewin, W. H. G. 1976, *IAU Circ.*, No. 2922.
[169] Lewin, W. H. G. 1977, *IAU Circ.*, No. 3078.
[170] Lewin, W. H. G. 1977, *Amer. Sci.*, **65**, No. 5, p. 605.
[171] Lewin, W. H. G. 1977, *MNRAS*, **179**, 43.
[172] Lewin, W. H. G. 1977, *Annals N.Y. Acad. Sci.*, **302**, 210. (*Proceedings of the 8th Texas Symposium on Relativistic Astrophysics*, Boston, Massachusetts, December 1976.)
[173] Lewin, W. H. G. 1978, *IAU Circ.*, No. 3193.
[174] Lewin, W. H. G. 1979, *In:* 'Advances in space exploration' (*Proceedings of IAU/COSPAR Symposium on X-Ray Astronomy*, Innsbruck, Austria, May 1978), eds. W. A. Baity and L. E. Petersen (Oxford: Pergamon), Vol. 3, p. 133.
[175] Lewin, W. H. G. 1980, *In:* 'Globular clusters' (*Proceedings of Globular Cluster Meeting*, held in Cambridge, England, August 1978), eds. D. Hanes & B. Madore (Cambridge University Press), p. 315.
[176] Lewin, W. H. G. & Clark, G. W. 1979, *In: Symposium on the Results and Future Prospects of X-Ray Astronomy*, 3-4 August 1979, Tokyo (Tokyo: ISAS, University of Tokyo), p. 3.
[177] Lewin, W. H. G. & Clark, G. W. 1980. *Annals N.Y. Acad. Sci.*, **336**, 451. (*Proceedings of the 9th Texas Symposium on Relativistic Astrophysics*, Munich, West Germany, December 1978.)
[178] Lewin, W. H. G., Cominsky, L. & van Paradijs, J. 1978, *IAU Circ.*, No. 3308.
[179] Lewin, W. H. G., Cominsky, L., Walker, A. R. & Robertson, S. C. 1980, *Nature*, **287**, 27.
[180] Lewin, W. H. G., Doty, J., Clark, G. W. *et al.* 1976, *Ap. J.*, **207**, L95.
[181] Lewin, W. H. G., Doty, J., Hoffman, J. A. & Li, F. K. 1976, *IAU Circ.*, No. 2984.
[182] Lewin, W. H. G. & Hoffman, J. A. 1977, *IAU Circ.*, No. 3079.
[183] Lewin, W. H. G., Hoffman, J. A. & Doty, J. 1976, *IAU Circ.*, No. 2994.
[184] Lewin, W. H. G., Hoffman, J. A. & Doty, J. 1977, *IAU Circ.*, No. 3039.
[185] Lewin, W. H. G., Hoffman, J. A. & Doty, J. 1977, *IAU Circ.*, No. 3087.
[186] Lewin, W. H. G., Hoffman, J. A., Doty, J. *et al.* 1977, *Nature*, **267**, 28.
[187] Lewin, W. H. G., Hoffman, J. A., Doty, J. *et al.* 1976, *MNRAS*, **177**, 83P.
[188] Lewin, W. H. G., Hoffman, J. A., Doty, J. *et al.* 1977, *IAU Circ.*, No. 3075.
[189] Lewin, W. H. G., Hoffman, J. A., Marshall, H. *et al.* 1978, *IAU Circ.*, No. 3190.
[190] Lewin, W. H. G. & Joss, P. C. 1977, *Nature*, **270**, 211.
[191] Lewin, W. H. G., Li, F. K., Hoffman, J. A. *et al.* 1976, *MNRAS*, **177**, 93P.
[192] Lewin, W. H. G., Marshall, H. & Cominsky, L. 1978, *IAU Circ.*, No. 3211.
[193] Lewin, W. H. G., McClintock, J. E. & Ricker, G. R. 1971, *Ap. J.*, **169**, L17.
[194] Lewin, W. H. G., van Paradijs, J., Cominsky, L. & Holzner, S. 1980, *MNRAS*, **193**, 15.
[195] Li, F. 1976, *IAU Circ.*, No. 2936.
[196] Li, F. K. & Clark, G. W. 1977, *IAU Circ.*, No. 3095.
[197] Li, F. K. & Clark, G. W., private communication.
[198] Li, F. K., Joss, P. C., McClintock, J. E. *et al.* 1980, *Ap. J.*, **240**, 628.
[199] Li, F. & Lewin, W. H. G. 1976, *IAU Circ.*, No. 2983.
[200] Li, F. K., Lewin, W. H. G., Clark, G. W. *et al.* 1977, *MNRAS*, **179**, 21P.
[201] Liang, E. P. T. 1977, *Ap. J.*, **211**, L67 and *Ap. J.*, **218**, 243.
[202] Lightman, A. P., Press, W. H. & Odenwald, S. F. 1978, *Ap. J.*, **219**, 629.
[203] Liller, M. H. & Carney, B. W. 1978, *Ap. J.*, **224**, 383.

[204] Liller, W. 1976, *IAU Circ.*, No. 2929.
[205] Liller, W. 1976, *IAU Circ.*, No. 2936.
[206] Liller, W. 1977, *Ap. J.*, **213**, L21.
[207] Liller, W. 1979, *IAU Circ.*, No. 3366.
[208] Maraschi, L. & Cavaliere, A. 1977, *In: Highlights in Astronomy*, Vol. 4, Part I, ed. E. A. Müller (Dordrecht: Reidel), p. 127.
[209] Margon, B. 1980, *IAU Circ.*, No. 3478.
[210] Margon, B. & Whitter, K. B. 1978, *IAU Circ.*, No. 3246.
[211] Markert, T., Backman, D. E., Canizares, C. R. et al. 1975, *Nature*, **257**, 32.
[212] Markert, T. H., Backman, D. & McClintock, J. 1976, *Ap. J.*, **208**, L115.
[213] Markert, T. H., Bradt, H. V., Clark, G. W. et al. 1975, *IAU Circ.*, No. 2765.
[214] Markert, T., Canizares, C. R. & Clark, G. W. et al. 1977, *Ap. J.*, **218**, 801.
[215] Marshall, F. J. 1979, *IAU Circ.*, No. 3336.
[216] Marshall, H. & Lewin, W. H. G. 1978, *IAU Circ.*, No. 3208.
[217] Marshall, H., Li, F. & Rappaport, S. 1977, *IAU Circ.*, No. 3134.
[218] Marshall, H. L., Ulmer, M., Hoffman, J. A. et al. 1979, *Ap. J.*, **227**, 555.
[219] Mason, K. O., Bell-Burnell, S. J. & White, N. E. 1976, *Nature*, **262**, 474.
[220] Mason, K. O., Middleditch, J., Nelson, J. E. & White, N. E. 1980, *Nature*, **287**, 516.
[221] Matilsky, T., Bradt, H., Buff, J. et al. 1976, *Ap. J.*, **210**, L127.
[222] Matsuoka, M., Inoue, H., Koyama, K. et al. 1980, *Ap. J.*, **240**, L137.
[223] McClintock, J. E. 1977, *IAU Circ.*, No. 3084.
[224] McClintock, J. E. 1977, *IAU Circ.*, No. 3088.
[225] McClintock, J. E., Canizares, C. & Backman, D. E. 1978, *Ap. J.*, **223**, L75.
[226] McClintock, J. E., Canizares, C. R., Bradt, H. V. et al. 1977, *Nature*, **270**, 320.
[227] McClintock, J. E., Canizares, C., Li, F. K. & Grindlay, J. 1980, *Ap. J.*, **235**, L81.
[228] McClintock, J. E., Canizares, C. R., van Paradijs, J. et al. 1979, *Nature*, **279**, 47.
[229] McClintock, J. E., Rappaport, S., Nugent, J. & Li, F. K. 1977, *Ap. J.*, **216**, L15.
[230] McClintock, J. E., Remillard, R. A., and Margon, B. 1981, *Ap. J.*, **243**, 900.
[231] Middleditch, J., Mason, K. O., Nelson, J. & White, N. 1981, *Ap. J.*, **244**, 1001.
[232] Milgrom, M. 1978, *Astron. Ap.*, **67**, L25.
[233] Mook, D., Kurtz, M., Weed, J. & Johns, M. 1978, *IAU Circ.*, No. 3251.
[234] Murakami, T., Inoue, H., Koyama, K. et al. 1980, *Ap. J.*, **240**, L143.
[235] Murakami, T., Inoue, H., Koyama, K. et al. 1980, *Pub. Astron. Soc. Japan*, **32**, 543.
[236] Murdin, P., Penston, M. J., Penston, M. V. et al. 1974, *MNRAS*, **169**, 25.
[237] Neugebauer, M. & Tsurutani, B. T. 1978, *Ap. J.*, **226**, 494.
[238] Oda, M. 1979, *IAU Circ.*, No. 3349.
[239] Oda, M. 1979, *IAU Circ.*, No. 3366.
[240] Oda, M. 1979, *IAU Circ.*, No. 3392.
[241] Makishima, K. et al. 1981, *Ap. J.*, **244**, L79.
[242] Koyama, K. et al. 1981, *Ap. J.*, **247**, L27.
[243] Oke, J. B. 1977, *Ap. J.*, **217**, 181.
[244] Oke, J. B. & Greenstein, J. L. 1977, *Ap. J.*, **211**, 872.
[245] Ostriker, J. P., Rees, M. J. & Silk, J. 1970, *Ap. Lett.*, **6**, 179.
[246] Ostriker, J. P., Spitzer, L. & Chevalier, R. A. 1972, *Ap. J.*, **176**, L51.
[247] Paczynski, B. & Jaroszynski, M. 1978, *Acta Astron.*, **28**, 111.
[248] Parsignault, D. & Grindlay, J. E. 1978, *Ap. J.*, **225**, 970.
[249] Pedersen, H., Lub, J. et al. 1982, *Ap. J.*, **263**, 325.
[250] Pedersen, H., Oda, M., Cominsky, L. et al. 1979, *IAU Circ.*, No. 3399.
[251] Peterson, C. J. 1976, *Astron. J.*, **81**, 617.
[252] Peterson, C. J. & King, I. R. 1975, *Astron. J.*, **80**, 427.
[253] Pramesh Rao, A. & Venugopal, V. R. 1980, *Bull. Astron. Soc. India*, **8**, 41.
[254] Press, W. H. & Teukolsky, S. A. 1977, *Ap. J.*, **213**, 183.
[255] Proctor, R. J., Skinner, G. K. & Willmore, A. P. 1978, *MNRAS*, **185**, 745.
[256] Rappaport, S., Joss, P. C. & Webbink, R. 1982, *Ap. J.*, **254**, 616.
[257] Rosenbluth, M. N., Ruderman, M., Dyson, F. et al. 1973, *Ap. J.*, **184**, 907.
[258] Ruderman, M. 1972, *Ann. Rev. Astron. Ap.*, **10**, 427.

[259] Sato, S., Kawara, K., Kobayashi, Y. *et al.* 1980, *Nature*, **286**, 688.
[260] Schreier, E., Levinson, R., Gursky, H. *et al.* 1972, *Ap. J.*, **172**, L79.
[261] Schwarzschild, M. & Harm, R. 1965, *Ap. J.*, **142**, 855.
[262] Seitzer, P., Smith, G. & Ross, B. 1979, *IAU Circ.*, No. 3372.
[263] Seitzer, P., Tuohy, I. R., Mason, K. O. *et al.* 1979, *IAU Circ.*, No. 3406.
[264] Shapiro, S. L. 1977, *Ap. J.*, **217**, 281.
[265] Share, G., Wood, K., Yentis, D. *et al.* 1978, *IAU Circ.*, No. 3190.
[266] Silk, J. & Arons, J. 1975, *Ap. J.*, **200**, L131.
[267] Svestka, J. 1976, *Ap. Space Sci.*, **45**, 21.
[268] Swank, J. H., Becker, R. H., Boldt, E. A. *et al.* 1977, *Ap. J.*, **212**, L73.
[269] Swank, J. H., Becker, R. H., Boldt, E. A. *et al.* 1978, *MNRAS*, **182**, 349.
[270] Swank, J. H., Becker, R., Pravdo, S. & Serlemitsos, P. J. 1977, *IAU Circ.*, No. 2963.
[271] Swank, J., Becker, R., Pravdo, S. *et al.* 1976, *IAU Circ.*, No. 3000.
[272] Swank, J. H., Becker, R. H., Pravdo, S. H. *et al.* 1976, *IAU Circ.*, No. 3010.
[273] Swank, J. H., Eardley, D. M. & Serlemitsos, P. J. 1979, preprint.
[274] Taam, R. E. 1980, *Ap. J.*, **241**, 358.
[275] Taam, R. E. & Picklum, R. E. 1978, *Ap. J.*, **224**, 210.
[276] Taam, R. E. & Picklum, R. E. 1979, *Ap. J.*, **233**, 327.
[277] Takagishi, K., Nagareda, K., Matsuoka, M. *et al.* 1978, *ISAS Research Note 60*, preprint CSR-P-78-31.
[278] Tananbaum, H., Chaisson, L. J., Forman, W. *et al.* 1976, *Ap. J.*, **209**, L125.
[279] Tananbaum, H., Gursky, H., Kellogg, E. M. *et al.* 1972, *Ap. J.*, **174**, L143.
[280] Tawara, Y. 1980, Ph.D. Thesis Univ. of Tokyo, *ISAS Research Note 103*.
[281] Thomas, R. M., Duldig, M. L., Haynes, R. F. *et al.* 1979, *MNRAS*, **187**, 299.
[282] Thorstensen, J., Charles, P. & Bowyer, S. 1978, *IAU Circ.*, No. 3253.
[283] Thorstensen, J., Charles, P. & Bowyer, S. 1978, *Ap. J.*, **220**, L131.
[284] Thorstensen, J., Charles, P., Bowyer, S. *et al.* 1979, *Ap. J.*, **233**, L57.
[285] Thorstensen, J., Charles, P. A., Bowyer, S. *et al.* Talks presented at the HEAD/AAS Meeting, San Diego, California, September 1978 and at the HEAD/AAS Meeting, Cambridge, Massachusetts, January 1980.
[286] Ulmer, M. P., Hjellming, R. M., Lewin, W. H. G. *et al.* 1978, *Nature*, **276**, 799.
[287] Ulmer, M. P., Lewin, W. H. G., Hoffman, J. A. *et al.* 1977, *Ap. J.*, **214**, L11.
[288] van den Heuvel, E. P. J. 1977, *Annals N.Y. Acad. Sci.*, **302**, 14.
[289] van Horn, H. M. & Hansen, C. J. 1974, *Ap. J.*, **191**, 479.
[290] van Paradijs, J. 1978, *IAU Circ.*, No. 3197.
[291] van Paradijs, J. 1978, *Nature*, **274**, 650.
[292] van Paradijs, J. 1979, *Ap. J.*, **234**, 609.
[293] van Paradijs, J. 1980, *IAU Circ.*, No. 3487.
[294] van Paradijs, J., Cominsky, L. & Lewin, W. H. G. 1978, *IAU Circ.*, No. 3294.
[295] van Paradijs, J., Cominsky, L. & Lewin, W. H. G. 1979, *MNRAS*, **189**, 387.
[296] van Paradijs, J., Joss, P. C., Cominsky, L. & Lewin, W. H. G. 1979, *Nature*, **280**, 375.
[297] van Paradijs, J. & Lewin, W. H. G. 1978, *Nature*, **276**, 249.
[298] van Paradijs, J., Rybicki, G. & Lamb, D. Q. 1980, Talk presented at the HEAD/AAS Meeting, Cambridge, Mass., January 1980, abstract in *Bull. Amer. Astron. Soc.*, **11**, 788.
[299] van Paradijs, J., Verbunt, F., van der Linden, T. *et al.* 1980, *Ap. J.*, **241**, L161.
[300] Wallace, R. K. & Woosley, S. E. 1981, *Ap. J. Suppl.*, **45**, 339.
[301] Watson, M. G. 1976, *MNRAS*, **176**, 19P.
[302] Wheeler, J. C. 1977, *Ap. J.*, **214**, 560.
[303] Whelan, J., Ward, M. J., Allen, D. A. *et al.* 1977, *MNRAS*, **180**, 657.
[304] White, N. E. & Burnell, S. 1977, *IAU Circ.*, No. 3067.
[305] White, N. E., Mason, K. O., Carpenter, G. F. & Skinner, G. K. 1978, *MNRAS*, **184**, 1P.
[306] Willmore, A. P., Mason, K. O., Sanford, P. W. *et al.* 1974, *MNRAS*, **169**, 7.
[307] Wilson, A. M., Carpenter, G. F., Eyles, C. J. *et al.* 1977, *Ap. J.*, **215**, L111.
[308] Woosley, S. E. & Taam, R. E. 1976, *Nature*, **263**, 101.

[309] Wyckoff, S. 1979, *IAU Circ.*, No. 3386.
[310] 1978, *IAU Circ.*, No. 3243.
[311] Makishima, K. *et al.* 1981, *Ap. J.*, **247**, L23.
[312] Grindlay, J. 1980, *IAU Circ.*, No. 3506.
[313] Terzan, A. 1971, *Astron. Ap.*, **12**, 477.
[314] Sztajno, M., Basinska, E., Cominsky, L., Marshall, F. J. & Lewin, W. H. G. 1983, *Ap. J.*, in press.
[315] Kudryashov, A. D. & Érgma, E. V. 1980, *Acta Astron.*, **30**, 453; and 1980, *Sov. Astron. Lett.*, **6**, 375.
[316] Giacconi, R., Gursky, H., Kellogg, E. *et al.* 1971, *Ap. J.*, **167**, L67.
[317] Apparao, K. M. V. & Chitre, S. M. 1979, *Ap. Space Sci.*, **63**, 125.
[318] Brecher, K., Morrison, P. & Sadun, A. 1977, *Ap. J.*, **217**, L139.
[319] Joss, P. C. & Rappaport, S. 1977, *Nature*, **265**, 222.
[320] Cowley, A. P., Crampton, D. & Hutchings, J. B. 1979, *Ap. J.*, **231**, 539.
[321] Crampton, D. & Cowley, A. P. 1980, *Astron. Soc. Pac.*, **92**, 147.
[322] Lodenquai, J., Canuto, V., Ruderman, M. & Tsuruta, S. 1974, *Ap. J.*, **190**, 141.
[323] Canuto, V. 1970, *Ap. J.*, **159**, 641.
[324] Rappaport, S., Buff, J., Clark, G. *et al.* 1976, *Ap. J.*, **206**, L139.
[325] Pounds, K. Talk presented at the HEAD/AAS Meeting, San Diego, California, September 1978.
[326] Mason, K. O., Middleditch, J., Nelson, J. E. *et al.* 1980, *Ap. J.*, **242**, 109.
[327] Lewin, W. H. G., van Paradijs, J., Oda, M. & Pounds, K. 1979, *IAU Circ.*, No. 3334.
[328] Lewin, W. H. G., Cominsky, L. & Oda, M. 1979, *IAU Circ.*, No. 3420.
[329] Lewin, W. H. G. & Cominsky, L. 1979, *IAU Circ.*, No. 3428.
[330] Gursky, H. 1973, Lecture presented at the NASA Advanced Study Institute on Physics of Compact Objects, Cambridge, England, July 1973.
[331] Grindlay, J. & Hertz, P. 1981, *Ap. J.*, **247**, L17.
[332] Charles, P. A., Thorstensen, J. R., Bowyer, S. *et al.* 1980, *Ap. J.*, **237**, 154.
[333] Pedersen, H., van Paradijs, J., Motch, C. *et al.* 1982, *Ap. J.*, **263**, 340.
[334] Walter, F. M., White, N. E. & Swank, J. 1981, *IAU Circ.*, No. 3611.
[335] Motch, C. & Ilovaisky, S. A. 1981, *IAU Circ.*, No. 3609.
[336] McClintock, J. & Petro, L. 1981, *IAU Circ.*, No. 3615.
[337] van Paradijs, J., Pedersen, H. & Lewin, W. H. G. 1981, *IAU Circ.*, No. 3626.
[338] Basinska, E. M., Sztajno, M., Cominsky, L., Marshall, F. J. & Lewin, W. H. G. 1983, submitted to *Ap. J.*
[339] Hoshi, R. 1981, *Ap. J.*, **247**, 628.
[340] Oda, M. & Tanaka, Y., private communication.
[341] van Paradijs, J. 1981, *Astron. Ap.*, **101**, 174.
[342] Ohashi, T., Inoue, H., Koyama, K. *et al.* 1982, *Ap. J.*, **258**, 254.
[343] Murdin, P., Allen, D. A., Morton, D. C., Whelan, J. A. J. & Thomas, R. M. 1980, *MNRAS*, **192**, 709.
[344] van Paradijs, J., Pedersen, H. & Lewin, W. H. G. 1981, *IAU Circ.*, No. 3626.
[345] Pedersen, H., van Paradijs, J. & Lewin, W. H. G. 1981, *Nature*, **294**, 725.
[346] Faulkner, J., Flannery, B. F. & Warner, B. 1972, *Ap. J.*, **175**, L79.
[347] Warner, B. 1976, *In:* 'Structure and evolution of close binary systems', *IAU Symposium No. 73*, eds. P. Eggleton, S. Mitton & J. Whelan (Dordrecht: Reidel), p. 85.
[348] Walter, F. M. *et al.* 1982, *Ap. J.*, **253**, L67.
[349] Oda, M. & the Hakucho Team 1981, *IAU Circ.*, No. 3624.
[350] Kahn, S., Grindlay, J., Halpern, J. & Ladd, E. 1981, *IAU Circ.*, No. 3624.
[351] Grindlay, J. 1981, *IAU Circ.*, No. 3620.
[352] Hjellming, R. M. & Wade, C. M. 1971, *Ap. J.*, **168**, L21.
[353] Tarenghi, M. & Reina, C. 1972, *Nature Phys. Sci.*, **240**, 53.
[354] Thorstensen, T., Charles, P. A. & Bowyer, S. 1980, *Ap. J.*, **238**, 964.
[355] Taam, R. E. 1981, *Ap. J.*, **247**, 257.

[356] Wallace, R. K., Woosley, S. E. & Weaver, T. A. 1982, *Ap. J.*, **258**, 696.
[357] Nomoto, K. 1981, *In: IAU Symposium No. 93, Fundamental Problems in the Theory of Stellar Evolution*, eds. D. Sugimoto, D. Q. Lamb & D. M. Schramm (Dordrecht: Reidel), p. 295.
[358] Michel, F. C. 1977, *Ap. J.*, **216**, 838.
[359] Inoue, H., *et al.* 1981, *Ap. J.*, **250**, L71.
[360] Marshall, H. L. 1982, *Ap. J.*, **260**, 815.
[361] Cominsky, L. R. 1981, PhD. thesis, Massachusetts Institute of Technology.
[362] Lewin, W. H. G. 1982, *Proceedings of Workshop on Accreting Neutron Stars*, eds. W. Brinkmann & J. Trümper, *MPE Report*, **177**, October 1982.
[363] Bradt, H. & McClintock, J. 1983, *Annual Reviews of Astronomy and Ap.*, Vol. 21.
[364] Lawrence, A., Cominsky, L., Engelke, C. *et al.* 1983, *Ap. J.*, **273**.
[365] Kunieda, H. *et al.* 1982, submitted to *Pub. Astron. Soc. Japan*.
[366] Lawrence, A., Cominsky, L., Lewin, W. H. G. *et al.* 1983, *Ap. J.*, **267**, 301.
[367] Lewin, W. H. G. & Clark, G. W., 1982, *In: Galactic X-ray sources*, eds. P. Sanford, P. Laskarides & J. Salton (John Wiley and Sons, New York), p. 319.
[368] Lewin, W. H. G. & Joss, P. C. 1981, *Space Sci. Rev.*, **28**, 3.
[369] Cowley, A. P., Crampton, D. & Hutchings, J. B. 1982, *Ap. J.*, **256**, 605.
[370] Hut, P. & Verbunt, F. (1982) preprint.
[371] Koyama, K., Inoue, H., Makishima, K. *et al.* 1981, *Ap. J.*, **247**, L27.
[372] White, N. E. & Swank, J. H., *Ap. J.*, **253**, L61.
[373] Tawara, Y., Hayakawa, S., Kunieda, H., Makino, F. & Nagase, F. 1982, *Nature*, **299**, 38.
[374] Leahy, D. A., Darbro, W., Elsner, R. F. *et al.*, 1983, *Ap. J.*, **266**, 160.
[375] Hertz, P. & Grindlay, J. E., 1983, *Ap. J.* (in press).
[376] Paczyński, B., 1983, *Ap. J.*, **267**, 315.
[377] Taam, R. E., 1982, *Ap. J.*, **212**, 825.

3

X-RAY EMISSION FROM NORMAL GALAXIES

Knox S. Long
Colombia Astrophysics Laboratory, Colombia University, New York, NY

Leon P. Van Speybroeck
Harvard-Smithsonian Center for Astrophysics, Cambridge, Mass.

3.1 Introduction

The 2-10 keV X-ray luminosity of the Galaxy is approximately 2×10^{39} erg s^{-1}. X-ray surveys in this energy range are complete above a luminosity of 5×10^{36} erg s^{-1}; 26 galactic sources exceeding this threshold have been detected (Clark *et al.* 1978). Most of these sources have luminosities in the range 10^{37}-10^{38} erg s^{-1}, close to the Eddington luminosity limit for a 1 M$_\odot$ object; they are primarily neutron stars or black holes in mass-exchanging binary systems (see Chapters 5, 8 and 9, this book). The number of sources in the Galaxy with luminosity of order 10^{36} erg s^{-1} is poorly known. Blumenthal & Tucker (1974), based on the increase in flux in the galactic plane, estimate that the number can be no greater than 100. All of the compact binary X-ray sources are time variable; some are visible only a small portion of the time and are classified as transients. The sources are concentrated in the galactic plane and within 60° of the galactic center.

Bright, compact X-ray sources can be divided into two broad categories: those associated with massive, extreme Population I companions, such as Cen X-3 and Cyg X-1, and those identified with lower-mass, older population stars, such as Cyg X-2, Sco X-1, and the globular cluster sources. The number of sources located in globular clusters is greater than expected for their number of stars, which has lead to suggestions that the globular cluster sources might not be classical X-ray binaries, but possibly massive black holes located at the center of the cluster (e.g., Bahcall & Ostriker 1975, Grindlay 1978). However, accurate determination of cluster source positions (Grindlay 1981), as well as the success of nuclear flash models in explaining X-ray bursts in globular cluster sources (e.g., Joss 1978, Lewin & Clark 1980, Lewin & Joss 1981), indicate that these systems are not radically different from other luminous X-ray sources in the Galaxy.

The center of our Galaxy is not bright at X-ray wavelengths. Recent imaging observations with the Einstein Observatory show a complex region of source emission within 20′ of the galactic center. Based on an assumed spectrum with $kT = 5$ keV and $N_H = 6 \times 10^{22}$, Watson et al. (1981) estimate that the luminosity of the galactic center diffuse emission is about 2.2×10^{36} erg s^{-1} (0.9–4.5 keV); the galactic nucleus itself may be a source with a luminosity of 1.5×10^{35} erg s^{-1}, the uncertainty being due to positional accuracies of order 40″ in this crowded field.

Luminous X-ray sources are not common objects; perhaps one in 10^9 stellar systems exhibits this phenomenon. Consequently, it is useful to study these objects in other galaxies to improve statistical estimates, to better define the extremes of the phenomenon, to determine the dependence of the probability of X-ray source formation upon the stellar population and galactic morphology, and possibly to discover new classes of these rare objects. Prior to the launch of the Einstein Observatory, X-ray emission had been detected in only three other normal galaxies (LMC, SMC and M31). In the case of the Magellanic Clouds, five relatively permanent sources were well studied both at X-ray and optical wavelengths. These sources are similar to the bright sources which exist in the Galaxy; in several cases they show pulsations and/or eclipses. The luminosities of these sources are near the upper end or above the luminosity distribution of the galactic X-ray sources. Clark et al. (1978) suggest that these sources are superluminous (above the Eddington limit for a neutron star with spherical accretion). Various explanations for their luminosity have been proposed, including the suggestion that the mass accretion rate in these galaxies can be higher because radiative pressure is reduced due to lower metal abundance. The LMC also was known to contain the remarkable X-ray transient A 0538−66 which has a peak luminosity in excess of 10^{39} erg s^{-1} (Skinner et al. 1980). For M31, only the total X-ray luminosity was known (Margon et al. 1974). The Einstein Observatory, which utilized a focusing X-ray telescope, was roughly 1000 times more sensitive than previous X-ray experiments. As a result, detailed X-ray surveys of nearby external galaxies could be performed with sensitivities comparable to the Uhuru and SAS surveys of the Galaxy. The results obtained from observations of normal galaxies with the Einstein Observatory are the subject of this chapter.

Giacconi et al. (1979) have described the Einstein Observatory. Two imaging instruments could be placed in the focal plane of the telescope: an imaging proportional counter (IPC) and a microchannel plate detector (HRI for high resolution imager). The IPC had a higher quantum efficiency and a larger energy bandpass, but it had poorer angular resolving power (FWHM of 1.4′ as compared to 4″). There are difficulties in comparing the Einstein surveys with previous ones because of the different bandpasses of the experiments. We will adopt the uniform conversion factor from IPC counts to X-ray flux (0.5–3.0 keV) of 2.56×10^{-11} erg cm^{-2} per count; this corresponds to an assumed 10 keV thermal bremsstrahlung source and absorption resulting from $N_H = 5 \times 10^{20}$ cm^{-2}. The uncertainties in luminosity result primarily from deviations from these assumptions, and are approximately ± 30 per cent for $2\,\text{keV} \leqslant kT \leqslant 10\,\text{keV}$ and $N_H < 2.5 \times 10^{21}$. These systematic errors dominate the statistical errors in all cases and, consequently, errors have not been assigned to individual observations.

3.2 Detailed observations of sources in nearby galaxies
The LMC

The Large Magellanic Cloud, the largest satellite of the Galaxy with a mass of approximately $10^{10}\,M_\odot$, is an irregular barred spiral. Optical photographs of the LMC are dominated by the diffuse optical bar, which is comprised principally of low-mass stars. An exceedingly active star formation region surrounded by an extensive HI complex is located at the eastern end of the bar. This region, along with the spiral arms most visible in HI maps, comprise the Population I portions of the LMC.

Long, Helfand & Grabelsky (1981) conducted a survey of the LMC with the imaging instrumentation on the Einstein Observatory; it is complete above a luminosity of 4×10^{35} erg s^{-1}. Initial imaging proportional counter observations were followed by HRI observations of many of the stronger sources. In the initial survey, 97 sources were detected ranging in luminosity from 10^{35} erg s^{-1} to 2×10^{38} erg s^{-1}. The positions of these sources are indicated on an optical photograph of the LMC in Figure 3.1. Some of the sources detected in this survey (<25) are foreground stars or background active galaxies not associated with the LMC. Most, however, must be in the cloud itself. Nearly half of the sources found in the survey are concentrated around the 30 Doradus HI complex. About 20 sources lie inside the portion of the LMC defined by the optical bar, but this cannot necessarily be taken as an association with Population II objects since an HI arm of the LMC overlies the bar (McGee & Milton 1966).

Most of the sources whose nature is understood are SNR. Twenty-five have been so identified – based on high-resolution X-ray observations or on positional coincidences with SNR which were previously known from radio and optical observations. The relatively high percentage of SNR identifications is not surprising because of the low-energy bandpass of the experiment and the small absorption in the direction of the LMC.

Five sources in addition to LMC X1-4 detected with the HRI appear to be pointlike. They, as well as sources showing evidence of variability, are candidates for compact objects, but it is as yet unclear how many of these are located in the LMC. Optical observations performed by M. Pakull of four of the HRI error boxes indicate that the counterpart for three of the sources are likely to be galactic foreground stars. In one of these cases, the optical candidate is a

Fig. 3.1. X-ray source positions plotted on an optical photograph of the LMC (courtesy of P. Hodge, University of Washington). North is at the top, east is left in this and subsequent illustrations of this chapter. The 30 Doradus complex of OB associations contains nearly half the X-ray sources (Long, Helfand & Grabelsky 1981).

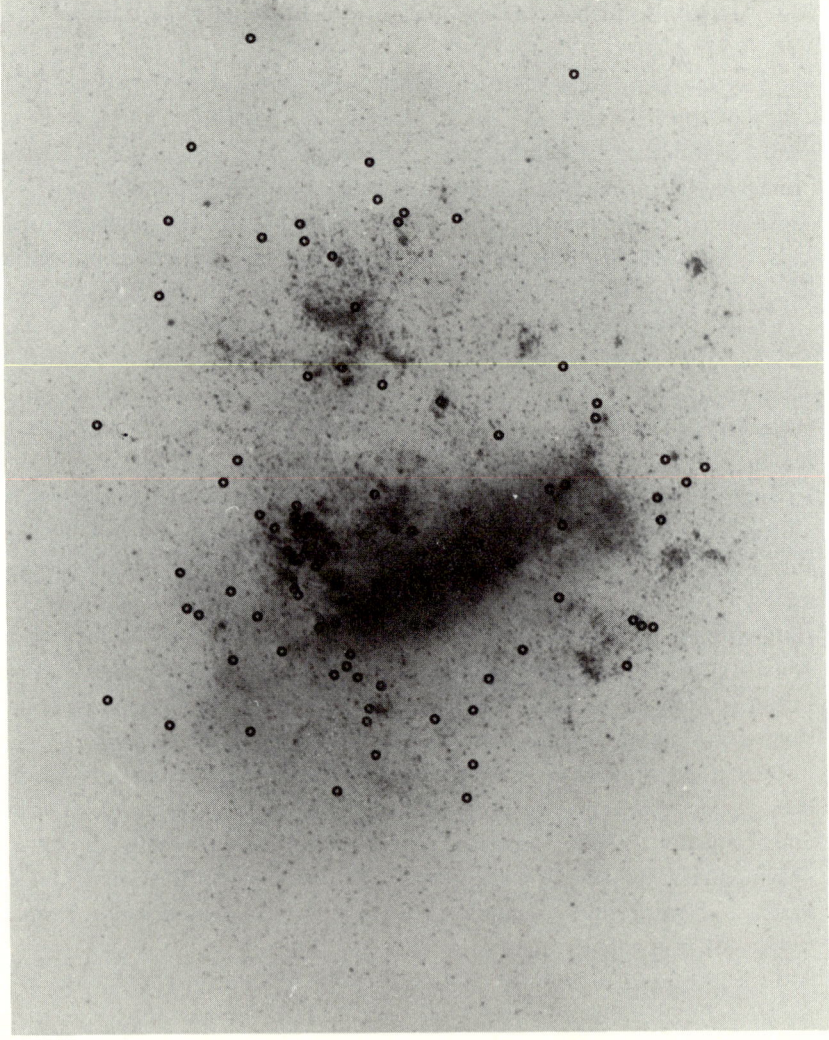

$m_v = 6.8$ RS CVn star. The remaining object has as a candidate a UV excess object with variable He II and Balmer emission.

The luminosities of most of the LMC sources are considerably less than those of strong galactic X-ray binaries. Some are likely to be low-luminosity accreting systems similar to 4U 2129 + 47 (Thorstensen et al. 1979); others are most likely SNRs. Long, Helfand & Grabelsky (1981) attempted to identify source classes in the LMC by comparing the positions of X-ray sources with those in various source catalogues. With the exception of catalogues with a strong SNR contribution, the only correlation which appeared was with massive stars in the LMC, and this correlation was rather weak. There is a strong concentration of sources around the 30 Doradus nebula, including a source which, within the accuracy of the IPC, is coincident with R Doradus itself. The region around 30 Doradus also shows extended emission. This may be due to a large number of low-luminosity sources or hot diffuse gas. Such emission is similar to that observed in the Orion nebula (Ku & Chanan 1979) or Eta Carinae (Seward et al. 1979).

The SMC

A second companion of the Galaxy, the Small Magellanic Cloud, is an irregular galaxy about $\frac{1}{10}$ as massive as the LMC but with significantly more hydrogen per unit mass. Seward & Mitchell (1981) have used the IPC to map the SMC to a limiting sensitivity of 3×10^{35} erg s^{-1}; their survey is essentially complete above 10^{36} erg s^{-1}. They detected 26 sources of which six are known not to be associated with the SMC and one additional source probably also is a foreground star. Followup observations with the HRI have shown that the second brightest source, after the binary SMC X-1, is a previously unrecognized SNR; Seward & Mitchell conclude that four other sources which exhibit soft X-ray spectra also are likely to be SNR. The remaining 13 sources are not distributed uniformly; a region of high source density probably includes six or seven sources associated with the SMC, of order four additional uncatalogued QSOs would be expected at this sensitivity level, and the remaining two or three sources cannot be assigned either to the SMC or background with any degree of confidence. The total number of sources associated with the SMC is 12-15. A detailed comparison of the SMC and LMC emission requires further identification of the SMC sources, but the relative numbers of bright binaries, SNR, and weak unidentified sources, appear to scale approximately as the ratio (~ 8) of the optical luminosities of the two galaxies.

M31

M31 is the only large galaxy which is near enough for the detailed study of sources with intensities comparable to those in our own Galaxy; M31 is well studied at other wavelengths, and is similar enough to our Galaxy to permit meaningful comparisons. It is somewhat larger than our Galaxy, and therefore offers a statistical advantage in the study of rare objects, such as binary X-ray sources. An initial survey was performed in January 1979 with the IPC; this

Fig. 3.2. M31 X-ray source positions plotted upon an optical photograph (*a*) and upon the HI map of Emerson (1976) (*b*). The outer sources are associated with spiral arm features or globular clusters. The large number of sources near the nucleus of the galaxy is not apparent in these presentations (van Speybroeck & Bechtold 1980).

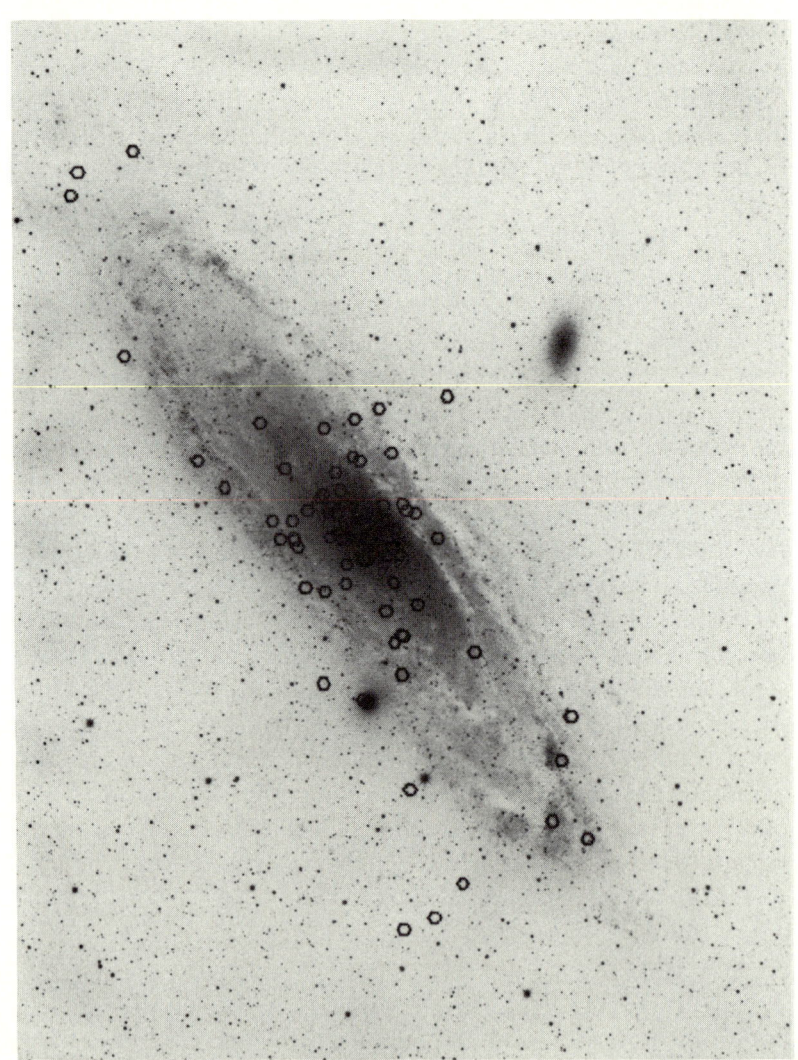

covered most of the visible galaxy, with a limiting sensitivity of $\sim 10^{37}\,\text{erg}\,\text{s}^{-1}$ (van Speybroeck *et al.* 1979, van Speybroeck *et al.* 1983). The observations of the central part of M31 were repeated 6 months later with about $\frac{1}{2}$ of the exposure time. These images contain isolated sources in the outer portions of the galaxy and a confused region surrounding the nucleus. The central region has been observed and resolved into individual sources with the HRI on three occasions separated by 6 mth intervals. Many of the IPC source positions also have been observed with the HRI, and positions accurate to about 5 arcsec have been obtained in these cases.

Ninety point sources have been detected. The source positions are overlaid upon a visible light photograph of the galaxy in Figure 3.2(*a*). There is a large

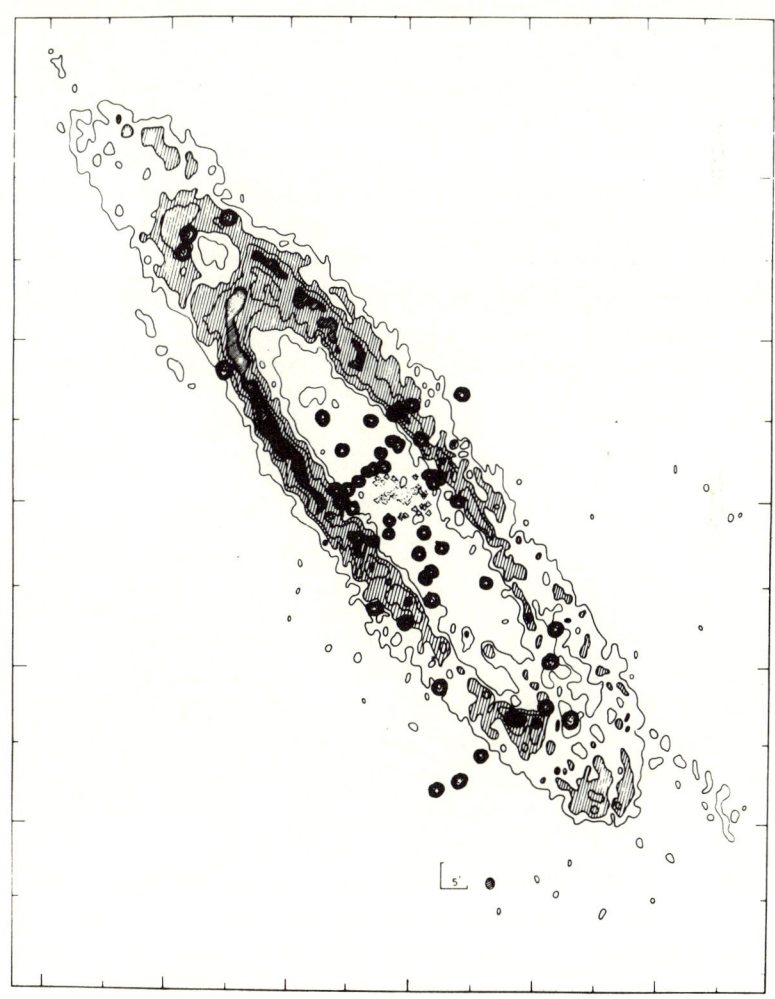

group of sources associated with the bulge (not easily seen in this figure) and a general association with the spiral arms of the galaxy. There also are two sources which may be associated with M32 and a few sources which do not seem to be associated with the galaxy. The apparently isolated source to the NW, approximately halfway between M31 and NGC 205, actually is a globular cluster of M31. The sources to the extreme SW are not located sufficiently accurately to secure identifications, but one also may be an M31 globular cluster. We also expect approximately one foreground star or background quasar in fields of this sensitivity. The three sources to the extreme NE seem to be located in a rather optically-faint portion of the disk. The comparison of X-ray source positions and the H1 map of Emerson (1976) in Figure 3.2(b) is more illustrative; here, the three NE sources are seen to be associated with a pronounced HI feature in that region, and there is a general correlation between the non-bulge X-ray sources and the HI. This is not surprising because we expect X-ray sources to be associated with young, massive stars, and in M31 the indicators of Population I activity, such as OB associations, HI, HII, CO, and dust, are centered in an arm about 9 kpc from the nucleus (Emerson 1974, 1976, 1978; Baade & Arp 1964; Berkhuijsen 1977; Combes et al. 1977(a), (b)).

Some sources are identified with individual objects in M31. There are, for example, two possible coincidences with the supernova remnant list of D'Odorico, Dopita & Benvenuti (1979); these have X-ray luminosities of about 2×10^{37} erg s^{-1}, comparable to the brighter SNR in the LMC. The supernova of 1885 is not detected at the 10^{37} erg s^{-1} level.

The largest number of source identifications are globular clusters. Nineteen of the M31 globular clusters listed by Sargent et al. (1977) or Battistini et al. (1980) have been detected with the HRI; the position errors in these cases are small enough to be confident of almost all identifications. The positions of five additional sources observed only with the IPC also are consistent with globular clusters, but in this case, the position errors would result in approximately three spurious associations. The 19 HRI detections are listed in Table 3.1. In our Galaxy, 13 X-ray sources have been detected among 130 optically-discovered globular clusters (Grindlay 1981). The Sargent et al. (1977) and Battistini et al. (1980) lists include 325 globular clusters within 27.5 arcmin of the centers of the M31 fields observed, and 24 are detected (including perhaps three spurious associations), which is consistent with the fraction found in our Galaxy. However, the globular clusters in M31 are more luminous in X-rays; the largest and mean X-ray luminosities among the 13 optically-discovered X-ray-emitting globular clusters in our Galaxy are 5.2 and 0.68×10^{37} erg s^{-1} respectively; the comparable numbers for M31 X-ray-emitting globular clusters are 26.4 and 4.7×10^{37} erg s^{-1}. In fact, only two of the globular clusters in our Galaxy would have been detected in M31, whereas, if the populations are the same, we would expect nine. We thus are left with two possibilities (or combinations thereof).

(1) The globular clusters in M31 are brighter X-ray sources than those in our Galaxy.

(2) The probability of a globular cluster including a bright X-ray source is significantly higher in M31 than in our Galaxy, and we are observing the high-luminosity tail of the X-ray distribution in M31. This perhaps is the more likely explanation; the globular clusters of M31 have a higher metallicity than those of our own Galaxy (Harris & Racine 1979, and references therein), and perhaps therefore contain more neutron stars.

The X-ray sources in globular clusters in our Galaxy occur primarily in the more centrally condensed globular clusters (e.g., Clark 1975; Grindlay 1977). Van Speybroeck et al. (1981) have noted that the brighter globular clusters in M31 contain X-ray sources more frequently than optically-fainter clusters. Conjectured relations between the probability that a globular cluster contains an X-ray source and the optical luminosity can be tested by performing a Kolmogorov test. The statistic in this test is the maximum distance between the

Table 3.1. *M31 X-ray globular clusters*

Globular cluster		X-ray luminosity ($\times 10^{37}$)
Sargent et al. (1977)	Battistini et al. (1980)	
198	143	1.7
200	148	2.3
–	144	2.2
–	146	2.0
–	153	4.5
148	86	3.1
213	158	3.2
144	82	2.9
52	5	9.4
322	–	4.4
307	–	26.4
280	225	5.6
235	185	1.5
108	45	5.7
192	135	10.7
199	147	1.0
264	213	1.9
169	107	0.5
185	127	0.5

(Less than one chance coincidence is expected based on a $10''$ error radius.)

(normalized) cumulative number of X-ray source detections and the cumulative sum of the conjectured function of luminosity. The test has been performed for various power law models, as shown in Figure 3.3 and summarized in Table 3.2. The globular clusters in the sample are the 270 in the lists of Sargent *et al.* (1977) or Battistini *et al.* (1980) which are within 27.5 arcmin of the center of an IPC field and for which optical magnitudes are known. The 22 X-ray sources include five IPC detections and 17 HRI detections. The test rejects at the 95 per cent confidence level those models in which the probability of a globular cluster containing a luminous X-ray source is proportional to the luminosity raised to a power either smaller than 0.386 or larger than 1.157. Including only the more certain HRI detections would change the permitted region to $0.75 < \alpha < 1.34$, so we will consider the accepted value of α to lie between 0.386 and 1.34.

These rough results already reject source formation models in which the probability of X-ray source formation is independent of the luminosity, or else scales as L^2 or higher powers. For example, binary or three-body capture pro-

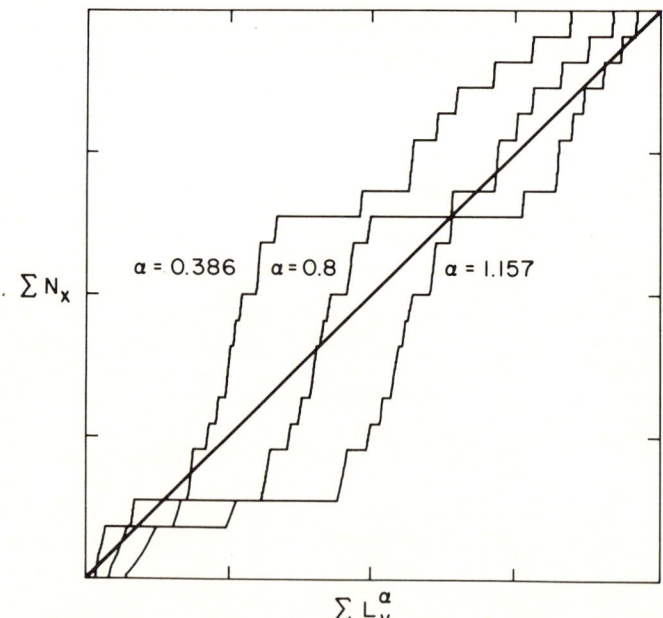

Fig. 3.3. Kolmogorov test of the hypothesis that the probability of a globular cluster containing an X-ray source is proportional to a power of the luminosity. The globular clusters in the M31 X-ray fields with known magnitudes are sorted in order of decreasing luminosity (L); the horizontal steps are proportional to L^α, and the vertical steps occur at X-ray detections. The hypothesis has an acceptable fit for $\alpha \approx 0.8$; $\alpha < 0.386$, or $\alpha > 1.157$ are rejected at the 5 per cent confidence level (van Speybroeck *et al.* 1981).

cess models neglecting the correlation between luminosity and other cluster parameters would yield source formation probabilities proportional to L^2 or L^3 and are rejected. Realistic binary capture models depend upon the velocity and density distributions, which unfortunately have not been measured for the M31 globular clusters. Jones (1977), Markert et al. (1977), and Clark et al. (1978) have emphasized the similarities between the spectra of globular cluster X-ray sources and the sources commonly associated with the galactic bulge. There are 35 point sources in M31 within the 1000-pc bulge radius quoted by Morton, Andereck & Bernard (1977), and it is interesting to compare this with the number that would be expected from the luminosity of the bulge assuming that the same X-ray source formation probability correlation with luminosity observed for M31 globular clusters also obtains for the bulge. The expected number of bulge X-ray sources under these assumptions is given in column 4 of Table 3.2. A model of X-ray source formation probability scaling approximately as $L^{0.6}$ would yield an acceptable fit to the globular cluster data and also provide the observed number of X-ray sources in the bulge of M31. The X-ray source formation probability actually must depend upon the density and velocity distributions – see, for example, Press & Teukolsky (1977); the luminosity dependence given here must be a consequence of an unfortunately unknown statistical correlation between luminosity and other globular cluster parameters. The X-ray data indicate that, if the same mechanism is responsible for the formation of both

Table 3.2. *Kolmogorov test of hypothesis*

	P (X-ray source) αL_V^α		
α	D	$P(D)$	N_{bulge}[a]
0.0	0.429	0.001	0.08
0.3	0.334	0.021	2
0.386	0.302	0.048	5
0.4	0.297	0.054	6
0.5	0.259	0.130	16
0.6	0.220	0.280	45
0.7	0.179	0.533	125
0.8	0.171	0.591	344
0.9	0.207	0.346	938
1.0	0.245	0.174	2.5×10^3
1.1	0.283	0.077	6.8×10^3
1.157	0.305	0.046	1.2×10^4
1.2	0.321	0.030	1.8×10^4
1.5	0.434	0.001	3.2×10^5
2.0	0.593	0.0001	3.3×10^7

[a] Assuming equivalent M31 bulge magnitude of $B = 5.14$.

M31 bulge and globular cluster sources, then the density and velocity parameters of the M31 bulge are not vastly different from what would be expected from a hypothetical globular cluster with the bulge luminosity.

It should be noted that the cluster data themselves do not exclude models in which the X-ray sources evolve independently $(P(x) \sim L^1)$, and thus yield an X-ray source probability proportional to the number of objects in the cluster, or models in which a cluster member interacts with an external object. These models, however, would yield the same rate of X-ray source production per object in similar stellar populations, such as the bulge of M31, which is not observed – the X-ray source production per star in globular clusters being some 100 times as great as that in the bulge of M31. This also provides another argument against the bulge of M31 and similar systems being formed largely from the debris of disrupted, evolved globular clusters.

If the globular cluster source formation process does not take place efficiently in the bulge, then the bulge sources must be created by some other mechanism – perhaps disruption of a limited number of globular clusters. For example, 350 globular clusters disrupted by tidal interactions within the bulge would contribute about 1 per cent of the bulge light and provide the observed number of X-ray sources. This scenario may be more viable than scaling the bulge source production as $L^{0.6}$ because it also results in the total X-ray luminosity of galaxies scaling as L (if the globular cluster numbers scale as galaxy luminosity), which, as will be discussed below, is approximately the relationship observed in a large sample of normal galaxies. More information on the formation of the low-mass binary X-ray sources in the globular clusters is given in Chapter 8 of this book.

The spatial distribution of X-ray globular clusters in our Galaxy is more concentrated towards the bulge than the total population (Grindlay 1981). A similar trend in M31 is noted (van Speybroeck *et al.* 1981). The 325 globular clusters of Sargent *et al.* (1977) or Battistini *et al.* (1980), which fall in the M31 HRI observation fields can be divided into three groups as in Table 3.3. The three categories are relatively rich in objects located within the bulge, disk, and halo respectively, the OB associations of Baade & Arp (1964) being used as tracers of the disk population. The probability of a globular cluster containing a strong X-ray source is significantly larger if the globular cluster is located within the denser regions of the galaxy. We do not have a satisfactory explanation for this effect. The higher metallicity of clusters near the center of M31 (Harris & Racine 1979, and references quoted therein) may imply a higher density of neutron stars among the inner globular clusters. Two other possibilities which have been suggested, assuming that the X-ray source consists of a collapsed object in a binary system, are tightening of the binary orbit due to collisions with objects outside the cluster, and exchange collisions between putative but unobserved globular cluster binaries and single external objects (to provide either the collapsed object or a new source of gas).

The central region of M31 has been observed with the HRI (~3" resolution) on three occasions; the sum of these is shown in Figure 3.4. The central region of M31 has a very high density of X-ray sources; there are 19 resolved sources plus ten diffuse or confused regions with $L_x > 10^{37}$ within two arcmin of the nucleus of M31. The distribution of sources near the center of M31 also appears to be elongated along the EW axis.

The luminosity distribution of the X-ray sources in M31 is shown in Figure 3.5(c). The mean luminosity for all detected sources is 3.6×10^{37} erg s^{-1}; the lower cutoff corresponds to the detection threshold and is not real, although the contribution of sources below the threshold must be relatively small because the sum of the individual sources detected is comparable to previous measurements of the total X-ray luminosity of M31. The globular cluster categorization is based upon optical identifications. The two remaining categories are an attempt to separate disk and inner bulge sources, for, in our own Galaxy, the bulge sources are more luminous and softer than the disk sources and presumably are a different population (Jones 1977, Markert et al. 1977, Clark et al. 1978). The inner bulge sources in M31 are brighter than the outer sources, as indicated in Figure 3.5. These inner bulge X-ray sources are more concentrated than the optical bulge of M31; Morton, Andereck & Bernard (1977) give an optical bulge radius of approximately 5 arcmin. The 16 X-ray sources between 2' and 5' from the bulge, however, have a mean luminosity of 2.2×10^{37} erg s^{-1}, or more similar to the outer sources (mean $L_x = 2.7 \times 10^{37}$) than to the inner bulge group (mean

Table 3.3. *Correlation, M31 globular clusters and OB associations*[a]

Globular cluster group	Globular[b] clusters	X-ray globular clusters	Fraction
(1) $R < 10'$ from nucleus	73	10	0.137
(2) $R > 10'$ from nucleus and five or more OB associations within 5'	154	10	0.065
(3) $R > 10$, from nucleus and fewer than five OB associations within 5'	98	3	0.031

Note: The probability of obtaining ten or more of 23 objects in ($\frac{73}{325}$) of the sample is 0.020. The probability of obtaining three or fewer of 23 objects in $\frac{98}{325}$ of the sample is 0.0522.

[a] Baade & Arp 1964.
[b] Sargent et al. 1977, or Battistini et al. 1980.

$L_x = 4.8 \times 10^{37}$). The inner sources thus appear to comprise a different population from the outer disk group and have X-ray luminosities comparable to the M31 globular cluster sources. The extent and location of this group of sources is similar to the 400 pc region in which Rubin & Ford (1971) find evidence of gas expansion velocities of order 100 km s^{-1} superimposed upon a rapid rotation. This inner region thus may constitute a distinct structure within M31.

The sources in this inner region appear to lie approximately along an EW axis, but it is not certain if the sources actually form a dense, quasi-spherical population or if, instead, the sources are located in the plane of the disk. If they are a spherical population, then the source density is much greater than near the nucleus of our own Galaxy. A 2 arcmin radius in M31 corresponds to approximately 2.3° at the center of our Galaxy; there is only one persistent source (and three known transients) within 2.3° of the nucleus of our own Galaxy which could have been detected in M31, whereas there are 19 point sources and ten diffuse or confused regions within 2' of the nucleus of M31. If the M31 sources form a spherical population, then they constitute a structure very different from any found in our own Galaxy.

Fig. 3.4. HRI exposure of the center of M31. Three exposures totalling 75 000 s have been merged in these data. The bright source 1 arcmin east of the brightest source in the field is coincident with the nucleus of M31. The brighter sources appear to be organised along an approximately east–west axis (van Speybroeck *et al.* 1981).

Alternatively, the X-ray sources near the center of M31 may not be a spherical population, but instead may lie largely within the plane of the galaxy. This geometry is suggested by the bulge sources in our own Galaxy, which certainly lie in the plane. In this view, the 'bulge' sources would delineate bars in both galaxies, and the X-ray sources would become tracers of a dynamically-important population – perhaps the remnants of the approximately 350 globular clusters which may have been necessary for their formation.

The HRI observations include a source well within the 5 arcsec position errors of the M31 nucleus; the source density is such that there is approximately a 10 per cent probability of an ordinary X-ray source being located in this region by chance. The luminosities observed during the three observations were $10^{38}\,\mathrm{erg\,s^{-1}}$, $<10^{37}\,\mathrm{erg\,s^{-1}}$, and $5\times 10^{37}\,\mathrm{erg\,s^{-1}}$. This source may be associated

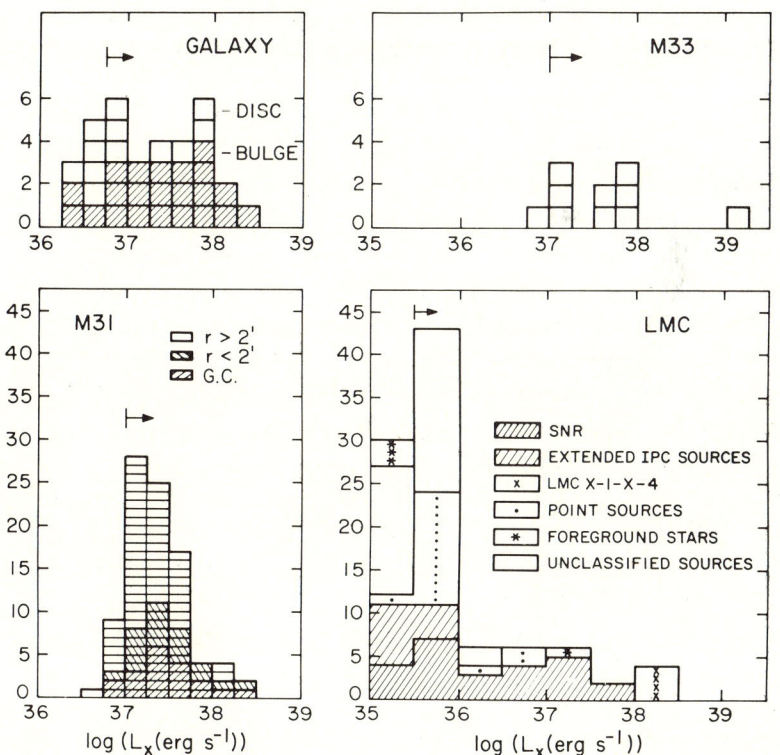

Fig. 3.5. X-ray luminosity distributions in our Galaxy (Clark et al. 1978), M33 (Long et al. 1981), M31 (van Speybroeck et al. 1981), and the LMC (Long, Helfand & Grabelsky 1981). The survey completeness thresholds are indicated. The bright source in M33 is associated with the nucleus of that galaxy. All other source luminosities are consistent with the Eddington limit for neutron stars, assuming pure hydrogen gas and spherical accretion.

with the otherwise inactive nucleus of M31; it also may be an ordinary X-ray source or a transient. Variations of a factor of 10 are unusual for ordinary non-eclipsing sources; sources in the bulge of our own Galaxy are not observed to eclipse, and time variations were not observed in this source during the approximately one-day observations. Transient X-ray sources normally do not have a duty cycle of $\frac{2}{3}$ (admittedly a low-precision estimate in this case) so it is probable that the nucleus of M31 is an X-ray source with a luminosity of order 10^{38} erg s^{-1}.

M33

M33 is a local group Sc which has well-defined, open spiral arms and a mass which is about $\frac{1}{4}$ that of the Galaxy. The IPC image of this galaxy (Long et al. 1981b) shows ten discrete sources associated with M33. These are indicated on an optical photograph of the galaxy in Figure 3.6. The luminosities of nine of these sources are comparable to those of compact binary sources in the Galaxy. None of the source positions corresponds to the position of suggested SNR in M33. The weakest source has a luminosity which is comparable to or slightly larger than the strongest SNR in the LMC. At least one of the sources varied between two observations, indicating that it must be compact. All of the sources except one lie within the main HI complex of M33. Several can be associated with individual Population I features. None is associated with as strong a radio source as Cyg X-3.

The central source undoubtedly is associated with the nucleus of M33. HRI observations (Markert et al. 1979) resolve the central IPC source into two components: the dominant one is located less than 3 arcsec from the optical nucleus, which is within the measurement errors. The HRI observations (Markert 1980) also show that the X-ray luminosity of the M33 nucleus varies, the two observations differing by 40 per cent. The luminosity during the IPC observations (Long et al.) was 2×10^{39} erg s^{-1}. The nucleus of M33 is unprepossessing at optical wavelengths; it is semi-stellar, smaller than a typical globular cluster (Gallagher 1980, Walker 1964). It is not detected at radio wavelengths. Its velocity dispersion is less than 30 km s^{-1}.

M101

M101, the nearest unobscured Sc supergiant galaxy, is the prime distance calibrator for galaxies of this type. It was observed twice, in January and June 1979, with the IPC. The two images superimposed on a PSS photograph of M101 in Figure 3.7(e) and (f) reveal at least five distinct sources or source complexes which appear to be associated with the galaxy. There also are several other sources outside the optical image of the galaxy, most of which are

Fig. 3.6. The positions of ten X-ray sources are superimposed on this photograph of M33 obtained with the Palomar Schmidt through an Hα interference filter. The circles, which are 2 arcmin in diameter, correspond approximately to 90 per cent confidence contours. Many of the X-ray sources are associated with individual HI features or other Population I markers.

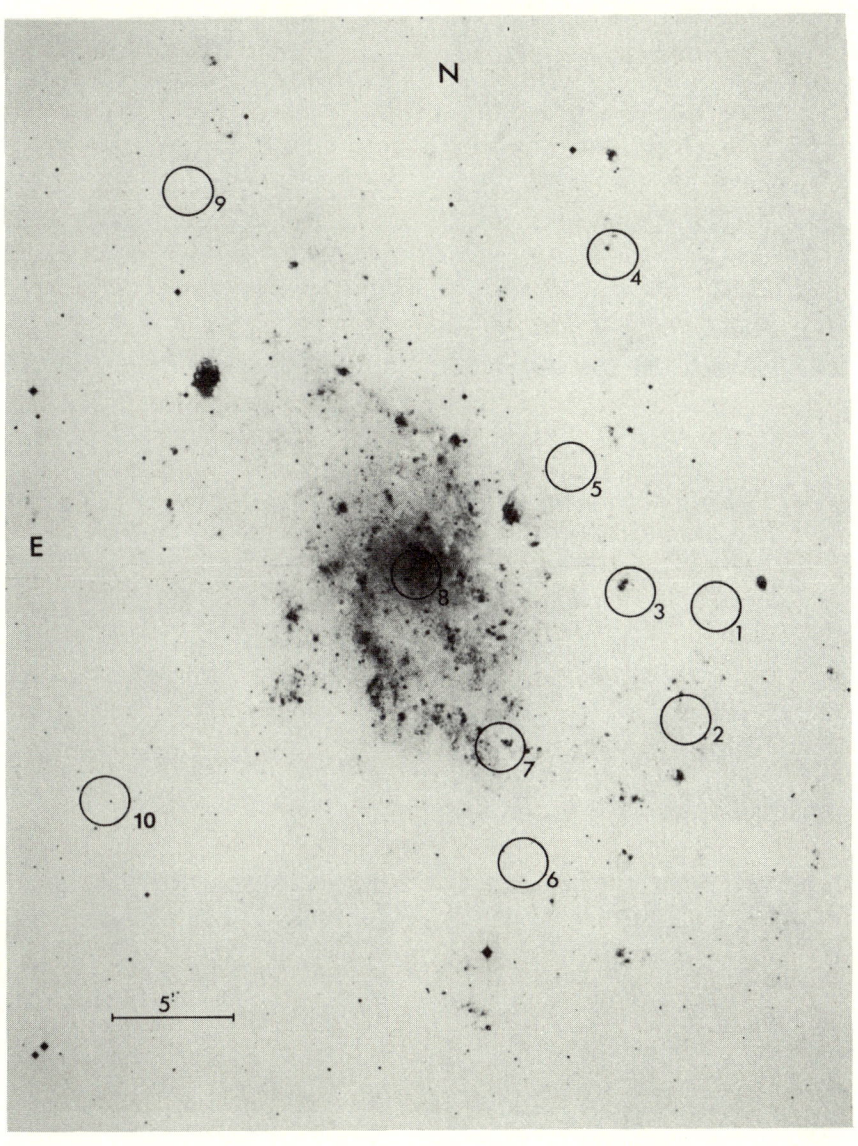

Fig. 3.7. IPC contours superimposed upon optical photographs of: (a) NGC 247; (b) NGC 2403; (c) NGC 3031 = M81; (d) NGC 5236 = M83; (e) and (f) NGC 5457 = M101 (exposures taken at 6 mth intervals). The apparently bright sources north west of NGC 2403 (b) and east of M81 (c) are not associated with the galaxies.

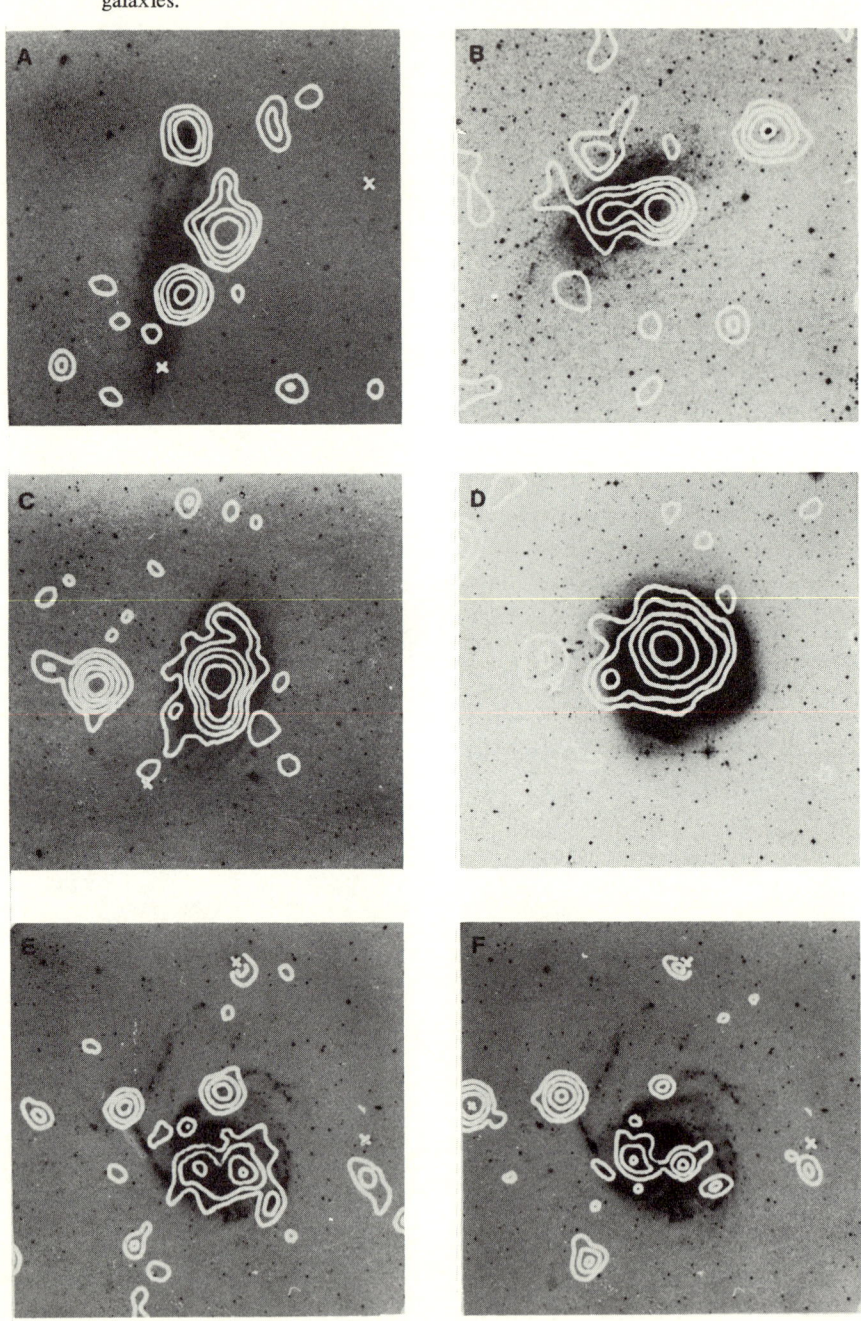

probably not located near M101. All of the sources in the outer portions of M101 can be associated with prominent spiral arms of the galaxy. Two of these are variable. At a distance of 7.2 Mpc (Sandage & Tammann 1975), individual sources (or source complexes) have luminosities ranging from 4×10^{38} erg s^{-1} to 2×10^{39} erg s^{-1}, substantially greater than individual sources in the Galaxy.

Optically, the nucleus of M101 is very weak. Unlike M33, the nucleus of M101 is also weak at X-ray wavelengths (less than about 2×10^{38}).

NGC 247

NGC 247 is a relatively small Sc galaxy in the Sculptor group. Three sources were clearly detected within the optical galaxy; the luminosities were in the range $10^{38} - 3 \times 10^{38}$ erg s^{-1}. None of the detected sources is coincident with the nucleus. One is coincident with a very prominent H II region. The IPC contours are overlaid upon the optical image in Figure 3.7(a).

M81 and M100

M81 is an Sb galaxy with narrow emission lines and a very large Ne II/Hα ratio at its nucleus. The IPC contours near M81 (NGC 3031) are shown in Figure 3.7(c) (the strong source to the east of the galaxy probably is not associated with M81). The galaxy is extended in the IPC data but appears to include a strong nuclear component. The HRI data (Elvis & van Speybroeck 1982) show an unresolved ($\Delta\theta < 2''$) source with a luminosity of order 7×10^{39} erg s^{-1} coincident with the nucleus. Peimbert & Torres-Peimbert (1981) have observed faint Mg II and Hα lines with widths of 10 200 km s^{-1} and 5300 km s^{-1} respectively. The X-ray and optical properties of the M81 nucleus are approximately those of a Seyfert galaxy nucleus about two orders of magnitude fainter than any previously observed. Peimbert & Torres-Peimbert conclude that if the broadening is caused by circular motions, then a central object, possibly a black hole, with a mass between 1×10^6 and $1.8 \times 10^7 M_\odot$ is required.

M100 also may represent a transition case between 'normal' and Seyfert galaxies. Palumbo *et al.* (1981) report that the nucleus of M100 has an X-ray luminosity of 1.5×10^{40} erg s^{-1} and that a preliminary analysis of IUE and VLA data indicate a rather complex optical and radio structure.

N 253

N 253 is a bright nearby Sc in the Sculptor group seen at high inclination. The IPC observation of this galaxy shows six discrete sources. The sources away from the nucleus have luminosities ranging from about 5×10^{37} erg s^{-1} to 4×10^{38} erg s^{-1} assuming the distance is about 3.4 Mpc.

The emission is dominated by a source at the galactic center, which is confused in the IPC image. The nuclear region has an X-ray luminosity of about 2×10^{39} erg s^{-1}. Much of this emission appears diffuse in the HRI, and no strong point source coincides with the nucleus. The other properties of the nucleus of N 253 place it at the extreme active end for 'normal' galaxies. It is a strong radio source. It exhibits non-rotational velocities (Ulrich 1978). The velocity full width of the radio recombination line H102α is nearly 400 km s^{-1} (Seaquist & Bell 1977). The Ne II line emission is strongly concentrated at the nucleus with $\frac{1}{3}$ confined within 6" (Beck, Lacy & Gaballe 1979). The density in the central 300 pc based on the Ne II rotation curve is 9 M$_\odot$ pc^{-3} in comparison to 120 M$_\odot$ pc^{-3} in our Galaxy (Oort 1977). Rieke *et al.* (1980) have reviewed the extensive measurements of the nuclear properties of NGC 253. The X-ray source positions are indicated in Figure 3.8.

3.3 The survey of normal galaxies

Shorter observations of a much larger sample of galaxies, with the Einstein Observatory reveal only gross features concerning the X-ray emission.

Fig. 3.8. X-ray source positions in NGC 253. The galaxy also has bright diffuse emission from its nuclear region. The individual sources have luminosities between 5×10^{37} and 4×10^{38} erg s^{-1}, assuming a distance of 3.4 Mpc.

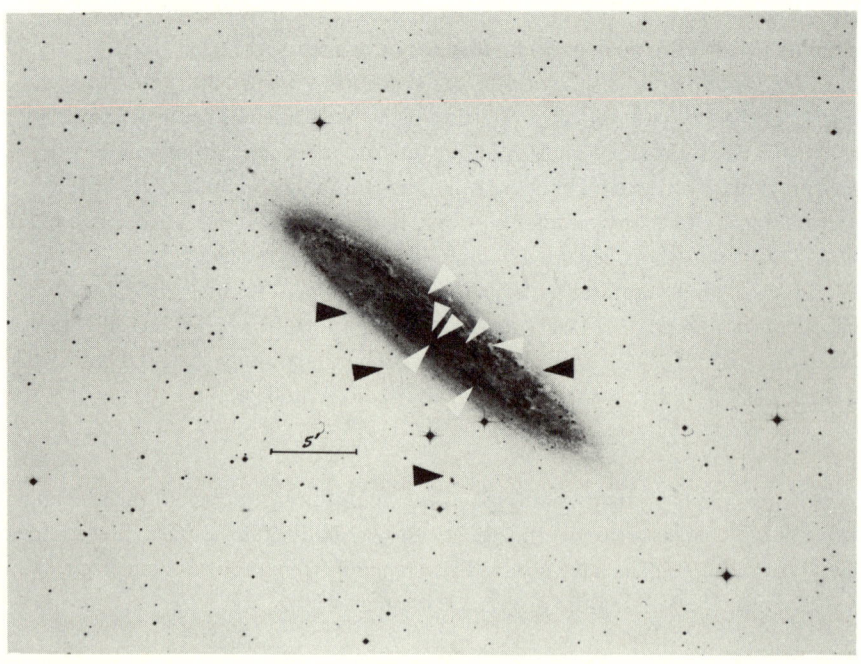

Typical sensitivities obtained are of order $2 \times 10^{-13}\,\mathrm{erg\,cm^{-2}\,s^{-1}}$. These results are presented in more detail in Long et al. (1983). The X-ray luminosities of these galaxies have been obtained by summing the excess flux in the individual images inside of the D(25) diameters as listed in the *Second Reference Catalogue of Bright Galaxies* (de Vaucouleurs et al. 1976; hereafter RC2). Detections of 51 galaxies have been obtained in this manner; 3σ upper limits have been obtained on 19 others. These results are listed in Table 3.4. Type was taken from RC2 (column 6), B is the corrected face on total magnitude (RC2, column 18) when given; otherwise the asymptotic magnitude B_T (column 6) is used; $B-V$

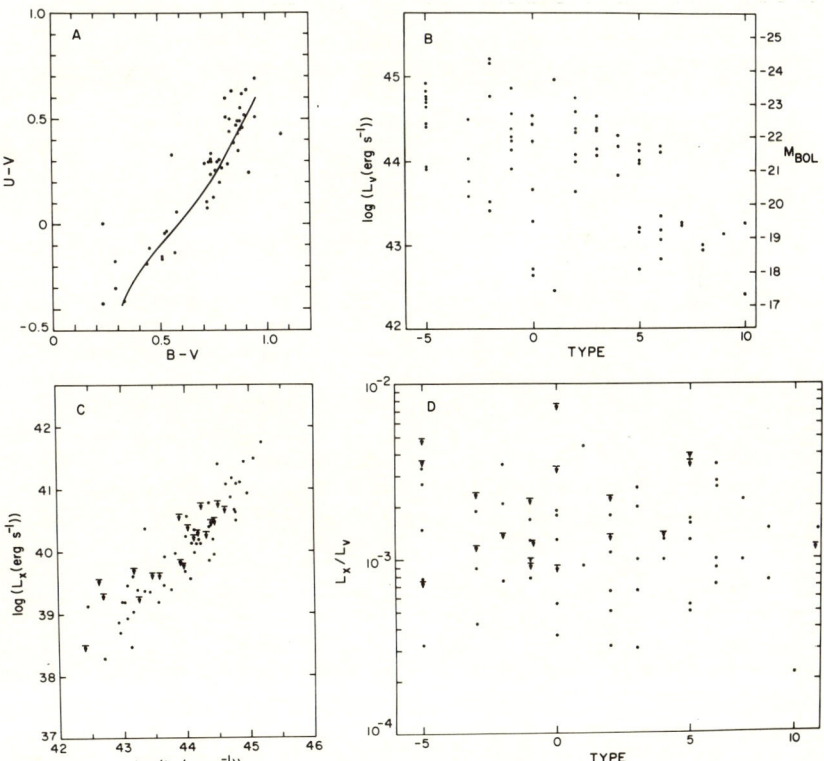

Fig. 3.9. The normal galaxy survey. The colors of the sample are shown in (a) together with the locus of normal galaxy colors taken from Figure 3.1(a) of Larson & Tinsley (1978). The visible luminosity v. galaxy-type distribution of the sample is shown in Figure 3.9(b). The early galaxies (negative type) tend to be more luminous than the late galaxies in this sample. The distribution L_x (0.5–3.0 keV) v. L_v is shown in Figure 3.9(c); L_x is approximately proportional to L_v with a spread of about a decade in L_x at any visible luminosity value. The ratio of L_x to L_v does not have a strong dependence on the type of galaxy (Figure 3.9(d)).

Table 3.4. *Normal galaxy X-ray survey*

Name	Type	B	$B-V$	$U-V$	Dist. MPC	L_V	L_X	L_X/L_V
NGC 224	3	3.59	0.74	0.34	0.690	0.12E 45	0.36E 40	0.32E-04
NGC 247	7	8.85	0.52	−0.04	3.400	0.16E 44	0.16E 40	0.10E-03
NGC 253	5	7.40	0.82	0.39	3.400	0.95E 44	0.48E 40	0.51E-04
NGC 520	0	11.55	0.72	0.11	46.720	0.34E 45	0.64E 41	0.19E-03
NGC 524	−1	11.24	0.88	0.65	51.900	0.73E 45	0.12E 42	0.17E-03
NGC 598	6	5.79	0.44	−0.18	0.720	0.11E 44	0.83E 39	0.74E-04
NGC 628	5	0.48	0.51	−	3.400	0.90E 43	0.50E 39	0.56E-04
NGC 720	−5	10.89	0.86	0.47	38.580	0.53E 45	0.15E 42	0.28E-03
NGC 936	−1	10.77	0.88	0.49	26.640	0.30E 45	<0.65E 41	<0.22E-03
NGC 1073	5	11.23	0.45	−0.11	26.440	0.10E 45	0.17E 41	0.17E-03
NGC 1097	3	9.91	0.92	0.25	24.540	0.60E 45	0.12E 42	0.21E-03
NGC 1300	4	10.70	0.58	0.06	30.400	0.25E 45	<0.35E 41	<0.14E-03
NGC 1313	7	8.95	−	−	4.820	0.18E 44	0.41E 40	0.22E-03
NGC 1316	−2	9.39	0.83	0.44	32.640	0.14E 46	0.31E 42	0.22E-03
NGC 1332	−3	10.84	0.81	−	30.400	0.31E 45	0.61E 41	0.19E-03
NGC 1350	2	10.97	0.77	0.30	29.280	0.24E 45	<0.34E 41	0.14E-03
NGC 1380	−2	10.70	1.07	0.43	29.280	0.58E 45	0.45E 41	0.77E-04
NGC 1398	2	10.28	0.87	−	30.400	0.59E 45	0.31E 41	0.52E-04
NGC 1533	−3	11.58	0.89	0.46	11.200	0.25E 44	0.23E 40	0.91E-04
NGC 1559	6	10.36	0.29	−0.17	20.500	0.13E 45	0.37E 41	0.29E-03
NGC 1574	−3	10.99	0.74	0.30	19.980	0.11E 45	<0.26E 41	<0.24E-03
LMC	9	0.14	0.55	−	0.055	0.13E 44	0.31E 39	0.23E-04
NGC 2366	10	10.67	0.38	−	3.250	0.25E 43	<0.32E 39	<0.13E-03
NGC 2403	6	8.30	0.37	−	3.250	0.22E 44	0.24E 40	0.11E 03
NGC 2763	6	12.26	0.51	−0.15	10.680	0.68E 43	0.24E 40	0.36E-03
NGC 2775	2	10.85	0.78	0.31	19.300	0.12E 45	0.14E 41	0.12E-03
NGC 2835	5	10.95	−	−	10.680	0.14E 44	<0.53E 40	<0.37E-03
NGC 2841	3	9.58	0.73	0.30	12.020	0.14E 45	0.95E 40	0.68E-04
NGC 2848	5	12.64	0.57	−0.13	10.680	0.52E 43	<0.21E 40	<0.41E-03
NGC 2859	−1	11.36	0.85	−	29.260	0.19E 45	<0.20E 41	<0.10E-03
NGC 2903	4	9.05	0.53	−0.03	7.600	0.68E 44	0.94E 40	0.14E-03
NGC 3031	2	7.24	0.81	0.60	3.250	0.99E 44	0.18E 41	0.18E-03
NGC 3077	0	10.28	0.72	0.08	3.250	0.52E 43	0.19E 39	0.37E-04
NGC 3368	2	9.79	0.78	0.20	15.980	0.22E 45	0.71E 40	0.33E-04
NGC 3377	−5	10.85	0.79	0.27	15.980	0.83E 44	<0.64E 40	<0.77E-04

Table 3.4. (*Continued*)

Name	Type	B	$B-V$	$U-V$	Dist. MPC	L_V	L_X	L_X/L_V
NGC 3489	−1	10.81	0.74	0.34	15.980	0.80E 44	<0.78E 40	<0.98E-04
NGC 3585	−5	10.64	0.87	0.43	24.740	0.28E 45	0.90E 40	0.32E-04
NGC 3593	0	11.24	0.64	–	15.980	0.46E 44	0.86E 40	0.19E-03
NGC 3628	3	9.47	0.64	–	15.980	0.24E 45	0.62E 41	0.26E-03
NGC 3818	−5	12.45	0.83	0.50	31.660	0.80E 44	<0.40E 41	<0.50E-03
NGC 3923	−5	10.74	0.91	0.64	32.700	0.49E 45	0.76E 41	0.16E-03
NGC 4236	8	9.32	0.25	–	3.250	0.86E 43	0.66E 39	0.77E-04
NGC 4244	6	9.66	0.24	0.01	5.000	0.15E 44	0.11E 40	0.73E-04
NGC 4251	−2	11.25	–	–	18.440	0.32E 44	<0.45E 40	<0.14E-03
NGC 4382	−1	9.82	0.81	0.51	20.260	0.36E 45	0.46E 41	0.13E-03
NGC 4449	10	9.51	0.33	−0.36	5.000	0.17E 44	0.25E 40	0.15E-03
NGC 4459	−1	11.07	0.88	0.45	20.780	0.14E 45	<0.18E 41	<0.13E-03
NGC 4579	3	10.33	0.76	0.26	21.740	0.24E 45	0.25E 41	0.10E-03
NGC 4594	1	8.74	0.84	0.63	19.260	0.92E 45	0.87E 41	0.95E-04
NGC 4636	−5	10.29	0.89	0.46	25.440	0.43E 45	0.12E 42	0.27E-03
NGC 4638	−3	11.74	0.78	–	20.260	0.58E 44	0.25E 40	0.43E-04
NGC 4643	0	11.25	0.87	0.49	25.440	0.17E 45	<0.59E 41	<0.35E-03
NGC 4649	−5	9.62	0.95	0.69	21.860	0.67E 45	0.25E 42	0.37E-03
NGC 4697	−5	9.96	0.87	0.35	26.140	0.59E 45	0.46E 41	0.79E-04
NGC 4753	0	10.43	0.85	0.39	22.740	0.28E 45	0.16E 41	0.57E-04
NGC 4826	2	8.96	0.75	0.13	5.000	0.44E 44	0.29E 40	0.67E-04
NGC 5068	6	10.53	–	–	7.900	0.11E 44	0.30E 40	0.26E-03
NGC 5078	1	12.00	–	–	7.900	0.29E 43	0.13E 40	0.46E-03
NGC 5101	0	11.58	–	–	7.900	0.43E 43	<0.34E 40	<0.78E-03
NGC 5102	−3	9.86	0.58	0.19	7.900	0.37E 44	<0.45E 40	<0.12E-03
NGC 5236	5	7.85	–	–	7.900	0.13E 45	0.23E 41	0.17E-03
NGC 5253	0	10.37	0.29	−0.30	7.900	0.19E 44	<0.18E 40	<0.96E-04
NGC 5457	6	7.96	0.40	–	7.200	0.15E 45	0.14E 41	0.91E-04
NGC 5532	−2	12.64	0.90	0.52	142.120	0.16E 46	0.56E 42	0.36E-03
NGC 5566	2	10.73	0.71	0.29	34.580	0.38E 45	<0.51E 41	<0.13E-03
NGC 5866	−1	10.39	0.74	0.31	19.400	0.17E 45	0.14E 41	0.80E-04
NGC 5898	−5	12.09	0.95	0.51	42.560	0.26E 45	0.33E 41	0.13E-03
NGC 5907	5	10.08	0.56	0.33	18.400	0.16E 45	<0.22E 41	<0.14E-03
NGC 6744	4	8.30	–	–	10.380	0.15E 45	0.16E 41	0.11E-03
NGC 7793	8	9.36	0.51	−0.16	3.400	0.10E 44	0.16E 40	0.16E-03

and $U-V$ are the corrected, face on values of RC2, column 23, if available; otherwise the asymptotic values in columns 19 and 20 were used. The distance is taken from Sandage & Tammann (1975) if available, then from de Vaucouleurs (1979), and finally from the redshift given in RC2 using $H_0 = 50$. L_v was calculated using 2.48×10^{-5} erg s^{-1} for $m_{bol} = 0$; m_{bol} is taken as m_v plus a small correction term corresponding to main sequence stars whenever $B-V$ is available. The X-ray flux (0.5–3.0 keV) was calculated using a constant conversion coefficient of 2.56×10^{-11} erg cm^{-2} per count.

These data also are summarized in Figure 3.9. The galaxy colors are shown in Figure 3.9(a); the solid line, taken from Larson & Tinsley (1978), is representative of normal galaxies. The optical luminosity v. type distribution of the sample is shown in Figure 3.9(b). In general, the early galaxies contained in the sample are more luminous than the later ones. The plot of L_x v. L_v for these galaxies, together with 3σ upper limits for those galaxies not detected, is shown in Figure 3.9(c); this is the principal result of the survey. The total X-ray luminosity (0.5–3.0 keV) divided by the total visible luminosity is 1.9×10^{-4}. Although the survey spans approximately three decades in intrinsic luminosity and includes both early and late galaxies, in almost all cases, the ratio of X-ray to optical luminosity is included in the decade 4×10^{-5}–4×10^{-4}. This ratio is substantially independent of the galaxy type, at least for normal galaxies, as shown in Figure 3.9(d).

We have carried out correlation analyses of these galaxies with other measured properties: optical magnitude, color, radio flux, mass (of hydrogen and total mass). The determination of the significance of other correlations is hampered by the fact that there exists an overwhelming correspondence between X-ray flux and visual magnitude, or equivalently between X-ray luminosity and optical luminosity, as is shown in Figure 3.9(c).

There is a small correlation in this sample indicating that ellipticals may have a higher X-ray–optical luminosity ratio than spirals; it is not clear whether this effect, if it exists, is an intrinsic property of ellipticals or of large galaxies; in particular, the emission of large galaxies in clusters frequently is dominated by the hot diffuse gas trapped in the gravity potential well of the galaxy rather than either galactic-type compact sources or galactic nuclei. (Two cases, M86 and NGC 4472, where this was known to be true, were not included in the sample.) Fabbiano, Feigelson & Zamorani (1982) have analysed data from relatively short exposures of 27 galaxies selected on the basis of unusual morphology or colors, evidence for interactions, etc. Thirteen of these galaxies were detected. The ratios of X-ray to visible luminosity of these galaxies substantially overlap the results given here, although tending towards the high side of the distribution. Two galaxies in the Fabbiano set, NGC 3125, which has an emission-line nucleus, and

NGC 3991, which has radio emission comparable to Seyfert galaxies and is a member of an interacting system with NGC 3994 and NGC 3995, have L_x/L_v of order 2.5×10^{-3}. Similarly, M82 has a ratio of about 1.4×10^{-3}. These very peculiar galaxies have ratios of L_x/L_v which exceed those of the normal galaxies studied here by a decade or less. Almost all galaxies which do not have active nuclei, or emit by gravity trapped hot gas, appear to fall into a relatively narrow range of L_x/L_v regardless of the type or size of the galaxy.

3.4 Very bright non-nuclear sources

The Eddington limit for a solar-mass object is approximately 1.3×10^{38} erg s^{-1}, and for the most massive neutron stars allowed by current models approximately three times this, assuming pure hydrogen gas and spherical accretion. The X-ray sources in the Galaxy are, in general, consistent with these limits, the brightest disk sources, Cyg X-3 and Cir X-1, having X-ray luminosities of approximately 1×10^{38} and 2×10^{38} respectively (Bradt, Doxsey & Jernigan 1979), and the brightest bulge source, 1758−250 having an X-ray luminosity of 2.9×10^{38}. This also is true of the sources found in M31 and M33, excluding the nucleus (Figure 3.5). The most luminous objects found in M31 emit approximately 2.5×10^{38} erg s^{-1} in X-rays and, thus, are similar to objects found in our own Galaxy. Transient sources somewhat brighter than this have been observed (e.g., 1742−289, 6×10^{38}) in the galaxy. Clark et al. (1978) observed that the sources in the Magellanic Clouds are considerably brighter than typical galactic sources. LMC X-4 and SMC X-1, for example, have been observed as bright as 5 and 6×10^{38} erg s^{-1}, respectively, and the recurrent transient A 0538−66 has a peak luminosity of 1.5×10^{39} erg s^{-1} (0.5-20 keV) during outbursts (Skinner et al. 1980).

The observations of normal galaxies with the Einstein Observatory have yielded a number of additional very bright objects. The galaxy NGC 2403 includes a source, apparent at the west edge of the galaxy in Figure 3.7(b), which has a luminosity of 1.0×10^{39} erg s^{-1}, assuming a distance of 3.25 Mpc. The galaxy M101 includes three point sources brighter than 10^{39} erg s^{-1}; this is a particularly important case because the observation was repeated and variability demonstrated, so that we know that single point objects probably are responsible for the X-ray emission, and also because the probability of obtaining three serendipitous sources not associated with obvious foreground stars or background quasars in the area of one galaxy is remote. The three X-ray source luminosities obtained during the two observations, in units of 10^{39} erg s^{-1}, are: 1.3-2.4, 1.4-0.4, and 1.4-1.2. There also is an apparently extended feature having a total yield of 2.7×10^{39} erg s^{-1}, which probably is the sum of a number of sources. Two more dramatic examples include one in NGC 6744 with

$L_x = 6.2 \times 10^{39}$, and an object reported by Palumbo et al. (1981) in M100 which, if associated with that galaxy, would have $L_x = 1.0 \times 10^{40}$. Palumbo et al. chose to interpret this object as a background quasar because of its high luminosity for a galactic X-ray source, but the number of other objects now observed with significant fractions of this emission makes the assignment of this object to M100 more viable.

These sources all are found in the outer disks of spiral galaxies; the most beautiful case illustrated here is M101 (Figure 3.7(e) and (f)), in which the association of these very bright objects with the well-delineated spiral features is apparent. We do not know of any similar objects found in elliptical galaxies which are not associated with the nuclei or other active features (e.g., bright radio knots in M87, Cen A). This may, however, be a selection effect; there are fewer early galaxies in our sample, and most of these are located at greater distances than the spirals which have been observed. Such objects also would not have been detected in the bulges of the spiral galaxies; the emission would have been assigned incorrectly to the nucleus or to the aggregate of the bulge sources; future observations of these galaxies to detect significant variability or to resolve the individual bulge sources may produce examples of very bright objects in these Population II regions as well.

We also must note that there is a correlation between the brightness of these sources and the distances assigned to the galaxy. Such a correlation is expected for a magnitude limited sample, such as this one, but also would be found if (a) the objects actually are foreground or background objects, or (b) the distance scales are incorrect.

Proper estimation of the probability of chance associations requires a detailed study of the exposures used to detect these sources and the areas assigned to the outer parts of the galaxies studied. We have not completed such an analysis; our impression is that the incidence of these objects is significantly higher than the chance rate, at least in the case of M101. The question of the distance scale is, of course, both significant and difficult; if the Hubble constant is 100 rather than the 50 used here, then the brightest of these objects will be reduced to $1\text{-}2.5 \times 10^{39}$ erg s^{-1}, still considerably in excess of the Eddington limit for neutron stars. It also must be emphasized that these luminosities refer to a bandwidth much narrower than typical temperatures for bright X-ray sources, and the total emission must be several times that stated here – thus, partially offsetting the 'loss' due to a possible revision of the distance scale.

3.5 Summary

The increased sensitivity of the Einstein Observatory has permitted the X-ray detection of more than 50 normal galaxies. For the nearest galaxies, studies comparable to the Uhuru survey of the Galaxy were performed. In spiral galaxies, source populations have been identified with Population I features, such as H II regions and spiral arms; other sources, at least in M31, are clearly Population II objects. Some preliminary conclusions can be drawn:

(1) Despite various morphological differences, the X-ray luminosity of normal galaxies is proportional to their optical luminosity, revealing no strong dependence on galaxy type. Other correlations exist, but at present they can either be attributed to the more significant correlation with L_{opt} or to selection effects in the sample. This ratio can be used to place limits on the fraction of the X-ray background which arises from emission from normal galaxies. The integrated brightness of extragalactic background light is $1 \pm 1.2\,S_{10}$ at 5100 Å (Dube, Wickes & Wilkinson 1977). If all of this flux were made up by emission from normal stars in galaxies, and if galaxies have X-ray luminosities approximately 1.9×10^{-4} of their optical luminosity, then X-ray emission from normal galaxies would comprise 6 ± 7 per cent of the diffuse background [$2.6 \times 10^{-8}\,\mathrm{erg\,cm^{-2}\,s^{-1}\,sr^{-1}}$ (0.5–3.0 keV)]. This limit is comparable to upper limits calculated from source population studies (van Paradijs 1978). It does not rule out the possibility that X-ray emission from normal galaxies at earlier epochs is greater if the star formation rate or initial mass function was different. The correlation between X-ray luminosity and optical luminosity in its current form does not indicate significant differences in the ratio of X-ray and visible luminosities for Population I and Population II systems, for similar ratios are obtained in different galaxy types having differing percentages of Population I and Population II features. Fabbiano, Feigelson & Zamorani (1982) have argued that higher X-ray to optical luminosities in a sample of blue galaxies is directly traceable to higher star formation rates there. Although this may be correct, it is not yet proven in our opinion because the X-ray–optical luminosity ratios of nearby normal galaxies, which are otherwise similar, range over an order of magnitude.

(2) In those galaxies where a substantial number of individual sources can be detected with sensitivity comparable to surveys of the Galaxy, the number of sources which have been detected are comparable to the number of sources one would expect on mass considerations. In the 2–11 keV range, 26 sources exist in our Galaxy with luminosities exceeding $5 \times 10^{36}\,\mathrm{erg\,s^{-1}}$, the completeness limit of galactic surveys (Clark *et al.* 1978). Twenty of these have luminosities in excess of $10^{37}\,\mathrm{erg\,s^{-1}}$. In M31 there are 80 point sources with luminosities in excess of $10^{37}\,\mathrm{erg\,s^{-1}}$, and in M33 there are nine. While caution should be exercised in comparing results in different energy windows, the relative number of sources in M33, the Galaxy and M31 appear to be approximately consistent with the ratios of the masses of the galaxies $3.9 \times 10^{10}\,M_\odot$, $1.3 \times 10^{11}\,M_\odot$ and $3.1 \times 10^{11}\,M_\odot$ (van den Bergh 1975).

(3) There are substantial numbers of X-ray sources in the Magellanic Clouds with luminosities in the range 10^{35}-10^{36} erg s^{-1}, lower than most X-ray binaries but higher than known uncollapsed stellar systems. The only known classes of objects which have X-ray emission in this luminosity range are SNR, and a limited number of low-luminosity binary systems exemplified by 4U 2129+47 (Thorstensen et al. 1979).

(4) About seven X-ray sources with luminosities of at least 10^{39} erg s^{-1} in the 0.5-3.0 keV band have been found in the arms of nearby spiral galaxies. These sources maintain average luminosities in excess of the Eddington limit for neutron stars at least for the 5000 s typically required for these observations.

(5) The X-ray properties of the nuclei of normal galaxies are not presently predictable on the basis of their optical and radio properties. The wide, optical emission lines observed in M81, together with its X-ray emission and the high ratio of X-ray to visible luminosity observed in M100, suggest that the Seyfert phenomena may exist at low levels in otherwise normal galaxies.

Acknowledgements

We wish to thank W. Ku, G. Fabbiano, and J. Schwarz for access to unpublished results and useful discussions. We also thank D. Schwartz and D. Helfand for a critical reading of the manuscript. This work was supported at Columbia by NASA contract NAS 8-30753 and at SAO by Smithsonian funds and NASA contract NAS 8-30751.

References

Baade, W. & Arp, H. 1964, *Ap. J.*, **139**, 1027.
Bahcall, J. & Ostriker, J. 1975, *Nature*, **256**, 23.
Bahcall, N. A. & Hausman, M. A. 1976, *Ap. J. (Lett.)*, **207**, L181.
Battistini, P., Bonoli, F., Braccesi, A., Fusi Pecci, F., Malagnini, M. L. & Marano, B. 1980, *Astron. Ap. Suppl.*, **42**, 357.
Beck, S. C., Lacy, J. H. & Geballe, T. R. 1979, *Ap. J.*, **231**, 28.
Berkhuijsen, E. M. 1977, *Astron. Ap.*, **57**, 9.
Blumenthal, G. R. & Tucker, W. H. 1974, *Ann. Rev. Astron. Ap.*, **12**, 1.
Bradt, H. V., Doxsey, R. E. & Jernigan, J. G. 1979, *In: X-ray Astronomy*, eds. W. A. Baity & L. E. Peterson (Oxford: Pergamon).
Clark, G. W. 1975, *Ap. J. (Lett.)*, **199**, L143.
Clark, G., Doxsey, R., Li, F., Jernigan, F. G. & van Paradijs, J. 1978, *Ap. J. (Lett.)*, **221**, L37.
Coombes, F., Encrenaz, P. J., Lucas, R. & Weliachew, L. 1977a, *Astron. Ap.*, **55**, 311.
Coombes, F., Encrenaz, P. J., Lucas, R. & Weliachew, L. 1977b, *Astron. Ap.*, **61**, L7.
de Vaucouleurs, G., de Vaucouleurs, A. & Corwin, H. G. 1976, *Second Reference Catalogue of Bright Galaxies* (Austin: University of Texas Press) (RC2).
de Vaucouleurs, G. 1979, *Ap. J.*, **227**, 729.
D'Odorico, S., Dopita, M. A. & Benvenuti, P. 1979, *Astron. Ap. Suppl.*, **40**, 67.
Dube, R. R., Wickes, W. C. & Wilkinson, D. T. 1977, *Ap. J. (Lett.)*, **215**, L51.
Elvis, M. & van Speybroeck, L. 1982, *Ap. J. (Lett.)*, **257**, L51.

Emerson, D. T. 1974, *MNRAS*, **169**, 607.
Emerson, D. T. 1976, *MNRAS*, **176**, 321.
Emerson, D. T. 1978, *Astron. Ap.*, **63**, L29.
Fabbiano, G., Feigelson, E. & Zamorani, G. 1982, *Ap. J.*, **256**, 397.
Faber, S. M. & Jackson, R. E. 1976, *Ap. J.*, **204**, 668.
Fomalont, E. B. 1968, *Ap. J. Suppl.*, **15**, 203.
Gallagher, J. S. 1980, private communication.
Giacconi, R. *et al.* 1979, *Ap. J.*, **230**, 540.
Gottesman, S. T., Lucas, R., Weliachew, L. & Wright, M. C. H. 1976, *Ap. J.*, **204**, 699.
Grindlay, J. E. 1977, *Highlights of Astron.*, **4**, 111.
Grindlay, J. E. 1978, *Ap. J.*, **221**, 234.
Grindlay, J. *Proceedings of the High Energy Astrophysics Division of the AAS Meeting on X-ray Astronomy*, Cambridge, ed. R. Goacconi (Dordrecht: Reidel), 1981, Vol. 87, p. 79.
Harris, W. E. & Racine, R. 1979, *Ann. Rev. Astron. Ap.*, **17**, 241.
Jones, C. 1977, *Ap. J.*, **214**, 856.
Joss, P. C. 1978, *Ap. J. (Lett.)*, **225**, L123.
Ku, W. H. M. & Chanan, G. A. 1979, *Ap. J. (Lett.)*, **234**, L59.
Larson, R. B. & Tinsley, B. M. 1978, *Ap. J.*, **219**, 46.
Lewin, W. H. G. & Clark, G. W. 1980, *Ann. N.Y. Acad. Sci.*, **336**, 451.
Lewin, W. H. G. & Joss, P. C. 1981, *Space Sci. Rev.*, **28**, 3.
Long, K. S., Fabbiano, G., Ku, W. H. M., Schwarz, J. & van Speybroeck, L. 1983, paper in preparation.
Long, K. S., Helfand, D. J. & Grabelsky, D. A. 1981a, *Ap. J.*, **248**, 925.
Long, K. S., D'Odorico, S., Charles, P. A. & Dopita, M. A. 1981b, *Ap. J. (Lett.)*, **246**, L61.
Margon, B., Bowyer, S., Cruddace, R., Heiles, C., Lampton, M. & Troland, T. 1974, *Ap. J. (Lett.)*, **191**, L117.
Markert, T. H., Canizares, C. R., Clark, G. W., Hearn, D. R., Li, F. K., Sproti, F. & Winkler, P. F. 1977, *Ap. J.*, **218**, 801.
Markert, T. H., Kriss, G. A., Canizares, C. R., McClintock, J. E. & Winkler, P. F. 1979, *Bull. AAS*, **11**, 633.
Market, T. H. 1980, private communication.
McGee, R. X. & Milton, J. A. 1966, *Australian J. Phys.*, **19**, 343.
Morton, D. C., Andereck, C. D. & Bernard, D. A. 1977, *Ap. J.*, **212**, 13.
Oort, J. H. 1977, *Ann. Rev. Astron. Ap.*, **15**, 295.
Palumbo, G. G. C., Maccacaro, T., Panagia, N. G., Vettolani, G. & Zamorani, G. 1981, *Ap. J.*, **247**, 484.
Peimbert, M. & Torres-Peimbert, S. 1981, *Ap. J.*, **245**, 845.
Press, W. H. & Teukolsky, S. A. 1977, *Ap. J.*, **213**, 183.
Rieke, G. H. & Low, F. J. 1975, *Ap. J.*, **197**, 17.
Rieke, G. H., Lebofsky, M. J., Thompson, R. I., Low, F. J. & Tokunaga, A. T. 1980, *Ap. J.*, **238**, 24.
Rubin, V. & Ford, W. K., Jr. 1971, *Ap. J.*, **170**, 25.
Sandage, A. & Tammann, G. A. 1975, *Ap. J.*, **196**, 313.
Sargent, W. L. W., Schechter, A., Boksenberg, A. & Shortridge, K. 1977, *Ap. J.*, **212**, 326.
Seaquist, E. R. & Bell, M. B. 1977, *A and A*, **60**, L1.
Seward, F. D., Forman, W. R., Giacconi, R., Griffiths, R. E., Harnden, F. R., Jr, Jones, C. & Pye, J. P. 1979, *Ap. J. (Lett.)*, **234**, L55.
Seward, F. D. & Mitchell, M. 1981, *Ap. J.*, **243**, 736.
Skinner, G. K., Shulman, S., Share, G., Evans, W. D., McNutt, D., Meekins, J., Smathers, H., Wood, K., Yentis, D., Byram, E. T., Chubb, T. A. & Friedman, H. 1980, *Ap. J.*, **240**, 619.
Thorstensen, J., Charles, P., Bowyer, S., Briel, U. G., Doxsey, R. E., Griffiths, R. E. & Schwartz, D. A. 1979, *Ap. J. (Lett.)*, **233**, L57.

Ulrich, Marie-Helene 1978, *Ap. J.*, **219**, 414.
van den Bergh, S. 1975, *Ann. Rev. AA*, **13**, 217.
van Paradijs, J. 1978, *Ap. J.*, **226**, 586.
Van Speybroeck, L., Epstein, A., Forman, W., Giacconi, R., Jones, C., Liller, W. & Smarr, L. 1979, *Ap. J. (Lett.)*, **234**, L45.
Van Speybroeck, L. & Bechtold, J. *Proceedings of the High Energy Astrophysics Division of the AAS Meeting on X-ray Astronomy*, Cambridge, ed. R. Giacconi (Dordrecht: Reidel), 1980.
Van Speybroeck *et al.* 1983, article in preparation.
Walker, M. F. 1964, *Astron. J.*, **69**, 744.
Watson, M. G., Willingale, R., Grindlay, J. E. & Hertz, P. 1981, *Ap. J.*, **250**, 142.

4

ACCRETING DEGENERATE DWARFS IN CLOSE BINARY SYSTEMS

France A. Córdova
University of California, Los Alamos National Laboratory

Keith O. Mason
*University College London
Mullard Space Science Laboratory*

4.1 Introduction

Interacting binary systems thought to contain accreting white dwarfs include classical novae, recurrent novae, dwarf novae, novalike objects, the magnetic variables, and possibly also some symbiotic and Mira variables. The designation cataclysmic variable can be reasonably extended to include all of these objects on the basis of their outburst behavior or their spectral and photometric similarities to the outbursting types. In this review, however, we discuss only those systems with short orbital periods ($P_{orb} < 1$ d) for which there is reasonable evidence that the accreting component is a degenerate dwarf. This excludes the symbiotic and Mira variables which have orbital periods of hundreds of days or more and which contain a blue star of low luminosity whose exact nature is in doubt [24, 182]. Close binaries containing an accreting neutron star rather than a white dwarf are discussed in Chapter 5 of this book.

Cataclysmic variables exhibit some or all of the following characteristics: a flat or blue optical spectral distribution, broad emission or absorption lines of hydrogen and helium, rapid variability, marked aperiodic changes in optical brightness, and low-luminosity X-ray emission ($L < 10^{34}$ erg s^{-1}). Most of the known cataclysmic variables have been catalogued in the optical region of the spectrum, but a few have been detected first at X-ray energies.

Crawford & Kraft [52] pointed out that the important properties of the cataclysmic variables might be largely determined by the transfer of material from the companion star to the compact component of the binary. Warner & Nather [276] and Smak [212] elucidated a model for this transfer in which the companion star fills its Roche-limiting surface and transfers material through the inner Lagrangian point into an accretion disk surrounding the central star. At the point of intersection of the mass stream and the disk a shock is formed. This produces a 'bright spot' on the disk that manifests itself as a prominent hump in the optical light curve of those systems (e.g. many dwarf novae during their quiescent states) in which the emission from the rest of the disk is weak. Often, however, the accretion luminosity of the disk is the principal component of the optical and ultraviolet spectrum. The companion star may contribute significantly to the luminosity of the system in the red and infrared, while the degenerate, accreting star and the interface between the inner disk and this star may contribute to the far ultraviolet luminosity. This standard model for cataclysmic variables must now be modified to accommodate the recent discovery of relatively strong magnetic fields in some systems. The presence of these fields may alter the geometry of the accretion flow and in some cases may even prevent a disk from forming.

The optical observations and theory of cataclysmic variables have been reviewed most recently by Robinson [195] and Warner [270]. An overview of white dwarfs in general and a historical perspective on cataclysmic systems are given by Liebert [122] and Payne-Gaposchkin [182] respectively.

In the present article, we shall concentrate on the advances made in the past few years. Cataclysmic variables have now been observed at all wavelengths from the infrared to the far ultraviolet (10μ–1100 Å), and in the X-ray band from 100 eV to >20 keV. The data spanning this large spectral range have made it possible to model the continuum and line-emitting regions, determine as a function of energy the relative contributions from the various components of the systems, and gauge how these are affected by the outbursts of the stars. High speed photometric observations have been extended into the X-ray band and, together with optical measurements, have provided insights into the source of the pulsations and other rapid variability found in cataclysmic binaries. The ultraviolet and X-ray observations confirm that most of the accretion luminosity is emitted at energies above the optical band. Infrared and optical spectroscopy and photometry have allowed a more accurate determination of the nature of the companion star and the geometry of the binary systems, while optical polarimetry has demonstrated the importance of magnetic fields in governing the accretion flow in some systems. These data have greatly increased the usefulness of cataclysmic binaries as a laboratory for the study of the mass accretion process and the physics of compact stars.

4.2   Classification of cataclysmic binaries

The various subtypes of novae are classified on the basis of their outburst properties. The classical (or ex-) novae have the longest recurrence times; by definition they have been seen to erupt only once. The energy emitted in a nova outburst exceeds 10^{44} erg. In contrast the dwarf novae have outbursts lasting a few days that recur at intervals of weeks or months, and the energy liberated in such an outbursts is $\sim 10^{38}$-10^{39} erg. Stars with outbursts separated by more than about 10 yr are generally classified as recurrent novae. Many of the well-studied members of this class contain giant companions which dominate the spectrum during quiescent states. Thus, even though the outburst amplitudes of recurrent novae (4-9 mag) are somewhat less than those of classical novae (8-15 mag), the bolometric luminosity of the recurrent novae and classical novae eruptions are similar. The distinction between the various types of nova based on their outburst morphology is blurred by the presence of a few straddlers: among them, the classical nova GK Per which has displayed recurrent outbursts of amplitude 2-4 mag on a timescale of a few years, and WZ Sge which has recurrent outbursts about every 30 yr but whose photometric and spectroscopic behavior is otherwise similar to the dwarf novae [270].

Stars that have not been observed to erupt but which otherwise have the optical characteristics of the novae are termed novalike objects. This category probably contains a number of historical novae whose outbursts were not recorded, and also some stars in permanent or extended outbursts of the kind experienced by dwarf novae [280]. The magnetic variables (or AM Her stars) were also originally classified as novalike or irregular variables [114] but now form a distinct subclass of cataclysmic binary. Some of the magnetic stars and also some novalike objects (e.g., TT Ari and MV Lyr) exhibit high and low brightness states lasting for months or years.

The various subclasses of cataclysmic variable have been subdivided still further on the basis of their detailed outburst behavior. Ex-novae are generally classified as either fast or slow depending on the characteristic decay time of their outbursts [180]. Dwarf novae are subtyped as U Gem, SU UMa, or Z Cam variables. Each of these subtypes has two, morphologically different, kinds of outbursts. The U Gem stars often alternate long and short outbursts, but the spread in the total energy of these eruptions is less than a factor of 4. The SU UMa stars, in contrast, exhibit 'supermaxima' that contain $\gtrsim 8$ times the energy of their normal outburst [255]. The superoutbursts occur much less frequently than the normal eruptions, but with a higher degree of regularity. It is not known whether the two types of outburst have the same origin. In a few stars superoutbursts sometimes begin during a normal outburst [11, 12, 164]. In the third category of dwarf novae, the Z Cam stars, the most common type of eruption lasts for only a few days. Yet occasionally a Z Cam star will fail to return to its quiescent light level after an outburst; instead it will remain in a 'standstill' condition at an intermediate brightness, sometimes for years.

From the evidence collected by Warner [270] it appears that the time-averaged energy emitted by a given dwarf nova may be constant, regardless of the mode of outburst behavior. This is further supported by the ΔE-Δt relationship which appeared in a first form as a period–amplitude relationship [115] and was reinvestigated for the dwarf nova SS Cygni by Brecher, Morrison & Sadun [26]. This relationship shows a 'relaxation oscillator' behavior in which the energy of an outburst ΔE is proportional to the time Δt to the following outburst (see also Chapter 2, Section 2.4.3.3 of this book). The time required for a given dwarf nova to decline to its quiescent level after an outburst appears to be relatively constant. Bailey [3] has shown that this decay time is related to the orbital period of the system. Mattei [136] has fit a linear relation to data on 15 dwarf novae and finds $T_{\text{decline}} = 8.9\, P_{\text{orb}} + 0.3$, where the quantities are expressed in days. Such a relationship is expected in a model where the decay reflects the time it takes to empty a disk whose size is limited by the Roche lobe of the accreting star.

4.3 Dynamical properties

The properties of cataclysmic variables with known orbital periods are summarized in Table 4.1. Included are X-ray sources suspected of being cataclysmic variables and a few objects that are probably related. Fast, linear, electronic detectors now allow spectroscopic measurements with a time resolution of a few minutes and wavelength resolution of order 1 Å, increasing the accuracy with which orbital elements can be determined (e.g. [296, 30, 232, 263]. The spectral signature of the mass-donating star can usually be discerned in the visual range for systems with orbital periods >6 h, consistent with the expected period–luminosity relationship for main sequence stars [270]. Observations in the red and near-infrared are required to obtain information on cooler companions in shorter period orbits. In eclipsing systems the contribution of the secondary can be differentiated by observing the eclipse of the disk by this star [124, 8]. The spectral type of the secondary can then be determined from its spectral distribution. The mass, size and luminosity of the secondary has been estimated from the mass–radius and spectral-type–luminosity relationships for main sequence stars. There are indications, however, that many of the systems that have been studied contain secondaries whose radii are larger than that of a main sequence dwarf with the same mass (e.g. U Gem [261]; AE Aqr [167]; BV Cen [15]; DQ Her [215]; Stepanyan's Star [298]; and possibly Z Cha [8]). In addition, Wade [261, 263] finds that the red star in U Gem is overluminous for its spectral type. There are several possible reasons for these empirical results. First, the mass transfer process can modify the effective temperature of the secondary star [289, 262]. Secondly, synchronous rotation of the secondary will lessen the hydrostatic pressure in its core, thus reducing its nuclear luminosity [162]. A third possibility is evolution of the secondary off the main sequence. Systems with periods greater than about 8 h are expected to have evolved secondaries, but nuclear evolution may also be important in shorter period systems if the secondary was already somewhat evolved at the onset of mass transfer [289]. For these reasons masses and luminosities which are calculated under the assumption of a particular mass-radius relationship must be regarded as unreliable [263, 187].

The measured orbital periods of cataclysmic variables range from 17 min to $16\frac{1}{2}$ h. There are several striking features in the period distribution. One is the absence of systems with periods between 2 and 3 h. Whyte & Eggleton [289] conclude that the period gap is probably not due to chance. The reason for the gap is not understood although a number of explanations have been put forward. One possibility, based on evolutionary arguments, is that there really are no cataclysmic binaries with such orbital periods. Alternatively, the binaries may evolve through tnis period range very rapidly or in a detached or inactive state [289, 187, 199].

Excluding AM CVn and G61-29, in which both the primary and secondary stars are thought to be degenerate [278, 147], there appears to be a minimum orbital period of 80 min among the cataclysmic binaries. This has been interpreted as marking the point in the evolution of the companion when its mass falls below the value at which nuclear burning can take place, $M = 0.085\,M_\odot$ [157, 187]. Up to this point the binary period has been decreasing due to gravitational radiation losses. When hydrogen burning ceases, however, the binary period begins to increase because of mass transfer driven by the inverse mass-radius relationship of the now degenerate secondary. This scenario is only valid for stars which are initially fully convective and which have periods $\lesssim 2$ h at the onset of mass transfer. Clearly, a different evolutionary path must be invoked to explain AM CVn and G61-29.

There is some evidence from the period distribution that there may be a relationship between the orbital parameters and the subtype of cataclysmic system. For instance, the classical novae, with the exception of GK Per, appear to have periods exclusively between 3 and 5 h. The SU UMa stars, a subtype of dwarf nova, generally have periods less than 2 h, while most of the other dwarf novae have periods >3 h. Current theoretical studies suggest that the outburst properties of these stars could be related to their evolutionary state, so that correlations between subclass and orbital period might be expected in certain cases. Scenarios for evolution through various subclasses have been outlined by Whyte & Eggleton [289].

152

TABLE 1: ORBITAL PARAMETERS

Star	Subclass	V (mag.)	P orbital (hours)	K_1 (km s^{-1})	K_2 (km s^{-1})	i (degrees)	Secondary (Spectral Type)	D (pc)	Ref.
GK Per	cN	10.2-14.0	16.43:	34+12	140+11		K2V-IVp	480	(155)(139)(63)(21)
BV Cen	dN	10.5-13.2	14.63	156-6	154+9	61+5	G5-G8IV-V	450	(15)(257)
V1668 Cyg (=Nova Cyg 78)	cN		10.56:					3.6 Kpc	(225)(28)
V Sge	?		12.34	320:	85:			2.75 Kpc:	(77)
AE Aqr	N1	9.5-13.9	9.88	129+7	163+6	58+ 4	K2-K5IV-V	150;70	(181)(167)(247)(33)
RU Peg	dN	9.8-11.5	8.90	94+3	121+2		G8IVn		(108)(230)
BT Mon	cN	9.0-12.9	8.01						(197)
Lanning 10	?	16	7.71	240+17			F5-G0	1 Kpc	(241)(87)
AC Cnc	dN?	~14.5	7.2						(116)
EM Cyg	dN	13.8-16.3	6.98	170+10;188+7	135+3	~65	K2V	320	(97)(231)
Z Cam	dN	11.9-14	6.96	137	193	54	K7		(193)(103)
SS Cyg	dN	10.2-14.5	6.60	90+2;118+8	153+2;120+6	37+3	K5V	125+25	(232)(50)(102)
RW Sex	N1	8.1-12.0	5.93:	74+5					(48)
2A1822-371	n.s.?	10.6	5.57	70		>70	>600	1-5 Kpc	(1332)(286)(130)(288)
RW Tri	N1	15.4-16.4	5.57	<150		82	M0V	180	(264)(23)(214)(58)
2A0526-328 (=TV Col)	?	12.6-13.4	5.49(5.18)	116+8:					(145)(94)
4U2129+47	n.s.?	13.5	5.25			>70		1-2 Kpc:	(248)(286)
1E0643-1648	dN	17-18.5	~5.2						(93)
RX And	dN	10-13	5.08	78	410				(108)
V3885 Sgr	N1	10.3-13.5	4.94:	159+21					(49)
T Aur	cN	10.4	4.91	154+34		60		830	(139)(265)(20)
UX UMa	N1	14.8	4.72			~70	M4V	214	(266)(183)(60)
DQ Her	cN	13.8	4.65	136+5		>77;~90	M3V	420	(183)(92)(55)(296)(215)(297)
BD Pav	cN?	14.6	4.30						(10)
SS Aur	dN	12.4->16	4.33	~85	450:				(110)
U Gem	dN	10.5-14.8	4.25	104;137+8	283+15	67+8	M5	76+30	(213)(261)(263)(229)
HR Del	cN	8.8-14.5	4.08:		98:	42+5		810	(90)(284)
H2215-086	N1?	12.4	4.0						(173)
WW Cet	dN	13.5	3.83	120					(110)
LX Ser	?	12.2-15.5	3.80	162:		75		500	(298)(85)
(=Stepanyan's Star)									
H2252-035	MV?	~14.0	3.59						(172)
		~13.3							(284)(252)

Star	Subclass	Mag range	Period	i (°)	Spectrum	T (K)	Refs
V1500 Cyg	cN	>21	3.35			430	(284)(109)
TT Ari	N1	9.3-14.5	3.30:	65+5		1400	(284)(165)
MV Lyr	N1	12-18	3.21	25−4			(51)(217)
AM Her	MV	12.8-14.5	3.09	<140	M5V	320	(295)(297)
TU Men	dN	11-17	2.8		M4V	75±10	(228)
YZ Cnc	dN	10.2-14.6	2.21(s)	12:			(164)
AN UMa	MV	14-16.5	1.91	40±20			(203)
PG 1550+191	MV	15.0-15.8	1.90				(123)
E0139-681	MV	14.9-16.4	1.83				(1)(254)
WX Hyi	dN	11.6-14.2	1.80	40+10:			(210)
AY Lyr	dN	12.6-18.3	1.81(s)	67+6			(164)(89)
Z Cha	dN	11.9-15.3	1.79	87+14	late MV	125±20	(192)(8)(256)
VW Hyi	dN	8.5-14.0	1.78	78+14	79+2		(210)
HT Cas	dN	12.6-16.9	1.77	115+6	60+10		(171)(299)
E1013-477	MV?	~17	1.72		~78		(135)
E1405-451	MV	~15.5	1.69				(134)
VV Pup	MV	14-17	1.67				(203)
EX Hya	dN	10.0-13.5	1.64	68+9	66:	76-190	(27)(211)
EK TrA	dN	12.0-14.5	1.56(s)		>70		(258)
OY Car	dN	11.7-16	1.51	~125;110±10	78+3	150:	(191)(5)(259)
V436 Cen	dN	11.9-15.5	1.50	118±10			(65)(209)
V2051 Oph	dN	16.5	1.50		2.8 mag. eclipse		(22)
WZ Sge	dN	7.6-15.5	1.36	<38	75-85		(112)
2A0311-227 (=EF Eri)	MV	15.5	1.35		M4 or later		(204)(272)
GP Com (=G61-29)	N1;td	15.8	0.78	14.6+3.7:			(147)
AM CVn (=HZ29)	N1;td	14.2	0.29				(278)

Footnotes

(1) Subclass: cN = classical nova; rN = recurrent nova; dN = dwarf nova; N1 = nova-like object; MV = magnetic variable; td = twin-degenerate; n.s. = neutron star as accreting object.

(2) A colon indicates an uncertain value.

(3) "s" in the period column indicates a superhump period; no orbital period has been determined for these cases.

4.4 X-ray emission

X-ray emission is now an established property of cataclysmic variables. The measurements made to date are summarized in Table 4.2. At least two distinct kinds of X-ray component have been detected in these stars. The most common has a relatively hard spectral distribution with a characteristic temperature of order 10 keV; ultra-soft X-ray-EUV emission has also been detected from a few systems.

4.4.1 Hard X-rays

The hard X-ray component has been detected at a sensitivity level of $\sim 10^{29} (d/100 \text{ pc})^2 \text{ erg s}^{-1}$ from about 70 per cent of the more than 50 cataclysmic variables that have been observed in the energy range 0.1-4.5 keV with the imaging detectors on the Einstein Observatory [43, 18, 17, 42]. Less sensitive all-sky surveys have shown, however, that few cataclysmic variables emit at a level above $5 \times 10^{31} (d/100 \text{ pc})^2 \text{ erg s}^{-1}$ [281, 41].

Hard X-ray emission has been detected from stars representative of each of the five main subclasses of cataclysmic variable (dwarf nova, classical nova, recurrent nova, novalike object and magnetic variable). This emission, in many cases, is variable by at least a factor of 2 on timescales greater than a minute [42, 17, 99]. The X-ray flickering is discussed further in Section 4.7. There are indications that the absolute X-ray luminosity and the fraction of the total luminosity emitted in the X-ray band may be a function of a star's subtype. Becker & Marshall [18] find a relationship between the present X-ray luminosity of classical novae and the rate of decline of the nova outburst such that the fast novae have a higher X-ray luminosity. The inclusion of additional data shows, however, that no simple linear relationship exists [308] and further work is necessary to establish the reality of the suggested correlation. Córdova & Mason [42] note that X-rays are detected from six highly inclined dwarf novae during quiescence (HT Cas, U Gem, Z Cha, EX Hya and WZ Sge) but not from three comparatively nearby classical novae and novalike objects (DQ Her, T Aur and UX UMa) with a similar range of inclinations. This may reflect differences in the structure of the innermost accretion disk region among the subclasses. Córdova & Fenimore [42] also find that the average ratio of the hard X-ray to visual band flux is higher for dwarf novae in quiescence than dwarf novae during outburst. Typical values for F_x (0.1-4 keV)/F_v (5000-6000 Å) are ~ 1 and $\lesssim 0.06$ respectively, when this X-ray flux is calculated for the hard component only. The classical novae, recurrent novae and novalike objects have F_x/F_v ratios that are less, on average, than for dwarf novae during quiescence, but F_x/F_v can be as high as ~ 10 for the magnetic variables (e.g. AM Her and EF Eri).

The difference in F_x/F_v between dwarf novae in outburst and in quiescence might come about because the large increase in the optical brightness of these stars during outburst is accompanied by a reduction or, in some cases, only a moderate increase in the hard X-ray flux. SS Cyg and U Gem are the only dwarf novae, however, whose hard X-ray emission has been monitored extensively during optical outburst. The hard X-ray flux of SS Cyg is found to increase initially at the beginning of outburst by a factor of 2-3 and then to decrease below the quiescent level [190, 234]. Another short-lived hard X-ray flare is seen at the end of the outburst. The behavior of U Gem is different; its hard X-ray flux increases by a factor of 5-10 near the beginning of the outburst and probably remains high throughout the eruption [236, 43, 131].

The hard X-ray spectrum of SS Cyg has been measured at various stages in its outburst cycle by Swank [233]. Two spectral components are required to fit the data. The cooler can be represented by a thermal bremsstrahlung model with a temperature of ~ 0.6 keV and does not change in temperature or strength during optical outburst. The second, hotter emission spectrum has a characteristic temperature of ~ 8 keV during optical quiescence, and ~ 5 keV during the outburst. During the flare at the end of the outburst the temperature rises to ~ 20 keV. A two-component, hard X-ray spectrum has also been found in EX Hya during quiescence (temperatures of 0.6 keV and ~ 5 keV; [233; see also [111]).

Some of the best data on the hard X-ray spectrum of a cataclysmic variable has been obtained on the magnetic variable AM Her which has been measured from 2 to 150 keV [200]. Over this large energy range a single component bremsstrahlung distribution does not adequately fit the data, but a model in which part of the X-ray flux is reflected back from the surface of the white dwarf, with a higher effective absorption than the radiation seen directly, yields an acceptable fit with a temperature of ~ 30 keV. Similar departures from a simple, optically thin spectrum could be produced by Compton scattering in the accretion column above the X-ray source [95].

Iron line emission is observed at ~ 6.7 keV in the spectrum of both SS Cyg and EX Hya [127, 234, 233]. Iron line emission has also been seen in the X-ray spectra of the magnetic variables AM Her and EF Eri [200, 285].

4.4.2 Soft X-rays

A second type of X-ray emission has been observed from a few cataclysmic binaries that include the magnetic variables AM Her, AN UMa, EF Eri (2A 0311−227), and VV Pup [75, 74, 177, 179] and two dwarf novae, SS Cyg and U Gem [188, 126, 129, 44, 46]. The characteristic temperature of this emission is only a few tens of eV, and as such it is only barely detectable above

TABLE 2: X-Ray Observations

(q = quiescence)
(o = outburst)
(s = standstill)

Star	Subclass	Energy Band (keV)	kT (keV)	X-Ray Flux** (erg cm⁻²s⁻¹)	D (pc)	Lx (erg s⁻¹)	V⁺⁺ (mag)	Refs.
RX And	dn(q)	0.1-4	[10]	7.3 (-12)	130	1.3 (31)	13.6	(17)
AE Aqr	N1	0.1-4	[10]	7.0 (-12)	430+	1.6 (32)	11.8	(168)
V603 Aql	cN	0.1-4	[10]	7.5 (-12)			11.3	(18)
V794 Aql	?	0.1-4	[10]	5.1 (-12)			14.5	(243)
TT Ari	N1	0.1-4	[10]	0.8-2.7 (-11)			~10.5	(43)(42)
T Aur	cN	0.1-4	[10]	<1.8 (-13)	800+	<1.3 (31)	14.8	(18)
Z Cam	dN(s)	0.1-4	[10]	1.6 (-12)			11.5	(17)
SY Cnc	dN(o)	0.1-4	[10]	<3.8 (-13)			12.5	(42)
YZ Cnc	dN(q?)	0.1-4	[10]	1.4 (-12)			[14.0]	(43)
AC Cnc	dN?	0.1-0.5	[10]	<0.6-6.2 (-11)				(151)
SV CMi	dN(q?)	0.1-4	[10]	<2.7 (-13)			15.3]	(42)
AM CVn	N1;td	0.1-4	[10]	<2.2 (-13)			~14.4	(17)
HT Cas	dN(q)	0.1-4	[10]	2.2 (-12)			[16.8]	(42)
BV Cen	dN(q)	0.1-4	[10]	3.8 (-12)	450	8.7 (31)	13.1	(42)
V436 Cen	dN(q)	0.1-4	[10]	6.2 (-12)			15.1	(42)
WW Cet	dN(q?)	0.1-4	[10]	7.6 (-12)			[13.6]	(17)
Z Cha	dN(o)	0.1-4	[10]	6.5 (-13)	125	1.2 (30)	13.3	(17)
"	"(q)	0.1-4	[10]	1.4 (-12)	"	2.5 (30)	15.1	(17)
2A0526-328	?	2-6	*	3 (-11)			[13.5]	(206)
G61-29	N1;td	0.1-4	[10]	4.3 (-12)			[15.8]	(170)
T CrB	rN	0.1-4	[10]	2.2 (-13)	1500	5.6 (31)	10.0	(43)
EM Cyg	dN(q)	0.1-4	[10]	2.9 (-12)	320	3.4 (31)	13.2	(17)
SS Cyg	dN(q)	2-20	~20	1.6(-10)	150	4.1 (32)	11.9	(127)
"	"(o,rise)	2-18	[20]	4.8 (-10)	"	1.2 (33)	~10.0	(190)
"	"(o,rise)	2-25	[20]	3.2 (-10)	"	8.1 (32)	~8.5	(234)
"	"(o,peak)	2-25	~7	4 (-11)	"	1.0 (32)	~8.5	(234)
"	"(o,peak)	0.15-0.5	[20]	3.2 (-10)	"	8.1 (32)	11.0	(234)
"	"(o,peak)	0.4-4	0.03	4.5 (-11)	"	1.2 (32)	~8.5	(44)
"	"(q)	0.4-4	0.56	7.1 (-12)	"	1.8 (31)	~12	(233)
"	"(o,peak)	0.4-4	~8	4.7 (-11)	"	1.2 (32)	~12	(233)
"	"(o,peak)	0.4-4	0.56	7.1 (-12)	"	1.8 (31)	~8.5-10.5	(233)
"	"(o,decline)	0.4-4	~5	1.8 (-12)	"	4.6 (31)	~8.5-10.5	(233)
V1500 Cyg	cN	0.4-4	~20	1.4 (-10)	1400+	3.5 (32)	~11	(91)
HR Del	cN	0.5-3	-	<1.6 (-13)	810+	<3.6 (31)	>15.1	(91)
AB Dra	dN (o,rise)	1-4	-	7 (-13)		5.2 (31)	11.8	(42)
AH Eri	dN(q)	0.1-4	[10]	2.7 (-12)			12.9	(42)
U Gem	dN(q)	0.1-4	[10]	7.6 (-13)	75	2.0 (30)	[17.5]	(43)(54)
"	"(o)	2-25	~5	2.4-3.8 (-11)	"	~2.0 (31)	14.2	(236)
"	"(o)	2-25	~4	1.5 (-11)	"	9.5 (30)	8.8	(234)
"	"(o)	0.15-0.5	[10]	3.2 (-10)			10.0	(129)
AH Her	dN(o)	0.1-4	[10]	3.6 (-13)		2.0 (32)	9.0	(43)
"	"(o,decline)	0.1-4	[10]	<1.6 (-13)			12.0	(42)
AM Her	MV	0.1-0.3	-	~3.3 (-11)	75	2.1 (31)	12.7	(75)
"	"	2-60	~35	7.6 (-10)	"	4.8 (32)	14.8	(235)
"	"	2-6	>30	6.6(-11)	"	4.2 (31)	13.0	(54)
"	"	0.17-0.5	0.025	0.7-7.6 (-10)	"	0.45-4.8 (32)	12.8	(250)
"	"	2-150	~35	4.1(-10)	"	2.6 (32)	12.6-13.3	(200)
DQ Her	cN	0.1-4	[10]	<1.2 (-13)	420	<2.4 (30)	~13	(43)
V 446 Her	cN	0.1-4	[10]	<2.2 (-13)			14.8	(170)
V 533 Her	cN	0.1-4	[10]	<1.6 (-13)	1000-1500	<2-4 (31)	[18.8]	(43)
EX Hya	dn(q)	0.1-4	[10]	6.9-7.8(-11)	100-200?	0.8-3.5 (32)	[16.5]	(17)
"	"	0.7-2	~4.5	8.6(-11)	"	1-4 (32)	~13.0	(40)
						1-4 5,6 (31)	~13.0	(233)

157

Name	Type	Band (keV)	Flux**		Distance	m_v	Ref.
λ Leo						14.4	(18)
AY Lyr	dN(o)	0.18-0.5	<0.43			12.4	(42)
"	dn(q/o)	0.1-4	1 (-11)			13.1-18	(37)
LL Lyr	dn (q)	0.1-4	<2.0(-13)			18.3	(42)
MV Lyr	dN(q)	0.18-0.43	<2.0(-13)			17.1	(42)
"	N1	0.1-4	4 (-12)			[13.5]	(128)
"	"	0.1-4	3.1 (-12)			12.6	(17)
"	"	0.1-4	1.4 (-12)			[14.3]	(17)
RS Oph	rN	0.1-4	<1.8 (-13)			11.6	(42)
V841 Oph	cN	0.1-4	5.1 (-13)	5.7 (31)	990+	[13.3]	(18)
CN Ori	dN(o)	0.1-4	<2.4(-13)			[12.7]	(17)
GK Per	cN(o)	2-18	2 (-10)	5.2 (33)	480+	12.6	(107)
"	cN(q)	0.1-4	4.8-8.4 (-12)	1.3-2.2 (32)	"	~13.3	(43)(18)
KT Per	dN(o)	0.1-4	4.6 (-13)			12.2	(43)
RU Peg	dN(q)	2-10	1.9 (-11)			[12.6]	(233)
RR Pic	cN	0.1-4	8.4 (-13)	2.4 (31)	500+	~12.0	(18)
CP Pup	cN	0.1-4	1.6 (-12)	9.1 (31)	710+	~15.0	(18)
VV Pup	MV	0.5-4.5	1.8 (-11)?			?	(179)
"	"	0.1-0.5	3(-11)			?	(179)
T Pyx	rN	0.1-4	<1.6 (-13)			[15.2]	(42)
WZ Sge	dN outburst decline)	0.1-4	5.0 (-12)			14.4	(17)
V1017 Sgr	rN	0.1-4	4.9 (-13)			13.7	(42)
V1059 Sgr	cN	0.1-4	3.8 (-13)			[16.5]	(18)
V1223 Sgr	MV?	2-6	6.6 (-11)			~13.2	(224)
V3885 Sgr	N1	0.1-4	1.2-2.4 (-12)	1.6 (32)	1930+	~10.4	(42)(43)
U Sco	rN	0.1-4	<2.0 (-13)			18.3	(42)
LX Ser	?	0.1-4	2.4 (-13)			~14	(239)
RW Sex	N1	0.1-4	1.9 (-12)			~10.6	(42)
SU UMa	dn(q)	0.1-4	1.6 (-11)	2.0 (30)	~200	~14.0	(38)
UX UMa	N1	0.1-0.4	4.3 (-13)			~13.8	(17)
AN UMa	MV	0.1-4	2.8 (-11)			14.5	(74)
"	"	0.1-4	7.2 (-13)			16.3	(244)
TW Vir	dN(q)	0.1-4	2.0 (-12)			[16]	(42)
CK Vul	cN	0.1-4	<1.8 (-13)	<1.6 (30)	280+	>16.5	(18)
Lanning 10	?	0.1-4	<1.1 (-13)	<1.2 (31)	1 Kpc	~14.5	(241)
E0139-681	MV	0.1-4	1-2 (-11)			[14.9-16.4]	(1)
2A0311-227	MV	0.18-0.5	0.025			15.5	(31)
"	"	0.5-60	18			15.5	(36)(285)
1E0643.0-1648	dN	0.1-6	variable			10-13	(34)
E1013-477	MV?	0.1-4	4.0 (-13)			[17]	(135)
E1405-451	MV	0.1-4	0.4-5 (-11)			[15.0-16.5]	(98)
E1551 + 718	MV	0.1-4	9.2 (-13)			~12-17	(133)
2A 1822-371	n.s?	2-11	1.5-7 (-10)	>1 (36)	1-5 Kpc	15.4-16.4	(25)
"	"	0.5-50	1.0 (-9)	"	"	15.4-16.4	(288)
4U 2129+47	n.s?	2-11	2-6 (-10)	0.2-0.3 (35)	1-2 Kpc?	17-18.5	(25)
H2215-086	N1?	0.1-4	1.9 (-12)				(173)
H2252-035	MV?	0.5-50	6.7 (-11)			13.3	(287)(69)

Footnotes: All data in brackets are estimated values from the best available data.

* Crab-like spectrum assumed (e.g. see ref. 25)

** Flux in bandpass given in column 3. Exponent is in parenthesis. The conversion factor for the Einstein bandpass (0.1-4 keV) is 2.7 x 10^{-11} erg cm^{-2} s^{-1} per IPC ct s^{-1} for an assumed spectrum with kT = 10 keV and N$_H$ = 1 x 10^{20} cm^{-2}.

+ distance from ref. 284; other distances are given in references in this Table or Table 1.

++ in cases where V mag is not supplied by references indicated, it was contributed by the AAVSO or the Var. Star Sect., N.Z. R.A.S. Some values for V have been updated by these organizations since the original X-ray papers appeared.

the energy threshold of current low-energy X-ray detectors (which are typically sensitive above ~0.1 keV). Most of the energy in a spectrum with this temperature is radiated in the largely unsurveyed extreme ultraviolet (EUV) range. The soft X-ray emission of SS Cyg and U Gem has been detected only during their optical outbursts. The 0.18–0.5 keV flux of U Gem, for example, increases by more than a factor of 100 between quiescence and outburst [129, 131]. This does not necessarily imply, however, that the luminosity in the EUV–soft X-ray component as a whole increases by this amount since the flux detected above ~0.1 keV is very sensitive to the temperature. Ultra-soft X-ray emission has been searched for in many other cataclysmic variables, and in particular in more than 20 other erupting dwarf novae, but has not been found [41, 42, 45]. From the information in Table 4.2 it is clear that the observed soft X-ray luminosity of AM Her, U Gem and SS Cyg is comparable to, or exceeds that emitted as hard X-rays when both emissions are present. The dominance of the soft X-ray component is even more striking when it is considered that most of the energy contained within it may be radiated in the EUV.

A blackbody spectral model of the soft emission has been favored over an optically thin bremsstrahlung model because of theoretical arguments (see discussion later in this section). Observations with the objective grating spectrometer on the Einstein X-ray satellite confirm that there is no detectable line emission in the soft X-ray spectrum of AM Her, further favoring the blackbody model [76]. The low-energy spectrum of AM Her and probably also SS Cyg is significantly absorbed by interstellar or circumstellar matter ($N_H \sim$ a few $\times\ 10^{20}$ H atoms cm^{-2} [250, 44]). The absorbing column to U Gem appears to be the least among these three stars ($N_H < 2.5 \times 10^{20}$ H atoms cm^{-2} [46]). The amount of absorption will be critical in determining whether cataclysmic binaries can eventually be observed in the EUV, and the total flux in this band unambiguously measured. Constraints on the temperature of the soft X-ray–EUV component can be derived from observations in the ultraviolet; this is discussed further in Section 4.5.

Another characteristic of the soft X-ray emission is its variability. The soft X-ray emission of AM Her is modulated with the 3.09 h binary period; the flux of this star can also vary substantially in an apparently random manner on timescales of hours. Variability by up to a factor of 10 on a similar timescale has been observed in U Gem [129, 131].

Quasi-periodic pulsations have been detected in the soft X-ray flux of SS Cyg and U Gem with characteristic periods of ~9 s and ~25 s, respectively [44, 46]. In addition, flaring or bursting behavior has been detected in U Gem on a timescale of ~20 s when no obvious pulsations were present [131]. The short-timescale variability of the soft X-ray flux is one of the strongest indications that it is produced near the white dwarf. No pulsations were detected in the hard X-ray emission of SS Cyg at the time when they were observed at soft X-ray wavelengths [44]. The short-timescale variability of the soft X-ray flux may thus be unrelated to that of the hard X-rays.

AM Her is also variable on a timescale of tens of seconds. The variability in this case is best described by shot noise models, although instances of apparently periodic pulse trains have also been noted [250].

4.4.3 Models

Two alternative accretion geometries have been considered for producing both hard and soft X-radiation from deep in the potential well of the degenerate dwarf. In one case the accretion disk is assumed to penetrate to the stellar surface. X-radiation is produced in a boundary layer between the disk and the degenerate dwarf where the rapidly orbiting material of the inner disk is forced to dissipate its excess Keplerian energy [185]. In the second geometry, matter falls radially onto the degenerate star and releases its kinetic energy in a strong shock near the stellar surface. Pseudo-radial accretion will occur if the accreting matter is channelled by the magnetic field of the compact star. The shocked material can then cool through cyclotron interactions with the magnetic field as well as by emitting bremsstrahlung radiation [119, 105].

In both types of model the relative amount of hard and soft X-radiation will depend on the accretion rate. At rates below about 10^{16} g s^{-1} gas heated in a boundary layer can expand above the plane of the disk before cooling and form a hot, hard X-ray-emitting 'corona' around the compact star [186, 251]. Pringle & Savonije [186] suggest that the boundary layer is heated by shocks, but these must be strong for the gas to attain the high temperatures measured in cataclysmic variables (the maximum temperature would result if all the Keplerian energy were released in a single, strong shock, and is given by $T_s \sim 2.3 \times 10^8$ K $(M_*/M_\odot)(R_*/10^9 \text{cm})^{-1}$ K). This is difficult to achieve in the strongly sheared flow of the inner disk whose geometry would tend to favor production of a large number of cooler, oblique shocks. To attain high temperatures Pringle & Savonije suggest a two-stage process in which gas that is initially mildly shocked in the boundary layer expands into the path of, and interferes with, gas still circulating in the inner disk. Tylenda [251], however, argues that turbulent viscosity will be a more efficient mechanism than shocks for dissipating energy in the boundary layer and that this mechanism can account for the observed high temperatures without resorting to complicated flow geometries.

In contrast, at high accretion rates ($\gtrsim 10^{16}$ g s^{-1}) the density in the boundary layer region becomes such that the free-free cooling timescale of the gas is much shorter than the adiabatic expansion time, so that the material cools before it can expand above the disk. Radiation emitted in the boundary layer is then absorbed by the disk and thermalized, resulting in a much softer X-ray emission spectrum ($T \sim 10^5$-10^6 K [185]) than at low accretion rates. The observations, however, particularly those of U Gem, dictate that some high-temperature gas must remain since hard X-rays are still seen during outburst.

In the case of a magnetically controlled flow onto the compact star, a strong shock is formed in the accretion column at a distance above the stellar surface that depends on the velocity of the flow and the cooling rate of the shocked gas. The ratio of soft to hard X-radiation is determined by the relative importance of bremsstrahlung and cyclotron cooling in the post-shock region. At high accretion rates or low magnetic field strength bremsstrahlung dominates and energy is radiated primarily as hard X-rays. At low accretion rates or high field strength the balance shifts towards cyclotron cooling and radiation at energies below the X-ray band [119, 105]. About half the energy radiated in the accretion column is intercepted by the surface of the degenerate star. Conduction of energy by electrons may also be important in heating the stellar surface [106]. The ultra-soft X-ray component observed in a number of magnetic variables is identified in this model as the high-energy tail of blackbody radiation from the heated surface of the degenerate dwarf.

Often the dominant accretion mechanism operating in a given system is clear. For example, the detection of strong optical polarization and the geometry inferred from the light curves of the AM Her stars point to pseudo-radial accretion, while the observation of doubled emission lines or a disklike continuum in many stars favors disk accretion. However, there may exist intermediate cases of cataclysmic variables which are not generally recognized as being magnetic, but which nevertheless contain degenerate dwarfs that have appreciable magnetic fields. In general, the distance above the stellar surface at which magnetic forces will disrupt the disk will depend on the field strength and the accretion rate [304]. For certain combinations of these parameters systems may exist that have partial accretion disks which are disrupted by the magnetic field of the central star. An expression for the innermost radius of the disk in terms of the magnetic field strength and the mass accretion rate is given by Ghosh & Lamb [305]. cf. their eq. 11. As emphasized by Córdova, Mason & Nelson [43], transitions between pseudo-radial and disk accretion are possible in stars whose accretion rates vary with time (e.g. the dwarf novae).

Although much of the X-ray emission of cataclysmic variables probably originates in the vicinity of the degenerate dwarf, there may be some contribution to the X-ray flux from other parts of the binary system. Possible sources include the corona of the companion star, or a similar coronal region associated with the disk. The detection by Swank of a spectral component with a temperature of ~0.6 keV in SS Cyg that is unchanged in outburst requires at least one X-ray-emitting component that is unaffected by the outburst process. The luminosity in this component (1.8×10^{31} erg s^{-1}) is consistent with an extrapolation of the rotation-dependent coronal X-ray flux for isolated dwarf stars with periods greater than 12 h (cf. [267]). The quiescent hard X-ray flux of U Gem is also consistent with this relationship [42]. It should be cautioned, however, that relationships derived for isolated or detached binary stars may not necessarily be applicable to stars in interactive binary systems whose convective layers may be altered by the mass transfer process. Pringle [185] has argued that the shock on the accretion disk that gives rise to the 'bright spot' is unlikely to produce observable X-radiation in the manner proposed by Warner [268] because of the high degree of absorption expected in disk material surrounding it.

4.5 The ultraviolet and optical spectrum
4.5.1 Disk emission

In many cataclysmic binaries the high angular momentum of the matter that streams off the companion causes the formation of a differentially rotating disk around the accreting star. If the accretion rate is constant and the disk optically thick, each elemental annulus of the disk will emit a spectrum that can be approximated by a blackbody whose temperature is a function of radius. The temperature in the disk is then given by [208, 16] (see also Chapter 10 of this book for a detailed discussion of disk models);

$$T(R) = T_*(R/R_1)^{-3/4}[1 - (R_1/R)^{1/2}]^{1/4},$$

where

$$T_* = 4.1 \times 10^4 (\dot{m}/10^{16} \text{g s}^{-1})^{1/4}(M_1/M_\odot)^{1/4}(R_1/10^9 \text{ cm})^{-3/4} \text{ K}. \quad (4.1)$$

Here R_1 is the inner radius of the disk, $\dot{m}$ is the mass accretion rate and M_1 is the mass of the white dwarf. The maximum temperature in the disk, $T_{\max} = 0.488 T_*$, occurs at a radius $49/36 R_1$. The minimum temperature, T_{out}, occurs at the outer edge of the disk, R_{out}. The total luminosity of the disk is

$$L_d = 6.7 \times 10^{32} (\dot{m}/10^{16} \text{ g s}^{-1})(M_1/M_\odot)(R_1/10^9 \text{ cm})^{-1} \text{ erg s}^{-1}. \quad (4.2)$$

For frequencies, ν, such that $kT_{\text{out}} \ll h\nu \ll kT_{\max}$, Lynden-Bell [125] has shown that the spectrum of a steady-state disk will have the form $F_\nu \propto \nu^{1/3}$ (where $F_\nu \propto \nu^\alpha$ implies $F_\lambda \propto \lambda^{-(2+\alpha)}$). When $h\nu < kT_{\text{out}}$, the disk spectrum will approach a Rayleigh–Jeans form with $F_\nu \propto \nu^2$, while at frequencies such that $h\nu \sim kT_{\max}$, the spectrum decays exponentially.

Recent observations extending from the infrared to the ultraviolet show that the steady-disk model can successfully reproduce the spectral distribution of some types of cataclysmic variables, in particular the dwarf novae in outburst and the novalike 'disk' stars (i.e. the UX UMa variables [271]). In these systems the disk dominates the ultraviolet and visual continuum and is responsible for the shallow, broad absorption lines in the spectrum. By fitting a disk model to the data, the mass accretion rate, the size of the disk, and conditions at the outer disk edge (temperature and opacity) can be estimated (cf. [15, 59]). In Figure 4.1(a) we show the spectrum of the dwarf nova SS Cyg during outburst as an example of a star with a substantial accretion disk. The ultraviolet data were obtained by Holm [83] and the optical data were taken from Kiplinger [102]. The solid curve in Figure 4.1(a) is an optically thick disk spectrum with parameters $T_* = 1.6 \times 10^5$ K and $R_{out}/R_1 = 80$. This spectrum, which has been dis-

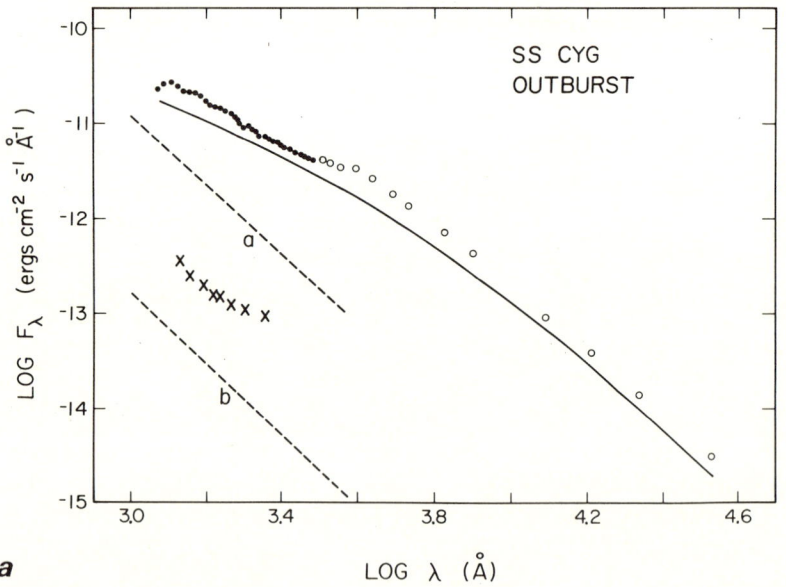

Fig. 4.1. (a) The spectrum of SS Cygni during outburst. The solid circles are data taken by Holm with the IUE satellite on 14 June 1978; the open circles are from Kiplinger [102] and were taken during a different outburst, but at a similar brightness level. The reddening correction applied by Kiplinger to his data has been removed and the data have been corrected by a few tenths of a magnitude so that the flux at 5500 Å corresponds to that measured with the IUE Fine Error Sensor. The solid line is the best-fit optically thick, steady state disk model, which has been displaced downward from the data by $\text{Log} F_\lambda = 0.2$. The dashed lines are the Rayleigh–Jeans tails of blackbody spectra with temperatures of 15 eV (curve a) and 20 eV (curve b), normalized so as to produce the soft X-ray flux observed simultaneously with the ultraviolet outburst data. The crosses represent the far ultraviolet spectrum of SS Cyg taken during optical quiescence by Fabbiano et al. [54].

placed downward from the data by $\log F_\lambda = 0.2$ for clarity, reproduces well the gross shape of the continuum. Assuming the above value of T_* and making use of Equations (4.1) and (4.2), we find that $\dot{m} = 2.3 \times 10^{18} (M_1/M_\odot)^{-1} (R_1/10^9 \text{cm})^3$ g s^{-1} and $L_d = 1.6 \times 10^{35} (R_1/10^9 \text{cm})^2$ erg s^{-1}. The observed 1100–30 000 Å flux is 3.3×10^{-8} erg cm^{-2} s^{-1}, so that at an assumed distance of 150 pc, $R_1 = 7.5 \times 10^8$ cm, consistent with the expected radius of a degenerate dwarf. The parameters derived from the successful application of the steady-disk model to several other stars are summarized in Table 4.3.

Various authors have attempted to improve on this simple blackbody model. Mayo, Wickramasinghe & Whelan [137] calculate disk spectra using model atmospheres for the optically thick regions of the accretion disk. The models, which accommodate various values of R_{out}/R_1 and mass accretion rate, give good agreement with the UBV colors and hydrogen line profiles of dwarf novae in outburst and the novalike disk stars. These authors also consider the effects of inclination on the observed spectrum, and Pacharintanakul & Katz

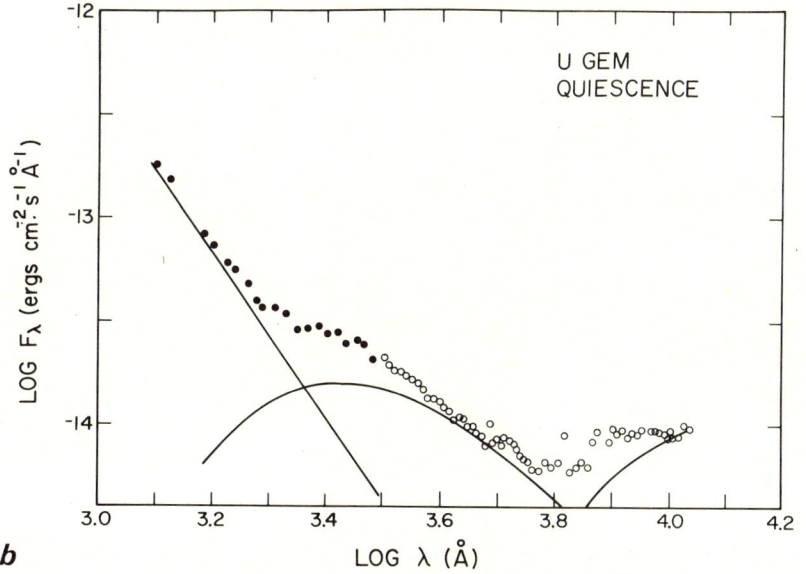

Fig. 4.1. (b) The spectrum of U Gem during quiescence. The solid circles are data taken by Wade with the IUE satellite on 12 February 1980: the open circles are optical and near-infrared spectrophotometry obtained on 11 March 1977 by Oke & Wade. Shown is the average of two optical–IR measurements made in the same phase interval of the binary orbit as the long wavelength IUE observations ($\phi = 0.65$–0.92). The short wavelength IUE data were taken during the phase interval $\phi = 0.18$–0.63 but, unlike the long wavelength data, do not change with orbital phase (Fabbiano et al. [54]). The spectrum has been approximated by three blackbody components (solid lines) as described in the text.

TABLE 3: STEADY – DISK MODEL PARAMETERS

Star	SUBCLASS	T_*	T_{out}	R_{out}/R_1	$\dot{m}_{16}$*	Refs.
SS Cyg	dN (outburst peak)	1.6×10^5 K	6300K	~80	60	(this paper)
EX Hya	dN (quiescence)	7×10^4	2200	~100	1	(15)(59)(211)
VW Hyi	dN (outburst decline)	$>1.6 \times 10^5$	<5400	>90	180?	(15)
RW Tri	N1	1.6×10^5	3700	~160	8	(58)(59)
UX UMa	N1	8.4×10^4	6500	~30	50	(59)(60)

*$\dot{m}$ is given in units of 10^{16} g s^{-1}. For EX Hya, RW Tri and UX UMa, $\dot{m}$ was derived by fitting the IR light curves. For SS Cyg and VW Hyi, Eq. 4.5 of Ref. 15 was used, assuming $M_1 = 1.3 M_\odot$ and $R_1 = 7 \times 10^8$ cm (see text) for SS Cyg, and $M_1 = 0.8 M_\odot$ and $R_1 = 8.5 \times 10^8$ cm (i.e. the radius of the degenerate star) for VW Hyi. The value of $\dot{m}$ for VW Hyi, in particular, is very uncertain (see Ref. 15).

[154] show how reprocessing in a steady-state disk around a degenerate dwarf can significantly increase the infrared flux depending on the disk thickness and curvature. Blanketed model atmospheres for early-type stars [117] can provide good fits to the UV and optical data on cataclysmic variables for effective temperatures $T_{eff} \sim 1.5\text{-}1.7 \times 10^4$ K and $\log g \sim 4$. This is because of the flatness of the observed spectrum which may (possibly fortuitously) mimic a stellar atmosphere [90, 84, 70, 39].

The opacity of the disk in the cooler, outer regions has been examined by Williams [292] and Frank & King [59]. Williams treats the case of steady, highly supersonic turbulence ($\alpha > 1$ in the alpha-disk model of Shakura & Sunyaev [208]) and finds that the optical emission lines may be produced in the optically thin regions of the outer disk which do not fall below a temperature of ~ 6000 K.

Williams & Ferguson [312] also suggest that the He I emission lines are formed with the Balmer lines in the outer disk, and that the helium abundance is much larger, by number, than the hydrogen abundance. They compute the continuum fluxes for model disks with enhanced helium. Frank & King show that the viscosity may actually be much less than unity, thus giving a higher density in the disk. Molecular opacity may then be important near R_{out} so that the disk could be optically thick out to much lower temperatures. The authors relate measurements of T_{out} (and thence R_{out}) and the mass accretion rates of three systems, EX Hya, RW Tri and UX UMa (cf. Table 4.3), to the viscosity, and find that $\alpha = 0.1$, 0.6 and 20 respectively.

Tylenda [311] has considered the continuous as well as line radiation from accretion disks with comparatively low rates of mass transfer ($<10^{16}$ g s^{-1}) which are optically thin in their outer regions. His models differ from those of Williams [292] mainly in the adoption of a much lower viscosity ($\alpha \sim 1$).

4.5.2 Other sources of emission: contributions from the companion star, bright spot, and white dwarf

While disk models adequately represent the spectrum of a number of cataclysmic variables from the ultraviolet to the infrared, there are many cases in which they fit over only a limited range of the spectrum [238]. Kiplinger [102, 103] finds that the steady, optically thick disk model does not fit the observed spectrum of SS Cygni during quiescence nor the spectra of either SS Cyg or Z Cam during the decline from outburst. Mayo, Wickramasinghe & Whelan [137] find that the dwarf novae in quiescence and old novae in general do not lie in the steady-disk region of the color–color diagram.

The deviation from a steady-disk distribution often takes the form of an optical and/or IR excess (although this is sometimes difficult to assess quantitatively if the data in different energy bands were not taken simultaneously). This excess may be due in part to the contribution from the bright spot on the outer edge of the disk or from the companion star. A striking example is the quiescent spectrum of the dwarf nova U Gem. This is shown in Figure 4.1(b) where we have combined unpublished ultraviolet, optical and near-infrared spectrophotometry of the star obtained by Oke & Wade [152]. Figure 4.1(b) is similar to the spectrum of U Gem reported by Fabbiano et al. [54], but here we have included long-wavelength ultraviolet data (1900-3000 Å) to complete the spectral coverage, and the spectrum corresponds to an orbital phase when the bright spot is prominent. We have approximated the composite 1100-10 000 Å continuum of U Gem with three blackbody spectral components. These can be identified, with emission from the companion star which dominates in the red and infrared (color temperature $T_{BB} \sim 2400$ K); the bright spot on the accretion disk which is observed in the visual and near ultraviolet ($T_{BB} \sim 10\,000$ K), and whose light is modulated with the binary orbit; and a far UV component, that probably originates near the degenerate star, whose slope ($F_\lambda \propto \lambda^{-4}$) is consistent with the Rayleigh-Jeans tail of a hot blackbody ($T_{BB} \gtrsim 1.2 \times 10^5$ K; see also Fabbiano et al. [54]). There is no component in the spectrum that can readily be identified with the spectrum of an optically thick accretion disk.

The far ultraviolet blackbody spectrum observed from U Gem during quiescence, like the ultra-soft X-ray emission found in this star during optical outburst, suggests that an important part of the luminosity of the system may be emitted in the EUV range between 0.01 and 0.1 keV. Similar hot ultraviolet emission has been identified in the quiescent state spectrum of SS Cyg [54] and the magnetic variable AM Her [189] both of which also evidence ultra-soft X-ray emission (Section 4.4). The far UV data taken on U Gem during optical quiescence can be combined with limits set on the quiescent X-ray flux above 0.18 keV [129] to constrain the temperature of the EUV spectrum. The temperature is found to lie in the range $1.2-1.8 \times 10^5$ K (10-15 eV) if the blackbody spectral model is adopted. Fabbiano et al. [54] obtain similar estimates for the EUV temperature of both U Gem and SS Cyg; the temperature of AM Her, however, may be somewhat higher if the far UV and soft X-ray components in this star are one and the same [189]. Assuming that the far ultraviolet emission of U Gem is the tail of a hot blackbody, its total luminosity is inferred to be $1.5-4 \times 10^{33}$ erg s^{-1} at a distance of 75 pc, although only $\sim 6 \times 10^{31}$ erg s^{-1} of this, the fraction longward of 1100 Å, is observed. The luminosity in the postulated EUV blackbody is considerably in excess of the $\sim 4 \times 10^{31}$ erg s^{-1} found in the 10 000 K (0.8 eV) component that dominates the optical spectrum.

Further information on the nature of the EUV emission can be obtained from the spectrum of SS Cyg. Holm's ultraviolet data on that star (Figure 4.1(a)) were acquired simultaneously with HEAO-1 soft X-ray measurements ([44]; see Section 4.4). The Rayleigh–Jeans tails of blackbody spectra with temperatures of 15 eV and 20 eV, normalized to produce the observed soft X-ray signal, are shown in Figure 4.1(a). The lower limit to the X-ray temperature during outburst is not well constrained by the HEAO-1 data alone, but the temperature cannot be much less than ~15 eV if such a blackbody spectrum is not to violate the observed far UV flux measured simultaneously. Also plotted in Figure 4.1(a) are the far ultraviolet SS Cyg data of Fabbiano et al. [54] that were taken during optical quiescence. These data illustrate the approximately λ^{-4} slope of the spectrum at the shortest wavelengths. If a quiescent-state, hot, blackbody component is also assumed to be responsible for the soft X-ray emission found during outburst, some inferences concerning changes in the EUV emission between outburst and quiescence can be made. These will depend on the temperature, T_x, assumed for the soft X-ray emission during outburst. If kT_x is, for instance, 20 eV, the far UV flux due to this component must increase by about a factor of 5 when the star goes from optical outburst to quiescence; this must be accompanied by a reduction in the temperature of the emission to account for the disappearance of the soft X-rays above 0.1 keV during quiescence. If, instead, kT_x is 15 eV, the far UV flux in this component during outburst exceeds that during quiescence by about a factor of 10. This is comparable to the lower limit on the ratio of outburst to quiescent soft X-ray flux given by Córdova et al. [44]. Thus a change in the normalization without any change in the temperature of a 15 eV spectrum would be consistent with the present outburst and quiescent data on SS Cyg.

If the soft X-ray emission during outburst is produced in the boundary layer where the inner disk encounters the degenerate dwarf ([185]; see Section 4.4), its luminosity will correspond to the flow of Keplerian energy of matter in the innermost disk orbit. By the virial theorem, this will approximately equal the energy already lost in passage through the disk. Setting the luminosity of the EUV–soft-X-ray component of SS Cyg equal to the disk luminosity during outburst, L_d (calculated above), implies a blackbody temperature of ~18 eV, and an emitting area approximately equal to the surface area of the white dwarf.

Fabbiano et al. [54] discuss steady nuclear burning of accreted matter in the envelope of the degenerate dwarf as a possible source of the inferred EUV luminosity in U Gem, SS Cyg, and AM Her (see also [200] on AM Her). Steady hydrogen burning by way of proton–proton reactions has been discussed by Starrfield, Truran & Sparks [223]. It has been shown [61, 163] that for a given CNO abundance there is a maximum steady luminosity (L_{max}), and hence a maximum accretion rate, at which steady burning is possible. For normal solar abundances $L_{max} \sim 5 \times 10^{-2} L_\odot$ (or $\dot{m} \sim 5 \times 10^{13}\,\mathrm{g\,s^{-1}}$). The observed luminosity in the EUV component of U Gem is at least five times greater than this value. The theoretical investigations show that, for a 1 $M_\odot$ degenerate dwarf, steady nuclear burning can proceed at such a luminosity if the CNO abundance is $\sim 1.5 \times 10^{-3}$ of solar. This would require that the CNO diffusion timescale in U Gem be $\sim 10^4$ yr, as compared to theoretical estimates of 10^6-10^7 yr [57]. The discrepancy is even greater for SS Cyg and AM Her which are believed to have higher accretion rates than U Gem during quiescence.

The continuum spectra of the magnetic variables bear little resemblance to that of stars with accretion disks. Raymond et al. [189] interpret their far UV spectra of AM Her in terms of a two-component continuum, one with a slope of λ^{-1} which is always present, and a second λ^{-4} component (discussed above) that dominates in the far UV and is modulated with the binary period. The UV continuum of EF Eri is also consistent with $F_\lambda \propto \lambda^{-4}$ [35].

4.5.3 Ultraviolet emission lines

In general, the magnetic variables (e.g. AM Her [189]; EF Eri [35]) have strong emission line-to-continuum ratios when compared with the systems that have disks (cf. [238]). Strong line emission is found, too, in AE Aqr and DQ Her [96, 120], two stars which may also have magnetic fields (see Section 4.6). The ratios of the ultraviolet emission lines of AE Aqr have been investigated by Jameson, King & Sherrington [96] who find that the lines from the high ionization states (He II $\lambda 1640$, C IV $\lambda 1549$, N V $\lambda 1240$, and Si IV $\lambda 1400$) are produced in a 'chromospheric' region which is photo-ionized by radiation from the hot, central parts of the accretion disk. He II may be a recombination line, but all the other lines appear to be excited collisionally. The Mg II $\lambda 2800$ line, in contrast, is probably formed in the optically thin outer disk, and is ionized collisionally rather than by photo-ionization.

Dwarf novae in quiescence also have UV emission lines of C IV, N V, Si IV and He II. Szkody [238] has compared the collisional-excitation ratios given by Jameson, King & Sherrington [96] for these high-excitation lines to the spectra of several dwarf novae and has found the data consistent with photo-ionization by a source with a blackbody temperature $T_{BB} \sim 5 \times 10^4 - 1 \times 10^5$ K.

An interesting discovery that has come from recent UV observations is that many dwarf novae in outburst [73, 39], and also some novalike objects [307, 66, 70], exhibit P Cygni-like UV line profiles or, alternatively, absorption lines whose central cores are shifted to shorter wavelengths. These lines betray the presence of an outflowing wind and have edge velocities as high as 5000 km s^{-1}, about the escape velocity from the surface of the white dwarf. The present authors have compared the UV spectra of a number of such stars and conclude that velocity shifted absorption lines are most evident in systems that are viewed at relatively low inclination [39]. Thus it is argued that the wind is emitted preferentially in a direction perpendicular to the disk plane. Support for the existence of a wind is found from an analysis of UV data on two high inclination eclipsing systems, the novalike stars UX UMa and RW Tri. Holm, Panek & Schiffer [306] find that the UV emission lines of UX UMa are not eclipsed along with the continuum, suggesting that they are formed in a region more extended than the inner disk. A similar result is found for RW Tri by the present authors. Estimates of the mass loss rate have been made by comparing the line profiles measured in low inclination C Vs with those which are formed in the stellar winds of normal O and B stars. Values of order $10^{-11} M_\odot$ are derived, which is about 10^{-2}–10^{-3} of the mass accretion rate onto the white dwarf.

4.6 Magnetic fields in cataclysmic variables

There are seven cataclysmic variables, AM Her, VV Pup, AN UMa, EF Eri, PG 1550+191, E 1405−451 and E 0139−681, in which strong optical polarization has been measured, indicative of a substantial magnetic field. One other star, E 1013−477, is probably also of this type [135] although no polarization measurements have yet been made of it. The recognition of this class of star followed the suggested identification of the optical variable AM Her with the hard X-ray source 3U 1809+50 by Berg & Duthie [19]. Tapia [245] observed unprecedented strong linear and circular polarization in the red light of that star, and its identification as the X-ray source was confirmed by the observation of a common 3.1 h periodicity in the soft X-ray [75], hard X-ray [235], optical photometric [240], radial velocity [47, 184] and polarization measurements [245]. Although AM Her remains the most widely studied of the magnetic variables (see [32]), similar behavior has been observed in the other stars of this type (e.g. [113, 68, 203, 204]). The accretion geometry of these systems is believed to be completely controlled by the magnetic field of the compact star, which funnels material from the mass-donating companion into an accretion column above the magnetic poles and may prevent any accretion disk from forming. Much of the observed radiation probably originates in these accretion columns. The rotation period of the compact star is believed to be synchronously locked to the orbital period of the binary.

The orbital light curves of the magnetic variables can show a great deal of structure. Major causes of the light curve modulation probably include: obscuration of the small emitting region by the accretion column and by the body of the degenerate dwarf star itself as it rotates; variations in the emissivity of polarized light as a function of angle with respect to the magnetic field lines; heating of one face of the companion star; and the production of radiation at different wavelengths at different heights in the accretion column. An energy dependence of the light curves would also be expected if there were more than one active pole accreting at different rates (e.g. because of their alignment with respect to the companion star), or if the centroid of the dipole field were displaced with respect to the center of mass of the degenerate star, resulting in a different surface magnetic field strength at each pole [105, 121].

The optical linear polarization curve of the magnetic variables takes the form of a narrow spike of full width between 10 and 20 per cent of the orbital period, depending on the star. The maximum amount of light that is polarized is typically a few per cent, and no linearly polarized light is detected outside this narrow pulse. The linear polarization spike can provide an excellent fiducial point for measuring the orbital ephemerides of the magnetic variables, but it is not an immutable feature [7, 142]. The peak of this spike coincides with a time of zero circular polarization. In contrast to the linear polarization, the percentage of circularly polarized light varies smoothly throughout the cycle [245, 246, 113, 6]. The shape of the orbital circular polarization curve of AM Her evolves with time [4], but does not appear to be greatly affected by transitions from high to low brightness states [121]. The orbital polarization variations of the magnetic variables might be due to the angular dependence of Faraday depolarizing effects and high-harmonic cyclotron emission in the radiating region [226, 140, 301]. In this model the linear polarization spike is seen in viewing directions that are perpendicular to the field lines that permeate the emitting volume, while the maximum circular polarization is seen when the emitting region is viewed in a direction parallel to the field lines.

The magnetic field strength of the degenerate dwarf has been measured accurately in two stars, VV Pup and AM Her. A value of 3×10^7G is derived for VV Pup based on cyclotron features observed in its optical spectrum [253, 227, 290]. The observation of Zeeman splitting of lines in the low-state spectrum of AM Her provides an estimate of between 1.3 and 2.2×10^7G for its surface field [202, 121].

The ultraviolet and optical spectra of the magnetic variables are characterized by strong emission lines (cf. Section 4.4; [47, 67]). In the optical region, narrow and broad components of the lines have been identified whose velocities vary out of phase with one another [47, 67, 203, 204]. The narrow component lags the broad component by 25°, 49°, 55° and 135° respectively for EF Eri, AN UMa, VV Pup, and AM Her. The narrow emission lines most probably arise on the heated face of the companion, or in low-velocity gas near the inner Lagrangian point, whereas the broad component is formed in higher-velocity material that is further down the accretion funnel [295, 203, 204]. The narrow-line component of AM Her was still present in the low state of that star, but the broad component was essentially absent [121].

The fact that the compact star in the AM Her systems is phase-locked with the binary orbital motion suggests that the torque exerted by the magnetic field on the companion star is sufficient to overcome the tendency for the compact star to be spun up by accreted material. Chanmugam & Dulk [302] show that if the magnetic white dwarf is formed spinning rapidly, and the companion star acts as a unipolar inductor, the large electromagnetic torque acting on the white dwarf can synchronize its rotation in $\sim 10^4$ yr. This timescale is much shorter than the $\sim 10^{10}$ yr required for ohmic dissipation from the magnetic interaction of the two stars to bring about synchronization [100].

Stars with weaker fields or larger accretion rates may not be able to maintain synchronism. Possible systems of this type among the cataclysmic variables are DQ Her, V533 Her and AE Aqr whose relatively stable periodic optical oscillations (of order tens of seconds) may reflect the rotation period of a magnetized degenerate dwarf [14, 118, 101, 166, 167, 168]. Hutchings *et al.* [94] have suggested that 2A0526−328 may be a system in which the orbital period and rotation period of the degenerate star differ slightly. They identify their spectroscopic period of 5.49 h with orbital motion and the photometric period of 5.18 h (discovered by Motch) [145] with white dwarf rotation.

Another example of a non-synchronously rotating compact object may be the X-ray star H 2252−035 which was optically identified with an object having a 3.6 h binary period [69, 172]. H 2252−035 exhibits an optical periodicity of 14.3 min [172] and an X-ray periodicity of 13.4 min [287]. It is thought that the X-ray pulsations originate from a magnetic, accreting star and are reprocessed by the companion star [172] or a bulge on the outer disk caused by the impact of the gas stream from the companion [72]. Because of the orbital motion, the reprocessing region sees the X-ray beam every 14.3 min, as opposed to every 13.4 min for a sidereal observer. It is not clear whether the accreting object in H 2252−035 is a white dwarf or a neutron star [287, 72]. The X-ray source 4U 1849−31 (=V1223 Sgr) has an optical periodicity of 13.2 min and may be similar to H 2252−035 [224].

4.7 Short timescale variability

The temporal behavior of cataclysmic binaries includes periodic modulation associated with orbital motion and rotation, and flickering and pulsations reflecting the mass transfer process and the dynamics of matter near the surface of the accreting star.

4.7.1 Orbital modulation

A variation on the timescale of the binary period can arise for a variety of reasons.

(1) A common phenomenon is an extended hump in the visual light curve. In many stars (e.g. U Gem) the hump is due to anisotropic emission from the mass transfer bright spot on the rim of an optically thick accretion disk. A qualitatively similar orbital hump is found in some magnetic variables (e.g. VV Pup [277]). The emission in this case originates in the vicinity of the magnetic polar cap associated with the degenerate star (Section 4.6).

(2) In highly inclined systems the optical light of the bright spot, the disk, and/or the central star is often eclipsed by the companion. Near-infrared observations of RW Tri [124] and Z Cha [8] reveal, in addition to the primary eclipse, a secondary eclipse which is interpreted as the occulation of the red dwarf companion by the disk.

(3) Ellipsoidal variations resulting from distortion of the red dwarf in the gravitational field of the degenerate star have been detected in the near-infrared light curves of U Gem [60], EM Cyg [97] and possibly also Z Cha [8]. The analyses of these variations provide data on the orbital elements of the binary system and the size and luminosity of the component stars and the accretion disk.

(4) 'Superhumps' having periods a few per cent longer than the binary orbit are observed during the superoutbursts of SU UMa stars. The amplitude of the superhumps (0.2-0.3 mag) is similar in all SU UMa systems and must therefore be insensitive to inclination. Papaloizou & Pringle [159] suggest that the difference between the superhump and the orbital period is due to apsidal motion in a slightly eccentric ($e \sim 10^{-4}$) binary orbit. In the model of Vogt [256] the difference in periods is due to apsidal motion in an eccentric ($e \sim 0.6$) ring which surrounds the inner accretion disk during superoutbursts: a 'super-bright spot' is formed at the point of impact of the gas stream and the eccentric ring. The ring may be formed by a brief episode (10^3 s) of enhanced mass transfer from the companion.

(5) A photometric variation in EX Hya with a period of 67 min, which is ~ 2 per cent longer than two-thirds of the orbital period, has been discovered by Vogt, Krzeminski & Sterken [259] who suggest that it is due to a variation in the mass transfer rate caused by pulsation of the companion star. This pulsation could be a toroidal mode which is excited by tidal interaction due to a favorable commensurability between the orbital and pulsation periods if the orbit is slightly eccentric [160]. An alternative model is that the 67 min modulation is associated with the rotation period of the white dwarf [275]. The X-ray emission of EX Hya is also modulated with the 67 min period, and the modulation is most pronounced at energies below ~ 2 keV [111].

4.7.2 Flickering

Aperiodic activity on timescales of a few minutes is observed in both the optical and X-ray light of most cataclysmic variables [270, 42]. In some stars there is evidence that the optical flickering is associated with the bright spot on the accretion disk. A notable example is U Gem in which the flickering amplitude is found to be much reduced during the eclipse of the bright spot (cf. Figure 2 in [270]). The bright spot is not the only possible source of rapid light variation. The AM Her stars show marked aperiodic optical and X-ray variability, which is thought to arise in a magnetically controlled accretion column [242, 250]. Simultaneous X-ray and optical data obtained by the authors and their colleagues on the novalike variable TT Ari show that marked variations in the optical light are correlated with similar variability in the X-ray band [99]. Since hard X-ray emission (i.e., $kT > 2$ keV) can probably not be produced in the outer accretion disk, the implication is that the flickering source at both wavelengths has its origin close to the degenerate dwarf. For the dwarf novae HT Cas [171] and possibly OY Car [260] there is independent evidence that the optical flickering may be associated with the inner-disk region.

4.7.3 Rapid oscillations

Pulsations have been observed with a wide variety of coherence properties in all subtypes of cataclysmic binary. The periods, amplitudes, and coherence values of the pulsations detected are listed in Table 4.4. Generally the periods are of order tens of seconds. The amplitudes are only a fraction of a percent in the optical band, but higher-amplitude pulsations have been detected at X-ray frequencies.

Because of the wide spectrum of pulse coherence observed among the different stars, it is doubtful that a single mechanism is responsible for the oscillations in all systems. The most stable pulsations have $Q \equiv \dot{P}^{-1}$ in the range 10^{11}–10^{13}. It has been suggested that these oscillations reflect the rotation period of a magnetized, degenerate dwarf star whose emission is beamed in a manner analogous to some rotating neutron stars. Examples include the classical novae DQ Her [101], and V533 Her [166], and the novalike object AE Aqr [167]. Hard X-ray pulsations with the same period as the optical pulsations may have been detected in AE Aqr [168], but no X-ray emission has been detected from either DQ Her or V533 Her [43].

Two optical oscillations have been seen in the long period dwarf nova WZ Sge. A pulsation of period 27.87 s usually dominated the power spectrum before the 1978 outburst of the star, although a second pulsation at 28.97 s was occasionally present by itself or simultaneously with the primary pulsation [198, 169]. The second pulsation at the period of 28.97 s dominated for an interval of at least two months in observations made seven months after the outburst [144]. Both periods have displayed large secular increases: $\dot{P} = 8 \times 10^{-12}$ s s^{-1} for the 27.87 s pulsation during the interval 1976-8; and $\dot{P} = 2 \times 10^{-10}$ s s^{-1} for the

TABLE 4: RAPID PERIODICITIES

Star	Subclass λ band*	Period	Coherence**	Refs.
RX And	dN/V	36s	QP	(194,237)
AE Aqr	N1/V	33.078s	$\dot{p}<2.5\times10^{-14}$	(167,170)
AE Aqr	N1/V	~18, ~36s	QP	(167)
AE Aqr	N1/hx	33s	?	(168)
TT Ari	N1/V	12, 32, 500-1100s	QP	(216,99)
TT Ari	N1/hx	9, 12, 32s	?	(99)
Z Cam	dN/V	16.0-18.8s	c	(193)
SY Cnc	dN/V	23.3-33.0s	c	(143,171,194)
YZ Cnc	dN/V	75-95s	QP	(171)
YZ Cnc	dN/hx	227s	?	(42)
AM CVn	td/V	26.3s	QP	(171)
HT Cas	dN/V	20.2-20.4s	c	(171)
HT Cas	dN/V	~100s	QP	(171)
V436 Cen	dN/V	19.5-20.1s	c	(274)
Z Cha	dN/V	27.7s	c	(269)
EM Cyg	dN/V	16.6-21.2s	c	(149,171)
SS Cyg	dN/V	8.5-10.9s	c,QP	(80,86,174)
SS Cyg	dN/V	32-36s	QP	(171,196)
SS Cyg	dN/sx	~9, ~12s	~20 cycles	(44,46)
U Gem	dN/V	73, 146s	QP	(196)
U Gem	dN/sx	20-30s	1-2 cycles	(46)
AH Her	dN/V	24.0-38.8s	c,QP	(81,82,171)
AH Her	dN/V	~100s	QP	(171)
AM Her	MV/V	various	QP	(242)
AM Her	MV/sx	20-60s	QP	(250)
DQ Her	cN/V	71.07s	$\dot{p}=-8.7\times10^{-13}$	(176)
V533 Her	cN/V	63.63s	$\dot{p}<2.5\times10^{-13}$	(166,170)
VW Hyi	dN/V	24-32s	QP	(274)
VW Hyi	dN/V	88,413s	QP	(71,274)
X Leo	dN/V	~160	QP	(143)
CN Ori	dN/V	24.3-25.0s	c	(274)
RU Peg	dN/V	11.6-11.8s	c	(175)
RU Peg	dN/V	~51s	QP	(175,196)
GK Per	cN/V	~380s	QP	(171)
KT Per	dN/V	22.0-29.2s	QP	(148,194)
KT Per	dN/V	82,147s	QP	(196)
RR Pic	cN/V	20-40s	QP	(273)
RW Sex	N1/V	620s, 1280s?	QP	(78)
WZ Sge	dN/V	27.87s	$\dot{p}=8\times10^{-12}$	(169,198)
WZ Sge	dN/V	28.97s	$\dot{p}=4\times10^{-10}$	(144,169)
V1223 Sgr	MV?/V	13.2m	c	(224)
V3885 Sgr	N1/V	29.0s	QP	(79)
UX UMa	N1/V	28.5-30.0s	$\dot{p}\sim1\times10^{-5}$	(146)
2A0311-227	MV/V	4-7m	QP	(291)
2A0311-227	MV/sx,hx	4-7m	QP	(177)
E1405-451	MV/V	1-3s	QP	(134)
H2215-086	N1?/V	20.9m	c	(173)
H2252-035	MV?/V	14.3m	c	(172)
H2252-035	MV?/hx	13.4m	c	(287)

*V = visual; sx = soft X-ray (0.1-0.5 keV); hx = hard X-ray (>0.5 keV); td = twin degenerate.
**QP = quasi-periodic; c = 'coherent'.

28.97 s pulsation after the 1978 outburst. Erratic phase jitter of 30°-60° has been seen in the 27.87 s oscillation on a probable timescale of 1-3 wk [169]. It is difficult to account for the presence of more than one stable period in a rigid-rotator model. Non-radial pulsations of the degenerate dwarf have been suggested as an alternative mechanism [198]. The upper limit on the pulsed fraction of the hard X-ray flux in WZ Sge is 15 per cent [17].

Compared to the pulsations described above, the optical oscillations that have been observed during the visual outbursts of many dwarf novae are much less stable in period ($\dot{P} = 10^{-5}$-10^{-6} s s^{-1}) and phase (shifts of up to 90° cycle^{-1}). Two novalike objects, UX UMa [146] and V3885 Sgr [79] and the ex-nova RR Pic [273] may sometimes exhibit like pulsations. The oscillation periods range from several seconds to hundreds of seconds, and the amplitudes are of order 0.1 per cent. Soft X-ray pulsations with similar characteristics but much higher amplitudes (~15-30 per cent on average) have been observed in the dwarf novae SS Cyg and U Gem during optical outbursts (see also Section 4.4). In SS Cyg X-ray pulsations have been detected both times that the source was observed, once at a period P ~9 s and once at P ~12 s [44, 46]. In U Gem, enhanced soft X-ray emission has been seen during all three (pointed) outburst observations conducted to date, but in only one of these were soft X-ray pulsations (P ~25 s) detected [46, 131].

Robinson & Nather [196] have proposed classifying the dwarf nova pulsations into 'coherent' oscillations with periods in the range 7-40 s and comparative stability over the few-hour length of the typical observing run, and 'quasi-coherent' oscillations with periods from 30 to 400 s and stability over only a few pulsation cycles. In SS Cyg and RU Peg, both types of oscillations have been observed simultaneously [196]. During the course of a single outburst the 'coherent' periodicity can drift smoothly in period as the luminosity changes, reaching a minimum value near the time of maximum light; variations in period of up to 16 per cent have been observed over a single outburst, and 35 per cent between different outbursts [82, 171]. There is also evidence that the period varies with orbital phase in some stars (e.g. V436 Cen and VW Hyi [274]). The period of the 'quasi-coherent' oscillation can differ by factors of 2-5 between different observations of the same star [196].

The X-ray observations and some recent optical observations suggest that there may be no clear dichotomy in the coherence of the pulsations among the dwarf novae. The X-ray data on SS Cyg show that the level of coherence can change by a factor of 3 on a timescale of an hour, and by a factor of 10 between different outbursts [46]. The range in coherence values for the soft X-ray pulsations of SS Cyg overlaps that of the optical 'coherent' and 'quasi-coherent' oscillations. The coherence of the optical pulsation in both SS Cyg and AH Her was found to be a function of outburst phase, being higher near optical maximum [80, 81]: in SS Cyg the coherence decreased by more than a factor of 7 during the outburst. This suggests that the pulsations in the dwarf novae become unobservable toward the end of an outburst because the coherence decreases to the point where they degenerate into random flickering.

The high amplitude of the pulsed soft X-rays in SS Cyg and U Gem indicates that the soft X-ray–EUV component is the primary pulsed emission in these stars. No pulsations have been detected in the hard X-ray component of either SS Cyg or U Gem during any outburst phase [234, 233]. The modelling of the phase shifts of the optical pulses seen during the eclipses of highly inclined variables such as Z Cha, DQ Her, UX UMa, and HT Cas suggests that the optical oscillations in these stars are due to reprocessing in the inner accretion disk of a high-energy pulsed component originating close to the degenerate star [269, 30, 183, 171]. Multi-color optical observations of AH Her [81] and SY Cnc [143] indicate that the optical pulsation spectrum is much bluer than the disk spectrum. Middleditch & Córdova [143] find that no single blackbody spectrum or combination of blackbody spectra can fit the colors of the pulsations of SY Cnc.

Various models have been proposed for the dwarf nova pulsations: These include (1) rotating hot spots in the disk or on the surface of the degenerate star, (2) magnetic and hydrodynamic instabilities in the accretion flow, and (3) nonradial pulsations in the disk or surface layer of the degenerate star. These have been reviewed by Patterson [171] and Córdova & Mason [303] who conclude that neither the present observations nor the theoretical models are detailed enough to exclude most of the proposals. The pulsation phenomenon can be described mathematically by: (1) a number of superposed, periodic pulse trains that interfere with each other to produce apparent shifts in the phase of the resultant oscillation [196, 80, 46]; (2) a strictly periodic clock whose phase is disturbed in a random manner [44]; and (3) the superposition of a number of oscillations with closely spaced periods [158, 161]. These physically disparate models can produce essentially indistinguishable light curves and Fourier spectra from data sets of the length, coherence and statistical precision afforded by currently available observations. Longer optical and X-ray observing runs to more sensitively measure the coherence of the oscillations, and pulse timing of the high-amplitude X-ray observations through an outburst to more accurately constrain the noise process, may clarify the origin of the pulsation phenomenon.

4.8   The outburst
4.8.1   Classical novae

The idea that the outburst of a classical nova is a thermonuclear runaway in the accreted envelope of the degenerate star was originally proposed by

Schatzman ([201]; see also [109]). Hydrodynamic modelling can reproduce the observed gross features of the ejection process following such a nuclear explosion [64, 221, 219, 249]. Recent studies suggest that the effects of non-radial accretion and rapid rotation of the central star could also be important [53, 104, 218]. The theoretical studies have led to predictions of various properties of the outburst (cf. [222]). The computations indicated that the ejected shells of fast novae would have enhanced carbon, oxygen and nitrogen (CNO) abundances relative to solar abundances, and that N would be enhanced relative to C and O. The CNO cycle was shown to be important in defining a post-outburst extended phase of shell burning at constant luminosity and in governing the speed class of novae. These predictions have been verified by recent observations of the composition of nova shells ([56, 249] and examples therein), and observations of the constant luminosity phase in recent novae (e.g. FH Ser: [62]; Nova Cygni 1975: [294]; Nova Vul 1976: [150]; Nova Cygni 1978: [225]).

Truran [249] and Shara, Prialnik & Shaviv [207] show that in addition to the CNO abundance, the mass of the white dwarf is important in determining whether the outburst is fast or slow: faster outbursts will occur on more massive white dwarfs. The observation that the mass of the degenerate star in DQ Her may be relatively low ($\sim 0.5\,M_\odot$ [296, 215]) would then resolve the question of why this nova had such a slow outburst development when it was so rich in CNO, since it appears to be extremely difficult to produce a fast outburst on a low-mass white dwarf, regardless of the CNO abundance.

4.8.2 Recurrent novae

The origin of the outbursts of the recurrent novae is less clear. Webbink [282] has suggested that the outbursts of the recurrent nova T CrB are caused by enhanced mass accretion, based on an examination of its outburst light curve. Williams *et al.* [293] conclude, however, that there is no obvious preference for either an accretion event or a thermonuclear event as the cause of the outburst of U Sco. Its spectral evolution differs from that of other recurrent novae (e.g. T CrB and RS Oph). Yet the outburst of U Sco also differs from that of the classical novae in a number of respects. The optical and UV data on the June 1979 eruption of that star show ejection at high velocities ($\sim 7500\,\text{km s}^{-1}$) of a helium-rich shell whose mass ($\sim 10^{-7}\,M_\odot$) was very small compared to the typical mass of a classical nova shell [9, 293]. Although the nova ejecta is rich in N, the combined abundance of carbon, nitrogen and oxygen is essentially solar. Furthermore, the constant luminosity phase that is characteristic of classical novae was not observed. Williams *et al.* [293] (see also [310]) discuss various ways in which a classical nova outburst might nevertheless produce the observed properties of U Sco. They also note that the differences among recurrent nova outbursts may be explained by differences in their companion stars. The giant companions of T CrB and RS Oph may modify the development of the ejecta; U Sco, on the other hand, probably has a low-mass, helium-rich companion [283, 293, 310].

Data on the outburst of the long period erupting variable WZ Sge favors an enhanced mass accretion episode. This star erupted in December 1978 after about 32 yr of minimal activity [138]. Its UV spectrum was more similar to that of the dwarf novae than to that of the classical or recurrent novae, and there was no evidence of mass loss [220]. The appearance of 'superhumps' (Section 4.7) in the visual light curve on the outburst decline has encouraged its reclassification as a dwarf nova of the SU UMa type rather than a recurrent nova ([178] and references therein).

4.8.3 Dwarf novae

Observations of the quasi-periodic dwarf nova outbursts indicate that the major source of enhanced luminosity during the eruption is the accretion disk. The models that have been proposed to explain an increased mass flux through the disk invoke either unstable mass accretion within the disk [153, 88], or an instability in the companion star [16]. There are a number of observations which have been interpreted as supporting the latter model (e.g. [256, 178]). Furthermore, Bath & Pringle [13] have considered the time-dependent evolution of an accretion disk following a burst of mass transfer from the companion star and demonstrate that they can reproduce the gross morphology of a dwarf nova outburst in this way.

Observational support for a disk instability comes from measurements of the quiescent-state luminosity of U Gem. Paczynski & Schwarzenberg-Czerny [156] have noted a discrepancy between the luminosity of the bright spot on the disk and that of the disk itself, which may imply that the rate of mass transfer from the companion is much higher than the accretion rate through the disk. The calculations of Meyer & Meyer-Hofmeister [141] illustrate that, for certain values of the surface density of material in a disk, two values of viscosity (accretion rate) are possible depending on whether the disk structure is convective or radiative. They suggest that the quiescent-outburst cycle of the dwarf novae is caused by transitions between the convective (low-viscosity) and radiative (high-viscosity) states. In the model of Meyer & Meyer-Hofmeister, the double-valued nature of the surface density function is caused by a change in the ionization structure of the gas as the temperature increases. Cannizzo, Ghosh & Wheeler [300] argue that a more significant transition is related to the onset of convection in an optically thin, cold (3000 K) torus. Matter transferred from the companion star accumulates in the cold torus until a critical surface density is reached. Beyond this value, no cold, steady-state solution exists, and the torus heats up until a hot, stable solution is obtained; this is accompanied by an increase in the viscosity. Smak [309] has also constructed a vertically integrated, steady-state convective model of disk accretion, which differs mainly in its treatment of the viscosity.

There are a number of other observations that are relevant to the question of the origin of the dwarf nova outburst. Smak [212] and Warner [269] deduced that the disk expands during the initial phase of the outburst, by studying changes in the eclipse light curves of U Gem and Z Cha. The spectral distribution also reddens early in the outburst (see [137] and examples therein; [13]) and the entire continuum brightens (cf. [71]). The UV emission lines that are prominent during quiescence do not appear to change during the rise to outburst in the one instance studied (RX And [238]), suggesting that the UV line-emitting region is not affected during the initial outburst rise. During the outburst the optical and UV emission lines go into absorption (see Section 4.5.3 regarding the latter). There are also dramatic changes in the hard and soft X-ray components during the outbursts of dwarf novae (Section 4.4) and various quasi-periodic phenomena appear (rapid oscillations and orbital-related humps: see Section 4.7). Because of the observational difficulties associated with trying to detect the early stage of a dwarf nova outburst, it is unclear at what point the various changes observed in different spectral bands happen, or exactly how they are correlated with each other.

4.9 Conclusions

Substantial advances continue to be made in the understanding of cataclysmic binary systems. We expect that much of the impetus for future work will come from improved and more extensive observations in the ultraviolet and X-ray bands, regions of the spectrum that have only recently become available for the study of cataclysmic variables. There is growing evidence, too, that these stars emit a substantial flux in the EUV between ~100 and 1000 Å, and it is hoped that the interstellar absorption column to at least a few stars will be sufficiently low to enable their spectra and fluxes to be measured in this range. The application of advanced techniques in the optical and infrared region of the spectrum will also be important, particularly in determining basic data such as the orbital parameters and the masses of the component stars. Simultaneous observations over a wide range of wavelengths are essential in modelling the emission from these variable systems. The voluminous observational data that have been gathered on cataclysmic variables have, to paraphrase Carroll [29], filled our heads with ideas..., but to a large extent we still do not exactly know they are. Outstanding questions include: the mechanism behind the dwarf nova outburst; the cause of the pulsations and other rapid variability; the origin of the strong EUV emission inferred in some stars; the influence of the magnetic field of the degenerate dwarf in determining the accretion geometry, the physics of the white dwarf envelope, and the appearance of the system; and, on a more fundamental level, the evolutionary history of cataclysmic variables (discussed further in Chapter 8, Section 3 of this book) and the relationship of the various subclasses to each other and to the binaries that contain accreting neutron stars rather than degenerate dwarfs.

Acknowledgements

The authors are grateful to A. Holm, I. Tuohy and R. Wade for communicating unpublished data; and to K. Jensen, W. Priedhorsky, E. L. Robinson, S. Starrfield, R. Wade, R. Webbink, and R. Williams for their comments on a preliminary version of the manuscript. This work was supported by NASA Grants NAGW-44 and NAS 6-5022B and the US Department of Energy.

References

[1] Agrawal, P. C., Rao, A. R., Riegler, G. R., Pickles, A. J. & Visvanathan, N. 1982, *I.A.U. Circ.*, No. 3649.
[2] Angel, J. R. P. 1977, *Ap. J.*, **216**, 1.
[3] Bailey, J. 1975, *J. Brit. Astron. Ass.*, **86**, 30.
[4] Bailey, J. & Axon, D. J. 1981, *MNRAS*, **194**, 187.
[5] Bailey, J. & Ward, M. 1981, *MNRAS*, **194**, 17P.
[6] Bailey, J., Hough, J. M. & Axon, D. J. 1980, *Nature*, **285**, 306.
[7] Bailey, J. A., Jones, D. H. P., Parkes, G. E. & Mason, K. O. 1978, *MNRAS*, **184**, 73P.
[8] Bailey, J., Sherrington, M. R., Giles, A. B. & Jameson, R. F. 1981, *MNRAS*, **196**, 121.
[9] Barlow, M. *et al.* 1981, *MNRAS*, **195**, 61.
[10] Barwig, H. & Schoembs, R. 1981, *Inf. Bull. Var. Stars*, No. 2031.
[11] Bateson, F. M. 1977, *N.Z. J. Sci.*, **20**, 73.
[12] Bateson, F. M. 1978, *MNRAS*, **184**, 567.
[13] Bath, G. T. & Pringle, J. E. 1981, *MNRAS*, **194**, 967.
[14] Bath, G. T., Evans, W. D. & Pringle, J. E. 1974, *MNRAS*, **166**, 113.
[15] Bath, G. T., Pringle, J. E. & Whelan, J. A. J. 1980, *MNRAS*, **190**, 185.
[16] Bath, G. T., Evans, W. D., Papaloizou, J. & Pringle, J. E. 1974, *MNRAS*, **169**, 447.
[17] Becker, R. H. 1981, *Ap. J.*, **251**, 626.
[18] Becker, R. H. & Marshall, F. E. 1981, *Ap. J.*, **244**, L93.
[19] Berg, R. A. & Duthie, J. G. 1977, *Ap. J.*, **211**, 859.
[20] Bianchini, A. 1980, *MNRAS*, **192**, 127.
[21] Bianchini, A., Hamzaoglu, E. & Sabbadin, F. 1981, *Astron. Ap.*, **99**, 392.
[22] Bond, H. & Webbink, R. 1977, private communication.
[23] Borne, K. D. 1977, *BAAS*, **9**, 556.
[24] Boyarchuk, A. A. 1969, In: *Nonperiodic Phenomena in Variable Stars*, ed. Detre, L. (Academic Press, Budapest), p. 395.
[25] Bradt, H. V., Doxsey, R. E. & Jernigan, J. G. 1979, 'Adv. Space Exploration' (*Proc. of IAU/COSPAR Symp. on X-Ray Astronomy*, Innsbruck, Austria), eds. Baity, W. A. & Peterson, L. E. (Pergamon, Oxford), Vol. 3, p. 3.
[26] Brecher, K., Morrison, P. & Sadun, A. 1977, *Ap. J.*, **217**, L139.
[27] Breysacher, J. & Vogt, N. 1980, *Astron. Ap.*, **87**, 349.

[28] Campolonghi, F., Gilmozzi, R., Guidoni, U., Messi, R., Natali, G. & Wells, J., 1980, *Astron. Ap.*, **85**, L4.
[29] Carroll, L. 1929, *Through the Looking Glass; and What Alice found there* (Dutton, New York).
[30] Chanan, G. A., Nelson, J. E. & Margon, B. 1978, *Ap. J.*, **226**, 963.
[31] Charles, P. A. & Mason, K. O. 1979, *Ap. J.*, **232**, L25.
[32] Chiappetti, L., Tanzi, E. G. & Treves, A. 1980, *Spa. Sci. Rev.*, **27**, 3.
[33] Chincarini, G. & Walker, M. F. 1981, *Astron. Ap.*, **104**, 24.
[34] Chlembowski, T., Halpern, J. P. & Steiner, J. E. 1981, *Ap. J.*, **247**, L35.
[35] Coe, M. J. & Wickramasinghe, D. T. 1981, *Nature*, **290**, 119.
[36] Cooke, B. A. *et al.* 1978, *MNRAS*, **182**, 489.
[37] Córdova, F. A. & Garmire, G. P. 1979, *Nature*, **279**, 782.
[38] Córdova, F. A. & Mason, K. O. 1980, *Nature*, **287**, 25.
[39] Córdova, F. A. & Mason, K. O. 1982, *Ap. J.*, **260**, 716.
[40] Córdova, F. A. & Riegler, G. R. 1979, *MNRAS*, **188**, 103.
[41] Córdova, F. A., Jensen, K. A. & Nugent, J. J. 1981, *MNRAS*, **196**, 1.
[42] Córdova, F. A. & Mason, K. O. 1983, *MNRAS*, submitted.
[43] Córdova, F. A., Mason, K. O. & Nelson, J. E. 1981, *Ap. J.*, **245**, 609.
[44] Córdova, F. A., Chester, T. J., Tuohy, I. R. & Garmire, G. P. 1980a, *Ap. J.*, **235**, 163.
[45] Córdova, F. A., Nugent, J. J., Klein, S. R. & Garmire, G. P. 1980b, *MNRAS*, **190**, 87.
[46] Córdova, F. A., Chester, T. J., Mason, K. O., Kahn, S. M. & Garmire, G. P. 1982, *Ap. J.*, submitted.
[47] Cowley, A. P. & Crampton, D. 1977, *Ap. J.*, **212**, L121.
[48] Cowley, A. P., Crampton, D. & Hesser, J. E. 1977a, *PASP*, **89**, 716.
[49] Cowley, A. P., Crampton, D. & Hesser, J. E. 1977b, *Ap. J.*, **214**, 471.
[50] Cowley, A. P., Crampton, D. & Hutchings, J. B. 1980, *Ap. J.*, **241**, 269.
[51] Cowley, A. P., Crampton, D., Hutchings, J. B. & Marlborough, J. M. 1975, *Ap. J.*, **195**, 413.
[52] Crawford, J. A. & Kraft, R. P. 1956, *Ap. J.*, **123**, 44.
[53] Durisen, R. H. 1977, *Ap. J.*, **213**, 145.
[54] Fabbiano, G., Hartmann, L., Raymond, J., Steiner, J., Branduardi-Raymont, G. & Matilsky, T. 1981, *Ap. J.*, **243**, 911.
[55] Ferland, G. J. 1980, *The Observatory*, **100**, 166.
[56] Ferland, G. J. & Shields, G. A. 1978, *Ap. J.*, **226**, 172.
[57] Fontaine, G. & Michaud, G. 1979, *Ap. J.*, **231**, 826.
[58] Frank, J. & King, A. R. 1981a, *MNRAS*, **195**, 227.
[59] Frank, J. & King, A. R. 1981b, *MNRAS*, **196**, 507.
[60] Frank, J., King, A. R., Sherrington, M. R., Jameson, R. F. & Axon, D. J. 1981, *MNRAS*, **195**, 505.
[61] Fujimoto, M. Y. & Truran, J. W. 1982, *Ap. J.*, **257**, 303.
[62] Gallagher, J. S. & Code, A. D. 1974, *Ap. J.*, **189**, 303.
[63] Gallagher, J. S. & Oinas, V. 1974, *PASP*, **86**, 952.
[64] Gallagher, J. S. & Starrfield, S. G. 1978, *Ann. Rev. Astron. Ap.*, **16**, 171.
[65] Gilliland, R. 1982, *Ap. J.*, **254**, 653.
[66] Greenstein, J. L. & Oke, J. B. 1982, *Ap. J.*, **258**, 209.
[67] Greenstein, J. L., Sargent, W. L. W., Boroson, T. A. & Boksenberg, A. 1977, *Ap. J.*, **218**, L121.
[68] Griffiths, R. E., Ward, M. J., Blades, J. C., Wilson, A. S., Chaisson, L. & Johnston, M. D. 1979, *Ap. J.*, **232**, L27.
[69] Griffiths, R. E. *et al.* 1980, *MNRAS*, **193**, 25P.
[70] Guinan, E. F. & Sion, E. M. 1982, *Ap. J.*, **258**, 217.
[71] Haefner, R., Schoembs, R. & Vogt, N. 1979, *Astron. Ap.*, **77**, 7.
[72] Hassall, B. J. M. *et al.* 1981, *MNRAS*, **197**, 275.
[73] Heap, S. R. *et al.* 1978, *Nature*, **275**, 385.
[74] Hearn, D. R. & Marshall, F. J. 1979, *Ap. J.*, **232**, L21.

[75] Hearn, D. R. & Richardson, J. A. 1977, *Ap. J.*, **213**, L115.
[76] Heise, J. & Brinkman, A. C. 1979, *In: Galactic X-ray Sources*, eds. Sandford, P. W., Laskarides, P. & Salton, J. (J. Wiley), p. 393.
[77] Herbig, G. H., Preston, G. W., Smak, J. & Paczynski, B. 1965, *Ap. J.*, **141**, 617.
[78] Hesser, J. E., Lasker, B. M. & Osmer, P. S. 1972, *Ap. J.*, **176**, L31.
[79] Hesser, J. E., Lasker, B. M. & Osmer, P. S. 1974, *Ap. J.*, **189**, 315.
[80] Hildebrand, R. H., Spillar, E. J. & Stiening, R. F. 1981a, *Ap. J.*, **243**, 223.
[81] Hildebrand, R. H., Spillar, E. J. & Stiening, R. F. 1981b, *Ap. J.*, **248**, 268.
[82] Hildebrand, R. H., Spillar, E. J., Middleditch, J., Patterson, J. J. & Stiening, R. F. 1980, *Ap. J.*, **238**, L.145.
[83] Holm, A. 1979, private communication.
[84] Holm, A. & Gallagher, J. 1974, *Ap. J.*, **192**, 425.
[85] Horne, K. 1980, *Ap. J.*, **242**, L167.
[86] Horne, K. & Gomer, R. 1980, *Ap. J.*, **237**, 845.
[87] Horne, K., Lanning, H. H. & Gomer, R. H. 1982, *Ap. J.*, **252**, 681.
[88] Hoshi, R. 1979, *Prog. Theor. Phys.*, **61**, 1307.
[89] Howarth, I. D. 1977, *J. Br. Astron. Ass.*, **88**, 79.
[90] Hutchings, J. B. 1979, *Ap. J.*, **232**, 176.
[91] Hutchings, J. B. 1980, *PASP*, **92**, 458.
[92] Hutchings, J. B., Cowley, A. P. & Crampton, D. 1979, *Ap. J.*, **232**, 500.
[93] Hutchings, J. B., Cowley, A. P. & Crampton, D. 1981, *I.A.U. Circ.*, 3585.
[94] Hutchings, J. B., Crampton, D., Cowley, A. P., Thorstensen, J. R. & Charles, P. A. 1981, *Ap. J.*, **249**, 680.
[95] Illarionov, A. F. & Sunyaev, R. A. 1972, *Sov. Astron. A. J.*, **19**, 38.
[96] Jameson, R. F., King, A. R. & Sherrington, M. R. 1980, *MNRAS*, **191**, 559.
[97] Jameson, R. F., King, A. R. & Sherrington, M. R. 1981, *MNRAS*, **195**, 235.
[98] Jensen, K., Nousek, J. & Nugent, J. 1982, *Ap. J.*, **261**, 625.
[99] Jensen, K. A., Córdova, F. A., Middleditch, J., Mason, K. O., Grauer, A., Horne, K. & Gomer, R. 1983, *Ap. J.*, 270, in press.
[100] Joss, P. C., Katz, J. I. & Rappaport, S. A. 1979, *Ap. J.*, **230**, 176.
[101] Katz, J. I. 1975, *Ap. J.*, **200**, 298.
[102] Kiplinger, A. L. 1979, *Ap. J.*, **234**, 97.
[103] Kiplinger, A. L. 1980, *Ap. J.*, **236**, 839.
[104] Kippenhahn, R. & Thomas, H.-C. 1978, *Astron. Ap.*, **63**, 265.
[105] King, A. R. & Lasota, J. P. 1979, *MNRAS*, **188**, 653.
[106] King, A. R. & Lasota, J. P. 1980, *MNRAS*, **191**, 721.
[107] King, A. R., Ricketts, M. J. & Warwick, R. S. 1979, *MNRAS*, **187**, 77P.
[108] Kraft, R. P. 1962, *Ap. J.*, **135**, 408.
[109] Kraft, R. P. 1964, *Ap. J.*, **139**, 457.
[110] Kraft, R. P. & Luyten, W. J. 1965, *Ap. J.*, **142**, 1041.
[111] Kruszewski, A., Mewe, R., Heise, J., Chlebowski, T., Van Dijk, W. & Bakker, R. 1981, *Sp. Sci. Rev.*, **30**, 221.
[112] Krzeminski, W. & Kraft, R. P. 1964, *Ap. J.*, **140**, 921.
[113] Krzeminski, W. & Serkowski, K. 1977, *Ap. J.*, **216**, L45.
[114] Kukarkin, B. V. et al. 1969, *General Catalogue of Variable Stars*, 3rd edn, Moscow.
[115] Kukarkin, B. V. & Parenago, P. P. 1934, *Var. Star. Bull. IV*, No. 8, p. 44.
[116] Kurochkin, N. E. & Shugarov, S. Yu. 1980, *The Astron. Circ.*, No. 1114.
[117] Kurucz, R. L., Peytremann, E. & Avrett, E. H. 1974, *Blanketed Model Atmospheres for Early-Type Stars* (Washington: Smithsonian Institution).
[118] Lamb, D. Q. 1974, *Ap. J.*, **192**, L129.
[119] Lamb, D. Q. & Masters, R. 1979, *Ap. J.*, **234**, L117.
[120] Lambert, D. L. & Slovak, M. H. 1981, *PASP*, **93**, 477.
[121] Latham, D. W., Liebert, J. & Steiner, J. E. 1981, *Ap. J.*, **246**, 919.
[122] Liebert, J. 1980, *Ann. Rev. Astron. Ap.*, **18**, 363.
[123] Libert, J., Stockman, H. S., Williams, R. E., Tapia, S., Green, R., Rautenkranz, D., Ferguson, D. H. & Szkody, P. 1982, *Ap. J.*, **256**, 594.
[124] Longmore, A. J., Lee, T. J., Allen, D. A. & Adams, D. J. 1981, *MNRAS*, **195**, 825.

[125] Lynden-Bell, D. 1969, *Nature*, **233**, 690.
[126] Margon, B., Szkody, P., Bowyer, S., Lampton, M. & Paresce, F. 1978, *Ap. J.*, **224**, 167.
[127] Mason, K., Córdova, F. & Swank, J. 1979, *In: (COSPAR) X-ray Astronomy*, eds. Baity, W. A. & Peterson, L. E. (Pergamon Press: Oxford and N.Y.), p. 121.
[128] Mason, K. O., Kahn, S. M. & Bowyer, C. S. 1979, *Nature*, **280**, 568.
[129] Mason, K. O., Lampton, M., Charles, P. & Bowyer, S. 1978, *Ap. J.*, **226**, L129.
[130] Mason, K. O., Middleditch, J., Nelson, J., White, N., Seitzer, P., Tuohy, I. & Hunt, L. 1980, *Ap. J.*, **242**, L109.
[131] Mason, K. O., Kahn, S. M., Córdova, F. A., Middleditch, J. & Bowyer, S. 1982*a*, in preparation.
[132] Mason, K. O., Murdin, P. G., Tuohy, I. R., Seitzer, P. & Branduardi-Raymont, G. 1982*b*, *MNRAS*, **200**, 793.
[133] Mason, K. O., Reichert, G., Bowyer, S. & Thorstensen, J. 1982*c*, *PASP*, **94**, 521.
[134] Mason, K. O., Middleditch, J., Córdova, F. A., Jensen, K. A., Reichert, G. A., Murdin, P. G., Clark, D. & Bowyer, S. 1983, *Ap. J.*, **264**, 575.
[135] Mason, K. O., Córdova, F. A., Middleditch, J., Reichert, G. A., Bowyer, S., Murdin, P. G., Clark, D. 1983, *PASP*, submitted.
[136] Mattei, J. 1981, reported at the Fifth Annual Santa Cruz Summer Workshop in Astron. & Ap.
[137] Mayo, S. K., Wickramasinghe, D. T. & Whelan, J. A. J. 1980, *MNRAS*, **193**, 793.
[138] McGraw, J. T. 1978, reported by J. Patterson 1978, *I.A.U. Circ.*, No. 3311.
[139] McLaughlin, D. B. 1960, *Stars and Stellar Systems*, Vol. 6, Stellar atmospheres, ed. Greenstein, J. L. (Chicago: University of Chicago Press), p. 585.
[140] Meggitt, S. M. A. & Wickramasinghe, D. T. 1982, *MNRAS*, **198**, 71.
[141] Meyer, F. & Meyer-Hofmeister, E. 1981, *Astron. Ap.*, **104**, L10.
[142] Michalsky, J. J., Stokes, G. M. & Stokes, R. A. 1977, *Ap. J.*, **216**, L35.
[143] Middleditch, J. & Córdova, F. A. 1982, *Ap. J.*, **255**, 585.
[144] Middleditch, J. & Nelson, J. 1980, private communication.
[145] Motch, C. 1981, *Astron. Ap.*, **100**, 277.
[146] Nather, R. E. & Robinson, E. L. 1974, *Ap. J.*, **190**, 637.
[147] Nather, R. E., Robinson, E. L. & Stover, R. J. 1981, *Ap. J.*, **244**, 269.
[148] Nevo, I. & Sadeh, D. 1976, *MNRAS*, **177**, 167.
[149] Nevo, I. & Sadeh, D. 1978, *MNRAS*, **182**, 595.
[150] Ney, E. & Hatfield, B. F. 1978, *Ap. J.*, **219**, L111.
[151] Nousek, J., Córdova, F. & Garmire, G. 1980, *Ap. J.*, **242**, 1107.
[152] Oke, J. B. & Wade, R. A. 1981, private communication.
[153] Osaki, Y. 1974, *PASJ*, **26**, 429.
[154] Pacharintanakul, P. & Katz, J. I. 1980, *Ap. J.*, **238**, 985.
[155] Paczynski, B. 1965, *Acta. Astron.*, **15**, 197.
[156] Paczynski, B. & Schwarzenberg-Czerny, A. 1980, *Acta. Astron.*, **30**, 127.
[157] Paczynski, B. & Sienkiewicz, R. 1981, *Ap. J.*, **248**, L27.
[158] Papaloizou, J. & Pringle, J. E. 1978, *MNRAS*, **182**, 423.
[159] Papaloizou, J. & Pringle, J. E. 1979, *MNRAS*, **189**, 293.
[160] Papaloizou, J. & Pringle, J. E. 1980*a*, *MNRAS*, **190**, 13P.
[161] Papaloizou, J. & Pringle, J. E. 1980*b*, *MNRAS*, **190**, 43.
[162] Papaloizou, J. C. B. & Whelan, J. A. J. 1973, *MNRAS*, **164**, 1.
[163] Papaloizou, J. C. B., Pringle, J. E. & MacDonald, J. 1982, *MNRAS*, **198**, 215.
[164] Patterson, J. 1979*a*, *Astron. J.*, **84**, 804.
[165] Patterson, J. 1979*b*, *Ap. J.*, **231**, 789.
[166] Patterson, J. 1979*c*, *Ap. J.*, **233**, L13.
[167] Patterson, J. 1979*d*, *Ap. J.*, **234**, 978.
[168] Patterson, J. 1980*a*, *Ap. J.*, **240**, L133.
[169] Patterson, J. 1980*b*, *Ap. J.*, **241**, 235.
[170] Patterson, J. 1981, private communication.
[171] Patterson, J. 1981, *Ap. J. (Suppl.)*, **45**, 517.

[172] Patterson, J. & Price, C. 1981, *Ap. J.*, **243**, L83.
[173] Patterson, J. & Steiner, J. E. 1983, *Ap. J.*, **264**, L61.
[174] Patterson, J., Robinson, E. L. & Kiplinger, A. L. 1978, *Ap. J.*, **226**, L137.
[175] Patterson, J., Robinson, E. L. & Nather, R. E. 1977, *Ap. J.*, **214**, 144.
[176] Patterson, J., Robinson, E. L. & Nather, R. E. 1978, *Ap. J.*, **224**, 570.
[177] Patterson, J., Williams, G. & Hiltner, W. A. 1981, *Ap. J.*, **245**, 618.
[178] Patterson, J., McGraw, J. T., Coleman, L. & Africano, J. L. 1981, *Ap. J.*, **248**, 1067.
[179] Patterson, J., Beuermann, K., Fabbiano, G., Lamb, D. Q., Raymond, J. C., Swank, J. M., White, N. E. & Horne, K. 1982, *Ap. J.*, submitted.
[180] Payne-Gaposchkin, C. 1957, *The Galactic Novae* (Amsterdam: North Holland).
[181] Payne-Gaposchkin, C. 1969, *Ap. J.*, **158**, 429.
[182] Payne-Gaposchkin, C. 1977, *Astron. J.*, **82**, 665.
[183] Petterson, J. A. 1980, *Ap. J.*, **241**, 247.
[184] Priedhorsky, W. P. 1977, *Ap. J.*, **212**, L117.
[185] Pringle, J. E. 1977, *MNRAS*, **178**, 195.
[186] Pringle, J. E. & Savonije, G. J. 1979, *MNRAS*, **187**, 777.
[187] Rappaport, S., Joss, P. C. & Webbink, R. F. 1982, *Ap. J.*, **254**, 616.
[188] Rappaport, S., Cash, W., Doxsey, R., McClintock, J. & Moore, G. 1974, *Ap. J.*, **187**, L5.
[189] Raymond, J. C., Black, J. H., Davis, R. J., Dupree, A. K., Gursky, H., Hartmann, L. & Matilsky, T. 1979, *Ap. J.*, **230**, L95.
[190] Ricketts, M. J., King, A. R. & Raine, D. J. 1979, *MNRAS*, **186**, 233.
[191] Ritter, H. 1980a, *Astron. Ap.*, **85**, 362.
[192] Ritter, H. 1980b, *Astron. Ap.*, **86**, 204.
[193] Robinson, E. L. 1973a, *Ap. J.*, **180**, 121.
[194] Robinson, E. L. 1973b, *Ap. J.*, **183**, 193.
[195] Robinson, E. L. 1976, *Ann. Rev. Astron. Ap.*, **14**, 119.
[196] Robinson, E. L. & Nather, R. E. 1979, *Ap. J. (Suppl.)*, **39**, 461.
[197] Robinson, E. L., Nather, R. E. & Kepler, S. O. 1982, *Ap. J.*, **254**, 646.
[198] Robinson, E. L., Nather, R. E. & Patterson, J. 1978, *Ap. J.*, **219**, 168.
[199] Robinson, E. L., Barker, E. S., Cochran, A. L., Cochran, W. D. & Nather, R. E. 1982, *Ap. J.*, **251**, 611.
[200] Rothschild, R. E. *et al.* 1981, *Ap. J.*, **250**, 723.
[201] Schatzman, E. 1949, *Ann. Ap.*, **12**, 281.
[202] Schmidt, G. D., Stockman, H. S. & Margon, B. 1981, *Ap. J.*, **243**, L157.
[203] Schneider, D. P. & Young, P. 1980a, *Ap. J.*, **240**, 871.
[204] Schneider, D. P. & Young, P. 1980b, *Ap. J.*, **238**, 946.
[205] Schneider, D. P., Young, P. & Shectman, S. A. 1981, *Ap. J.*, **245**, 644.
[206] Schwartz, D. A. *et al.* 1979, *Adv. Space Exploration*, **3**, 435.
[207] Shara, M. M., Prialnik, D. & Shaviv, G. 1980, *Ap. J.*, **239**, 586.
[208] Shakura, N. I. & Sunyaev, R. A. 1979, *Astron. Ap.*, **24**, 337.
[209] Schoembs, R. 1981, reported at the Fifth Annual Santa Cruz Summer Workshop in Astron. & Ap.
[210] Schoembs, R. & Vogt, N. 1981, *Astron. Ap.*, **97**, 185.
[211] Sherrington, M. R., Lawson, P. A., King, A. R. & Jameson, R. F. 1980, *MNRAS*, **191**, 185.
[212] Smak, J. 1971, *Acta Astron.*, **21**, 15.
[213] Smak, J. 1976, *I.A.U. Symp. 73*, eds. Eggleton, P. R., Mitton, S. & Whelan, J. A. J., p. 149.
[214] Smak, J. 1979, *Acta. Astron.*, **29**, 469.
[215] Smak, J. 1980, *Acta. Astron.*, **30**, 267.
[216] Smak, J. & Stepién, K. 1969, *Nonperiodic Phenomena in Variable Stars*, ed. Petre, L. (Academic Press: Budapest), p. 335.
[217] Smak, J. & Stepién, K. 1975, *Acta. Astron.*, **25**, 379.
[218] Sparks, W. M. & Kutter, G. S. 1979, *I.A.U. Colloq. No. 53, White Dwarfs and Variable Degenerate Stars*, eds. van Horn, H. & Weidemann, V. (University of Rochester Press: Rochester, NY), p. 294.

[219] Sparks, W. M., Kutter, G. S., Starrfield, S. & Truran, J. W. 1980a, *Highlights of Astronomy*, ed. Waymann, P. A. (IAU), vol. 5, p. 505.
[220] Sparks, W. M., Wu, C.-C., Holm, A. V. & Schiffer, F. H. III 1980b, *Highlights of Astronomy*, ed. Wayman, P. A. (IAU), vol. 5, p. 285.
[221] Starrfield, S. 1980, *Spa. Sci. Rev.*, **27**, 635.
[222] Starrfield, S., Truran, J. W. & Sparks, W. M. 1978, *Ap. J.*, **226**, 186.
[223] Starrfield, S., Truran, J. W. & Sparks, W. M. 1981, *Ap. J.*, **243**, L27.
[224] Steiner, J. E., Schwartz, D. A., Jablonski, F. J., Busko, I. C., Watson, M. G., Pye, J. P. & McHardy, I. M. 1981, *Ap. J.*, **249**, L21.
[225] Stickland, D. J., Penn, C. J., Seaton, M. J., Snijders, M. A. J. & Storey, P. J. 1981, *MNRAS*, **197**, 107.
[226] Stockman, H. S. 1977, *Ap. J.*, **218**, L57.
[227] Stockman, H. S., Liebert, J. & Bond, H. E. 1979, *I.A.U. Colloq. 53, White Dwarfs and Variable Degenerate Stars*, eds. van Horn, H. M. & Weidemann, V. (University of Rochester Press: Rochester, N.Y.), p. 334.
[228] Stolze, B. & Schoembs, R. 1981, *Inf. Bull. Var. Stars*, No. 2029.
[229] Stover, R. J. 1981a, *Ap. J.*, **248**, 684.
[230] Stover, R. J. 1981b, *Ap. J.*, **249**, 673.
[231] Stover, R. J., Robinson, E. L. & Nather, R. E. 1981, *Ap. J.*, **248**, 696.
[232] Stover, R. J., Robinson, E. L., Nather, R. E. & Montemayor, T. J. 1980, *Ap. J.*, **240**, 597.
[233] Swank, J. H. 1980, private communication.
[234] Swank, J. H. 1979, *I.A.U. Colloq. No. 53, White Dwarfs and Variable Degenerate Stars*, eds. van Horn, H. M. & Weidemann, V. (University of Rochester Press: Rochester, N.Y.), p. 135.
[235] Swank, J. H., Lampton, M., Boldt, E., Holt, S. & Serlemitsos, P. 1977, *Ap. J.*, **216**, L71.
[236] Swank, J. H., Boldt, E. A., Holt, S. S., Rothschild, R. E. & Serlemitsos, P. J. 1978, *Ap. J.*, **226**, L133.
[237] Szkody, P. 1976, *Ap. J.*, **207**, 190.
[238] Szkody, P. 1981a, *Ap. J.*, **247**, 577.
[239] Szkody, P. 1981b, *PASP*, **93**, 456.
[240] Szkody, P. & Brownlee, D. E. 1977, *Ap. J.*, **212**, L113.
[241] Szkody, P. & Crosa, L. 1981, *Ap. J.*, **251**, 620.
[242] Szkody, P., Córdova, F. A., Tuohy, I. R., Stockman, H. S., Angel, J. R. P. & Wisniewski, W. 1980, *Ap. J.*, **241**, 1070.
[243] Szkody, P., Crosa, L., Bothun, G. D., Downes, R. A. & Schommer, R. A. 1981a, *Ap. J.*, **249**, L61.
[244] Szkody, P., Schmidt, E., Crosa, L. & Schommer, R. 1981b, *Ap. J.*, **246**, 223.
[245] Tapia, S. 1977, *Ap. J.*, **212**, L125.
[246] Tapia, S. 1979, *IAU Circ.*, 3327.
[247] Tanzi, E. G., Chincarini, G. & Tarenghi, M. 1981, *PASP*, **93**, 68.
[248] Thorstensen, J., Charles, P., Bowyer, S., Briel, U. G., Doxsey, R. E., Griffiths, R. E. & Schwartz, D. A. 1979, *Ap. J.*, **233**, L57.
[249] Truran, J. W. 1981, Progress in Particle and Nuclear Physics, ed. Wilkinson, D. (Pergamon Press: Oxford), vol. 6, p. 177.
[250] Tuohy, I. R., Mason, K. O., Garmire, G. P. & Lamb, F. K. 1981, *Ap. J.*, **245**, 183.
[251] Tylenda, R. 1981, *Acta Astron.*, **31**, 267.
[252] Visvanathan, N. & Wickramasinghe, D. T. 1979, *Nature*, **281**, 47.
[253] Visvanathan, N., Hiller, J. & Pickles, A. 1982, *I.A.U. Circ.*, No. 3658.
[254] Vogt, N. 1975, *Astron. Ap.*, **41**, 15.
[255] Vogt, N. 1980, *Astron. Ap.*, **88**, 66.
[256] Vogt, N. 1981, *Ap. J.*, **252**, 653.
[257] Vogt, N. & Breysacher, J. 1980, *Ap. J.*, **235**, 945.
[258] Vogt, N. & Semeniuk, I. 1980, *Astron. Ap.*, **89**, 223.
[259] Vogt, N., Krzeminski, W. & Sterken, C. 1980, *Astron. Ap.*, **85**, 106.
[260] Vogt, N., Schoembs, R., Kreminiski, W. & Pedersen, H. 1981, *Astron. Ap.*, **94**, L29.
[261] Wade, R. A. 1979, *Ap. J.*, **84**, 562.

[262] Wade, R. A. 1980, Ph.D. thesis, California Institute of Technology.
[263] Wade, R. A. 1981, *Ap. J.*, **246**, 215.
[264] Walker, M. F. 1963a, *Ap. J.*, **137**, 485.
[265] Walker, M. F. 1963b, *Ap. J.*, **138**, 313.
[266] Walker, M. F. & Herbig, G. H. 1954, *Ap. J.*, **120**, 278.
[267] Walter, F. M. 1981, *Ap. J.*, **245**, 677.
[268] Warner, B. 1974a, *MNRAS*, **167**, 47P.
[269] Warner, B. 1974b, *MNRAS*, **168**, 235.
[270] Warner, B. 1976a, *I.A.U. Symposium 73, The Structure and Evolution of Close Binary Systems*, eds. Eggleton, P., Mitton, S. & Whelan, J. (Dordrecht: Reidel), p. 85.
[271] Warner, B. 1976b, *The Observatory*, **96**, 49.
[272] Warner, B. 1980, *MNRAS*, **191**, 43P.
[273] Warner, B. 1981, *MNRAS*, **195**, 101.
[274] Warner, B. & Brickhill, A. J. 1978, *MNRAS*, **182**, 777.
[275] Warner, B. & McGraw, J. T. 1981, *MNRAS*, **196**, 59P.
[276] Warner, B. & Nather, R. E. 1971, *MNRAS*, **152**, 219.
[277] Warner, B. & Nather, R. E. 1972, *MNRAS*, **156**, 305.
[278] Warner, B. & Robinson, E. L. 1972, *MNRAS*, **159**, 101.
[279] Warner, B. & Thackaray, A. D. 1975, *MNRAS*, **172**, 433.
[280] Warner, B. & van Citters, C. W. 1974, *The Observatory*, **94**, 116.
[281] Watson, M. G., Sherrington, M. R. & Jameson, R. F. 1978, *MNRAS*, **184**, 79P.
[282] Webbink, R. 1976, *Nature*, **262**, 271.
[283] Webbink, R. 1978, *PASP*, **90**, 57.
[284] Webbink, R. & Gallagher, J. 1980, private communication.
[285] White, N. 1981, *Ap. J.*, **244**, L85.
[286] White, N. & Holt, S. 1982, *Ap. J.*, **257**, 318.
[287] White, N. E. & Marshall, F. E. 1981, *Ap. J.*, **249**, L25.
[288] White, N. E., Becker, R. H., Boldt, E. A., Holt, S. S., Serlemitsos, P. J. & Swank, J. H. 1981, *Ap. J.*, **247**, 994.
[289] Whyte, C. A. & Eggleton, P. P. 1980, *MNRAS*, **190**, 801.
[290] Wickramasinghe, D. T. & Visvanathan, N. 1980, *MNRAS*, **191**, 589.
[291] Williams, G. & Hiltner, W. A. 1980, *PASP*, **92**, 178.
[292] Williams, R. E. 1980, *Ap. J.*, **235**, 939.
[293] Williams, R. E., Sparks, W. M., Gallagher, J. S., Ney, E. P., Starrfield, S. G. & Truran, J. W. 1981, *Ap. J.*, **251**, 221.
[294] Wu, C.-C. & Kester, D. 1977, *Astron. Ap.*, **58**, 331.
[295] Young, P. & Schneider, D. P. 1979, *Ap. J.*, **230**, 502.
[296] Young, P. & Schneider, D. P. 1980, *Ap. J.*, **238**, 955.
[297] Young, P. & Schneider, D. P. 1981, *Ap. J.*, **247**, 960.
[298] Young, P., Schneider, D. P. & Shectman, S. A. 1981a, *Ap. J.*, **244**, 259.
[299] Young, P., Schneider, D. P. & Shectman, S. A. 1981b, *Ap. J.*, **245**, 1035.
[300] Cannizzo, J. K., Ghosh, P. & Wheeler, J. C. 1982, *Ap. J.*, **260**, L83.
[301] Chanmugam, G. & Dulk, G. A. 1981, *Ap. J.*, **244**, 569.
[302] Chanmugam, G. & Dulk, G. A. 1982, *I.A.U. Colloq. No. 72, Cataclysmic Variables and Related Objects*, eds. Livio, M. & Shaviv, G. (Dordrecht: Reidel), in press.
[303] Córdova, F. A. & Mason, K. O. 1982, *Pulsations in Classical and Cataclysmic Variable Stars, JILA Conference*, Univ. of Colorado, Boulder), eds. Cox, J. P. & Hansen, C. J., p. 23.
[304] Ghosh, P. & Lamb, F. K. 1979, *Ap. J.*, **232**, 259.
[305] Ghosh, P. & Lamb, F. K. 1979, *Ap. J.*, **234**, 296.
[306] Holm, A. V., Panek, R. J. & Schiffer, F. H. 1982, *Ap. J.*, **252**, L35.
[307] Krautter, J., Klare, G., Wolf, B., Duerbeck, H. W., Rahe, J., Vogt, N. & Wargau, W. 1981, *Astron. Ap.*, **102**, 337.
[308] Mason, K. O. & Córdova, F. A. 1982, *COSPAR, Proceedings on High Luminosity X-ray Binaries and Variable and Globular Clusters* (Ottawa, Canada); to appear in *Advances in Space Exploration* (Pergamon Press Ltd).

[309] Smak, J. 1982, preprint.
[310] Starrfield, S. G., Sparks, W. M. & Williams, R. E. 1982, *Advances in Ultraviolet Astronomy: Four Years of IUE Research*, eds. Kando, Y., Mead, J. M. & Chapman, R. D. NASA Conf. Publ. No. 2238, p. 470.
[311] Tylenda, R. 1981, *Acta Astron.*, **31**, 127.
[312] Williams, R. E. & Ferguson, D. H. 1982, *Ap. J.*, **257**, 672.

5

OPTICAL OBSERVATIONS OF COMPACT GALACTIC X-RAY SOURCES

Jan van Paradijs
Astronomical Institute 'Anton Pannekoek', University of Amsterdam

> *Ik keek naar een ster*
> *en onder mij zakte de*
> *aarde langzaam weg*
> Job Degenaar

5.1 Introduction

In this chapter I will discuss the optical observations of bright galactic X-ray sources ($L_x \gtrsim 10^{35}$ erg s^{-1}) containing neutron stars (or perhaps black holes), which have contributed to our understanding of their structure. The peculiar system SS 433 (see Chapter 7, this book) and systems in which the X-ray emission is the result of accretion onto a white dwarf (cataclysmic variables and AM Her-type systems) are the subject of separate chapters in this book.

We distinguish the following two groups of bright galactic X-ray sources. They are called Class I and Class II X-ray sources (other authors call them Type I and Type II sources; e.g. Chapters 1, 8 and 9, this book).

- *Class I: massive X-ray binaries (MXRB)*
  The companion star is a massive ($\gtrsim 10$ M$_\odot$) early-type star which transfers mass to the compact object by stellar wind [108] or by incipient Roche-lobe overflow [418]. The X-ray emission of most MXRB is pulsed, due to the fact that the accretion takes place on a rotating neutron star with a strong non-aligned magnetic field, along which the inflowing matter is guided toward the magnetic poles. The binary character of many of these systems is evident from periodic eclipses of the X-ray source by the companion star, and from Doppler shifts of the pulse-arrival times, varying in phase with the X-ray eclipses (if present). Orbital parameter analyses of MXRB have yielded invaluable information on the masses of neutron stars (see Chapter 1, this book). The X-ray spectra of MXRB are generally 'hard', with values of kT in exponential fits larger than ~ 15 keV [216]. The optical luminosity of the MXRB is dominated by the bright companion star. Values of the ratio of optical to X-ray luminosities are typically $\gtrsim 1$ (see Section 5.3).

- *Class II: low-mass X-ray binaries (LMXB)*

These systems are composed of a neutron star and a low-mass (in general $M \lesssim 1$ M$_\odot$) star, which transfers matter by Roche-lobe overflow. The group of sources comprises the bright bulge sources, the X-ray bursters, globular cluster sources, soft X-ray transients, and the well-known source Sco X-1. This X-ray emission from these sources is generally soft, with values of kT in exponential fits less than ~10 keV [216]. In a number of well-studied LMXB systems a correlation has been found between the hardness of the X-ray spectrum and the X-ray intensity [18, 59, 296, 504, 507]. X-ray pulsations have been observed for only three of these systems (Her X-1, 4U 1626−67 and GX 1+4). The presence of X-ray bursts and X-ray pulsations appears to be a mutually exclusive property of X-ray sources; it has been suggested that the absence of pulsations is due to the decay or alignment with the rotation axis of the neutron-star magnetic field with age ([129]; Chapter 2, Sections 1 and 2, this book).

The optical counterparts of LMXB are intrinsically faint objects. The spectra of most of them show a few characteristic emission lines superposed on a rather flat continuum; such a spectrum is often indicated by the term 'Sco X-1-type' spectrum, after the first optically-identified X-ray source of this kind. We will also use this term, but the reader should not take this to imply that Sco X-1 can be considered a prototype of the Class II X-ray sources (see Sections 5.2.1.4 and 5.2.4.5). The optical continuum of LMXB is dominated by the emission from an accretion disk around the neutron star, which is predominantly the result of reprocessing of a fraction of the X-rays into optical photons in the disk. In a few cases a noticeable contribution of a normal star is present. (Recently, Grindlay [147], on the basis of positional coincidence, suggested G-type stars as optical counterparts of the X-ray burst sources GX 3+1, GX 17+2 and 4U 1916−05. I feel that these identifications are uncertain. For the source 4U 1916−05, subsequently an alternative identification with a very faint ($V \sim 23$) star has been proposed [36].)

X-ray eclipses are generally not observed for the LMXB systems. As proposed by Milgrom [331] this may be a selection effect, as a consequence of the large thickness of the accretion disk, which shields the companion star from X-rays. Those X-ray sources which, because of the large inclination of their orbital plane, would otherwise show eclipses are then unobservable from the earth throughout their orbit. However, recent observations of periodic X-ray and optical variability of the sources 2A 1822−37 and 4U 2129+47 (see Section 5.2.4.1.4) and of optical periodic variability of 4U/MXB 1735−44 and 4U/MXB 1636−53 (see Section 5.2.4.1.5) suggest that the companion stars are not completely shielded from X-rays.

In Figure 5.1 the sky distributions of the optically-identified Class I and II sources are shown. The Class I sources are spread along the galactic equator and show a narrow latitude distribution ($\langle |b^{II}| \rangle = 2.4°$). The Class II sources are somewhat concentrated toward the galactic center and have a wider latitude distribution ($\langle |b^{II}| \rangle = 9.2°$). This is consistent with the idea that these sources are related to young (Population I) and older (Population II and old disk population) objects respectively.

The reason why the galactic X-ray sources seem to fall into these two groups is that the only way in which a long-lived ($\gtrsim 10^4$ yr) bright X-ray source can be

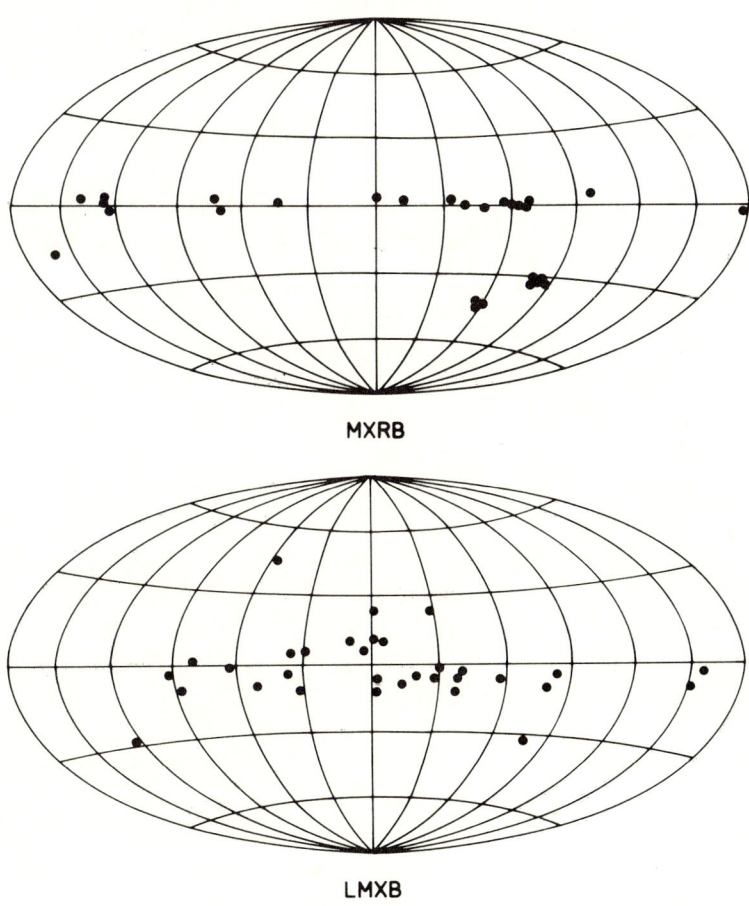

Fig. 5.1. Sky distributions (in galactic coordinates) of the optical counterparts of Class I and Class II X-ray sources (upper and lower panel respectively).

produced is either through stellar-wind (or incipient Roche-lobe overflow) mass transfer from a massive companion, or through Roche-lobe overflow of a low-mass star ([453] and Chapter 9, Section 9.3.1, this book).

Orbital periods have been reported for 18 LMXB, but some of these are highly uncertain. Well-established periods range from 41 min (4U 1626−67 [325]) to 9.8 d (Cyg X-2 [90]). Among the low-mass companion stars we find white dwarfs, late-type main-sequence stars and F–G-type giants. In the source GX 1+4 we find an M6 III companion star [107], which implies an orbital period of the order of one year. Clearly the LMXB sources comprise a variety of objects, and it is quite possible that several different evolutionary histories of close-binary systems exist which can produce an LMXB (see Chapter 8).

A large amount of data on optically-identified galactic X-ray sources and an extensive bibliography is contained in the catalogue of Bradt, Doxsey & Jernigan [40]. For much of the literature up to 1978 we refer to this work. An update of this catalogue is given by Bradt & McClintock [565]. Further reviews, covering various aspects of the optical properties of the bright galactic X-ray sources, can be found in [5, 13, 87, 182, 198, 241, 286, 541].

5.2 Class II sources: low-mass X-ray binaries

This section contains a discussion of the following topics:

- observational relations between the different types of Class II sources;
- their average optical properties;
- evidence for the existence of accretion disks in Class II sources;
- an outline of the basic structural properties of Class II sources;
- a discussion of the optical properties of some individual Class II sources.

5.2.1 Relations between different types of Class II sources

In Figure 5.2, relations and similarities are indicated between several types of Class II X-ray sources which may be distinguished, and the information each of them has contributed to our understanding of these objects. The following items deserve some attention.

5.2.1.1 Bulge sources – bursters

The single most important piece of information provided by the X-ray burst sources is that they contain neutron stars (see [271] and Chapter 2, this book). The burst sources can be considered a subset of the bulge sources. One main factor determining whether a bulge source is a burster or not may be the value of the persistent X-ray luminosity (equivalently: the accretion rate) [221, 475], but other factors, such as the magnetic field strength of the neutron star and the previous history of the accretion rate are possibly important in determining the burst behavior of a source [11, 225]. The burst sources have persistent X-ray luminosities less than a few tenths of the Eddington limit [475]. The luminosities of the non-bursting bulge sources are on the average higher than this value. This general correlation between burst activity and accretion rate is explained by some numerical calculations of thermonuclear flashes [221], and has been observed for some sources [79, 358]. (However, for other sources, no obvious relation between the variation of the persistent X-ray luminosity and burst activity is observed, see e.g. [270, 344, 362].)

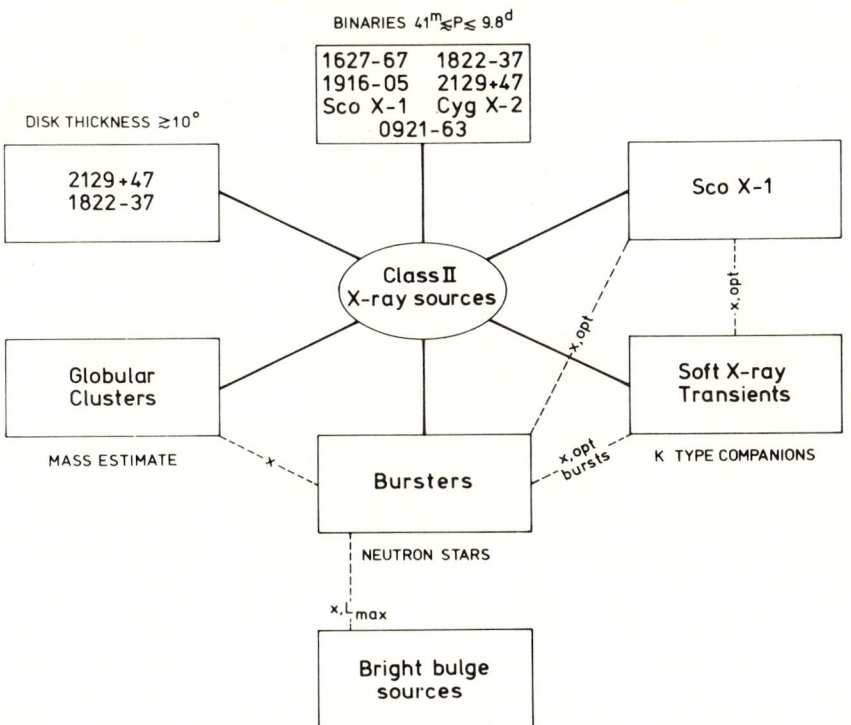

Fig. 5.2. In this scheme the relations between various types of Class II X-ray sources (rectangular boxes) are indicated by dashed lines. The abbreviations accompanying these lines have the following meaning: X – similar X-ray properties (softness of X-ray spectrum; intensity-spectral hardness relation); opt – similar optical properties (colors, spectral features); bursts – both types show the occurence of X-ray bursts; L_{max} – level of persistent X-ray emission is one of the main factors determining burst activity. Significant pieces of information which each of the various types have contributed are indicated by a short text accompanying the boxes (see Section 5.2.1).

5.2.1.2 Globular cluster X-ray sources – burst sources

Most of the globular cluster X-ray sources emit X-ray bursts [145] (see Table 2.1 in Chapter 2, this book). The average maximum luminosity of X-ray bursts from globular clusters (using distance estimates from optical studies) is about equal to that of the field-burst sources [469] (distances based on the 'standard-candle idea' [467] and an assumed spatial distribution symmetric around the galactic center [268]).

A fundamental contribution from the study of the globular cluster sources is a statistical mass determination from their positions relative to the cluster center [16]. Their masses turn out to lie in the range between 1.5 and 5.0 $M_\odot$ (90 per cent confidence limits) [145], which supports the idea that they are low-mass X-ray binaries. Because of the severe crowding of the error boxes of globular cluster sources no stellar optical counterpart has yet been discovered for any of them. Recent extensive reviews of the globular cluster X-ray sources have been given in [145, 264] (see also Chapter 2, Section 3, this book).

5.2.1.3 Soft X-ray transients

X-ray transients are undetectable for most of the time, then turn on and rise to a maximum X-ray brightness in typically a few days, and afterwards decay more slowly (timescale of weeks to months) into invisibility. Some transients have shown more than one outburst.

Similarly to the galactic X-ray sources in general, the transient sources fall into two groups: the 'hard' and the 'soft' X-ray transients, comparable to the Class I and II sources respectively [83]. Near maximum brightness the X-ray properties of the soft X-ray transients are similar to those of the persistent Class II sources. In several cases an optical counterpart has been detected, which simultaneously with the X-ray source showed a large brightness increase [28, 55, 141, 148, 347, 441]. The optical spectrum near maximum and during the initial decay of the outburst showed the 'Sco X-1'-type features (see Section 5.2.2). Soft X-ray transients have been of particular importance in the development of the LMXB model, based on the following observations.

- Several soft X-ray transients [245, 266, 267, 269, 283, 305, 344] have emitted X-ray bursts. This observation provides an additional link with the other Class II sources, and shows that these systems contain neutron stars. Furthermore, these observations indicate that the soft-transient phenomenon is the result of an accretion instability, making it comparable to dwarf-novae outbursts [222] (see Chapter 4, Section 8, this book).
- Late-type absorption features have been observed in the spectra of three (perhaps four) optical counterparts of soft X-ray transients, after

the X-ray intensity had reached a very low level [349, 364, 441, 476]. The optical counterpart had by then returned to a faint (approximately pre-outburst) brightness level. This observation indicates that in soft X-ray transients the neutron star is accompanied by a low-mass star. The spectral types (mid to late K-type) correspond to a companion-star mass between ~ 0.5 and $\sim 0.8\,M_\odot$.

The network of similarities between the different types of Class II sources suggests that the low-mass binary model holds for Class II sources in general.

5.2.1.4   *Class II sources – Sco X-1*

Sco X-1 is often considered an archetype for the Class II X-ray sources. An extensive review of the properties of this source has been given by Miyamoto & Matsuoka [337]. Its X-ray spectrum is soft ($kT \sim 5$ keV) and its optical spectrum is very similar to that of other Class II sources, in particular some burst sources [314]. A possible optical burst from Sco X-1 has been reported [309]. Sco X-1 is the first Class II X-ray source for which an orbital period (of 0.78 d) was established [88, 140]. However, in some important and intriguing aspects Sco X-1 differs from the other Class II sources. In particular, both the X-ray and optical brightness show flares on a timescale of minutes and less (see [337] and Section 5.2.4.5). Such flaring behavior is generally not observed in other well-studied Class II sources, in particular burst sources, some of which have been extensively monitored (both in X-rays and in optical brightness). Detailed correlation between X-ray and optical flaring has been observed on some occasions (see e.g. [53, 204, 394]), but often the X-ray emission varies appreciably without any change in optical brightness and vice versa [53, 204]. This is definitely different from e.g. the burst sources, for which a detailed correlation between the X-ray and optical intensity is observed during bursts [149, 154, 380]. This lack of general X-ray–optical correlation is an intriguing property of Sco X-1, which cannot be easily understood in the framework of the low-mass binary model (see also Section 5.2.4.5).

5.2.2   Average optical properties of Class II X-ray sources

In this section I will derive average values for the intrinsic colors, absolute visual magnitude and ratio of X-ray to optical luminosity of Class II X-ray sources (see also [470]). The sample to be discussed consists of the optical counterparts showing the characteristic 'Sco X-1'-type spectral features [314]. Cases in which a significant contribution to the optical flux from a normal star is present (as inferred from stellar absorption lines in the spectrum) are excluded from the sample, except if such a contribution is observable only during a fraction of a binary period.

The basic data (X-ray fluxes and UBV magnitudes), collected in Table 5.1, have been taken mainly from [40], except in cases where more recent information is available. Most sources emit a variable X-ray flux. If available, we have used simultaneous X-ray and optical data. For most sources such simultaneous data are not available. The statistically most satisfying estimate of the ratio of the X-ray to optical luminosity can be derived from the average value of the minimum and maximum X-ray fluxes quoted in [40]. For the transient sources we have used X-ray and optical data near maximum brightness, since then these sources are similar to their persistent counterparts. For sources which show eclipses of the optically-emitting region (e.g. 0921−63 [74] and 1822−37 [298]) we have taken the data at maximum. Interstellar extinction corrections, using the standard relations $A_V = 3 E_{B-V}$ and $E_{U-B} = 0.72 E_{B-V}$ were obtained from the original sources quoted in [40], and in many cases compared with more recent results, in particular the compilation of interstellar reddening data by Neckel & Klare [352].

5.2.2.1 Average color index

From Table 5.1 we obtain for the average, reddening-corrected, color indices $B - V$ and $U - B$ values of $+0.01 \pm 0.29$ (1 standard deviation) and -0.93 ± 0.20 (1 standard deviation) respectively. Part of the scatter in the distribution of the individual values of $B - V$ and $U - B$ is probably due to random errors in the reddening corrections (these are certainly not expected to be better determined than to ±0.1 mag). Thus the intrinsic colors of Class II X-ray sources are very similar. The average values of $U - B$ and $B - V$ are close to those of a flat energy distribution (F_ν = constant), for which (using Johnson's [211] absolute-flux calibration of the UBV system) $U - B = -1.0$ and $B - V = 0.2$.

5.2.2.2 Ratio of X-ray to optical luminosity

We will use the difference between the B-magnitude (corrected for interstellar extinction) and the 'X-ray magnitude' $m_X = -2.5 \log F_X (\mu \text{Jy})$ as a convenient parameter characterizing the ratio of X-ray to optical fluxes (see Table 5.1).

A histogram of the distribution of $B_0 - m_X$ is shown in Figure 5.3. It is clear that the values of $B_0 - m_X$ are mainly confined to a fairly narrow magnitude interval. It is furthermore striking that the sources which show easily recognizable evidence for periodic eclipses of the optical emission region (0921−63 [74] and 1822−37 [298]) or for a pronounced heating effect (2129+47 [318]) are all located at one extreme end of the $B_0 - m_X$ distribution. As discussed in Section 5.2.4.1 these sources are probably observed from near a critical inclination angle where the accretion disk just shields the central X-ray source. (The observed X-rays have been scattered from a highly-ionized corona above the inner parts of the accretion disk.) For this reason we have not used 0921–63, 1822–37 and 2129+47 in determining the average value of $B_0 - m_X$.

We then find an average $B_0 - m_X = 21.5 \pm 1.1$ mag (1 standard deviation). Part of the scatter around the mean value is due to errors in the interstellar reddening corrections ($\gtrsim 0.4$ mag) and to the non-simultaneity of the X-ray and optical observations of most sources in the sample. Since most of the optical emission probably originates in an accretion disk (see Section 5.2.3), projection effects will also give a finite contribution to the spread of the distribution of $B_0 - m_X$. For random disk orientations with an inclination cut off at $i \sim 75°$ (as required in Milgrom's [331] model of Class II sources) this projection effect introduces a scatter of ±0.35 mag.

It follows that the ratio of X-ray to optical fluxes of Class II X-ray sources is distributed within an interval of less than ±1 mag. Using the absolute-flux calibration of the UBV system [211], and taking into account that the optical energy distribution is approximately flat, we find for the average ratio of the energy flux emitted in X-rays (2–11 keV) and in optical radiation (3000–7000 Å) a value ~ 350.

5.2.2.3 Absolute visual magnitude

The absolute visual magnitudes can be estimated for the optical counterparts of X-ray burst sources and of some soft X-ray transients.

Fig. 5.3. Histogram of the 'color index' $B_0 - m_X$ (see Section 5.2.2.2). The three sources on the far left of the distribution are viewed from near a critical inclination angle at which the central X-ray source is just occulted by the accretion disk (see Section 5.2.4.1.4). The sources indicated by the dark squares show the presence of stellar absorption features in their optical spectrum, due to the companion of the neutron star.

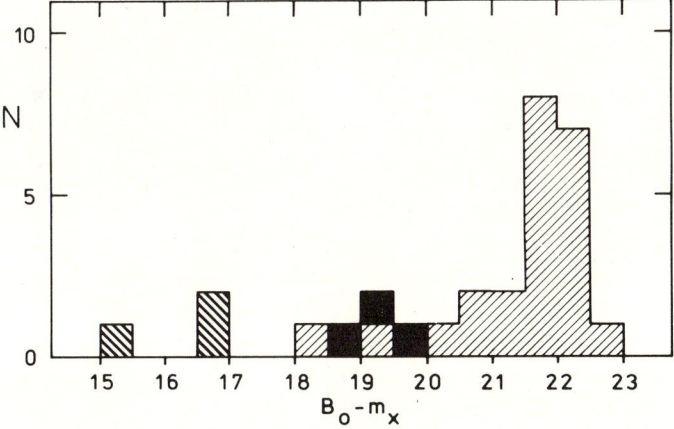

Table 5.1. *Low-mass X-ray binaries.*

Source	Optical counterpart	F_X (μJy)	var†	$V/B*$	B	V	U−B	E(B−V)	$B_0 - m_X$	Reference
0042+327[d]	#3 [58]	15 [131, 321]	T	19.3 [58]		0.6		0.2 [58]	22.0	
0142+614[d]	#3 [115]	5		~20				0.8 [352]	19.4[a]	
0521−720 LMC X-2	star [213, 371]	26		18.5* [371]				0.1	21.6	[374]
0614+091	V1055 Ori #D [115]	50 [295]	B?	18.5	0.3		−0.5	0.3	21.9	
0620−003	V616 Mon [28]	50 000	T	11.2	0.2		−0.8	0.4 [365, 525]	21.6	[127, 349, 502]
0918−549	star [147, 408]	10		19.6				0.4 [352]	20.9[a]	
0921−630	#5 [272]	2.3		15.3 [77]	0.4		−0.7	<0.3 [352]	15.4[b]	[44, 74, 91, 444, 561]
1254−690	#30 [142]	29	B	19.1	0.1			0.3 [352]	21.7	[299]
1455−315 Cen X-4	V822 Cen nova [55]	4 200	T, B	12.8 [55]	0.05		−0.9	0.0 [55]	21.9	[305, 476]
1524−617 TrA X-1	KY TrA #N [6]	800	T	17.5* [347]				0.7 [352]	22.0	
1543−624	#6 [6]	40		~20*				0.5 [352]	22.0	[315]
1556−605	star [63]	26		19.5*			−0.7	0.4 [352]	21.4	
1608−522	QX Nor nova [148]	1 100	T, B	I≲18.2				1.5 [148]		[344, 345]
1617−155 Sco X-1	V818 Sco [416]	19 000		12.2	0.2		−0.8	0.3 [517, 524]	21.9	[88, 140, 309, 314, 337, 394, 517]
1627−673	KZ TrA #4 [209]	20	P	18.5v [143]	0.1		−1.2	<0.3 [352]	20.7	[99, 143, 201, 205, 273, 317, 325, 389]
1636−536	V801 Ara	260	B	17.5	0.7		−0.7	0.6	21.8	[54, 362, 379−381, 547]

Source	Name	Number	Type	V [ref]		B−V	B₀−mX [ref]		References
1659−487 GX 339-4	V821 Ara #V [119]	400 [360]	T	15.5 [146]	0.8	−0.1	1.2 [119,352]	18.0	[144,146,199,342,359,554,555]
1705−250	V2107 Oph [141]	1200	T	15.9	0.6		0.5 [141,352]	22.2	
1728−247 GX1+4	V2116 Oph #GF [118]	125	P	19.0			1.7 [107]	19.1[c]	
1728−169 GX9+9	#DMB [118]	250		16.6	0.0	−0.5	0.3 [352]	21.4	
1735−444	V926 Sco #5 [209]	200	B	17.5	0.2	−0.8	0.3 [171]	22.3	[54,149,161,316,319]
1755−338	#6 [210]	60		19.3* [315]			0.3 [352]	22.6	
1822−371	V691 CrA #S [142]	9		15.3 [60]	0.1	−0.9	0.1 [301]	17.3[b]	[298,300,548,549]
1822−000	#F [147,408]	43		19.0 [147]			1.0 [352]	20.1[a]	
1937+049	MM Ser #D [117,443]	225	B	19.2* [443]		−0.6	0.7 [154,443]	22.3	[245,477]
1908+005 Aql X-1	V1333 Aql #5 [117,441]	1300	T, B	14.8 [59]	1.5	−0.4	0.7 [59]	21.3	
1916−053	star [36,117,147]	20	B	R = 23 [36]					[488,509]
1957+115	#M [117]	38		18.7	0.3	−0.6	0.3 [40]	22.0	
2030+407 Cyg X-3	V1521 Cyg IR object [22]	260					~6.5		
2129+470	V1727 Cyg [442]	4 [320]		16.6 [318]	0.4	−0.5	0.5 [442]	16.5[b]	[550,551]
2142+380 Cyg X-2	V1341 Cyg #10 [138]	480		14.7 [281]	0.5	−0.2	0.45 [78]	19.6[c]	[19,43,56,80,100,200,203,281,285,293]

† Variability type: T = transient, B = burster, P = pulsar.
[a] No B data available; assumed $(B-V)_0 = 0.0$
[b] Eclipses of the optically-emitting region; source not used in average $B_0 - m_X$ (see text).
[c] Stellar companion visible; source not used in average $B_0 - m_X$.
[d] Identification not firm (B. Margon, private communication).

For burst sources distances can be estimated using the idea that the average maximum luminosity L_{max} of X-ray bursts is the same for all sources ('standard-candle assumption' [467, 469]). This assumption is supported by the results of some numerical calculations of thermonuclear helium flashes [221] which indicate that L_{max} is approximately equal to the Eddington limit. Maximum luminosities of individual bursts from one source may differ by as much as a factor of ~7 [270, 344, 362]. However, the distribution of the maximum burst fluxes for one source have standard deviations of less than 35 per cent, and their average value probably is a meaningful quantity. It should be stressed here that the 'standard-candle assumption' deals with the average of the maximum X-ray burst luminosity, not with the values for individual bursts. Unfortunately, this point, apparently, is sometimes overlooked (see e.g. [304, 344, 361]). The strict use of the standard-candle assumption may give misleading results when applied to an individual source. However, observational results, in particular the implied uniformity of the blackbody radii of different burst sources [467] and the agreement of the average maximum burst luminosity for the 'field'-burst sources with the average results for bursts from globular cluster sources [469] indicate that the spread in the average value of L_{max} for different sources is

Table 5.2. *Absolute visual magnitudes of Class II X-ray sources*

A. *Distances from 'standard-candle assumption'*

Source	Distance (kpc)	M_V
Cen X-4	1.6	1.8
1636−53	5.9	1.9
1659−29	14.2	1.3
1735−44	9.4	1.7
Ser X-1	9.1	1.4
Aql X-1	5.0	−1.1

Average $\bar{M}_V = 1.2 \pm 1.1$

B. *Distances from spectral type of low-mass companions of soft X-ray transients*

Source	V_{max}	V_{min}	Spectral type	M_V (min)	M_V (max)
0620−00	11.2	18.3	K4 − 5 V	7.0	−0.1
Cen X-4	12.8	19	K3 − 7 V	7.3	1.1
Aql X-1	14.8	19.2	G7 − K3 V	5.9	1.5

Average M_V (max) = 0.8 ± 0.8

small enough to make the standard-candle assumption a meaningful tool to study the group characteristics of burst sources. (See also Chapter 2, Section 2.4.2.6, this book.)

In Table 5.2 the relevant data for optically-identified burst sources (including some transients) have been collected. Using distances for a standard-candle value of 3×10^{38} erg s^{-1} [469] we find an average absolute visual magnitude $M_V = 1.2 \pm 1.1$ (1 standard deviation).

Distances can be independently estimated for soft X-ray transients, which show a late-type stellar spectrum when the X-ray intensity has decreased to a very low value [349, 364, 441, 476], from the assumption that the companion star is on the main sequence. Using these distances we can determine the absolute visual magnitudes at maximum (when these systems are similar to their persistent counterparts). In Table 5.2 the relevant data have been collected. One finds for the average absolute magnitude of soft X-ray transients at maximum $M_V = 0.8 \pm 0.8$ (1 standard deviation), in good agreement with the value obtained for the X-ray burst sources.

It is a safe conclusion, that some important optical properties of Class II X-ray sources are distributed within fairly narrow ranges, and that their average values are meaningful quantities, that may be used in describing a 'typical' Class II source.

5.2.3 Accretion disks

5.2.3.1 Evidence for accretion disks in low-mass X-ray binaries

As a rule, absorption lines are not present in the optical spectra of Class II X-ray sources. Canizares, McClintock & Grindlay [54] have used this observation to put limits on the properties of any normal star in these systems, which can be expressed as an upper limit to its mass of ~ 2 M$_\odot$. The observation that the variation of the X-ray brightness of soft X-ray transients is accompanied by a large variation of their optical brightness (amplitude typically a factor of ~ 100) [28, 55, 141, 148, 347, 441] shows that almost all optical emission (except at most a few per cent) of soft X-ray transients near maximum is related to the X-ray emission. The overall similarity of soft X-ray transients at maximum to the other Class II X-ray sources suggests that this holds for all Class II sources.

The absence of large-amplitude periodic optical-brightness variations similar to those observed for Her X-1 (see Section 5.2.4.1.3) in Class II sources (except 4U 2129+47) shows that the optical emission, generally, cannot be the result of heating of one side of the companion star by X-rays.

Partial eclipses of the optically-emitting region have been observed for the sources 2A 1822−37 [298] and 2S 0921−63 [74, 77]. The phase relation of these eclipses relative to the X-ray light curve of 2A 1822−37 [510] and the radial-velocity and spectroscopic changes of 2S 0921−63 [91] indicate that in these systems the optical light source is centered on the X-ray source.

The analogy of the LMXB model to cataclysmic variables leads to the idea that the optical emission from the Class II sources originates in an accretion disk. Cataclysmic variables are close-binary systems, which differ from an LMXB in that they have a white dwarf instead of a neutron star accompanying a late-type main-sequence star. Light-curve analyses of these systems show that most of the optical emission originates in a region around the white dwarf. The optical spectrum contains strong emission lines whose profiles indicate that the emission region has a flat, rotating structure. (Note that the hydrogen emission lines in the spectrum of LMXB are generally much weaker than those in cataclysmic-variable spectra.) Theoretical calculations of particle trajectories have shown that the formation of a rotating, flattened gaseous body around a compact star, in which matter slowly spirals inward, is expected if mass transfer is taking place from a companion star through Roche-lobe overflow (see Chapter 10, this book). Extensive reviews of the properties of cataclysmic variables have been given in [412, 489], and Chapter 4, this book.

In cataclysmic variables a significant modulation of the optical brightness is often observed, due to the changing visibility of a 'hot spot' at the outer boundary of the disk, which contributes up to ~ 50 per cent of the optical light in these systems. For cataclysmic variables absolute magnitudes $M_V \sim +7$ (dwarf novae) to $\sim +4$ (novae at quiescence) have been derived [489], which shows that they are fainter than the optical counterparts of Class II X-ray sources by factors of ~ 150 and ~ 30 respectively. Since to first approximation the conditions at the outer boundary of an accretion disk are not very dependent on the nature of the central compact object, a hot spot in an LMXB is expected to have approximately the same luminosity as in cataclysmic variables. In view of their much larger brightness it is therefore not surprising that such hot-spot modulations have not been observed for Class II X-ray sources.

Accretion disks around white dwarfs radiate mainly through the conversion into heat of half the gravitational potential energy of matter spiralling inward. Most of the optical emission (3000–7000 Å) is generated at radial distances larger than $\sim 10^9$ cm. Therefore the optical brightness of the disk due to this internal heating is rather independent of the nature of the compact object. Calculations of the optical emission due to internal heating from accretion disks [311, 368] have shown that for accretion rates $\sim 5 \times 10^{-9}$ yr^{-1} (corresponding to an X-ray luminosity of $\sim 4 \times 10^{37}$ erg s^{-1}), the expected value of M_V is $\sim +5$. This internal heating of the accretion disk is completely insufficient to explain the optical emission of LMXB systems (cf. [378]).

Several lines of evidence indicate that the source of the optical emission of Class II X-ray sources is the absorption of a fraction of the X-rays by the accretion disk, and subsequent reradiation of this energy at larger wavelengths. As McClintock, Canizares & Tarter [312] have shown, the presence of the λ4640 emission line in the spectra of the optical counterparts of Class II X-ray sources is the result of the Bowen mechanism (see also [163, 226]). In this process other emission lines are also produced. Observations of the λ3500 region in the spectra of Her X-1 and A 0620−00 [291, 365] have confirmed the presence of these lines. This result indicates that reprocessing of a fraction of the X-rays does occur in at least some Class II X-ray sources.

In some cases the emission line profiles in the spectra of X-ray burst sources were split into two components [54]. This suggests that these lines are emitted from a flat, rotating region, and argues for an accretion disk as the site of the occurrence of the Bowen mechanism [172].

Simultaneous optical and X-ray observations of burst sources [149, 154, 380] have shown that optical bursts occur coincident with the X-ray bursts (see Section 5.2.4.2). They are delayed relative to the X-ray bursts by a few seconds [154, 316, 380]. The observed X-ray to optical burst flux ratios ($\sim 10^4$) and the optical burst delay further support the idea that the optical emission is the result of reprocessing of X-rays by matter in the vicinity of the X-ray source. The finite size of the reprocessing body causes a smearing of the optical burst compared to the X-ray burst.

The observed delay and smearing of an optical burst from 4U/MXB 1636−53 is consistent with the idea that the optical emission originates in an accretion disk [380]. A similar result was obtained by McClintock *et al.* [317] from simultaneous observations of correlated X-ray–optical pulsations from 4U 1626−67.

5.2.3.2 Disk properties

In view of the above arguments that the optical emission of Class II X-ray sources is dominated by reprocessing of X-rays in an accretion disk, the observed uniformity in their optical properties (see Section 5.2.2) suggests an underlying basic similarity in the structure of the disks in Class II X-ray sources. The nature of this similarity is as yet unclear.

An attractive starting point is to assume that the angular thickness of the disk, as seen from the neutron star, is the same for all Class II sources, i.e. the fraction of the X-ray luminosity (assumed isotropic) intercepted by the disk and reradiated as UV and optical photons, is a constant. This does not imply that the ratio of X-ray to optical luminosity is constant, since the bolometric correction of the disk emission will depend on the temperature distribution across the disk.

Fig. 5.4. Optical spectra of some Class II X-ray sources. Prominent emission lines are indicated (adapted from [54, 314]).

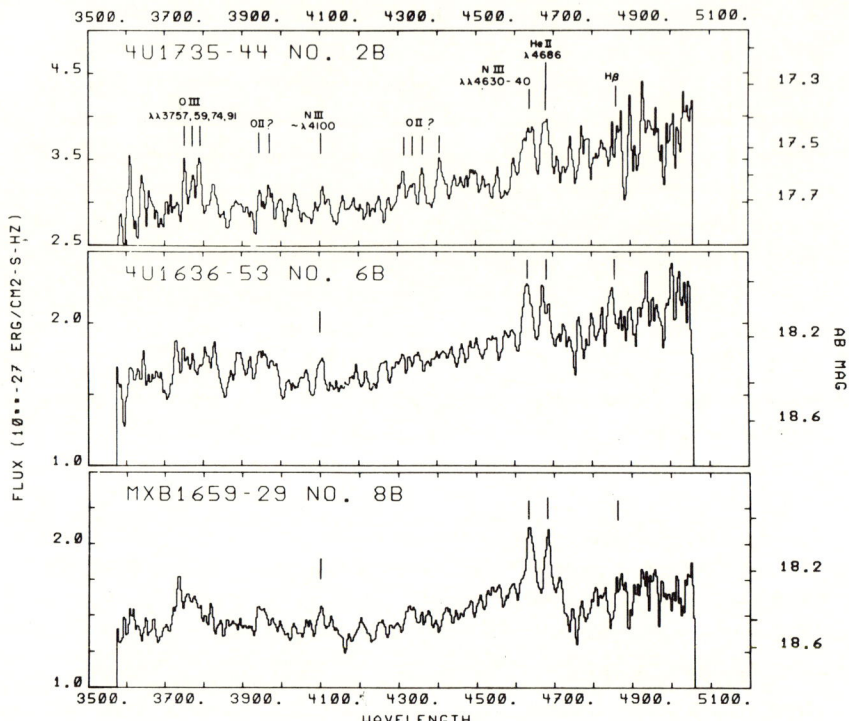

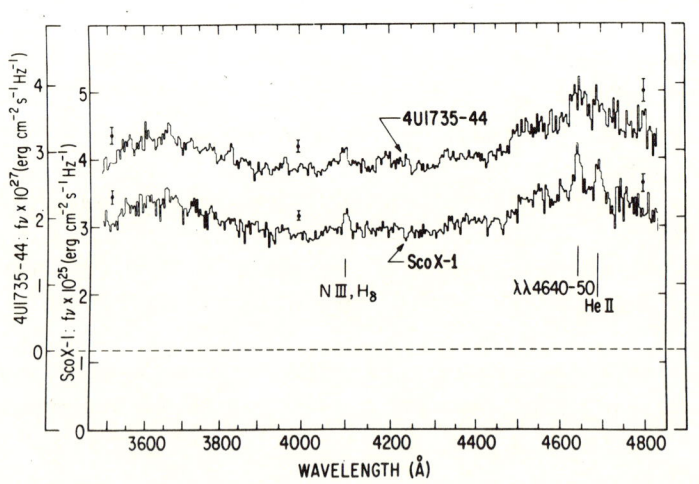

The latter depends on the luminosity of the X-ray source, and on the size of the disk. Disks probably have a diameter which is comparable to that of their Roche lobe. In view of the large range of orbital periods found for Class II X-ray sources (see Table 5.4) the sizes of their disks may be expected to show large differences.

On the other hand, the emission of the disks may be dominated by a hot inner region with a size and geometry which is similar for all Class II sources, in spite of the differences in their orbital separations. An argument for this possibility can be derived from the blackbody temperatures determined for sources with very different orbital periods of ~4 h (4U/MXB 1636–53 and 4U/MXB 1735–44), ~0.8 d (Sco X-1) and ~10 d (Cyg X-2), which are all approximately 30 000 K [78, 161, 380].

If the optical emission is the result of reprocessing of a fixed fraction of the X-ray luminosity, one expects that for blackbody temperatures near 30 000 K the optical- and X-ray-intensity variation is approximately related according to $F_{opt}(:) F_X^{\alpha}$, with $\alpha \sim 0.5$. Observed optical–X-ray variability can generally be described by such a power law, with values of α between ~0.4 and ~0.6 (see Sections 5.2.4.2 and 5.2.4.4), and the ratio of optical to X-ray luminosity is expected to vary according to $L_{opt}/L_X = F_{opt}/F_X (:) L_X^{-0.5}$.

The small spread in the distribution of M_V and the 'color index' $B_0 - m_X$ can then be understood as a consequence of an X-ray luminosity function of Class II sources which is confined to a restricted luminosity range. The observed spread of ±1 mag in M_V and in $B_0 - m_X$ would correspond to a range in X-ray luminosity of ±2 mag. Evidence for the existence of a high-luminosity cut off near ~10^{38} erg s^{-1} (~Eddington limit) was obtained by Margon & Ostriker [287]. A cut off in the luminosity function at low values of the X-ray luminosity could be the result of a selection effect against faint X-ray sources. However, arguing against such a selection effect are the results of recent *Einstein* observations of globular cluster sources [145], which indicate a real cut off or gap in their luminosity function: the globular cluster X-ray sources are either rather luminous ($\gtrsim 10^{36}$ erg s^{-1}) or very faint ($\lesssim 10^{34}$ erg s^{-1}).

Keeping in mind the above-made qualifying remarks, I will make estimates of the 'typical' thickness and size of accretion disks in Class II X-ray sources.

We can estimate the average thickness from the observed ratio of the optical (bolometric) flux f_{bol} to the X-ray flux f_X, making the assumption that all optical and ultraviolet flux is due to reprocessing of X-rays in the disk. If we approximate the disk, as seen by the observer, by a flat homogeneous plate, and assume that if the X-rays are emitted isotropically, the solid angle Ω subtended by the disk, as seen from the neutron star, is related to this ratio by the expression

$$f_{bol}/f_X = (1 - A) \cos i f(i) \Omega/2.$$

Here A is the X-ray albedo of the disk, i is the inclination of the orbital plane, and $f(i)$ is a limb-darkening factor.

Using a bolometric correction factor obtained from the far-UV energy distributions of Sco X-1 and 4U/MXB 1735−44 [161, 517], one finds that $B=0$ corresponds to a bolometric flux of $\sim 1.8 \times 10^{-4}$ erg cm^{-2} s^{-1} [470]. Integrating the X-ray flux between 2 and 11 keV, one finds that the observed average value $B_0 - m_X = 21.5$ corresponds to $f_X/f_{bol} = 50$ and

$$(1-A)\overline{\cos i\, f(i)}\, \Omega/2 = 0.020.$$

In Milgrom's [331] model (see Section 5.1) the minimum disk thickness required for the X-ray shielding depends on the angular size of the companion star, as seen from the neutron star. For Roche-lobe filling companions this is a function of the mass ratio $q = M_c/M_X$ only. Available estimates of the masses of neutron stars [403] and the companion stars [88, 90, 325, 349, 364, 441, 476] correspond to values of q between ~ 0.1 and ~ 0.7.

Using the following expressions, due to Paczynski [369], for the relative size of the Roche lobe,

$$R_L/a = 0.462(q/1+q)^{1/3} \quad \text{for } q < 0.8,$$
$$R_L/a = 0.38 + 0.2 \log q \quad \text{for } 0.3 < q < 20,$$

one finds that the disk should extend to $\sim 15°$ from the orbital plane (for comparison with theoretical predictions of standard disk models see Chapter 10, Section 10.3.2, this book). For a random distribution of disk orientations, with a corresponding inclination cut off at $i \sim 75°$, one obtains $\overline{\cos i} = 0.6$.

If the disk radiates isotropically we then find

$$(1-A)\Omega/2 = 0.033.$$

If the disk radiates similarly to a normal stellar atmosphere, $f(i)$ is approximately given by the expression for a grey atmosphere: $f(i) = 0.6 \cos i + 0.4$. Then $f(i) = 0.76$ and

$$(1-A)\Omega/2 = 0.043.$$

We thus find that the accretion disks extend to $<1.5/(1-A)°$ from the orbital plane. It appears that in order to have the disk effective in shielding the companion star from X-rays, the X-ray albedo A should be ~ 0.9, which is appreciably larger than the values ($A \sim 0.5$) calculated for X-rays perpendicularly incident on a stellar atmosphere [4, 279, 335]. This may be related to the large angle at which the X-rays reach the accretion disk. A discussion of possible systematic errors in the above determination of the disk thickness is given in [470].

We can use the observed average absolute visual magnitude $M_V = +1$ to make an estimate of the typical size of the emitting region ('disk') in Class II sources.

In this estimate it is assumed that the disk is a circular plate radiating as a black body with a temperature equal to the observed, spatially-averaged temperature. The energy distributions of Sco X-1, 4U/MXB 1735−44 and A 0620−00 can be fit with Planck functions with temperatures of 26 000–28 000 K [161, 525]. From the observed covariability of a coincident X-ray–optical burst a quiescent blackbody temperature of 28 000 K has been derived for the accretion disk in 4U/MXB 1636−53 [380]. Similar values have been derived from the observed X-ray and optical decay rates of soft X-ray transients (see Section 5.2.4.4).

From the expression $f_\lambda = \pi R^2 \cos i\, B_\lambda(T)/d^2$ we then derive (using $V = +1$ at $d = 10$ pc, $\cos i = 0.6$ and $T = 28\,000$ K) for this radius a value of 7×10^{10} cm.

A summary of the average properties of Class II X-ray sources is given in Table 5.3.

5.2.4 Optical variability of low-mass X-ray binaries

5.2.4.1 Orbital variations

Periodic variability in either X-ray or optical properties have been reported for 18 LMXB. A summary of the results is given in Table 5.4. In this table we have also included Her X-1, in spite of the fact that the relation (if any) between this source and a 'typical' LMXB system is not, as yet clear (see Chapter 8, Sections 8.2.1 and 8.5.1, this book).

5.2.4.1.1 Sco X-1

The binary character of this source had long been suspected because of similarities of its optical spectrum to those of certain cataclysmic variables, known to be close-binary systems [416]. However, for many years no direct evidence for a binary period could be detected, in spite of extended photometric and spectroscopic observations (see [53, 167, 340, 499] for reviews of early observations of Sco X-1). From an analysis of 1068 photographic plates taken over a time interval of 85 yr, Gottlieb, Wright & Liller [140] found a periodic

Table 5.3. *Average properties of low-mass X-ray binaries*

$B - V$	0.0
$U - B$	−0.9
M_V	1.0
F_X (2–11 keV)/F_{opt} (3000–7000 Å)	350
F_X (2–11 keV)/F_{bol} (UV + opt)	50
Disk thickness (degrees)	$1.5/(1-A)$
X-ray albedo (A)	0.9
Blackbody temperature	30 000 K
Size of optically-emitting region	7×10^{10} cm

modulation (with a single maximum per cycle) of ~0.2 mag in the optical brightness of Sco X-1, with a period of 0.787 313 d (see also [461, 523]). This period was later confirmed from radial-velocity observations of optical emission lines [88, 96]. The light- and radial-velocity curves of Sco X-1 are shown in Figure 5.5. From an orbital analysis, Crampton *et al.* [96] inferred probable masses $M_X \approx 1.3$ M$_\odot$ and $M_c \approx 1.0$ M$_\odot$, and an inclination angle $i \approx 30°$. Unless it underfills its Roche lobe substantially, the companion star has probably evolved off the main sequence [88].

The minimum in the optical light curve occurs when the emission line object is behind the companion star. This, together with the single-wave character of the light curve, suggests that the optical-brightness variation is due to X-ray heating of the companion star. However, the amplitude of the modulation is much smaller than expected [96] which suggests that the companion star is

Table 5.4. *Orbital periods reported for low-mass X-ray binaries*

Source	Period	Remark
1627−673	41 min	optical pulse timing [325]
1916−053	50 min	periodic X-ray dips [488, 509]
1636−536	~4 h	optical-brightness variation [379]
1735−444	4.3 h ? 2.9 h ?	optical-brightness and radial-velocity variation [319, 537]
2030+407 = Cyg X-3	4.8 h	X-ray-intensity variation [32]
2129+470	5.2 h	X-ray- and optical-intensity variation [318, 320, 442, 450]
1822−371	5.57 h	X-ray- and optical-intensity variation [298, 510, 548]
1455−315 = Cen X-4	8.2 h ?	X-ray-intensity variation [228]
1617−155 = Sco X-1	0.78 d	optical-brightness and radial-velocity variation [88, 140]
1908+005 = Aql X-1	1.3 d ?	X-ray- and optical-intensity variation [282, 493]
1656+354 = Her X-1	1.7 d	X-ray eclipses, Doppler shifts, optical-brightness and radial-velocity variation [14, 95, 437]
0614+091	5.2 d ?	X-ray-intensity variation [295]
0620−003	~7.8 d ??	X-ray- and optical-brightness variation [70, 120, 121, 278, 302, 411, 448, 495]
1702−363 = GX 349+2	8.7 d ??	X-ray-intensity variation [539]
1813−140 = GX 17+2	6.4 d ??	X-ray-intensity variation [539]
0921−630	9.0 d	optical-brightness and radial-velocity variation [7, 91]
2142+380 = Cyg X-2	9.8 d	optical-brightness and radial-velocity variation; X-ray-intensity variation [90, 100, 203, 293]
0042+327	11.2 d ??	X-ray-intensity variation [497]

strongly shielded from X-rays, most probably by the accretion disk. Another possible explanation for the brightness modulation is that perhaps the accretion disk is not axially symmetric (see also Section 5.2.4.1.5).

5.2.4.1.2 Cyg X-2 and 2S 0921−63

Cyg X-2 is one of the few LMXB for which spectral lines from the companion star are visible. Apart from the customary emission lines the optical spectrum shows an F2 III–IV star [90], with the absorption lines filled in by a strong UV continuum, probably originating from an accretion disk. Variability of the radial velocity and optical brightness of Cyg X-2 had been detected long ago, but no obvious periodicity was discovered (see [90] for references). Cowley, Crampton & Hutchings [90] showed that the radial velocity varies with a 9.843 d period (see also [100]). A tentative interpretation of the radial-velocity curve

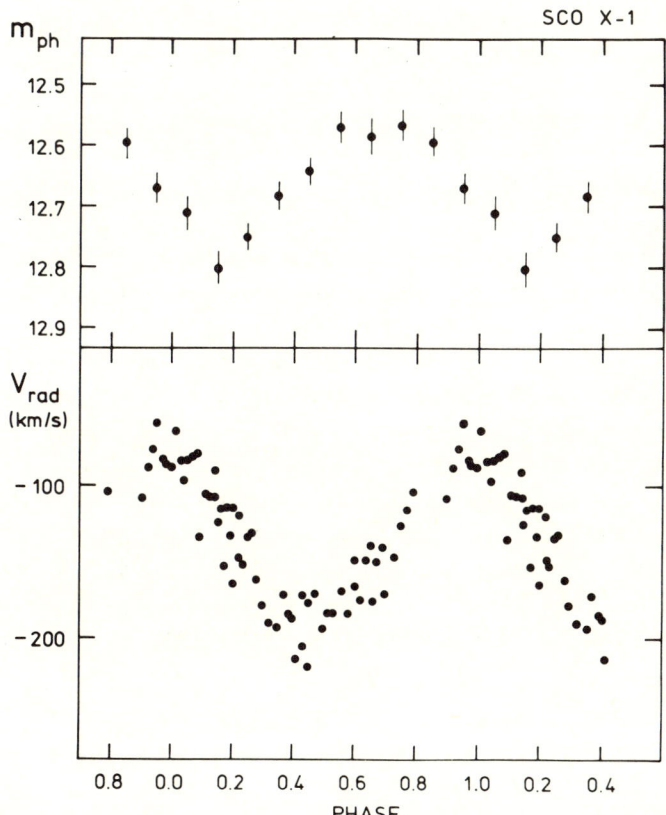

Fig. 5.5. Photographic light curve and He II λ4686 radial-velocity curve of Sco X-1 [88, 96, 523].

[90] is that for Cyg X-2 $1.3 \lesssim M_X \lesssim 1.8$ M$_\odot$, $0.5 \lesssim M_c \lesssim 1.1$ M$_\odot$, and $i \approx 75°$.

A 30 per cent modulation of the X-ray intensity of Cyg X-2 with the 9.84 d period has been reported [293]. This modulation appears only if the source is relatively faint [203]. The minimum in the X-ray light curve then occurs when the X-ray source is behind the companion star.

The UBV brightness of Cyg X-2 shows a double-wave variation with an amplitude of ~ 0.25 mag [90], indicating that the companion star is close to filling its Roche lobe. The light- and radial-velocity curves of Cyg X-2 are shown in Figure 5.6. During a bright (optical and X-ray) state of Cyg X-2 the U-light curve showed a single-wave modulation, indicating that then X-ray heating of the companion star was important. After transition to a lower X-ray and optical brightness the U-light curve reverted to the ellipsoidal type [90].

There is evidence for a phase-dependent change of the spectral type of the companion star, with later types preferentially occurring when the X-ray source is behind the companion [90]. By analogy to Her X-1 (see 5.2.4.1.3) this gives evidence that some X-ray heating of the companion star occurs in this system.

From the spectral type of the companion star and its apparent magnitude a distance of ~ 8 kpc is obtained. Cyg X-2 is likely to be a halo-population object [90].

Recurrent eclipses (~ 1 mag deep) in the optical light curve of 2S 0921−63, in phase with a smoother variation in $U-B$ and $B-V$ were reported in [74, 77] (see also [44]). Spectroscopic observations [91] showed that the radial velocity of the emission lines is also variable, with a period of 8.99 d. (This period fits the photometric variations.) There is a strong correlation between B and $B-V$, the source being redder at minimum. This is due to the improved visibility of the (red) companion star [77]. This agrees with the observation of late-type absorption features in the spectrum, which becomes stronger at minimum light [91, 444]. During the eclipses the He II λ4686 line does not disappear. This, together with the relative phasing of the photometric and radial-velocity variations suggest that the eclipses are due to a partial occultation of an accretion disk by the companion star and a bulge at the edge of the disk. The orbital solution for the radial-velocity curve [91] indicates that the companion star is an F-G giant with a mass of ~ 1 M$_\odot$ and a radius of ~ 7 R$_\odot$. The visibility of the spectral features of the companion star is consistent with the comparable optical luminosities of an F-G giant ($M_b \sim 1-2$) and of accretion disks in Class II X-ray sources ($M_b \sim 1$; see Section 5.2.2).

5.2.4.1.3 Her X-1

After the identification of HZ Her [171] as the optical counterpart of the eclipsing binary X-ray pulsar Her X-1 [274, 437] optical brightness variations with a period of 1.7 d were discovered by Bahcall & Bahcall [14]. The B-light curve (amplitude ~1.5 mag), shows a single minimum per orbital cycle. The color indices $B-V$ and $U-B$ are also variable with amplitudes of ~0.3 and ~1.0 mag respectively, the system being bluest when the X-ray source is in front of the companion star (see [15] for references on early observations of the light curve of HZ Her). The optical variability is mainly due to the fact that the side of the

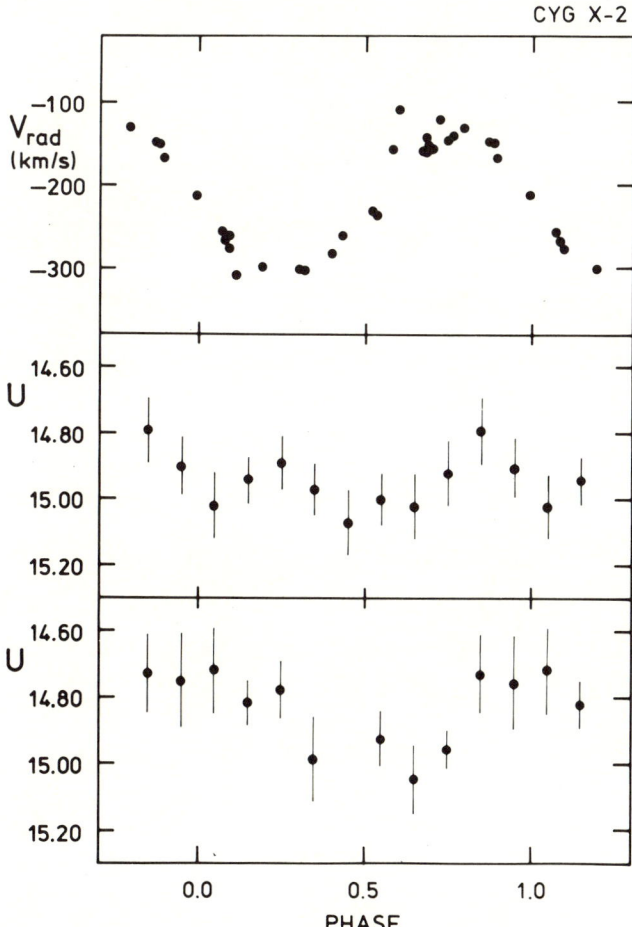

Fig. 5.6. Radial-velocity (metallic lines) and U-band light curve of Cyg X-2 [90]. The upper light curve is based on all observations from [19, 281], and shows an ellipsoidal variability. The lower light curve is based on observations made during a high state of Cyg X-2. The light curve then is single-peaked. This, together with the phasing of the light curve, indicates that X-ray heating contributed significantly to the optical brightness.

companion star facing the neutron star is brighter as the result of X-ray heating [14, 130, 280].

Magnitude estimates from photographic plates, taken during a time interval of 82 yr, revealed that during extended periods HZ Her is in an optically-low state, and then shows a double-wave light curve [217, 276, 500] (see however [125, 173]). *B*-light curves during the low state are shown in Figure 5.7.

Variations in the detailed shape of the light curve, varying in phase with the 35 d X-ray on-off light curve [139] were discovered by Boynton *et al.* [37] and Chevalier & Ilovaiski [68, 69] (see also [391]).

The most extensive photometric study of HZ Her has been made by Boynton and collaborators [37, 110, 137]. In spite of the large variation of the observed X-ray flux, the optical brightness averaged over an orbital period changes only little. A secondary minimum is present in the optical light curve, the strength of

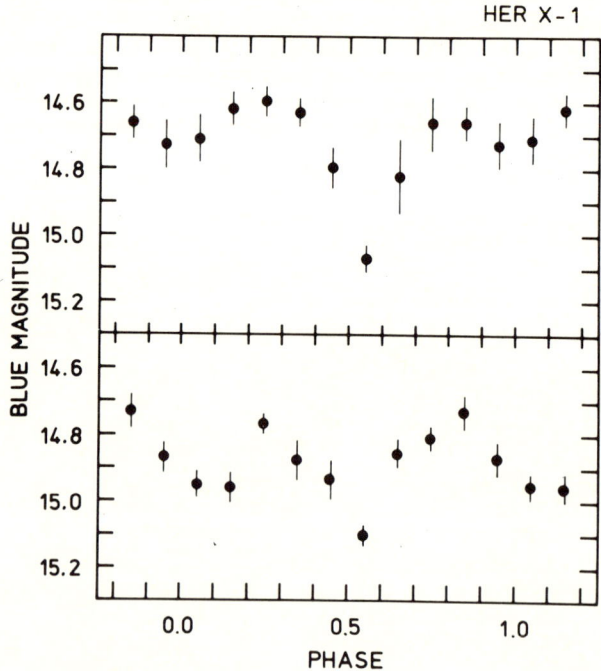

Fig. 5.7. *B*-magnitude light curve of HZ Her during extended low states according to [173] (upper curve) and [217, 276] (lower curve). The upper curve is based on data obtained between 1934 and 1940, and shows a single deep minimum, which can be ascribed to the occultation of luminous matter near the compact star. The lower curve contains data obtained between 1949 and 1956, and shows the ellipsoidal variation expected from a tidally-distorted, gravitationally-darkened star [500].

which varies appreciably during the X-ray on—off cycle. Also, a feature is present in the light curve, which drifts continuously toward earlier orbital phase throughout the 35 d on—off cycle, reappearing at the same orbital phase one on—off cycle later, see Figure 5.8.

Gerend & Boynton [137] showed that these variations can be explained by a phenomenological model of the Her X-1 system which contains a tilted preces-

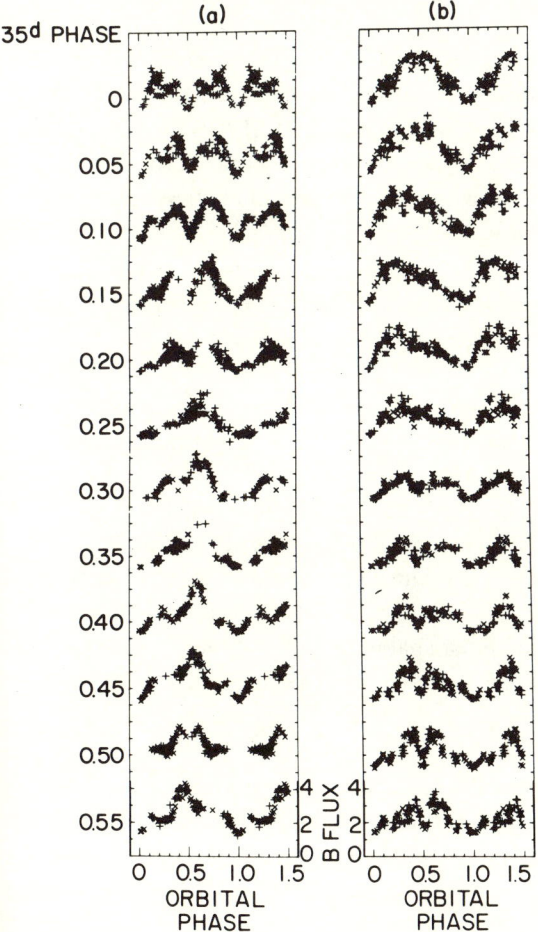

Fig. 5.8. Blue light curves of HZ Her during different phase bins in the 35 d X-ray on—off cycle [137]. The data in the curves labelled (a) have been folded modulo 1.7 d, and are a combination of data according to the symmetry rule $B(\phi, \psi) = B(-\phi, -\psi)$, where ϕ is the orbital phase and ψ is the phase in the 35 d cycle. The data in the curves labelled (b) have been sampled along a diagonal in the (ϕ, ψ) plane, and represent the light curve measured by an observer moving around the system in 35 d, together with the precessing disk.

sing accretion disk, which acts as a source of optical radiation, gives a variable shielding of the X-ray source (both as seen from the observer and the companion star), and occults part of the heated surface of the companion star. Independently, Petterson [396] calculated a theoretical model of a tilted accretion disk consisting of viscously-coupled precessing rings to explain the on-off X-ray behavior of Her X-1.

Crosa & Boynton [103] showed that the 35 d period, the occurrence of dips in the X-ray light curve, the preference of X-ray turn on to occur near orbital phases 0.3 and 0.7 and the shifting feature on the optical light curves are all closely related. They conclude that when the inner Lagrantian point enters the X-ray shadow the mass transfer rate is enhanced, leading to a bulge at the outer rim of the disk, circulating the neutron star in ~15 h. Subsequently, Boynton *et al.* [38] argued that the tilt mechanism resides somewhere in the mass transfer process.

Spectroscopic observations [33, 92, 94, 95] show that the spectral type of HZ Her varies in phase with the orbital period, between A5 and F0 at phase 0.0 (X-ray eclipses) and ~B0 at phase 0.5. Emission lines of He II λ4686 and C III/N III λ4640 are often present, but their strength varies irregularly on a short timescale, reaching, on the average, a maximum value near phase 0.5 A detailed spectrophotometric study of HZ Her has been published by Oke [363].

Because of the non-spherical shape of HZ Her and its non-uniform surface brightness the radial-velocity curve of the absorption lines is distorted (see Section 5.3.5). Analyses of absorption line radial-velocity curves (see Figure 5.9), taking this distortion into account [94, 95, 242], have yielded values of 1.0–1.5 $M_\odot$ and ~2.2 $M_\odot$ for the masses of the neutron star and the companion star respectively, in agreement with the results based on optical pulse-timing analysis [324] (see Section 5.2.4.3).

According to Crampton & Hutchings [95] the emission line radial-velocity curve of HZ Her varies approximately in antiphase with that of the absorption lines. They conclude that the emission features originate in a region located close to the neutron star. (This disagrees with the result of Koo & Kron [242] who find that the emission line and absorption line radial-velocity curves are roughly in phase, and that He II λ4686 emission is sometimes visible during X-ray eclipse.)

5.2.4.1.4 2A 1822−37 and 4U 2129+47

A 5.57 h period variability in the optical brightness of 2A 1822−37 was discovered by Mason *et al.* [298]. The amplitude of the modulation is ~1 mag, independent of wavelength. The light curve has an extended flat maximum level, and an asymmetric minimum which lasts for almost half the orbital cycle.

The X-ray intensity shows a 5.57 h sinusoidal modulation, with an additional dip in the intensity of ~30 min duration, occurring about 0.04 in phase after the optical minimum [510].

The optical spectrum contains emission lines of He II λ4686 and λ3815, C III/N III λ4640−50 and O III λ3765. Ca II H and K are present in absorption, and are probably of stellar origin [60, 548].

The optical-brightness modulation is due to the occultation of a luminous region, most probably an accretion disk, by the companion star [298].

The optical brightness and colors of 4U 2129+47 show a 5.2 h periodicity (see Figure 5.10) with an amplitude of ~1.5 mag in B, 0.25 mag in $B-V$ and 0.5 mag in $U-B$ [318, 442]. The light and color curves are remarkably similar to those Her X-1, indicating that the optical emission is dominated by the X-ray-heated surface of the companion star [318].

The optical spectrum shows He II λ4686 and C III/N III λ4640−50 in emission. Hβ and the Ca II H and K lines often appear in absorption; the strength of the latter lines is variable, indicating that at least part of them are of stellar origin [442, 550].

The X-ray luminosity of 4U 2129+47 also varies with the 5.2 h period [320, 450]. Contrary to the similarity of their optical light curves the X-ray light curve of 4U 2129+47 is very different from that of Her X-1.

Fig. 5.9. Radial-velocity curve of HZ Her from measurements of all absorption lines [94]. The solid line indicates the expected radial-velocity variation of the center of mass of HZ Her (K = 85 km s^{-1}; $V_0 = -60$ km s). The observed radial-velocity curve is distorted due to the non-sphericity and X-ray heating of HZ Her.

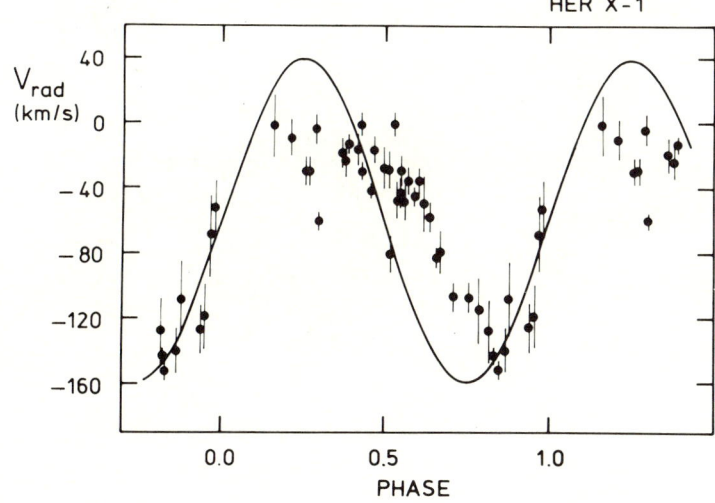

The observed X-ray light curves of 2A 1822−37 and 4U 2129+47 are the result of an eclipse by the companion star of an extended X-ray-scattering cloud [320, 510], presumably a highly-ionized corona above the central parts of the accretion disk (cf. [511]). The X-ray source itself is not directly observed, as it is shielded by the accretion disk. Modelling of the X-ray light curve yields inclinations between 70° and 78° and ~77° for 2A 1822−37 and 4U 2129+47 respectively [320, 510]. Thus these sources are very similar, in spite of the remarkable difference in their optical light curves.

These results imply that in these systems the disks have an angular thickness (as seen from the neutron star) of >10° from the orbital plane. If this value is

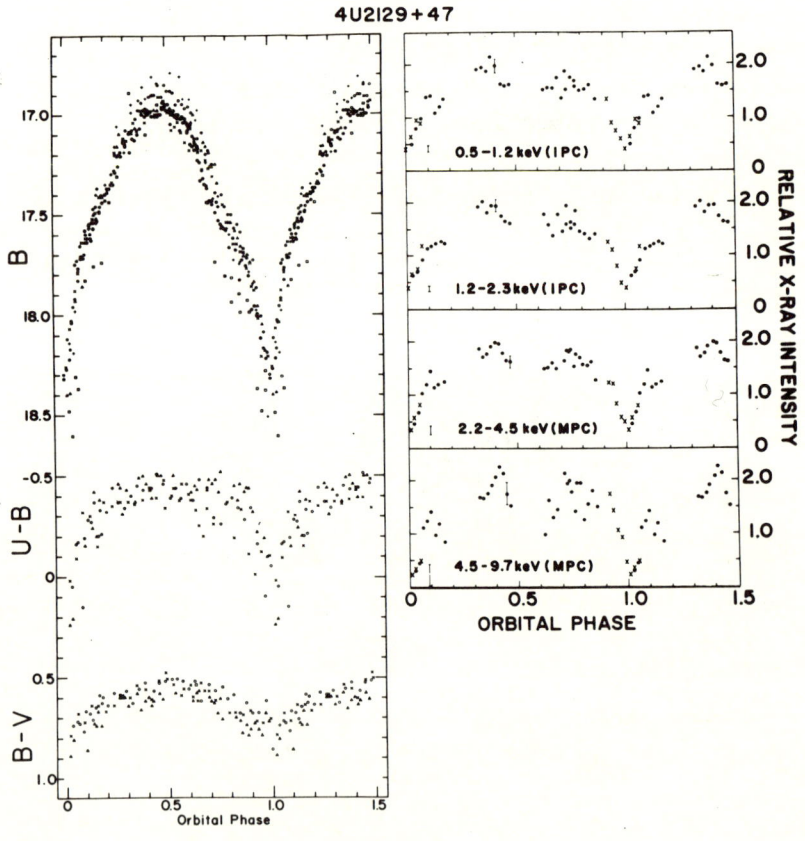

Fig. 5.10. Optical light and color curves, and X-ray light curves (for several energy intervals) of 4U 2129+47 [318, 320]. The shape of the X-ray light curve is independent of X-ray energy, and can be understood as the result of a partial eclipse of an extended region of X-ray emission by the companion star (see Section 5.2.4.1.4). The optical light and color curves are very similar to the average curves of HZ Her.

characteristic of Class II X-ray sources in general it confirms Milgrom's [331] model of these sources.

The observations of 2A 1822−37 and 4U 2129+47 indicate that the companion star in these systems is not completely shielded from X-rays, but that a polar cap is visible from the neutron star above the accretion disk. A small heating effect, such as has perhaps been observed in the optical light curves of Sco X-1 (Section 5.2.4.1.1), Cyg X-2 (Section 5.2.4.1.2), 4U/MXB 1636−53 and 4U/MXB 1735−44 (Section 5.2.4.1.5), may then be expected. However, with such a large disk thickness it is difficult to understand why the companion star of 4U 2129+47 undergoes strong X-ray heating, as is indicated by the shape of the optical light curve. Perhaps, similar to Her X-1, the disk in this system is tilted. Extended observations are necessary to detect a possible long-term modulation of the X-ray flux and of the shape of the optical light curve.

5.2.4.1.5 4U/MXB 1636−53 and 4U/MXB 1735−44

The persistent optical brightness of the burst sources 4U/MXB 1636−53 and 4U MXB 1735−44 is variable, with an amplitude of ∼20 per cent, and possible periods of ∼4 h and 4.3 h respectively [319, 379] (the light curve of 4U/MXB 1636−53 is shown in Figure 5.11). Interpreting these periods as orbital, the masses of the companion star (assuming, as usual, that they are Roche-lobe filling main-sequence stars) are ∼0.5 $M_\odot$ (see [489]). A possible 2.9 h period in radial-velocity variations has recently been reported for 4U/MXB 1735−44 [537].

As discussed by Pedersen, van Paradijs & Lewin [379], eclipses of the accretion disk by a companion star, or hot-spot occultations (analogous to some cataclysmic-variable light curves) can be excluded as causes of the observed brightness modulation of 4U/MXB 1636−53. The variability may be due to X-ray heating of a relatively small part of the companion star, not shielded by the accretion disk. In order to obtain a modulation of ∼20 per cent the disk thickness cannot be much larger than ∼6° from the orbital plane. As discussed above, disks may well be substantially thicker, and almost completely shield the companion star from X-rays. Perhaps the disk is not axially symmetric, e.g. being thicker at the side where matter is injected into it. The brightness modulation could then be the result of the azimuthal dependence of the disk brightness, in combination with the phase dependence of the angle at which different parts of the disk are viewed. Independent evidence for such a bulge at the rim of accretion disks has been obtained from observations of 2A 1822−37 [510], MXB 1916−05 [509], and MXB 1659−29 [84].

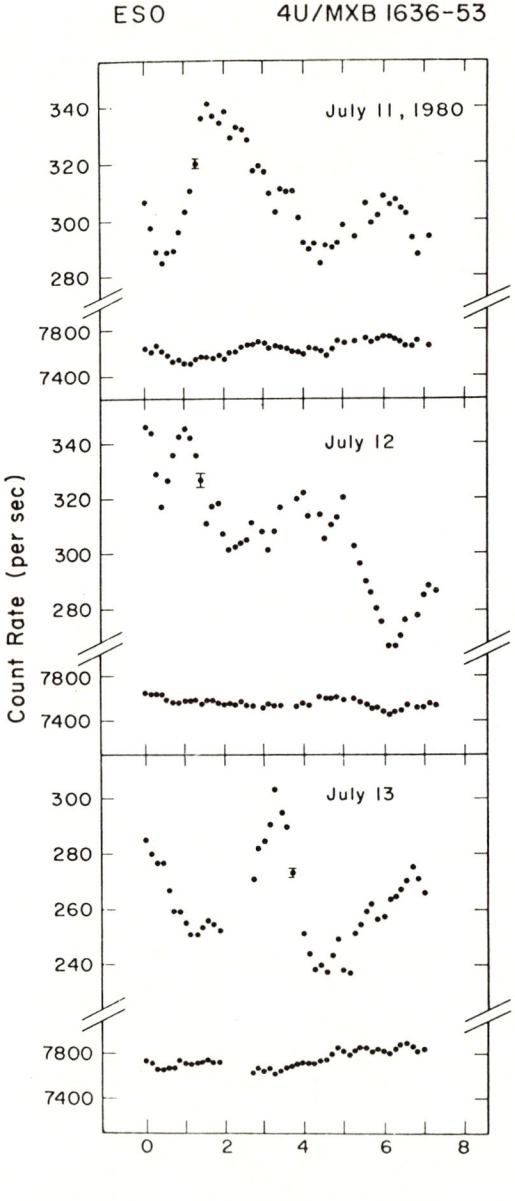

Fig. 5.11. Optical-brightness variation (3000–5500 Å) of 4U/MXB 1636–53 as observed during three consecutive nights [379]. The data are consistent with a periodic variability (P ~ 4 h). The approximate constant signal is that of a reference star.

5.2.4.1.6 4U 1626−67 and MXB 1916−05

An orbital period of 41 min has been determined by Middleditch et al. [325] from timing observations of optical pulsations from 4U 1626−67. These observations are described in Section 5.2.4.3.2.

A similar short orbital period has been found for the burst source MXB 1916−05 from periodic absorption dips in the X-ray light curve [488, 509]. The variations in the depth and duration of these dips suggest that these dips are due to a bulge at the edge of the accretion disk.

If the companions of 4U 1626−67 and MXB 1916−05 are Roche-lobe filling main-sequence stars their extremely short orbital periods correspond to companion-star masses $\lesssim 0.04\,M_\odot$. Such stars are too small to be able to ignite hydrogen. This suggests that the companions are degenerate objects. Several consistent solutions are possible (depending on the chemical composition of the degenerate star), with masses between ~ 0.01 and $\sim 0.05\,M_\odot$ [240, 509].

5.2.4.1.7 *Long orbital periods: soft X-ray transients and 2S 0614+091*

Possible periodic variability of the X-ray and optical brightness has been suggested for a number of soft X-ray transients.

The most extensive set of data has been obtained for A 0620−00. A period of 7.8 ± 0.7 d in the X-ray brightness variations was reported by Matilsky et al. [302]. The X-ray light curve is approximately sinusoidal and has a total amplitude of about a factor of 2. However, no evidence for such a modulation was found in X-ray observations by Watson, Ricketts & Griffiths [495], which partly overlap those of Matilsky et al.

An optical-brightness modulation of A 0620−00 with a possible period of 3.92 d was found by Duerbeck & Walter [120, 121]. However, from an independent analysis of their data, Matilsky et al. [302] were unable to reproduce this result. Chevalier, Ilovaisky & Mauder [70] reported an approximately sinusoidal modulation with an amplitude of ~ 20 per cent of the optical brightness of A 0620−00, with a period of 8.0 ± 0.2 d. Optical variability with a period of 7.39 ± 0.13 d, and a total amplitude of ~ 15 per cent was obtained by Tsunemi, Matsuoka & Takagishi [448]. These observations (which do not overlap) show a consistent phasing of the optical minima. However, observations by Lloyd, Noble & Penston [278] failed to show evidence for a periodicity near 7.5 d. Furthermore, photometric observations by Robertson, Warren & Bywater [411] showed two minima which are almost exactly out of phase with those reported by Chevalier, Ilovaisky & Mauder [70].

The X-ray flux of the soft X-ray transient H 1705−25 during the decay of the outburst showed a modulation on a timescale of about one week. This variability, however, is not strictly periodic [495].

Some evidence for a periodic modulation of the X-ray flux of Aql X-1 during the 1975 June outburst was found by Watson [493]. The period equals 1.3 ± 0.04 d, the total amplitude of the modulation is ~15 per cent. X-ray observations obtained during later outbursts failed to show evidence for this 1.3 d period [59, 245]. However, an optical-brightness modulation was found for Aql X-1 during outburst, with a period of 1.3 d and an amplitude of ~0.6 mag [282].

Kaluzienski, Holt & Swank [228] presented evidence for an 8.2 h periodicity in the X-ray flux of Cen X-4 during the 1980 outburst. This result is not supported by Hakucho observations obtained during the same period [305].

It would appear from the above that for none of the soft X-ray transients has the strictly periodic character of the X-ray- and optical-brightness variations been established. Perhaps the detected variability is due to an instability in the accretion flow, related to the transient character of the mass transfer. The periods suggested for A 0620−00 and Aql X-1 cannot be reconciled with the idea that the observed companion stars fill their Roche lobe. In particular, the period suggested for A 0620−00 implies a radius of the companion star of ~10 $R_\odot$, which would make it a giant star. This is ruled out by the spectroscopic data [349, 364]. Alternatively, the companion stars may substantially underfill their Roche lobe. This could be related to the transient character of their X-ray behavior.

Recently Marshall & Millit [295] reported that the X-ray intensity of the source 2S 0614+091 is variable with a period of ~5.2 d. They suggest that perhaps this long period is not orbital, but reflects the precession period of an accretion disk around the neutron star. In the two other systems where there may be such a precessing disk (Her X-1 and LMC X-4, see Sections 5.2.4.1.3 and 5.3.6.6) the ratio of the precession period to the orbital period is ~20. If this ratio applies to 2S 0614+091 the corresponding orbital period would be ~6 h.

5.2.4.2 Optical bursts

Optical bursts, in a number of cases correlated with X-ray bursts (see also Chapter 2, Section 2.4.2.7.4, this book) have been detected from the burst sources 4U/MXB 1735−44 [149, 316], MXB 1837+05 (=4U 1837+04=Ser X-1) [154] and 4U/MXB 1636−53 [380]. Optical bursts (without simultaneous X-ray observations) have been observed from the transient X-ray burst source Aql X-1 [477] (see Figure 5.12) and from the source 2S 1254−690, not previously known to be a burster [299]. Furthermore, a possible optical burst has been reported for Sco X-1 [309].

The optical bursts are delayed relative to the X-ray bursts by 1–3 s, and the ratio of the integrated optical to X-ray burst flux (corrected for interstellar extinction) is of the order of 10^{-4}. This value is too high by about six orders of magnitude to explain the optical emission as the long-wavelength extrapolation of the blackbody spectrum [170, 436] of the X-ray burst.

The observed energy ratio and the delay of the optical burst indicate that they are the result of the reprocessing of a fraction of the X-rays by matter in the vicinity of the X-ray source.

Because of the large and rapid variation of the X-ray signal the X-ray bursts can be considered to be a probe that illuminates the matter surrounding the neutron star, which heats up and responds by an increase of the optical radiation.

The delay of the optical burst is predominantly the result of flight-time differences between the directly-observed X-rays and the X-rays which are first absorbed and subsequently reemitted as optical photons. The delay due to the finite reprocessing time can be neglected [380]. Since the reprocessing body is of finite extent, different parts of it will give rise to different values of the delay, and consequently to a smearing of the optical signal. From a detailed comparison of the covariability of the optical and X-ray brightness in the burst one can deduce information on the location and geometry of the surroundings of the burst source.

A detailed modelling of the optical burst signal as thermally reprocessed X-rays in a body, approximated as a flat plate, was made by Pedersen et al. [380]. For one burst from 4U/MXB 1636−53, for which the data had sufficient signal-to-noise, they derived values for delay and smearing (3 ± 1 and $\lesssim 5$ s respectively) and for the variation of the (average) blackbody temperature of the reprocessing body during the burst (blackbody temperature outside bursts $\sim 28\,000$ K). For two other bursts with less signal-to-noise, only the delay could be determined. To within the accuracy of their determination they are equal to the above value.

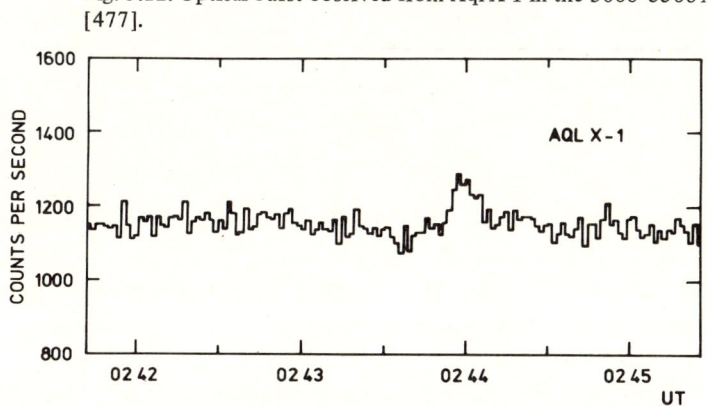

Fig. 5.12. Optical burst observed from Aql X-1 in the 3000–5500 Å passband [477].

Pedersen *et al.* [380] concluded that an accretion disk around the neutron star provides a plausible model for the reprocessing region. The projected area of the disk is $\sim 5 \times 10^{21}\,\mathrm{cm}^2$, corresponding to a projected linear size of ~ 1 lt s. From the observed delay a restriction can be put on the radius of the disk, which should be larger than the speed of light times half the delay. With the assumption that the radius of the accretion disk is between 75 and 100 per cent of the radius of the Roche lobe around the neutron star (see [380] for references), this in turn constrains the parameters of a consistent LMXB system (with a $1.4\,M_\odot$ neutron star). Pedersen *et al.* [380] find that the mass of the (Roche-lobe filling) companion star is between 0.5 and $2.0\,M_\odot$, which corresponds to a binary period between ~ 3 and $\sim 10\,\mathrm{h}$. These results are consistent with the possible orbital period suggested by the $\sim 4\,\mathrm{h}$ modulation of the persistent optical brightness of 4U/MXB 1636−53 [379] (see Section 5.2.4.1.5).

A total of 41 optical bursts were detected from the source 4U/MXB 1636−53 [381], five of which were observed in more than one passband simultaneously (see also [260]). (For most of these bursts there were no simultaneous X-ray data.)

The bursts show a great diversity in their profiles (see Figure 5.13) and their maximum fluxes and integrated fluxes range over a factor of ~ 8. A similar diversity in profile and energy has also been observed in the X-ray bursts from this source [362]. The integrated optical burst fluxes appear to be correlated with the waiting time since the previous burst. One particular burst was observed only 5.5 min after the previous one (see also [266, 344, 361] and Chapter 2, Section 2.4.2.5, this book, for similar short intervals between X-ray bursts). This suggests that in a thermonuclear flash not all fuel is consumed as first suggested by Lamb & Lamb [253] (see Chapter 2, Section 2.6.3.1, this book).

5.2.4.3 Optical pulsations

Optical pulsations have been discovered from Her X-1 and 4U 1626−67. These pulsations arise from the reprocessing of pulsed X-rays in the companion star and in the accretion disk, and provide a unique tool to study the orbital parameters of these systems.

5.2.4.3.1 Her X-1

The possible detection of optical pulsations from Her X-1 with the 1.24 s X-ray pulse period was reported by Lamb & Sorvari [252]. After a number of unsuccessful searches [81, 93, 134, 150, 346, 409] a definite detection of optical pulsations was made by Davidsen *et al.* [105] (see also [151, 323]). The most extensive set of data has been collected by Middleditch & Nelson [324]. The amplitude of the pulsations is phase dependent and reaches values up to 0.3 per

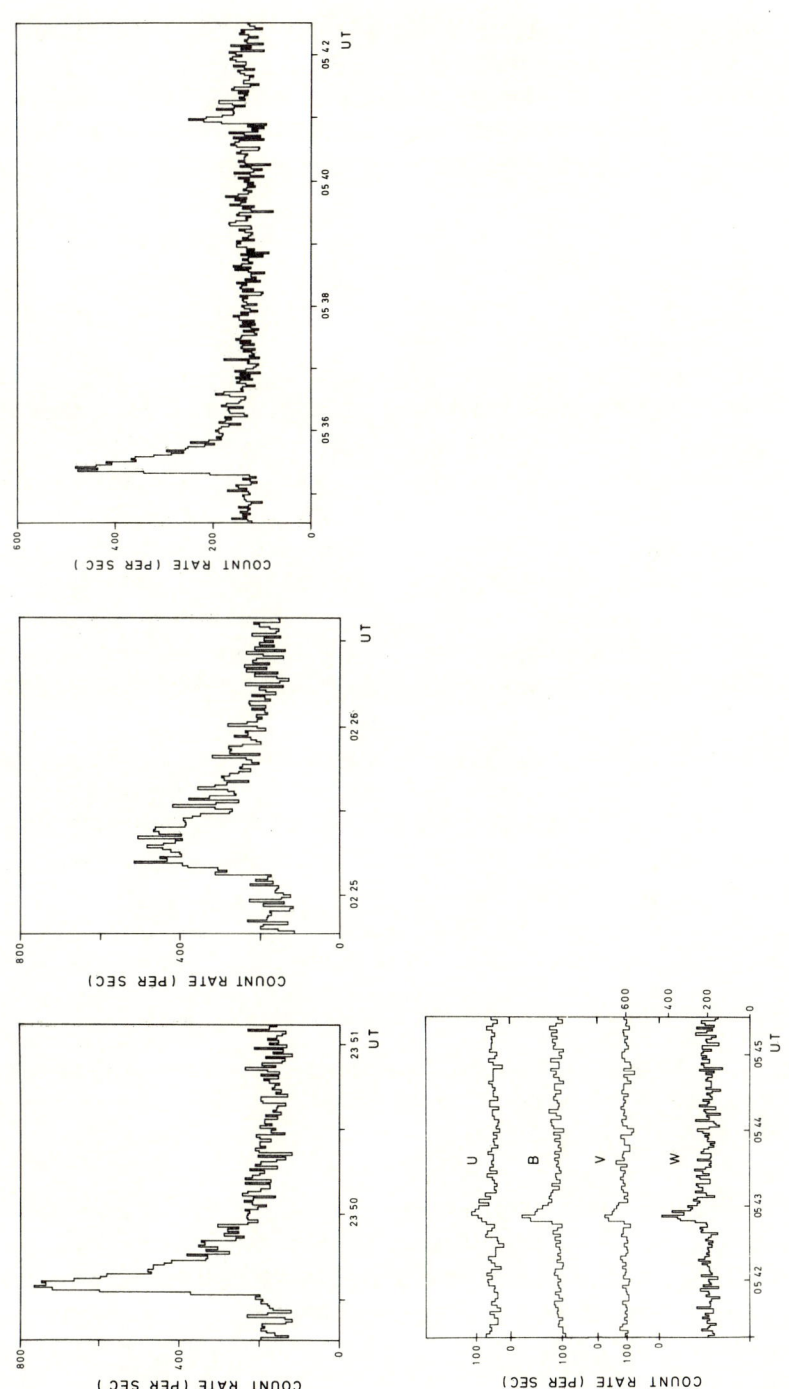

Fig. 5.13. Optical bursts observed from 4U/MXB 1636−53 [260, 381] (white light data, unless otherwise indicated).

cent. Observations of spectrally resolved optical pulsations show that they are primarily a broad-band phenomenon, arising from reprocessing of X-rays [106, 289, 353]. A detailed analysis of the optical pulsations of Her X-1 has been made by Middleditch & Nelson [324]. The orbital motion of the neutron star is known from the systematically varying Doppler shifts of the X-ray pulsations. Optical pulsations are present with the same Doppler-shift variations and in phase with the X-ray pulsations, but also optical pulsations are observed which show a different (but systematic) Doppler-shift variation. This is naturally explained by the fact that the first type of optical pulsations arises from reprocessing of X-rays in the accretion disk around the neutron star, whereas the second type arises from the surface of the companion star. (However, from near-simultaneous X-ray and optical observations, Joss *et al.* [224] conclude that the optical pulsations which are in phase with the X-ray pulsations arise not from the accretion disk, but from a region of enhanced density near the intersection of the accretion disk with an accretion stream from the companion star.) The pronounced variation with orbital phase of the strength of the optical pulsations indicates that the companion star is highly non-spherical, and is filling (or close to filling) its Roche lobe. As the pulsations arise from the X-ray illuminated side, the Doppler shifts do not represent the motion of the center of mass of the companion, but rather the motion of some weighted mean of the X-ray illuminated side. By assuming that the companion star fills its Roche lobe and the reprocessing time is constant across the surface one can calculate the correction factor. A comparison with the observed ratio yields $M_c/M_X = 1.68 \pm 0.10$. From the X-ray-eclipse duration one then derives for the inclination $i = 87° \pm 3°$, and $M_X = 1.30 \pm 0.14 M_\odot$ and $M_c = 2.18 \pm 0.11 M_\odot$. These values are consistent with the results obtained from the optical radial-velocity curve of HZ Her [94, 95, 242]. However, the high accuracy of this result may be illusory since no allowance has been made for the uncertainty in the rotational velocity of HZ Her [17].

5.2.4.3.2 4U1626−67

No periodic changes in the 7.68 s X-ray pulse period of this source due to Doppler shifts have been observed, and only an upper limit for the size of the neutron star orbit can be derived: $a_X \sin i < 0.04$ lt s [273, 400]. Optical pulsations at the X-ray pulse period were detected by Ilovaisky *et al.* [201] and Peterson *et al.* [389]. Extensive observations of the optical pulsations were made by Middleditch *et al.* [325]. Fourier spectra of the data, after subtraction of the ∼1000 s irregular variations (see [317]), show a fairly strong (∼3 per cent) optical modulation with the X-ray pulse period, interpreted as reprocessing from an accretion disk. Also, weaker (∼0.5 per cent) pulsations are seen with a 7.7012 s period. These are most readily explained as arising from reprocessing at the surface of a companion star, orbiting the neutron star in the same direction as the pulsar rotation, with a period of 2493 s. Connecting observations obtained during different nights with this tentative period, more accurately determined possible periods are 2492.32 and 2491.05 s.

Comparison of the phases of the optical pulsations arising from the disk and those from the companion star after correction for the effect of the orbital motion shows a residual phase difference which systematically varies. The average phase difference corresponds to the light-travel time from the neutron star to the companion, and equals 0.99 ± 0.32 lt s. The variation of the phase difference around this average is due to the variation of the distance from the companion to the observer, and yields $a_c \sin i = 0.31 \pm 0.08$ lt s. From the limit on $a_X \sin i$ it follows that $M_c/M_X < 0.1$.

As in Her X-1, account must be taken of the fact that a_c applies to some weighted mean of the companion-star surface, rather than its center of mass. A complication arises because this mean distance not only depends on the mass ratio but also on the fraction of the companion star which is shadowed by the accretion disk. However, the mere fact that we observe the 0.5 per cent optical pulsations set an upper limit to the thickness of the disk of $4.3°$. Detailed modelling then leads to the following binary parameters:

$a_c \sin i = 0.36 \pm 0.10$ lt s; $a = 1.14 \pm 0.40$ lt s; $i = 18^{+18°}_{-7}$;

$M_X = 1.8^{+2.9}_{-1.3} M_\odot$ and $M_c < 0.5 M_\odot$.

5.2.4.4 Optical observations of soft X-ray transients

Optical counterparts have been detected for six (perhaps seven) soft X-ray transients [28, 55, 141, 148, 347, 441]. In all cases where the optical observations were made near X-ray maximum the counterpart showed a coincident large increase of the optical brightness.

For four sources, more or less complete X-ray and optical light curves have been observed [55, 59, 305, 347, 502]. The decay time of the optical light curve is longer (by a factor of ~ 2) then that of the X-ray curve (see Figure 5.14). This can be understood as the result of a change of the spectral energy distribution of the light emitted by the accretion disk, whose effective temperature decreases with the amount of infalling X-rays. A progressively larger fraction of the total disk luminosity emerges in the visual region of the spectrum. If we make the assumption that the fraction of the X-rays intercepted by the disk does not change and that the disk radiates as a blackbody the observed decay rates of the X-ray and optical light curves provide an estimate of the blackbody temperature of the disk [127]. For the transients A 0620−00 and A 1524−61 blackbody temperatures near maximum of 34 000 and 27 000 K have been derived respectively [127, 347]. A similar analysis of the X-ray and optical light curve of Cen X-4 [55, 305] and Aql X-1 [59] yields temperatures of 35 000 and 12 000 K.

With the exception of the value obtained for Aql X-1 these results agree well with the blackbody temperatures derived from the shape of the ultraviolet and optical energy distributions of A 0620−00, Sco X-1 and 4U/MXB 1735−44 for which values near 28 000 have been derived [161, 517, 525]. However, it should be noted that in the case of Aql X-1 some doubt is justified about the applicability of this method to estimate the temperature: the B-light curve reached a maximum about one week after X-ray maximum; furthermore, the source was quite red at X-ray maximum ($B-V \sim 1.5$), and became much bluer during the initial decay part of the X-ray light curve [59].

The development of the optical spectra of the well-observed soft X-ray transients A 0620−00 and Cen X-4 during the X-ray outburst shows some common features ([55, 349, 476, 502] and references therein).

Initially, the optical emission consists of a blue continuum without obvious indications for the presence of emission lines. During a somewhat later stage weak emission lines (mainly He II $\lambda 4640$ and N III $\lambda 4640$) appear. The spectrum is then very similar to those of the steady Class II sources (see Figure 5.15). The

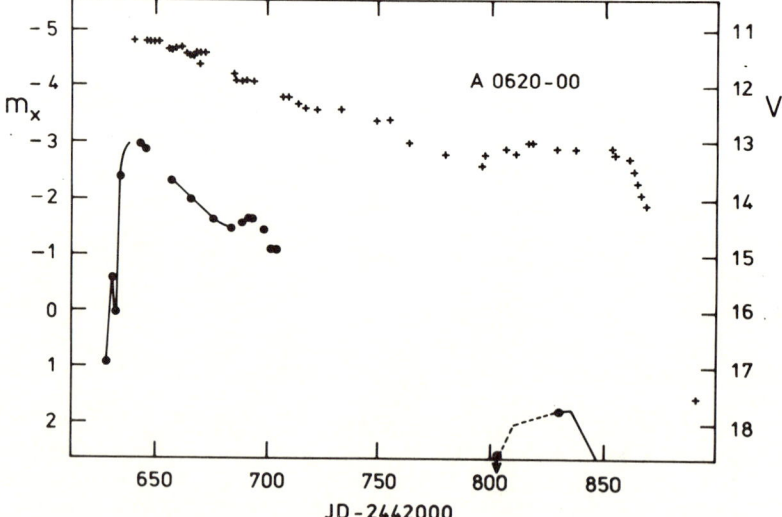

Fig. 5.14. X-ray and visual light curve of the soft X-ray transient A 0620–00 during the 1975–76 outburst. X-ray data are indicated by filled circles, visual data by crosses. The X-ray brightness is given in an arbitrary magnitude scale (left hand scale) (adapted from [502]).

Fig. 5.15. Optical spectra of the soft X-ray transient Cen X-4 [55, 476]. The upper panel shows spectra obtained near maximum of the 1979 May outburst, which are very similar to those of other Class II X-ray sources (cf. Figure 5.4). The lower panel shows the spectral energy distribution after the X-rays had decreased to a very low flux level. The Mg b feature around 5100 Å and the sharp break in the flux below 4400 Å indicate the presence of a K-type companion star.

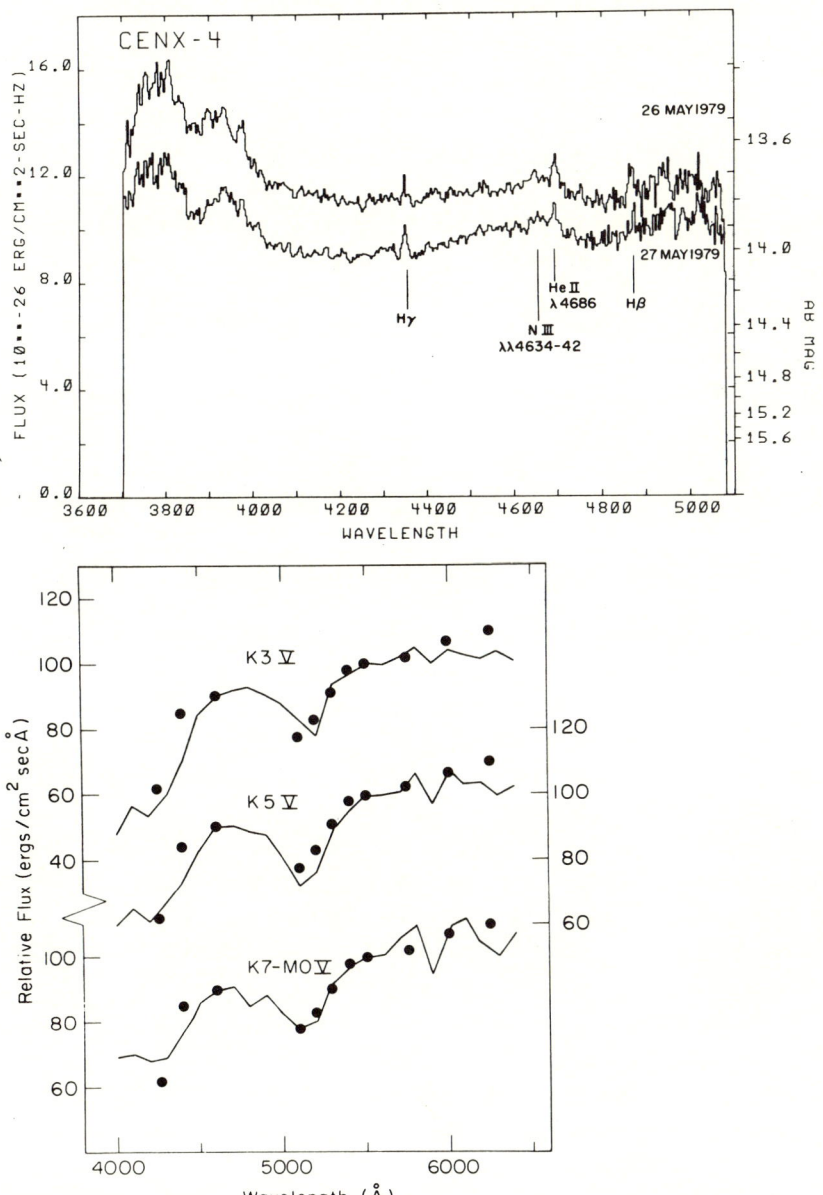

strength of these lines decreases during a later decay phase of the outburst. Balmer emission lines appear somewhat later, with increasing equivalent widths. The spectrum of H1705−25 during the initial decay also showed only He II λ4686 in emission [141].

During X-ray quiescence K-type spectral features of a companion star became visible in the spectra of A 0620−00, Cen X-4 and Aql X-1 [364, 441, 476]. Absolute magnitude estimates for the optical counterparts of A 1524−61 and H 1705−25 are consistent with the idea that in these systems also a late-type main-sequence companion is present [349]. Furthermore, a possible optical identification of the soft X-ray transient A 1742−28 with a K-type dwarf has been proposed [349].

The optical spectra of A 0620−00 and Cen X-4 also show strong Balmer emission lines during quiescence [349, 476, 502]. These emission lines are also present in the spectra of cataclysmic variables [489], and are thought to arise from the cool outer parts of an accretion disk, which is optically thin to continuum radiation, but optically thick in Balmer lines [516]. This suggests that during X-ray quiescence an accretion disk (or perhaps more appropriately an accretion ring) is present in soft X-ray transients. This idea is supported by observations of changes in the profiles of the Balmer lines in the spectrum of A 0620−00. These lines are very broad initially ($\sim$2500 km s^{-1} FWHM), showing a double structure in high-resolution spectrograms, and become narrower later on [349, 502]. During X-ray quiescence also Ca II H and K emission is present. Apparently the region dominating the line emission shifts toward the outer regions of the disk, which have a lower Keplerian velocity and are cool enough for Ca$^+$ to exist.

It is not clear as yet whether this structure is simply a left-over of the accretion disk formed during a previous mass transfer instability, or whether it shows that mass transfer continues at all times, the X-ray outburst being the result of an instability in a gradually forming disk.

In the case of dwarf novae the first possibility seems to be favored according to the numerical calculations of the evolution of accretion disks around white dwarfs after a burst of mass transfer from the companion star [20], which satisfactorily reproduce the outburst properties of these systems. However, arguments for the alternative possibility have also been presented [322].

It should be noted that the meagre evidence available suggests that the spectral behavior of Aql X-1 may be different. A spectrum obtained during X-ray quiescence [441] does not show the presence of Balmer emission lines.

5.2.4.5 *Non-periodic variability*

Although I believe that the LMXB model – in which the optical emission is dominated by reprocessing of X-rays in an accretion disk around the compact object - is correct, detailed observations of some sources indicate that this model certainly does not account for the large variety in observed optical behavior.

Particularly good examples are provided by the irregular brightness variations of GX 339−4 and Sco X-1.

The X-ray intensity of GX 339−4 shows long-term high and low states, a behavior also displayed in the optical brightness (see e.g. [146]). This indicates that most of the optical light is related to the X-ray emission, probably through reprocessing in an accretion disk. However, the source also undergoes X-ray intensity variations on a timescale of tens of milliseconds [566], similar to Cyg X-1 [357]. Recent observations by Motch, Ilovaisky & Chevalier [554] show that also the optical brightness of GX 339−4 is sometimes variable on this short timescale. In particular, they observed 20 ms quasi-periodic oscillations with an amplitude of ∼50 per cent. A 96 s interval of overlapping X-ray observations revealed that these 20 ms variations were also visible in X-rays; however, they were anticorrelated with the optical oscillations [555].

These results indicate that a major contribution to the optical brightness of GX 339−4 is sometimes emitted from within a distance of a few times 10^3 km from the compact object. As argued by Motch, Ilovaisky & Chevalier [554] and Fabian et al. [555], the optical emission cannot be the result of X-ray reprocessing. Fabian et al. propose a model in which the optical fluctuations are due to cyclotron emission from the inner disk region.

X-ray flaring of Sco X-1 was discovered by Lewin, Clark & Smith in 1967 [265]. Later observations showed that the X-ray brightness of Sco X-1 exhibits a very complex behavior (for a review of X-ray observations before 1977 see [337]). No periodicity has been observed in the X-ray flux variations.

Sco X-1 shows active and quiescent states with a duration between ∼1 and ∼5 d [53]. During the quiescent state the X-ray flux is variable on a timescale of about half a day. During active states brightness variations of up to a factor of 2 occur on a timescale of minutes. Petro et al. [394] recently reported rapid X-ray variability on a timescale of less than one second.

Optical variability was found as soon as the optical counterpart had been identified [416]. Extensive observations of Sco X-1 show that the B magnitude varies between $B \sim 13.6$ and $B \sim 12.2$ on a timescale of hours to days [337]. Optical flares of ∼0.2 mag lasting for ∼10 min and flickering (amplitude ∼2 per cent, timescale of minutes) is observed only when the source is optically bright ($B \lesssim 12.7$). Also, X-ray active states occur only when the source is above this optical threshold ([53, 337] and references therein). The optical-brightness histogram shows long-term changes [337]. The relative optical-brightness variations do not depend much on wavelength [341]. IUE observations have shown that this result extends into the far UV [517].

Between $B \sim 12.8$ and ~ 12.4 the X-ray and optical brightness (averaged over a few minutes) often show a correlation, approximately according to a power law: $F_{opt} (:) F_x^\alpha$ with $\alpha \sim 0.5$ (see e.g. [53]), similar to relations observed for other sources.

However, a general correlation between the X-ray and optical brightness of Sco X-1 during the active state does not exist. On the one hand, detailed correlation between optical and X-ray flares has been observed in a number of cases [53, 174, 239, 382, 394]. On the other hand, examples of X-ray flares have been observed without an optical counterpart, and vice versa [53, 204]. During the X-ray quiescent state the optical brightness may change by ~ 0.5 mag without any detectable variation in the X-ray flux.

A series of well-correlated X-ray–optical flares has been observed by Petro et al. [394] (see Figure 5.16). The optical flares are delayed with respect to the X-ray flares by less than ~ 3 s. Some of the X-ray flares have a rise time of less than one second, the corresponding optical rise times, however, are a factor of $\gtrsim 10$ larger. As argued by Petro et al. [394] (see also [53, 337]), this observation (together with the absence of a general correlation between X-ray and optical brightness)

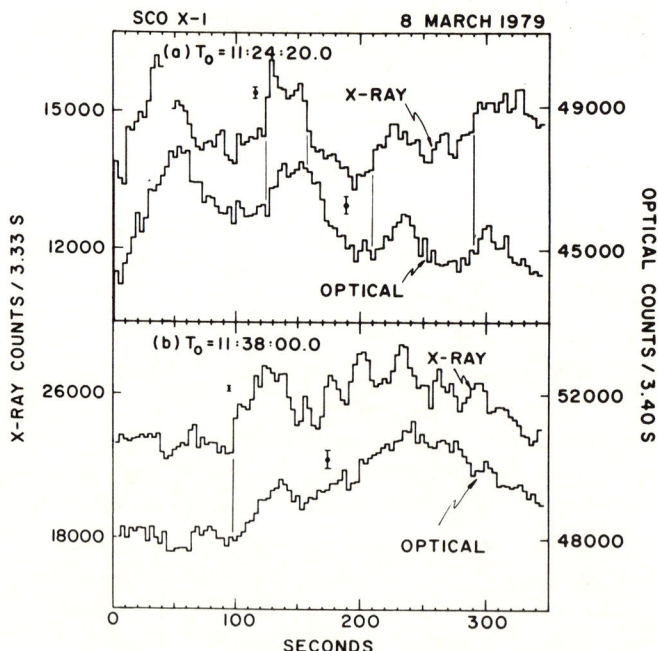

Fig. 5.16. Rapid correlated X-ray and optical variability of Sco X-1 (timescale less than 1 s), observed by Petro et al. [394].

rules out a thermal-bremsstrahlung model for the optical and continuum flux, as proposed on the basis of the overall shape of the electromagnetic spectrum ([337] and references therein).

The observed small delay of the optical flares observed by Petro *et al.* rule out that the optical emission in the flare is due to reprocessing of X-rays in the companion star, but is consistent with reprocessing in a region centered on the X-ray source. However, the observed smearing time of the optical signal ($\sim 20\,\text{s}$) is too large for the reprocessing to occur in an accretion disk [394]. This result rules against the simple accretion disk–X-ray reprocessing model, which has been quite successful in describing the properties of correlated optical X-ray bursts, and in providing a reasonable explanation of the overall optical properties of Class II sources.

It is, as yet, unclear whether the failure of this simple model is a peculiarity of Sco X-1, or whether its breakdown will show up in more extended observations of other Class II sources as well.

It is tempting to end this section with some speculation. If Sco X-1 turns out to be an exception, rather than a prototype of the Class II X-ray sources, this may perhaps be related to the existence of the double radio lobes associated with the X-ray source [41, 136, 168]. These suggest that matter has been (and perhaps still is) ejected perpendicularly to the accretion disk, presumably related to a supercritical accretion rate. If this is the case, matter near the X-ray source, not located in the accretion disk, may be present, which might provide an additional source of optical emission. Furthermore, the inner disk structure may well differ from that in other Class II sources and, e.g., provide a time-dependent shielding of the disk portions which in other sources are responsible for the optical emission through X-ray reprocessing.

5.3 Class I sources: massive X-ray binaries
5.3.1 Introductory remarks

As is discussed in Chapter 2 by Lewin & Joss, Class II X-ray sources (they call them galactic bulge sources) have been very effective in hiding their binary nature, and the establishment of the idea that they are low-mass binaries has been a gradual process. In contrast to this the discovery of massive X-ray binaries (MXRB) as an important class of bright galactic X-ray sources has been very prompt. The reason for this is that signatures of their binary character (presence of a normal stellar object, X-ray eclipses, binary motion of X-ray pulsars) are relatively easily observed. Their optical appearance is dominated by the emission from a rather normal star, and is generally not significantly influenced by the presence of the X-ray source. These stars can be fitted within an already existing astrophysical context, and much is known about their basic

Table 5.5. *Massive X-ray binaries*

Source	Optical counterpart	F_X (μJy)	var†	V	$B-V$	$U-B$	$E(B-V)$	Spectral type	L_{opt}/L_X	Orbital period (d)	Pulse period (s)	$V_{rot} \sin i$ (km s^{-1})	Reference
0050−727 SMC X-3	#4 [80]	<1-5	T	~14	−0.3	−1.0	0.03	O9 III-V [97]	20			~200 [97]	
0053−739 SMC X-2	#5 [80]	<1-7	T	16.0	−0.3	−0.5	0.03	B1.5 Ve [348]	1.2			200 [97]	[421]
0053+604	γ Cas	5-11		1.6-3.0 [338, 355]	−015	−1.08	0.05 [338]	B0.5 III-Ve [338]	220000			300-500 [176, 189, 338]	
0114+650	LSI+65° 010 #1 [115]	4		11.0	1.2	0.1	1.6 [24]	B0.5 IIIe	20000				
0115+634	V635 Cas #1 [212]	<2-250	T	14.5-16.3 [246, 326]	1.4	0.3	1.7	OB e [196]	3-10	24.3 [233, 405, 410]	3.61	~365 [196]	
0115−737 SMC X-1	Sk 160 [417]	2-57		13.3	−0.14	−0.98	0.03	B0 Ib	1.7	3.89	0.71	~200 [188]	[188, 401, 553]
0236+61	LSI+61° 303 #266 [46]	0.2 [26]		10.7	0.8		1.1	B1 Ib [102]	60000	26.45 [193, 438]		~200 [102]	
0352+309	X Per	11-37		6.0-6.6 [338, 355]	0.29	−0.82	0.5 [387, 558]	O9.5 III-Ve [338]	7500	0.93?, 581? 835 [135, 164, 186, 187, 328, 505, 560]		250-400 [180, 338, 505]	[559]
0532−664 LMC X-4	Sk-Ph [73]	4-60		14.0	−0.1	−1.1	0.1	O7 III-V [191]	2	1.40	13.5 [232, 544]	~170 [191]	[75, 395, 556]
0535−668	#Q [215]	1-180	T	13.1-14.9 [62]	0.1	−0.9	0.1	B2 IV+e [62, 195, 377, 533]	0.2	16.6 [427, 428]	0.069 [531]		[195, 377, 532-4]
0535+262	V725 Tau HDE 245770	3-2000	T	8.9-9.6 [355]	0.45	−0.54	0.8 [386]	09.7 IIe, B0 II-Ve [190]	25	>20 (110?) [492]	104	~300 [190]	
0538−641 LMC X-3	#13 [89]	1.7-44	T	16.9 [490]	−0.06	−0.66	0.1	B II-Ve [490]	0.1				
0540−697	#32 [89]	9-25		~14.5			0.1	OB [89, 194]					[372]
0544−665 LMC X-1	#1 [214]	1.8		15.4 [445]	−0.23	−0.96	0.1	B1 Ve [445]	25				[542]
0726−260	star	4.7		11.8 [425]	0.4	−0.7		B [425]	~200				

232

233

Source	Name	Range	T					Spectral type					Refs
0900−40 Vela X-1	GP Vel HD 77581	<28−280		6.9	0.47	−0.51	0.7 [124]	B0.5 Ib	1500	8.97	283	90−130 [3, 327, 513]	[406, 473]
1118−615	He 3-640 [207]	<5−70	T	12.2 [207]	0.96	−0.30	1.2 [207]	09.5 III-Ve [207]	140		405	~300 [207]	
1119−603	V779 Cen	<21−224		13.3	1.07	−0.04	1.4 [385]	06.5 II-III [175, 192, 343, 367]	80	2.09 [543]	4.84	250 [192]	[128]
Cen X-3	Krz [48]												
1145−619	Hen 715	7−1000	T	9.3 [355]	0.18	−0.81	0.25−0.49 [25, 160]	B0-1 Ve [197]	10	187.5? [492, 496]	292 [254, 508]	250−290 [160, 208]	[557]
1145.1−6141	HD 102567 star [197]	4−40		13.1 [564]	~2		1.7 [564]	B2 Iae [563]	1300	5.65? [564]	298 [254, 508]		
1223−624 GX 301-2	BP Cru WRA977 [39]	9−1000	T	10.8	1.76	0.42	1.8 [32, 159, 439]	B1-1.5 Ia [34, 159, 439]	55	41 [230, 492, 506, 552]	696		[535]
1258−613 GX 304-1	MMV # 2 [297]	6−51		14.7	1.8		2.1	B2 V ne [376]	120	>13 [313]	272	~600 [376]	
1417−624	# 7 [540]	8−22		17 [147]				Hα em. [147]		>15	17.6 [231]		
1516−569 Cir X-1	BR Cir # L [503]	5−3000	T	~19v [356, 503]	~3.6		~3.9	O I ? [503]	10	16.6 [227]			
1538−522	QV Nor # 12 [6]	8−30		14.4 [202]	1.9	0.6	2.2 [202]	B0 Iab [98]	450	3.73	529	~200 [98]	[109, 562]
1700−377	V884 Sco HD 153919	11−110		6.6	0.25	−0.72	0.52	06.5 f	3000	3.41	5820? [158, 249, 250, 303]	140−300 [85, 104, 178, 522]	
1907+09	star [424]	5−250		16.4 [424]	3.2		2.6	OB [424]	15	8.38 [294]			
1908+050	V1343 Aql SS433 [247, 432]	6 [491]		14v	2.1		2.7 [350]	Hα em, pec		13.1 [101]			[480], Chapter 7 of this book
1956+350 Cyg X-1	V1357 Cyg HD 226868	260−1320		8.9	0.78	−0.30	1.06 [446]	09.7 Iab	80	5.60		140 [30]	[536]
2206+543	star [425]	5.5 [491]		10 [425]				B [425]					

† T = transient.

properties (mass, luminosity, radius) by just looking at their spectrum. Within a few months after the discovery of the binary character of Cen X-3 [422] the basic outline of the evolution of MXRB systems had been established by van den Heuvel & Heise [455].

Whereas for the LMXB we have just reached the stage of uncovering their basic structure, much of the work on MXRB deals with comparatively detailed questions, and improvement of the accuracy of physically interesting parameters, such as the masses of the binary components.

In this section I will discuss optical observations of MXRB, in particular those related to the determinations of the masses of the components. These include observations of the variations of optical brightness, polarization and radial velocity.

An extensive discussion of the optical observations made before 1975 is given in [241]. Recent reviews of optical observations of MXRB, emphasizing their spectroscopic properties, have been given in [13, 182]. For a general discussion of the properties of massive early-type stars see [111].

5.3.2 Basic data

In Table 5.5 we have collected some basic data of X-ray sources with an early-type optical counterpart. It appears that most of the MXRB fall in either one of two groups, which may be characterized as follows.

(i) The primary star has a spectral type earlier than B2 and has evolved from the main sequence (luminosity class I, II and III). The orbital periods are generally less than 10 d. The primary is filling, or close to filling its Roche lobe (see Table 5.6).

(ii) The primary is a Be star, with a luminosity corresponding to a position close to the main sequence. The orbital periods (if known) tend to be long. The primary is underfilling its Roche lobe.

It was first proposed by Maraschi, Treves & van den Heuvel [284] that these two groups are distinguished by a different origin of the mass transfer process. In the first group mass transfer occurs through a strong stellar wind and/or by incipient Roche-lobe overflow [419, 420]. The mass transfer in the second group of objects is presumably related to the emission line character of the primary star, which is generally thought to be the result of their rapid rotation [429].

The evolutionary histories of these two types of MXRB have been discussed extensively in [407, 454]. A recent review on the properties of Be stars has been given in [429].

Most of the X-ray sources with Be-type companions are highly variable or transient [40, 407]. For a number of the hard X-ray transients recurrent outbursts at regular intervals have been observed [227, 427, 428, 492, 496]. In the

235

Table 5.6. *Orbital elements and other parameters of massive X-ray binaries.*

Source	K_{opt} (km s^{-1})	K_X (km s^{-1})	$a_{opt} \sin i$ (10^{11} cm)	$a_X \sin i$ (10^{12} cm)	M_{opt}/M_X	e_{opt}	e_X	ω_{opt}	ω_X	$R_L \sin i$ (10^{12} cm)	R† (10^{12} cm)	V_{rot}/V_{corot}	Ref.
SMC X-1	19	299	1.02	1.60	15.8	0:	<0.0007	–	–	1.06	1.6	1.0	[188, 401]
LMC X-4	38	465	0.75	0.92	12.0	0:	0:	–	–	0.59	0.9	0.6	[191, 232, 395, 544]
0900–40	21.8	273	2.69	3.37	12.6	0.14	0.092	355°	154°	2.18	1.7	0.5–0.7^b	[406, 473]
Cen X-3	24	415	0.69	1.19	17.3	0:	0.0008	–	–	0.79	0.9	0.9	[128, 192]
1538–52	33	323	1.70	1.66	9.8	0:	0:	–	–	1.06	2.0	1.0	[98, 109]
1700–37	19.0	–	0.88	–	21:	0.16	–	5°	–	1.25	1.8	0.5–1.1	[85, 104, 155, 178, 522]
Cyg X-1	72.7	–	5.55	–	3:	0.049^a	–	327°	–	1.19	1.9	0.9	[30]

: A colon indicates an assumed value.
† Stellar radius estimated from calibrations of M_V, bolometric correction and T_{eff} as a function of spectral type and luminosity class [27, 82, 481].
^{a} According to Bolton & Giess [31, 536] a reevaluation of the orbital elements of Cyg X-1 shows that e_{opt} is not significantly different from zero; the other orbital elements are unchanged.
^{b} Because of the eccentricity of the orbit the orbital angular velocity is not constant; at apastron the corotation factor becomes larger than 1.0.

case of GX 301−2 the recurrence period equals the orbital period found from Doppler-shift variations [230, 492, 506]. The periodic 16.6 d X-ray and optical outbursts of A 0535−66 [427, 428, 534] can be explained by a binary model, consisting of a neutron star in an eccentric orbit ($e \sim 0.7$) around a 12 M$_\odot$ B-type star. Optical and far-UV observations of this source [62, 533] show that near periastron the B star overfills its critical lobe, and a common envelope of approximately four times the original stellar size surrounds both binary components, accounting for the large increase in optical brightness. Attempts have been made to explain all recurrent hard X-ray transients with an eccentric-binary model, in which the accretion rate onto the neutron star changes periodically with the varying distance between the stars. These models often require extremely eccentric orbits ($e \gtrsim 0.9$) independent of the stellar-wind model used ([10] and references therein). Also in the case of 4U 0115+63 the recurrence interval is not related to the orbital period [405, 410]. Thus this model does not provide a general explanation for the hard X-ray transient phenomenon.

5.3.3 Mass determination: method

The determination of the masses of the components of a MXRB is, in principle, a straightforward procedure. The Doppler delay curve of the X-ray pulse-arrival times and the radial-velocity curve of the primary star yield a value of the mass function $f(M_{\text{opt}})$ and $f(M_X)$, respectively, where

$$f(M_X) = \frac{M_X \sin^3 i}{(M_X + M_{\text{opt}})^2} = (a_{\text{opt}} \sin i)^3 / P_{\text{orb}}^2$$
$$= 1.035 \times 10^{-7} K_{\text{opt}}^3 P_{\text{orb}} (1 - e^2)^{3/2} \, M_\odot$$

Fig. 5.17. Histogram of the ratio of bolometric (optical + UV) to X-ray luminosity of Class I X-ray sources.

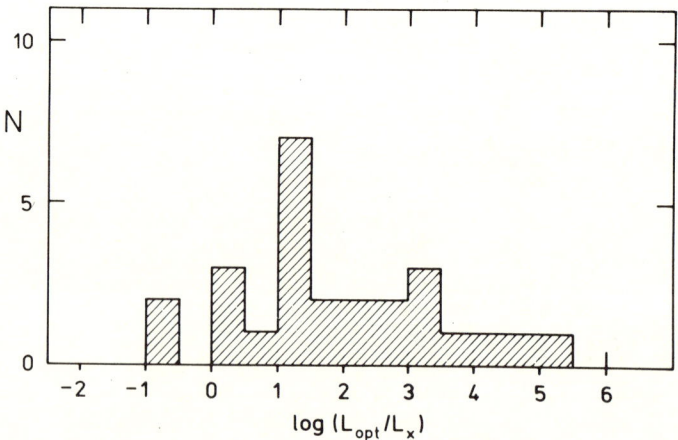

(and a similar expression for $f(M_{opt})$. Here K is the amplitude of the radial-velocity variation (in km s^{-1}), P_{orb} is the orbital period (in days) and e the eccentricity of the orbit. If both radial-velocity curves are available $M_{opt} \sin^3 i$ and $M_X \sin^3 i$ can be determined separately.

A complete mass determination can be made if the orbital inclination i is known. There are three possible methods available for an estimate of i. (See, however, the study of 4U 1626−67 for a different way to determine the orbital parameters, including the inclination [325].) I will discuss these here.

5.3.3.1 Duration of the X-ray eclipse

The principle of this method can be illustrated by consideration of a simple example. Let us suppose that the orbit is circular, and the primary spherical. Then the eclipse duration for a point-like X-ray source is given by the expression $(R/a)^2 = \cos^2 i + \sin^2 i \sin^2 (2\pi t_e/P_{orb})$. Here R is the radius of the primary, a is the separation between the two stars, i is the orbital inclination, t_e is half the eclipse duration and P_{orb} is the orbital period. If the relative size of the primary is known the inclination is a function of the eclipse duration only. In principle the radius of the primary follows from its spectral type and luminosity class. However, these estimates are not of sufficient accuracy to be useful (cf. [451]).

In general the primary will not be spherical, due to the gravitational perturbation of the companion star. In this case the relative size of the primary can be conveniently expressed as a function of the mass ratio $q = M_X/M_{opt}$ and a dimensionless potential parameter Ω [243, 248, 369]. An observed eclipse duration then specifies a relation between q, Ω and i [8, 402].

5.3.3.2 Polarization variations

Thomson scattering of originally unpolarized light in close-binary systems may yield a net linear polarization varying with orbital phase [47, 234, 330, 333, 414, 426]. This is due to the deviation from spherical symmetry of the system (deformation of the primary, presence of gas streams, e.g. an accretion disk). Under quite general conditions (see [333, 426]) the fundamental and first harmonic components (in orbital frequency) of the variation of the Stokes parameters Q and U describe ellipses in the (Q, U) plane. The eccentricity e of these ellipses are uniquely related to the inclination of the orbital plane by $e = \sin i$ for the fundamental and $e = \sin^2 i/(1 + \cos^2 i)$ for the first harmonic frequency. Thus observations of linear polarization variations offer a second, in principle very promising, way to determine the inclination of the orbital plane.

This method had been applied to several (non X-ray) close-binary systems, with results in agreement with those from analyses of their (Algol-type) light curves [413].

The practical use of this method in the case of MXRB is limited by significant variability of the linear polarization, not related to the orbital phase. Such variations may be due to changes in the number of scatterers (e.g. a variable mass flow). As a consequence it is necessary to extend the observations over a large number of orbital cycles.

Extensive polarimetric data have been collected for Cyg X-1 by Kemp and collaborators [237, 238]. The average linear polarization of this source varies in phase with the orbital period. The interpretation of these data, in particular the accuracy with which the inclination can be determined, is a debated issue [234, 426].

Less extensive observations have been reported for Vela X-1 and 4U 1700−37 [244, 468]. The linear polarization of these systems is variable, but the available data are too few to establish the origin of these variations.

5.3.3.3 Optical-brightness variations

The optical counterparts of MXRB with an evolved primary star undergo brightness variations in phase with the orbital period. The basic pattern in these light curves is a double-wave modulation of amplitude ~ 0.1 mag with two approximately equal maxima and two somewhat different minima.

The double-wave light curve ('ellipsoidal variation') is the result of the distortion of the primary star due to the gravitational attraction of the nearby compact star, and a non-uniform surface-brightness distribution ('gravity darkening'). In some systems there is evidence for heating of one side of the primary by X-rays, or for the presence of a luminous region near the compact star, presumably an accretion disk, which undergoes eclipses by the primary.

The distortion, and consequently the shape and amplitude of the light curve, depends on the relative size of the primary, i.e. on q and Ω. The observed amplitude depends, furthermore, on the inclination of the orbital plane.

Numerical calculations of ellipsoidal light curves of MXRB, some of which include complicating effects – such as X-ray heating of the primary, and deviation from corotation – have been presented in many papers [9, 165, 177, 248, 393, 435, 472, 501, 512, 515, 519, 530]. These calculations are based on the assumption that the shape of the primary star is given by a surface on which a potential function Ω, which includes the effects of gravitational and centrifugal forces, is constant [243, 277, 398]. Critical discussions of the various assumptions entering these calculations have been given in [9, 12, 434, 527].

I will limit myself here to a few remarks on attempts to incorporate the influence of radiation pressure on the shape of the primary star [423, 452]. These attempts have started from the idea that the radiation force, taken to be proportional to L/r^2 (L is the luminosity, r the radial distance to the stellar center) can be written as the gradient of a potential function (αr^{-1}). Obviously, by combining this 'radiative' potential with the gravitational and centrifugal potential a different set of equipotential surfaces is generated. In particular, the critical equipotential surface, on which the gradient of Ω becomes zero at some point, consists of two disconnected lobes.

Apart from the lack of spherical symmetry of the radiative flux, which invalidates the assumption that the radiation force is proportional to L/r^2, this approach generates equipotential surfaces which are difficult to interpret. In a stellar atmosphere the inward increase of the weight of the above-located layers is counterbalanced by gas pressure and radiation pressure together, and it is in this way that stellar-atmosphere models are constructed.

The critical surfaces defined by these modified potentials are related to the existence of a limit to the radiative flux (Eddington limit) above which the outward radiative force becomes stronger than the inward gravitational force, giving rise to an unstable situation, presumably resulting in a strong stellar wind [223]. This effect is independent of the deformation of the primary and occurs in single (spherical) stars as well. Since for the standard 'gravity-darkening law' the radiative flux, and thus approximately also the radiation force, is proportional to the gravity force, the Eddington limit is reached at each point of an equipotential surface simultaneously, without preference for e.g. the L_1 point. Of course, in more realistic descriptions of the primary star, the radiative flux near the L_1 point will not be exactly zero, and a locally enhanced stellar wind may occur near L_1, driving a mass flow to the compact star. Spectroscopic observations indicate that in some systems the stellar wind is indeed enhanced in the vicinity of the L_1 point [155].

5.3.4 Mass determination: results

Summarizing, there are three relations between q, i and Ω:

(q, i) from the two mass functions

$(q, i, \Omega)_1$ from the eclipse duration

$(q, i, \Omega)_2$ from the optical light curve

These, in principle, allow us to determine q, i and Ω separately, and hence the individual masses of the components of the MXRB.

For some sources this complete set of observations is not available. In such cases an additional relation between q, Ω and i has to be found. Normally one makes the assumption that the primary star fills its critical lobe, which uniquely relates q and Ω.

Optical radial-velocity curves of sufficient quality to be used in mass determinations have been published for seven sources. These curves have been collected in Figure 5.18.

Fig. 5.18. Radial-velocity curves of massive X-ray binaries. The data have been taken from the following sources: Vela X-1 [473]; 4U 1700−37 [155]; Cyg X-1 [2, 29, 30, 49, 50, 185, 430, 498]; Cen X-3 [192, 343]; SMC X-1 [188, 366]; 4U 1538−52 [98]; LMC X-4 [191, 395].

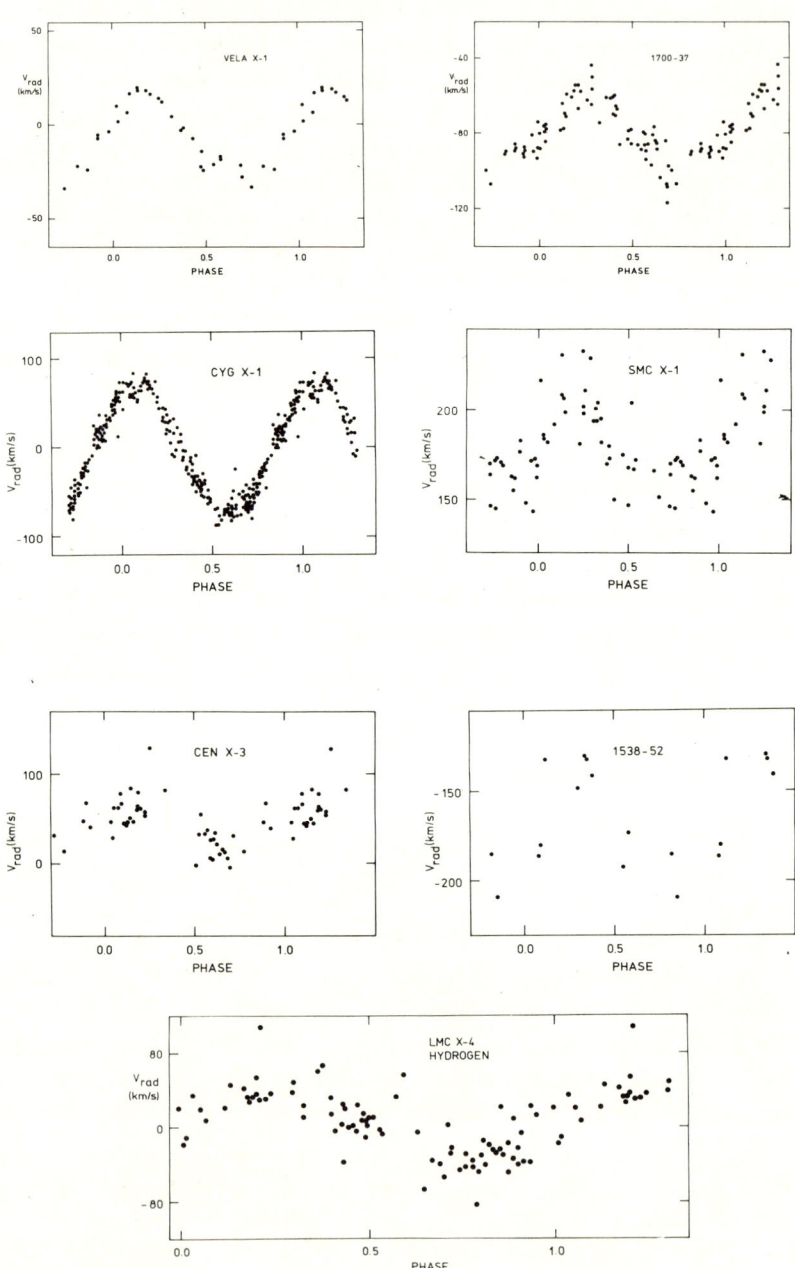

Optical observations of compact galactic X-ray sources 241

A complete solution for the masses (in the above sense) can be obtained for Vela X-1, SMC X-1, Cen X-3 and 1538−52. Recently, the source LMC X-4 has been added to this list, following the discovery of X-ray pulsations [232]. Partial information is available for 1700−37 (no X-ray pulsations observed) and for Cyg X-1 (no pulsations, no X-ray eclipses). A compilation of the results is given in Table 5.6.

5.3.4.1 Neutron-star masses

For an extensive discussion of the observed masses of neutron stars and the implications for the theory of stellar evolution we refer to Chapter 1 (see also [12, 403]). The results are not yet accurate enough to exclude significant differences between the masses of different neutron stars. However, the available results are consistent with the idea that all neutron stars have the same mass ($\sim 1.4\,M_\odot$).

5.3.4.2 Primary masses

A discussion of the observed masses of the primaries, with emphasis on the various sources of errors is given in Chapter 1, this book (see also [403]).

Of particular interest is the question of the 'overluminosity' of the primaries in MXRB. In previous studies [86, 112, 181] it has been concluded that the masses of these primaries, estimated from a comparison of their position in the HR-diagram with evolutionary tracks for stars of different masses, are significantly larger than their observed masses. Stated differently: the primaries are overluminous for their mass. The solution normally suggested for this discrepancy is that apparently the primaries have lost a large amount of mass during their previous evolution.

From the discussion by Rappaport & Joss it appears that there is indeed a tendency for the observed primary masses to be smaller than the evolutionary masses; however, with the exception of Cen X-3 the differences seem to be too small to conclude that there is a general overluminosity of the primaries of MXRB.

5.3.5 Distortions of the radial-velocity curves of massive X-ray binaries

It has been known for a long time that the line profiles of spectroscopic binaries often contain a contribution from gaseous streams in these systems [21, 415]. As a consequence the radial velocities obtained from these distorted line profiles may not reflect the orbital motion of the stellar center of mass.

The influence of gas streams on the spectroscopic and photometric behavior of MXRB has been inferred from several observations.

• Radial-velocity variations of the He II λ4686 emission line, out of phase with

the orbital motion of the primary have been ascribed to matter in between the primary and the X-ray source [98, 182, 185, 188, 191]. The phase dependence of Doppler-shifted Hα absorption features in Vela X-1 [23, 528] have similarly been associated with gas streams, possibly related to enhanced mass loss through the inner Lagrangian point or from a rapidly spinning disk [528].
- The light-curve behavior of SMC X-1 and LMC X-4 indicates the presence of an accretion disk [75, 76, 472].
- A significant amount of gas, moving in front of the X-ray source, has been postulated to explain the variability of the shape of the optical light curve of Vela X-1 [462].
- Intensity dips have been observed in the X-ray light curves of some MXRB [42, 57, 399, 449, 494]. These dips are often accompanied by an increased low-energy cut off, and are naturally explained by absorption of X-rays by gas streams. Although there is a preference for the dips to occur between phase 0.5 and 0.0 the phasing of the X-ray dips is highly variable, and there does not seem to be a stable flow pattern. An interpretation of the dips in terms of an accretion wake trailing the X-ray source has been given in [126, 206].

The overall impression one gains from these observations is that the gas flow in MXRB is extremely complex and variable, and insufficiently known to make an *ab initio* estimate of its effects on the radial-velocity curves.

Indirect evidence suggests that the effects are probably small. They are expected to be most severe for the hydrogen lines. However, in the best-studied systems (Vela X-1, 4U 1700−37 and Cyg X-1) the orbital parameters determined from the hydrogen lines do not differ significantly from those of photospheric lines [30, 155, 473].

In addition to the distortion of line profiles due to emission contributions from gas streams the following effects may play a role in systematically distorting the radial-velocity curves of the primaries.

- *Non-spherical shape of the primary*

Because of the relatively large size of the (approximately corotating) primary the radial-velocity variations of individual surface elements are much larger than the orbital velocity of its center of mass. Because of the tidal distortion of the primary, and the non-uniform temperature and gravity across its surface, the observed radial velocity, which is a spectrophotometric average over the stellar surface, may deviate from the orbital velocity of the primary's center of mass.

In order to estimate the effect on the observed radial-velocity curves, numerical calculations have been made (analogous to the light-curve calculations) in which the (appropriately Doppler-shifted) contributions of all visible surface elements to a line profile are combined [179, 471, 520]. These show that in some cases the orbital parameters may indeed be affected. The observed radial-velocity amplitude tends to be somewhat larger than the orbital velocity and spurious eccentricities may result (up to ~0.2) The apparent longitude ω of periastron is expected to be near 90°. The detailed results depend on the assumed temperatures of the primary, its rotational velocity and on the temperature sensitivity of the particular line considered.

Of the sources for which the orbital eccentricity is known from X-ray pulse timing data, only Vela X-1 has a sufficiently accurate optical radial-velocity curve to allow a study of the importance of this effect. The eccentricity of the orbit, as obtained from X-ray and optical data [404, 473], is in good agreement (see, however, [334]). A comparison of the observed and theoretical radial-velocity curves shows that the expected differences between the curves of different lines are not present [473]). Furthermore, the observed values of ω do not show a preference for a value near 90° (see also the discussion below). Thus, in the case of Vela X-1, the tidal distortion of the primary does not seem to influence the radial-velocity curve significantly.

- *X-ray heating*

A main effect of X-ray heating is the decrease of the temperature gradient and the formation of a temperature inversion in the outermost layers of the stellar atmosphere [4, 279, 335]. As a result absorption lines may be filled in, and may even revert into emission lines. (The extent to which this will occur can be estimated only through a detailed non-LTE calculation of the line formation in these atmospheres.) Thus X-ray heating will tend to shift the center of the light in the absorption lines away from the X-ray source, giving rise to an expected increase of the radial-velocity amplitude over the orbital velocity amplitude. Several complicating factors tend to make this qualitative argument too crude to be useful for an accurate assessment of the extent to which the radial-velocity curve is affected. The structure of the outer layers is rather strongly dependent on the amount of soft X-rays, about which little is known. Furthermore, the primary may be partly shielded from X-rays by an accretion disk. This possibility was suggested by Milgrom [329] to explain the absence of the theoretically expected large velocity amplitude for SMC X-1/Sk 160.

For Her X-1, X-ray heating of the companion star is so strong that the heated hemisphere is appreciably brighter than the back side of the star (see Section 5.2.4.1.3). In this case the average location of the absorption line forming region is shifted toward the center of mass of the system. As a consequence the observed radial-velocity curve has an amplitude which is substantially lower than that expected for the center of mass of HZ Her (see Figure 5.9). Modelling of the spectral line profiles, taking the distortion and heating of HZ Her into account, gives a reasonable explanation of the observed radial-velocity curve [94, 95, 242].

- *Non-isotropy of the stellar wind*

The distribution of the observed periastron longitudes has a marked concentration around $\omega = 0°$. As was shown by Milgrom [332] this may be the result of an enhanced absorption in a stellar wind near phases 0.0 and 0.5, making the radial velocity more negative near these phases. This introduces a spurious eccentricity of the orbit, with ω near 0°. The amplitude of the radial-velocity curve is not expected to be much affected. Observational evidence for an enhanced stellar wind near phases 0.0 and 0.5 has been obtained for 4U 1700−37 [155].

A possible way to distinguish such a spurious eccentricity from a real orbital one is by measuring the apsidal motion expected for an eccentric orbit. Such observations have been made for Vela X-1 [406]. The resulting upper limit on the apsidal motion for this system is consistent with the interpretation that the observed eccentricity is an intrinsic orbital property.

5.3.6 Optical light curves of massive X-ray binaries
5.3.6.1 Vela X-1/HD 77581

The average light curve of Vela X-1 [57, 133, 219, 393, 462, 478, 479, 529] is shown in Figure 5.19. Significant variability of the brightness occurs on timescales small compared to the orbital period. Often the shape of the light curve does not repeat from cycle to cycle, to the extent that sometimes an expected maximum does not occur [219]. According to Cherepashchuk [65] the shape of the light curve undergoes a periodic modulation, with a period of ~ 92 d.

Although an underlying ellipsoidal variation is evident, the shape of the light curve is definitely peculiar. The minimum near phase zero is shifted with respect to the X-ray eclipse ephemeris by ~ 0.1 cycle. Furthermore, the light curve does not show the expected symmetry: the maxima have different shapes, the one near phase 0.25 being much sharper than that near phase 0.75.

These peculiarities are probably related to the eccentricity of the orbit [404, 406, 473]. Due to the varying distance between the two stars, the distortion of the primary changes during the orbit, reaching a maximum at periastron near phase 0.17 [404].

It is not to be expected that a standard analysis of the light curve will yield a useful constraint on the inclination of the orbital plane. Perhaps the problem may be inverted: the mass ratio is known, and by making an assumption about the Roche-lobe filling factor at periastron the inclination can be determined from the eclipse duration. A detailed numerical calculation of the expected light curve, including the effect of the variation of both Ω and the corotation factor f

with phase might yield information on the dynamical behavior of the primary undergoing a variable gravitational perturbation.

Brightness modulations of HD 77581 at the X-ray pulse period, in a 30 Å wide passband centered on Hβ, were reported by Steiner [431]. These optical pulsations reached an amplitude of ~2 per cent, and were also observed during two X-ray

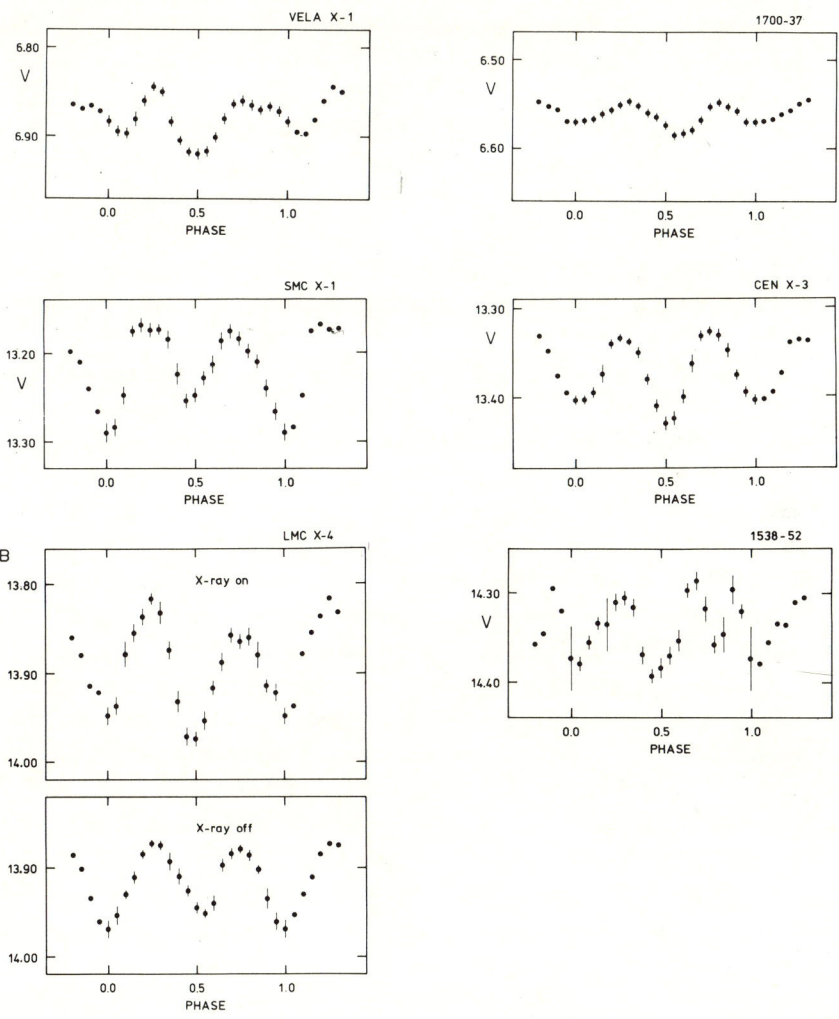

Fig. 5.19. Optical light curves of massive X-ray binaries. The data have been taken from the following sources: Vela X-1 [133, 393, 462, 478, 529]; 4U 1700—37 [132, 157, 384, 463, 464, 474]; SMC X-1 [383, 465, 466] and unpublished Walraven photometry; Cen X-3 [251, 307, 390] and unpublished Walraven photometry; 4U 1538—52 [202] and unpublished Walraven photometry; LMC X-4 [75].

eclipses. Broad-band optical variability of this amplitude cannot result from
X-ray heating [67]. Subsequent optical observations [57, 354, 440] have failed
to show the presence of both broad-band and Hβ pulsations. Upper limits derived
on the amplitude of any optical pulsations are an order of magnitude below the
value given by Steiner.

5.3.6.2 4U1700−37/HD153919

The optical light curve of this system [132, 157, 218, 339, 384, 459,
463, 464, 474] is shown in Figure 5.19. The residuals of individual observations
relative to the average light curve are much larger than the observational accuracy,
indicating intrinsic variability of the primary. Irregular variability, with amplitudes up to 0.05 mag has been observed in some apparently normal OB super-
giants [433]. The light curve of 4U1700−37 is dominated by a double-wave
modulation; however, the detailed shape is not well understood [474]. De Freitas
Pacheco [132] and Cherepashchuk & Khruzina [66] have proposed that the shape
of the light curve is modulated with a period of 32.5 d. Spectroscopic studies
[178] have shown that HD 153919 is one of the most extreme Of stars known,
with an extremely strong stellar wind. The layers where the optical continuum is
formed are already moving outward with a speed of $\sim$50 km s^{-1} [85]. In such
a situation equipotential surfaces may play a minor role in determining the shape
of the outer surface of the star.

The light curve is shifted with respect to the X-ray eclipse ephemeris by
$\sim$0.05 orbital cycle. Similarly, from the orbital solution for the optical radial-
velocity data [155], it follows that the center of X-ray eclipse is expected at
phase 0.06. This phase shift is perhaps due to additional strong absorption of
X-rays before the occultation by the primary stars [474]. If this is the case the
constraint on the inclination of the orbital plane from the observed eclipse
duration becomes rather uncertain.

According to Kruszewski and collaborators [249, 250] near phase 0.5 the
optical brightness of HD 153919 is modulated with a period of $\sim$95 min, decreasing on a e-folding timescale of $\sim$30 yr. A similar period was found earlier in the
X-ray flux by Matilsky, La Sala & Jessen [303]. However, Hammerschlag,
Henrichs & Shaham [158] have shown that such a period cannot be distinguished
from a spurious effect due to the particular window function of the X-ray data
sampling. The amplitude of the optical variability is sometimes as large as
0.02 mag, which is much larger than the modulation expected from X-ray heating.
Extended photometric observations made near phase 0.5 have failed to show the
reported 95 min variations [545, 546]. Thus the variations, reported by Kruszewski
et al., if real, must have been an ephemeral phenomenon.

5.3.6.3 SMC X-1/Sk 160

The light curve of SMC X-1 offers another example of possible complications, which limit the usefulness of this type of data in constraining the orbital parameters, but on the other hand may provide insight into other interesting physical phenomena. In Figure 5.19 we have plotted the average light curve, of SMC X-1 [51, 275, 383, 392, 457, 458, 465, 466].

Because of the large ratio of X-ray to optical luminosity, X-ray heating has a pronounced effect on the shape of the light curve. The minimum near phase 0.5 is partially filled in, and is less deep than the one near phase 0.0. There is a rather large variation of the colors with orbital phase.

The amplitude of the light curve is much too large to be explained by ellipsoidal variations. For the mass ratio of the system (M_{opt}/M_X between 12 and 22) the maximum expected amplitude is 0.065 mag [512]. According to van Paradijs & Zuiderwijk [472], the observed large amplitude is the result of a significant contribution to the optical brightness from a luminous region near the X-ray source, which is likewise eclipsed. This result is supported by observations of the radial-velocity and line-strength behavior of the He II λ4686 emission line [188] which show that it originates in the vicinity of the X-ray source.

These observations suggest the presence of an accretion disk in SMC X-1 implying that Roche-lobe overflow plays an important role in the mass transfer of this system. This result supports a similar conclusion, based on the inability of stellar-wind models to explain the large X-ray luminosity of SMC X-1 [86, 397]. It should be noted, however, that there are indications that the wind structure of early-type stars in the Magellanic Clouds is different from that of the galactic OB stars, in particular their terminal velocity may be significantly lower [183, 184, 538]. Because of the strong dependence of L_X on the wind velocity [256], part of the high X-ray luminosity of the Magellanic Cloud X-ray sources [80] may be related to a different wind structure.

5.3.6.4 Cen X-3

A comparison of the different published light curves [251, 307, 390] indicates that the shape of the light curve may not be stable. Values given for the amplitude of the brightness variation range between 0.10 and 0.15 mag. An average light curve, based on the above references and unpublished VBLUW photometry is shown in Figure 5.19. The amplitude of this average light curve (0.10 ± 0.01 mag) is somewhat larger than the values expected for the observed mass ratio ($M_{opt}/M_X = 17$ [192]), irrespective of the assumed corotation factor [512].

Perhaps in Cen X-3 an additional light source is present, similar to SMC X-1 [472]. Cen X-3 is one of the sources for which mass transfer is expected to occur by Roche-lobe overflow ([86, 256], and Chapter 9, this book).

Optical pulsations with an amplitude in excess of 0.1 per cent have not been detected [52, 259, 388].

5.3.6.5 Cyg X-1

This source is particularly interesting, since it is the only serious candidate for a black hole with a stellar mass. This result is based on the mass function obtained from the radial-velocity curve of the primary [30] and a primary mass expected for its spectral type ($\gtrsim 10\,M_\odot$), which give a lower limit to the mass of the compact star of $\sim 9\,M_\odot$ [111, 306, 370].

Trimble, Rose & Weber [447] suggested that the primary could be a low-mass ($\sim 0.5\,M_\odot$) helium star with an effective temperature and surface gravity similar to that of an OB supergiant. If this were true the distance of Cyg X-1 would be less than ~ 1 kpc. Distance estimates based on reddening and polarization studies [45, 64, 288, 290, 526] indicate that the distance is ~ 2.5 kpc, supporting the conclusion that Cyg X-1 contains a black hole.

A possible independent method to estimate the mass of the primary of Cyg X-1 was suggested by van den Heuvel (private communication). It is based on the observation that the ratio of the terminal velocity of the stellar wind of early-type stars to their escape velocity is approximately constant [1]. If this relation holds generally, an estimate of the terminal velocity (from high-resolution UV observations of resonance lines) would give an estimate of the escape velocity $V_{es} = (2GM/R)^{1/2}$. Together with the surface gravity $g = GM/R^2$ this would allow both M and R to be estimated separately.

Optical photometric data have been reported by many observers [152, 229, 258, 262, 263, 280, 482, 484, 485, 487]. The light curve shows the customary ellipsoidal variation. For the total amplitude values between 0.04 mag [484] and 0.06 mag [262] have been given. Analysis of the light curve [9, 30] yields a rather high inclination (30°–40°), consistent with the absence of X-ray eclipses.

Cyg X-1 shows long-term changes between states of high- and low-X-ray (2–10 keV) luminosity [357]. During transitions from a low to a high state the shape of the optical light curve undergoes a significant change [351, 483, 485].

Wilson & Fox [521] found evidence for long-term changes in the shape of the optical light curve of Cyg X-1; from a comparison with theoretical light curves, including the effects of eccentricity of the orbit [518], they conclude that the orbit of Cyg X-1 shows apsidal motion with a period of 4.5 yr. This period is much too short to be due to the finite mass distribution of the optical star and suggests the possible presence of a third perturbing body. Consistent mass-distance relations suggest a mass of the order of a solar mass and an orbital period of the order of a month for this third body. This is consistent with the results of a search for a third star in the Cyg X-1 system, based on possible variations of the systemic radial velocity [2].

In this regard the possible 39 d periodicity in the optical polarization detected by Kemp *et al.* [235, 236] is of interest (see also [113, 336, 486]).

5.3.6.6 LMC X-4

The optical brightness of LMC X-4 shows a double-wave variation in phase with the 1.4 d orbital period [72] (see Figure 5.19). The amplitude is variable, and reaches values which are too large to be understood from ellipsoidal variability only. Apart from regular X-ray eclipses LMC X-4 also shows a 30.5 d periodic on-off modulation of the X-ray intensity [257]. It was recently found [75, 76] that the shape and amplitude of the optical light curve are correlated with this X-ray on-off behavior. The amplitude varies between 0.09 mag during the X-ray off state to 0.16 mag during the on state. According to Chevalier *et al.* [75, 76] this behavior of the light curve can be understood in terms of a tilted precessing disk, analogous to Her X-1 (see Section 5.2.4.1.3).

5.3.6.7 4U 1538−52

The optical brightness of 4U 1538−52 shows a double-wave modulation, with an amplitude of ∼0.10 mag [202, 562] (see Figure 5.19). This amplitude is consistent with ellipsoidal variation at the mass ratio given for this system (M_{opt}/M_X between 7 and 15 [98]). There are indications that the maximum near phase 0.75 is variable.

5.3.6.8 4U 1223−62/Wra 977

Optical photometric observations of this source [35, 156, 456, 460] revealed variability of ∼0.1 mag on a timescale of days. Although several periods in the range between ∼10 and ∼25 d have been suggested, the variability is of a random character. No clear variability exists which is in phase with the 41 d orbital period [230, 373, 492, 506], except for a possible slight enhancement of the optical brightness near periastron passage [373].

Of particular interest are reports of optical variability, of 0.4 per cent amplitude, with a period which is consistent with the X-ray pulse period [71, 308, 310]. However, the X-ray luminosity of the source is too low by at least an order of magnitude to understand the observed optical pulsation amplitude in terms of a simple X-ray heating model [67]. We here encounter a similarly puzzling situation as for 4U 1700−37 (see Section 5.3.6.2).

Acknowledgements

I thank W. Lewin, B. Margon, C. Motch and F. Verbunt for their valuable comments on this paper. I thank I. Hoonhout and J. van Dijk for typing the manuscript, and M. Moesman for the preparation of the figures.

References

[1] Abbott, D. C. 1978, *Ap. J.*, **225**, 893.
[2] Abt, H. A., Hintzen, P. & Levy, S. G. 1977, *Ap. J.*, **213**, 815.
[3] Ammann, M. & Mauder, H. 1978, *Mitt. der Astron. Gesellschaft*, **43**, 219.
[4] Anderson, L. 1981, *Ap. J.*, **244**, 555.
[5] Apparao, K. M. V. & Chitre, S. M. 1976, *Sp. Sci. Rev.*, **19**, 281.
[6] Apparao, K. M. V., Bradt, H. V. *et al.* 1978, *Nature*, **271**, 225.
[7] Apparao, K. M. V., Naranan, S. *et al.* 1980, *Astron. Ap.*, **89**, 249.
[8] Avni, Y. 1976, *Ap. J.*, **209**, 574.
[9] Avni, Y. & Bahcall, J. N. 1975, *Ap. J.*, **197**, 675.
[10] Avni, Y. & Goldman, I. 1980, *Astron. Ap.*, **90**, 44.
[11] Ayasli, S. & Joss, P. C. 1982, *Ap. J.*, **256**, 637.
[12] Bahcall, J. N. 1978, *Ann. Rev. Astron. Ap.*, **16**, 241.
[13] Bahcall, J. N. 1978, In: *Physics and Astrophysics of Neutron Stars and Black Holes*, eds. R. Giacconi & R. Ruffini, North Holland (Amsterdam, New York, Oxford), p. 63.
[14] Bahcall, J. N. & Bahcall, N. A. 1972, *Ap. J. (Lett.)*, **178**, L1.
[15] Bahcall, J. N., Bahcall, N. A. *et al.* 1975, *PASP*, **87**, 141.
[16] Bahcall, J. N. & Wolf, R. A. 1976, *Ap. J.*, **209**, 214.
[17] Bahcall, J. N. & Chester, T. J. 1977, *Ap. J. (Lett.)*, **215**, L21.
[18] Basinska, E. *et al.* 1982, *Ap. J.* (in preparation).
[19] Basko, M. M., Goranskij, V. P. *et al.* 1976, *Peremm. Zvezdy*, **20**, 219.
[20] Bath, G. T. & Pringle, J. E. 1981, *MNRAS*, **194**, 967.
[21] Batten, A. H. 1973, *Binary and Multiple Systems of Stars*, Pergamon (Oxford, New York, Toronto, Sydney, Braunschweig).
[22] Becklin, E. E., Kristian, J. *et al.* 1972, *Nature Phys. Sci.*, **239**, 130.
[23] Bessell, M. S., Vidal, N. V. & Wickramasinghe, D. T. 1975, *Ap. J. (Lett.)*, **195**, L117.
[24] Bianchi, L. 1981, *Sp. Sci. Rev.*, **30**, 273.
[25] Bianchi, L. & Bernacca, P. L. 1980, *Astron. Ap.*, **89**, 214.
[26] Bignami, G. F. & Caraveo, P. A. 1981, *IAU Circular*, No. 3518.
[27] Böhm-Vitense, E. 1981, *Ann. Rev. Astron. Ap.*, **19**, 295.
[28] Boley, F., Wolfson, R. *et al.* 1976, *Ap. J. (Lett.)*, **203**, L13.
[29] Bolton, C. T. 1972, *Nature Phys. Sci.*, **240**, 200.
[30] Bolton, C. T. 1975, *Ap. J.*, **200**, 269.
[31] Bolton, C. T. & Giess, D. R. 1980, *IAU Symposium*, No. 88, p. 355.
[32] Bonnet-Bidaud, J. M. & van der Klis, M. 1981, *Astron. Ap.*, **101**, 299.
[33] Bopp, B. W., Grupsmith, G. & VandenBout, P. 1972, *Ap. J. (Lett.)*, **178**, L5.
[34] Bord, D. J. 1979, *Astron. Ap.*, **77**, 309.
[35] Bord, D. J., Mook, D. E. *et al.* 1976, *Ap. J.*, **203**, 689.
[36] Bowyer, S. & Clarke, J. 1981, *IAU Circular*, No. 3632.
[37] Boynton, P. E., Canterna, R. *et al.* 1973, *Ap. J.*, **186**, 617.
[38] Boynton, P. E., Crosa, L. & Deeter, J. 1980, *Ap. J.*, **237**, 169.
[39] Bradt, H. V., Apparao, K. M. V. *et al.* 1977, *Nature*, **269**, 21.
[40] Bradt, H. V., Doxsey, R. E. & Jernigan, J. G. 1979. In: *X-ray Astronomy* COSPAR Meeting, Innsbruck, May–June 1978 (eds. W. A. Baity & L. E. Peterson), p. 3.
[41] Braes, L. L. E. & Miley, G. K. 1971, *Astron. Ap.*, **14**, 160.
[42] Branduardi, G., Mason, K. O. & Sanford, P. W. 1978, *MNRAS*, **185**, 137.
[43] Branduardi, G., Kylafis, N. D. *et al.* 1980, *Ap. J. (Lett.)*, **235**, L153.

[44] Branduardi-Raymont, G., Corbet, R. et al. 1981, Sp. Sci. Rev., 30, 279.
[45] Bregman, J., Butler, D. et al. 1973, Ap. J. (Lett.), 185, L117.
[46] Brodskaya, E. J. & Shajn, P. F. 1958, Izw. Krim. Ap. Obs., 20, 299.
[47] Brown, J. C., McLean, I. S. & Emslie, A. G. 1978, Astron. Ap., 68, 415.
[48] Brucato, R. J., Kristian, J. & Westphal, J. A. 1972, Ap. J. (Lett.), 175, L137.
[49] Brucato, R. & Kristian, J. 1973, Ap. J. (Lett.), 179, L129.
[50] Brucato, R. J. & Zappala, R. R. 1974, Ap. J. (Lett.), 189, L71.
[51] Butler, C. J. & Byrne, P. B. 1973, Nature Phys. Sci., 243, 136.
[52] Canizares, C. R. & McClintock, J. E. 1974, Ap. J. (Lett.), 193, L65.
[53] Canizares, C. R., Clark, G. W. et al. 1975, Ap. J., 197, 457.
[54] Canizares, C. R., McClintock, J. E. & Grindlay, J. E. 1979, Ap. J., 234, 556.
[55] Canizares, C. R., McClintock, J. E. & Grindlay, J. E. 1980, Ap. J. (Lett.), 236, L55.
[56] Cathey, L. R. & Hayes, J. E. 1968, Ap. J. (Lett.), 151, L89.
[57] Charles, P. A., Mason, K. O. et al. 1978, MNRAS, 183, 813.
[58] Charles, P. A., Thorstensen, J. & Bowyer, S. 1978, MNRAS, 183, 29P.
[59] Charles, P. A., Thorstensen, J. R. et al. 1980, Ap. J., 237, 154.
[60] Charles, P. A., Thorstensen, J. R. & Barr, P. 1980, Ap. J., 241, 1148.
[61] Charles, P. A. & Thorstensen, J. R. 1981, IAU Circular, No. 3570.
[62] Charles, P. A., Booth, L. et al. 1981, Sp. Sci. Rev., 30, 423.
[63] Charles, P. A., Thorstensen, J. R. et al. 1981, Bull. AAS, 11, 720.
[64] Cheng, C. C., Phillips, K. J. H. & Wilson, A. M. 1974, Nature, 251, 589.
[65] Cherepashchuk, A. M. 1982, Astron. Zh., 59, 918.
[66] Cherepashchuk, A. M. & Khruzina, T. I. 1981, Astron. Zh., 58, 1226.
[67] Chester, T. J. 1979, Ap. J., 227, 569.
[68] Chevalier, C. & Ilovaisky, S. A. 1973, Nature Phys. Sci., 245, 87.
[69] Chevalier, C. & Ilovaisky, S. A. 1974, Astron. Ap., 35, 407.
[70] Chevalier, C., Ilovaisky, S. A. & Mauder, H. 1976, IAU Circular, no. 2957.
[71] Chevalier, C. & Motch, C. 1977, IAU Circular, No. 3146, 3152.
[72] Chevalier, C. & Ilovaisky, S. A. 1977, Astron. Ap., 59, L9.
[73] Chevalier, C. & Ilovaisky, S. A. 1978, ESO Messenger, No. 5, p. 4.
[74] Chevalier, C. & Ilovaisky, S. A. 1981, Astron. Ap., 94, L3.
[75] Chevalier, C., Ilovaisky, S. A. et al. 1981, Sp. Sci. Rev., 30, 405.
[76] Chevalier, C., Ilovaisky, S. A. et al. 1982, Astron. Ap. (in preparation).
[77] Chevalier, C. & Ilovaisky, S. A. 1982, Astron. Ap., 112, 68.
[78] Chiapetti, L., Maraschi, L. et al. 1981, Sp. Sci. Rev., 30, 287.
[79] Clark, G. W., Li, F. K. et al. 1977, MNRAS, 179, 651.
[80] Clark, G. W., Doxsey, R. et al. 1978, Ap. J. (Lett.), 221, L37.
[81] Cocke, W. J., Hintzen, P. et al. 1973, Nature Phys. Sci., 244, 137.
[82] Code, A. D., Davis, J. et al. 1976, Ap. J., 203, 417.
[83] Cominsky, L., Jones, C. et al. 1978, Ap. J., 224, 46.
[84] Cominsky, L. R. 1981, Ph.D. thesis, MIT.
[85] Conti, P. S. & Cowley, A. P. 1975, Ap. J., 200, 133.
[86] Conti, P. S. 1978, Astron. Ap., 63, 225.
[87] Cowley, A. P. 1979, Proc. NATO Advanced Study Institute on Compact X-ray Sources, Cape Sounion, Greece, June 1979.
[88] Cowley, A. P. & Crampton, D. 1975, Ap. J. (Lett.), 201, L65.
[89] Cowley, A. P., Crampton, D. & Hutchings, J. B. 1978, Astron. J., 83, 1619.
[90] Cowley, A. P., Crampton, D. & Hutchings, J. B. 1979, Ap. J., 231, 539.
[91] Cowley, A. P., Crampton, D. & Hutchings, J. B. 1981, Ap. J., 256, 605.
[92] Crampton, D. & Hutchings, J. B. 1972, Ap. J. (Lett.), 178, L65.
[93] Crampton, D. & Morbey, C. L. 1972, IAU Circular, No. 2418.
[94] Crampton, D. 1974, Ap. J., 187, 345.
[95] Crampton, D. & Hutchings, J. B. 1974, Ap. J., 191, 483.
[96] Crampton, D., Cowley, A. P. et al. 1976, Ap. J., 207, 907.
[97] Crampton, D., Hutchings, J. B. & Cowley, A. P. 1978, Ap. J. (Lett.), 223, L79.
[98] Crampton, D., Hutchings, J. B. & Cowley, A. P. 1978, Ap. J. (Lett.), 225, L63.

[99] Crampton, D. & McClure, R. D. 1979, *PASP*, **91**, 117.
[100] Crampton, D. & Cowley, A. P. 1980, *PASP*, **92**, 147.
[101] Crampton, D., Cowley, A. P. & Hutchings, J. B. 1980, *Ap. J. (Lett.)*, **235**, L131.
[102] Crampton, D. & Hutchings, J. B. 1978, *IAU Circular*, No. 3180.
[103] Crosa, L. & Boynton, P. E. 1980, *Ap. J.*, **235**, 999.
[104] Dachs, J. 1976, *Astron. Ap.*, **47**, 19.
[105] Davidsen, A., Henry, J. P. *et al.* 1972, *Ap. J. (Lett.)*, **177**, L97.
[106] Davidsen, A., Margon, B. & Middleditch, J. 1975, *Ap. J.*, **198**, 653.
[107] Davidsen, A., Malina, R. & Bowyer, S. 1977, *Ap. J.*, **211**, 866.
[108] Davidson, K. & Ostriker, J. P. 1973, *Ap. J.*, **179**, 585.
[109] Davison, P. J. N., Watson, M. G. & Pye, J. P. 1977, *MNRAS*, **181**, 73P.
[110] Deeter, J., Crosa, L. *et al.* 1976, *Ap. J.*, **206**, 861.
[111] De Jager, C. 1981, *The Brightest Stars* (Reidel, Dordrecht).
[112] De Loore, C. 1979, *IAU Symposium*, No. 83, p. 313.
[113] Dolan, J. F., Caraveo, P. *et al.* 1979, *Nature*, **280**, 126.
[114] Dorren, J. D. & Guinan, E. F. 1980, *IAU Symposium*, No. 88, p. 361.
[115] Dower, R. G., Apparao, K. M. V. *et al.* 1978, *Nature*, **273**, 364.
[116] Doxsey, R., Bradt, H. V. *et al.* 1973, *Ap. J. (Lett.)*, **182**, L25.
[117] Doxsey, R. E., Apparao, K. M. V. *et al.* 1977, *Nature*, **269**, 112.
[118] Doxsey, R. E., Apparao, K. M. V. *et al.* 1977, *Nature*, **270**, 586.
[119] Doxsey, R., Grindlay, J. *et al.* 1979, *Ap. J. (Lett.)*, **228**, L67.
[120] Duerbeck, H. W. & Walter, K. 1976, *Astron. Ap.*, **48**, 141.
[121] Duerbeck, H. W. & Walter, K. 1976, *In: X-ray Binaries*, NASA SP-389, p. 343.
[122] Dupree, A. K., Baliunas, S. L. & Lester, J. B. 1977, *Ap. J. (Lett.)*, **218**, L71.
[123] Dupree, A. K., Davis, R. J. *et al.* 1978, *Nature*, **275**, 400.
[124] Dupree, A. K., Gursky, H. *et al.* 1980, *Ap. J.*, **238**, 969.
[125] Dvorak, T. Z. 1976, *Inf. Bull. Variable Stars*, No. 1082.
[126] Eadie, G., Peacock, A. *et al.* 1975, *MNRAS*, **172**, 35P.
[127] Endal, A. S., Devinney, E. J. & Sofia, S. 1976, *Ap. Lett.*, **17**, 131.
[128] Fabbiano, G. & Schreier, E. J. 1977, *Ap. J.*, **214**, 235.
[129] Flowers, E. & Ruderman, M. 1977, *Ap. J.*, **215**, 302.
[130] Forman, W., Jones, C. & Liller, W. 1972, *Ap. J. (Lett.)*, **177**, L103.
[131] Forman, W., Jones, C. *et al.* 1978, *Ap. J. Suppl.*, **38**, 357.
[132] Freites Pacheco, J. A. de, Steiner, J. E. & Quast, G. R. 1974, *Astron. Ap.*, **33**, 131.
[133] Freites Pacheco, J. A. de & Quast, G. R. 1974, *Astron. Ap.*, **35**, 301.
[134] Frohlich, A. 1973, *Ap. Lett.*, **13**, 233.
[135] Galkina, T. S. 1980, *Izw. Krim. Ap. Obs.*, **61**, 77.
[136] Geldzahler, B. J., Fomalont, E. B. *et al.* 1981, *Astron. J.*, **86**, 1036.
[137] Gerend, D. & Boynton, P. 1976, *Ap. J.*, **209**, 562.
[138] Giacconi, R., Gorenstein, P. *et al.* 1967, *Ap. J. (Lett.)*, **148**, L129.
[139] Giacconi, R., Gursky, H. *et al.* 1973, *Ap. J.*, **184**, 227.
[140] Gottlieb, E. W., Wright, E. L. & Liller, W. 1975, *Ap. J. (Lett.)*, **195**, L33.
[141] Griffiths, R. E., Bradt, H. *et al.* 1978, *Ap. J. (Lett.)*, **221**, L63.
[142] Griffiths, R. E., Gursky, H. *et al.* 1978, *Nature*, **276**, 247.
[143] Grindlay, J. E. 1978, *Ap. J.*, **225**, 1001.
[144] Grindlay, J. E. 1979, *Ap. J. (Lett.)*, **232**, L33.
[145] Grindlay, J. E. 1980, Talk presented at the HEAD/AAS meeting, Cambridge, Mass., January 1980.
[146] Grindlay, J. 1981, *IAU Circular*, No. 3613.
[147] Grindlay, J. 1981, *IAU Circular*, No. 3620.
[148] Grindlay, J. E. & Liller, W. 1978, *Ap. J. (Lett.)*, **220**, L127.
[149] Grindlay, J. E., McClintock, J. E. *et al.* 1978, *Nature*, **274**, 567.
[150] Groth, E. J. & Nelson, M. R. 1972, *Ap. J. (Lett.)*, **176**, L111.
[151] Groth, E. J. 1974, *Ap. J.*, **192**, 517.
[152] Guinan, E. F., Dorren, J. D. *et al.* 1979, *Ap. J.*, **229**, 296.
[153] Gursky, H., Dupree, A. K. *et al.* 1980, *Ap. J.*, **237**, 163.
[154] Hackwell, J. A., Grasdalen, G. L. *et al.* 1979, *Ap. J. (Lett.)*, **233**, L115.

[155] Hammerschlag-Hensberge, G. 1978, *Astron. Ap.*, **64**, 399.
[156] Hammerschlag-Hensberge, G., Zuiderwijk, E. J. *et al.* 1976, *Astron. Ap.*, **49**, 321.
[157] Hammerschlag-Hensberge, G. & Zuiderwijk, E. J. 1977, *Astron. Ap.*, **54**, 543.
[158] Hammerschlag-Hensberge, G., Henrichs, H. & Shaham, J. 1979, *Ap. J. (Lett.)*, **228**, L75.
[159] Hammerschlag-Hensberge, G., de Loore, C. *et al.* 1979, *Astron. Ap.*, **76**, 245.
[160] Hammerschlag-Hensberge, G., van den Heuvel, E. P. J. *et al.* 1980, *Astron. Ap.*, **85**, 119.
[161] Hammerschlag-Hensberge, G., McClintock, J. E. & van Paradijs, J. 1982, *Ap. J. (Lett.)*, **258**, L1.
[162] Hardorp, J., Rohlfs, K. *et al.* 1959, *Luminous Stars in the Northern Milky Way I* (Hamburger Sternwarte-Warner and Swasey Observatory).
[163] Hatchett, S., Buff, J. & McCray, R. 1976, *Ap. J.*, **206**, 847.
[164] Henrichs, H. & van den Heuvel, E. P. J. 1977, *Astron. Ap.*, **54**, 817.
[165] Hill, G. & Hutchings, J. B. 1973, *Ap. Sp. Sci.*, **20**, 123.
[166] Hillditch, R. W. & Hill, G. 1974, *MNRAS*, **168**, 543.
[167] Hiltner, W. A. & Mook, D. E. 1970, *Ann. Rev. Astron. Ap.*, **8**, 139.
[168] Hjellming, R. M. & Wade, C. M, 1971, *Ap. J. (Lett.)*, **164**, L1.
[169] Hjellming, R., Hogg, D. & Hvatum, H. 1978, *IAU Circular*, No. 3180.
[170] Hoffman, J. A., Lewin, W. H. G. & Doty, J. 1977, *Ap. J. (Lett.)*, **217**, L23.
[171] Hoffmeister, C. 1941, *Kl. Verein. Beob. Bull.*, No. 24.
[172] Huang, S. S. 1972, *Ap. J.*, **171**, 549.
[173] Hudec, R. & Wenzel, W. 1976, *Bull. Astron. Inst. Chech.*, **27**, 325.
[174] Hudson, H. S., Peterson, L. E. & Schwartz, D. A. 1970, *Ap. J. (Lett.)*, **159**, L51.
[175] Humphreys, R. M. & Whelan, J. 1975, *Observatory*, **95**, 171.
[176] Hutchings, J. B. 1970, *MNRAS*, **150**, 55.
[177] Hutchings, J. B. 1974, *Ap. J.*, **188**, 341 (revision in *Ap. J.*, **201**, 413).
[178] Hutchings, J. B. 1974, *Ap. J.*, **192**, 677.
[179] Hutchings, J. B. 1977, *Ap. J.*, **217**, 537.
[180] Hutchings, J. B. 1977, *MNRAS*, **181**, 619.
[181] Hutchings, J. B. 1979, *IAU Symposium*, No. 83, 3.
[182] Hutchings, J. B. 1979, *Proc. NATO Advanced Study Institute on Compact X-ray Sources*, Cape Sounion, Greece, June 1979.
[183] Hutchings, J. B. 1980, *Ap. J.*, **235**, 413.
[184] Hutchings, J. B. 1982, *Ap. J.*, **255**, 70.
[185] Hutchings, J. B., Crampton, D. *et al.* 1973, *Ap. J.*, **182**, 549.
[186] Hutchings, J. B., Cowley, A. P. *et al.* 1974, *Ap. J. (Lett.)*, **191**, L101.
[187] Hutchings, J. B., Crampton, D. & Redman, R. O. 1975, *MNRAS*, **170**, 313.
[188] Hutchings, J. B., Crampton, D. *et al.* 1977, *Ap. J.*, **217**, 186.
[189] Hutchings, J. B. & Stoeckley, T. R. 1977, *PASP*, **89**, 19.
[190] Hutchings, J. B., Bernard, J. E. *et al.* 1978, *Ap. J.*, **223**, 530.
[191] Hutchings, J. B., Crampton, D. & Cowley, A. P. 1978, *Ap. J.*, **225**, 548.
[192] Hutchings, J. B., Cowley, A. P. *et al.* 1979, *Ap. J.*, **229**, 1079.
[193] Hutchings, J. B. & Crampton, D. 1980, *IAU Circular*, No. 3464.
[194] Hutchings, J. B., Cowley, A. P. & Crampton, D. 1980, *IAU Circular*, No. 3543.
[195] Hutchings, J. B., Cowley, A. P. & Crampton, D. 1981, *IAU Circular*, No. 3585.
[196] Hutchings, J. B. & Crampton, D. 1981, *Ap. J.*, **247**, 222.
[197] Hutchings, J. B., Crampton, D. & Cowley, A. P. 1981, *Astron. J.*, **86**, 871.
[198] Ilovaisky, S. A. 1979, *Proc. NATO Advanced Study Institute on Compact X-ray Sources*, Cape Sounion, Greece, June 1979.
[199] Ilovaisky, S. A. 1981, *IAU Circular*, No. 3586.
[200] Ilovaisky, S. A., Chevalier, C. *et al.* 1978, *Astron. Ap.*, **67**, 287.
[201] Ilovaisky, S. A., Motch, C. & Chevalier, C. 1978, *Astron. Ap.*, **70**, L19.
[202] Ilovaisky, S. A., Chevalier, C. & Motch, C. 1979, *Astron. Ap.*, **71**, L17.
[203] Ilovaisky, S. A., Chevalier, C. *et al.* 1979, *IAU Circular*, No. 3325.
[204] Ilovaisky, S. A., Chevalier, C. *et al.* 1980, *MNRAS*, **191**, 81.
[205] Ilovaisky, S. A., Chevalier, C. & Motch, C. 1981, *Sp. Sci. Rev.*, **30**, 415.

[206] Jackson, J. 1975, *MNRAS*, **172**, 483.
[207] Janot-Pacheco, E., Ilovaisky, S. A. & Chevalier, C. 1981, *Astron. Ap.*, **99**, 274.
[208] Janot-Pacheco, E., Chevalier, C. & Ilovaisky, S. A. 1982, *IAU Symposium*, No. 98, 151.
[209] Jernigan, J. G., Apparao, K. M. V. *et al.* 1977, *Nature*, **270**, 321.
[210] Jernigan, J. G., Apparao, K. M. V. *et al.* 1978, *Nature*, **272**, 701.
[211] Johnson, H. L. 1966, *Ann. Rev. Astron. Ap.*, **4**, 193.
[212] Johnston, M., Bradt, H. *et al.* 1978, *Ap. J. (Lett.)*, **223**, L71.
[213] Johnston, M. D., Bradt, H. V. *et al.* 1978, *Ap. J. (Lett.)*, **225**, L59.
[214] Johnston, M. D., Bradt, H. V. & Doxsey, R. E. 1979, *Ap. J.*, **233**, 514.
[215] Johnston, M. D., Griffiths, R. E. & Ward, M. J. 1980, *Nature*, **285**, 26.
[216] Jones, C. 1977, *Ap. J.*, **214**, 856.
[217] Jones, C., Forman, W. & Liller, W. 1973, *Ap. J. (Lett.)*, **182**, L109.
[218] Jones, C. & Liller, W. 1973, *Ap. J. (Lett.)*, **184**, L65.
[219] Jones, C. & Liller, W. 1973, *Ap. J. (Lett.)*, **184**, L121.
[220] Jones, C., Chetin, T. & Liller, W. 1974, *Ap. J. (Lett.)*, **190**, L1.
[221] Joss, P. C. 1978, *Ap. J. (Lett.)*, **225**, L123.
[222] Joss, P. C. 1979, *Comments on Ap.*, **8**, 109.
[223] Joss, P. C., Salpeter, E. E. & Ostriker, J. P. 1973, *Ap. J.*, **181**, 429.
[224] Joss, P. C., Li, F. *et al.* 1980, *Ap. J.*, **235**, 592.
[225] Joss, P. C. & Li, F. K. 1980, *Ap. J.*, **238**, 287.
[226] Kallman, T. & McCray, R. 1980, *Ap. J.*, **242**, 615.
[227] Kaluzienski, L. J., Holt, S. S. *et al.* 1976, *Ap. J. (Lett.)*, **208**, L71.
[228] Kaluzienski, L. J., Holt, S. S. & Swank, J. H. 1980, *Ap. J.*, **241**, 779.
[229] Kappelmann, N. & Mauder, H. 1978, *Mitt. der Astron. Gesellschaft*, **43**, 225.
[230] Kelley, R., Rappaport, S. & Petre, R. 1980, *Ap. J.*, **238**, 699.
[231] Kelley, R. L., Apparao, K. M. V. *et al.* 1981, *Ap. J.*, **243**, 251.
[232] Kelley, R. L., Jernigan, J. G. *et al.* 1981, *IAU Circular*, No. 3632.
[233] Kelley, R. L., Rappaport, S. *et al.* 1981, *Ap. J.*, **251**, 630.
[234] Kemp, J. C. 1980, *Astron. Ap.*, **91**, 108.
[235] Kemp, J. C., Herman, L. C. *et al.* 1977, *Nature*, **270**, 227.
[236] Kemp, J. C., Herman, L. C. & Barbour, M. S. 1978, *Astron. J.*, **83**, 962.
[237] Kemp, J. C., Barbour, M. S. *et al.* 1978, *Ap. J. (Lett.)*, **220**, L123.
[238] Kemp, J. C., Barbour, M. S. *et al.* 1979, *Ap. J. (Lett.)*, **228**, L23.
[239] Kestenbaum, H., Angel, J. R. P. *et al.* 1971, *Ap. J. (Lett.)*, **169**, L49.
[240] Kieboom, K. H. & Verbunt, F. 1981, *Astron. Ap.*, **95**, L11.
[241] Kondo, Y. (ed.) 1976, *Proc. Symposium on X-ray Binaries*, Goddard Space Flight Center, October 20-22, 1975, NASA SP-389.
[242] Koo, D. C. & Kron, R. G. 1977, *PASP*, **89**, 285.
[243] Kopal, Z. 1959, *Close Binary Systems*, Chapman and Hall (London).
[244] Korhonen, T. & Piirola, V. 1980, *Astron. Ap.*, **91**, 372.
[245] Koyama, K., Inoue, H. *et al.* 1981, *Ap. J. (Lett.)*, **247**, L27.
[246] Kriss, J., Cominsky, L. & Remillard, R. 1980, *IAU Circular*, No. 3543.
[247] Krumenaker, L. E. 1975, *PASP*, **87**, 185.
[248] Kruszewski, A. 1963, *Acta Astron.*, **13**, 106.
[249] Kruszewski, A. 1978, *Inf. Bull. Var. Stars*, No. 1424.
[250] Kruszewski, A., Surdej, J. *et al.* 1979, *Acta Astron.*, **29**, 481.
[251] Krzeminski, W. 1974, *Ap. J. (Lett.)*, **192**, L135.
[252] Lamb, D. Q. & Sorvari, J. M. 1972, *IAU Circular*, No. 2422.
[253] Lamb, D. Q. & Lamb, F. K. 1978, *Ap. J.*, **220**, 291.
[254] Lamb, R. C., Markert, T. H. *et al.* 1980, *Ap. J.*, **239**, 651.
[255] Lambert, D. L. & Tomkin, J. 1979, *Ap. J. (Lett.)*, **228**, L37.
[256] Lamers, H. G. J. L. M., van den Heuvel, E. P. J. & Petterson, J. A. 1976, *Astron. Ap.*, **49**, 327.
[257] Lang, F. L., Levine, A. M. *et al.* 1981, *Ap. J. (Lett.)*, **246**, L21.
[258] Lanning, H. H. 1975, *MNRAS*, **173**, 15P.
[259] Lasker, B. 1974, *Nature*, **250**, 308.

[260] Lawrence, A. et al. 1982 (in preparation).
[261] Lester, J. B. 1979, *Ap. J.*, **231**, 164.
[262] Lester, D., Nolt, I. G. & Radostiz, J. V. 1973, *Nature Phys. Sci.*, **241**, 125.
[263] Lester, D. F., Nolt, I. G. et al. 1976, *Ap. J.*, **205**, 855.
[264] Lewin, W. H. G. 1980, *In: Globular Clusters*, eds. D. Harris & B. Madore, Cambridge University Press, p. 315.
[265] Lewin, W. H. G., Clark, G. W. & Smith, W. B. 1968, *Ap. J. (Lett.)*, **152**, L55.
[266] Lewin, W. H. G., Hoffman, J. A. et al. 1976, *MNRAS*, **177**, 83P.
[267] Lewin, W. H. G., Hoffman, J. A. & Doty, J. 1976, *IAU Circular*, No. 2994.
[268] Lewin, W. H. G., Hoffman, J. A. et al. 1977, *Nature*, **267**, 28.
[269] Lewin, W. H. G., Hoffman, J. A. et al. 1978, *IAU Circular*, No. 3190.
[270] Lewin, W. H. G., van Paradijs, J. et al. 1980, *MNRAS*, **193**, 15.
[271] Lewin, W. H. G. & Joss, P. C. 1981, *Sp. Sci. Rev.*, **28**, 3.
[272] Li, F. K., van Paradijs, J. et al. 1978, *Nature*, **276**, 799.
[273] Li, F. K., Joss, P. C. et al. 1980, *Ap. J.*, **240**, 628.
[274] Liller, W. 1972, *IAU Circular*, No. 2415.
[275] Liller, W. 1973, *Ap. J. (Lett.)*, **184**, L37.
[276] Liller, W. 1976, *Proc. Symposium on X-ray Binaries*, Goddard Space Flight Center, October 20-2, 1975, NASA SP-389.
[277] Limber, D. N. 1963, *Ap. J.*, **138**, 1112.
[278] Lloyd, C., Noble, R. & Penston, V. 1977, *MNRAS*, **179**, 675.
[279] London, R., McCray, R. & Auer, L. H. 1981, *Ap. J.*, **243**, 970.
[280] Lyutyi, V. M., Sunyaev, R. A. & Cherepashchuk, A. M. 1973, *Sov. A. J.*, **17**, 1.
[281] Lyutyi, V. M. & Sunyaev, R. A. 1976, *Sov. A. J.*, **20**, 290.
[282] Lyutyi, V. M. & Shuganov, S. Yu. 1979, *Sov. A. J. (Lett.)*, **5**, 206.
[283] Makishima, K., Inoue, H. et al. 1981, *Ap. J. (Lett.)*, **244**, L79.
[284] Maraschi, L., Treves, A. & van den Heuvel, E. P. J. 1976, *Nature*, **259**, 292.
[285] Maraschi, L., Tanzi, E. G. & Treves, A. 1980, *Ap. J. (Lett.)*, **241**, L23.
[286] Margon, B. 1979, *In: X-ray Astronomy*, COSPAR Meeting, Innsbruck, May-June 1979 (eds. W. Baity & L. E. Peterson), p. 67.
[287] Margon, B. & Ostriker, J. P. 1973, *Ap. J.*, **186**, 91.
[288] Margon, B., Bowyer, S. & Stone, R. P. S. 1973, *Ap. J. (Lett.)*, **185**, L113.
[289] Margon, B., Davidsen, A. & Bowyer, S. 1976, *Ap. J. (Lett.)*, **208**, L35.
[290] Margon, B., Bowyer, S. & Kraft, R. 1975, *Nature*, **254**, 461.
[291] Margon, B. & Cohen, J. G. 1978, *Ap. J. (Lett.)*, **222**, L33.
[292] Margon, B. et al. 1978. *Nature*, **271**, 633.
[293] Marshall, N. & Watson, M. G. 1979, *IAU Circular*, No. 3318.
[294] Marshall, N. & Ricketts, M. J. 1980, *MNRAS*, **193**, 7P.
[295] Marshall, N. & Millit, J. M. 1981, *Nature*, **293**, 379.
[296] Mason, K. O., Charles, P. A. et al. 1976, *MNRAS*, **177**, 513.
[297] Mason, K. O., Murdin, P. G. et al. 1978, *MNRAS*, **184**, 45P.
[298] Mason, K. O., Middleditch, J. et al. 1980, *Ap. J. (Lett.)*, **242**, L109.
[299] Mason, K. O., Middleditch, J. et al. 1980, *Nature*, **287**, 516.
[300] Mason, K. O., Murdin, P. G. et al. 1982, *MNRAS*, **200**, 793.
[301] Mason, K. O. & Córdova, F. A. 1982, **255**, 603.
[302] Matilsky, T., Bradt, H. V. et al. 1976, *Ap. J. (Lett.)*, **210**, L127.
[303] Matilsky, T., La Sala, J. & Jessen, J. 1978, *Ap. J. (Lett.)*, **224**, L119.
[304] Matsuoka, M. 1980, *Proc. Symposium on Space Astrophysics*, Tokyo, 29 July 1980.
[305] Matsuoka, M., Inoue, H. et al. 1980, *Ap. J. (Lett.)*, **240**, L137.
[306] Mauder, H. 1973, *Astron. Ap.*, **28**, 473.
[307] Mauder, H. 1975, *Ap. J. (Lett.)*, **195**, L27.
[308] Mauder, H. 1976, *ESO Messenger*, No. 5, p. 3.
[309] Mauder, H. 1981, *ESO Messenger*, No. 24, p. 13.
[310] Mauder, H., Ammann, M. & Schulz, E. 1977, *Mitt. Astron. Gesellschaft*, **43**, 227.
[311] Mayo, S. K., Wickramasinghe, D. T. & Whelan, J. A. J. 1980, *MNRAS*, **193**, 793.
[312] McClintock, J. E., Canizares, C. R. & Tarter, C. B. 1975, *Ap. J.*, **198**, 641.

[313] McClintock, J. E. et al. 1977, *Ap. J. (Lett.)*, **216**, L15.
[314] McClintock, J. E., Canizares, C. R. & Backman, D. E. 1978, *Ap. J. (Lett.)*, **223**, L75.
[315] McClintock, J. E. et al. 1978, *IAU Circular*, No. 3251.
[316] McClintock, J. E., Canizares, C. R. et al. 1979, *Nature*, **279**, 47.
[317] McClintock, J. E., Canizares, C. R. et al. 1980, *Ap. J. (Lett.)*, **235**, L81.
[318] McClintock, J. E., Remillard, R. E. & Margon, B. 1981, *Ap. J.*, **243**, 900.
[319] McClintock, J. E. & Petro, L. D. 1981, *IAU Circular*, No. 3615.
[320] McClintock, J. E., London, R. A. et al. 1982, *Ap. J.*, **258**, 245.
[321] McHardy, I. M., Lawrence, A. et al. 1981, *MNRAS*, **197**, 893.
[322] Meyer, F. & Meyer-Hofmeister, E. 1981, *Astron. Ap.*, **104**, L10.
[323] Middleditch, J. & Nelson, J. 1973, *Ap. (Lett.)*, **14**, 129.
[324] Middleditch, J. & Nelson, J. 1976, *Ap. J.*, **208**, 567.
[325] Middleditch, J., Mason, K. O. et al. 1981, *Ap. J.*, **244**, 1001.
[326] Middleditch, J., Koski, A. & Burbidge, M. 1980, *IAU Circular*, No. 3510.
[327] Mikkelsen, D. R. & Wallerstein, G. 1974, *Ap. J.*, **194**, 549.
[328] Milgrom, M. 1976, *Astron. Ap.*, **53**, 321.
[329] Milgrom, M. 1977, *Astron. Ap.*, **54**, 725.
[330] Milgrom, M. 1978, *Astron. Ap.*, **65**, L1.
[331] Milgrom, M. 1978, *Astron. Ap.*, **67**, L25.
[332] Milgrom, M. 1978, *Astron. Ap.*, **70**, 763.
[333] Milgrom, M. 1979, *Astron. Ap.*, **76**, 338.
[334] Milgrom, M. & Avni, Y. 1976, *Astron. Ap.*, **52**, 157.
[335] Milgrom, M. & Salpeter, E. E. 1975, *Ap. J.*, **196**, 589.
[336] Milgrom, M. & Shaham, J. 1977, *Nature*, **270**, 228.
[337] Miyamoto, S. & Matsuoka, M. 1977, *Sp. Sci. Rev.*, **20**, 687.
[338] Moffatt, A. F. J., Haupt, W. & Schmidt-Kaler, T. 1973, *Astron. Ap.*, **23**, 433.
[339] Moffatt, A. F. J. & Dachs, J. 1977, *Astron. Ap.*, **58**, L5.
[340] Mook, D. E., Edwards, S. & Hiltner, W. A. 1972, *Ap. J. (Lett.)*, **177**, L63.
[341] Mook, D. E., Messina, R. J. et al. 1974, *Ap. J.*, **191**, 493.
[342] Motch, C., Ilovaisky, S. A. & Chevalier, C. 1981, *IAU Circular*, No. 3609.
[343] Mouchet, M., Ilovaisky, S. A. & Chevalier, C. 1980, *Astron. Ap.*, **90**, 113.
[344] Murakami, T., Inoue, H. et al. 1980, *Ap. J. (Lett.)*, **240**, L143.
[345] Murakami, T., Inoue, H. et al. 1981, *Publ. Astron. Soc. Japan*, **32**, 543.
[346] Murdin, P. 1972, *IAU Circular*, No. 2433.
[347] Murdin, P., Griffiths, R. E. et al. 1977, *MNRAS*, **178**, 27P.
[348] Murdin, P., Morton, D. C. & Thomas, R. M. 1979, *MNRAS*, **186**, 43P.
[349] Murdin, P., Allen, D. A. et al. 1980, *MNRAS*, **192**, 709.
[350] Murdin, P., Clark, D. H. & Martin, P. G. 1981, *MNRAS*, **193**, 135.
[351] Natali, G., Fabrianesi, R. & Messi, R. 1978, *Astron. Ap.*, **62**, L1.
[352] Neckel, T. & Klare, G. 1980, *Astron. Ap. Suppl.*, **42**, 251.
[353] Nelson, J. E., Chanan, G. A. & Middleditch, J. 1977, *Ap. J.*, **212**, 215.
[354] Nelson, J., Córdova, F. & Middleditch, J. 1979, *Ap. J.*, **229**, 294.
[355] Nicolet, B. 1978, *Astron. Ap. Suppl.*, **34**, 1.
[356] Nicolson, G. D., Feast, M. W. & Glass, I. S. 1980, *MNRAS*, **191**, 293.
[357] Oda, M. 1977, *Sp. Sci. Rev.*, **20**, 757.
[358] Oda, M. 1980, *IAU Circular*, No. 3506.
[359] Oda, M. 1981, *IAU Circular*, No. 3594.
[360] Oda, M. 1981, *IAU Circular*, No. 3616.
[361] Oda, M. 1981, ISAS Res. Note No. 162.
[362] Ohashi, T. 1980, ISAS Res. Note No. 141.
[363] Oke, J. B. 1976, *Ap. J.*, **209**, 547.
[364] Oke, J. B. 1977, *Ap. J.*, **217**, 181.
[365] Oke, J. B. & Greenstein, J. 1977, *Ap. J.*, **211**, 872.
[366] Osmer, P. S. & Hiltner, W. A. 1974, *Ap. J. (Lett.)*, **188**, L5.
[367] Osmer, P. S., Hiltner, W. A. & Whelan, J. A. J. 1975, *Ap. J.*, **195**, 707.
[368] Pacharintanakul, P. & Katz, J. I. 1980, *Ap. J.*, **238**, 985.

[369] Paczynski, B. 1971, *Ann. Rev. Astron. Ap.*, **9**, 183.
[370] Paczynski, B. 1974, *Astron. Ap.*, **34**, 161.
[371] Pakull, M. 1978, *IAU Circular*, No. 3313.
[372] Pakull, M. 1980, *IAU Circular*, No. 3472.
[373] Pakull, M. 1981, Ph.D. thesis, Univ. of Hamburg.
[374] Pakull, M. & Swings, J. P. 1979, *IAU Circular*, No. 3318.
[375] Pakull, M. & Parmar, A. N. 1981, *Astron. Ap.*, **102**, L1.
[376] Parkes, G. E., Murdin, P. G. & Mason, K. O. 1980, *MNRAS*, **190**, 537.
[377] Parmar, A. N., Branduardi-Raymont, G. & Murdin, P. 1981, *Sp. Sci. Rev.*, **30**, 433.
[378] Patterson, J., 1981, *Nature*, **292**, 810.
[379] Pedersen, H., van Paradijs, J. & Lewin, W. H. G. 1981, *Nature*, **294**, 725.
[380] Pedersen, H., Lub, J. et al. 1982, *Ap. J.*, **263**, 325.
[381] Pedersen, H., van Paradijs, J. et al. 1982, *Ap. J.*, **263**, 340.
[382] Pelling, M. R. 1973, *Ap. J.*, **185**, 327.
[383] Penfold, J. E., Warren, P. R. & Penny, A. J. 1975, *MNRAS*, **171**, 445.
[384] Penny, A. J., Olowin, R. P. et al. 1973, *MNRAS*, **163**, 7P.
[385] Penston, M. J. & Murdin, P. 1975, *MNRAS*, **172**, 377.
[386] Persi, P., Ferrari-Tonioli, M. et al. 1979, *MNRAS*, **187**, 293.
[387] Persi, P., Viotti, R. & Ferrari-Tonioli, M. 1977, *MNRAS*, **181**, 685.
[388] Peterson, B. A., Middleditch, J. & Nelson, J. 1975, *Ap. J. (Lett.)*, **195**, L31.
[389] Peterson, B. A., Wallace, P. et al. 1980, *MNRAS*, **190**, 33P.
[390] Petro, L. 1975, *Ap. J.*, **195**, 709.
[391] Petro, L. & Hiltner, W. A. 1973, *Ap. J. (Lett.)*, **181**, L39.
[392] Petro, L., Feldman, F. & Hiltner, W. A. 1973, *Ap. J. (Lett.)*, **184**, L123.
[393] Petro, L. & Hiltner, W. A. 1974, *Ap. J.*, **190**, 661.
[394] Petro, L., Bradt, H. V. et al. 1981, *Ap. J. (Lett.)*, **251**, L7.
[395] Petro, L. D. & Hiltner, W. A. 1982, *Astron. J.* (submitted).
[396] Petterson, J. A. 1975, *Ap. J. (Lett.)*, **201**, L61.
[397] Petterson, J. A. 1978, *Ap. J.*, **224**, 625.
[398] Plavec, M. 1958, *Mem. Soc. Roy. Sci. Liège*, **20**, 411.
[399] Pounds, K. A., Cooke, B. A. et al. 1975, *MNRAS*, **172**, 473.
[400] Pravdo, S. H., White, N. E. et al. 1979, *Ap. J.*, **231**, 912.
[401] Primini, F., Rappaport, S. & Joss, P. C. 1977, *Ap. J.*, **217**, 543.
[402] Rappaport, S. 1979, *Proc. NATO Advanced Study Institute on Compact X-ray Sources*, Cape Sounion, Greece, June 1979.
[403] Rappaport, S. A. & Joss, P. C. 1980, Talk presented at the HEAD/AAS meeting, Cambridge, Mass., January 1980, preprint MIT CSR/HEA-80-24.
[404] Rappaport, S., Joss, P. C. & McClintock, J. E. 1976, *Ap. J. (Lett.)*, **206**, L103.
[405] Rappaport, S., Clark, G. W. et al. 1978, *Ap. J. (Lett.)*, **224**, L1.
[406] Rappaport, S., Joss, P. C. & Stothers, R. 1980, *Ap. J.*, **235**, 570.
[407] Rappaport, S. A. & van den Heuvel, E. P. J. 1982, *IAU Symposium*, No. 98, 327.
[408] Reid, C. A., Johnston, M. D. et al. 1980, *Astron. J.*, **85**, 1062.
[409] Richer, H. B. & Isherwood, B. C. 1972, *IAU Circular*, No. 2431.
[410] Ricketts, M. J., Hall, R. et al. 1981, *Sp. Sci. Rev.*, **30**, 399.
[411] Robertson, B. S. C., Warren, P. R. & Bywater, R. A. 1976, *Inf. Bull. Var. Stars*, No. 1173.
[412] Robinson, E. L. 1976, *Ann. Rev. Astron. Ap.*, **14**, 119.
[413] Rudy, R. J. 1979, *MNRAS*, **186**, 473.
[414] Rudy, R. J. & Kemp, J. C. 1978, *Ap. J.*, **221**, 200.
[415] Sahade, J. & Wood, F. B. 1978, *Interacting Binary Stars* (Pergamon, Oxford).
[416] Sandage, A., Osmer, P. et al. 1966, *Ap. J.*, **146**, 316.
[417] Sanduleak, N. 1968, *Astron. J.*, **73**, 246.
[418] Savonije, G. J. 1978, *Astron. Ap.*, **62**, 317.
[419] Savonije, G. J. 1979, *Astron. Ap.*, **71**, 352.
[420] Savonije, G. J. 1980, *Astron. Ap.*, **81**, 25.
[421] Schlosser, W. & van Paradijs, J. 1979, *Astron. Ap.*, **75**, 112.
[422] Schreier, E., Levinson, R. et al. 1972, *Ap. J. (Lett.)*, **172**, L79.

[423] Schuerman, D. W. 1972, *Ap. Sp. Sci.*, **19**, 351.
[424] Schwartz, D. A., Griffiths, R. E. *et al.* 1980, *Astron. J.*, **85**, 549.
[425] Schwartz, D. A. *et al.* 1982, *Bull. A.A.S.*, **13**, 834.
[426] Simons, J. F. L., Aspin, C. & Brown, J. C. 1980, *Astron. Ap.*, **91**, 97.
[427] Skinner, G. K. 1980, *Nature*, **288**, 141.
[428] Skinner, G. K., Shulman, S. *et al.* 1980, *Ap. J.*, **240**, 619.
[429] Slettebak, A. 1979, *Sp. Sci. Rev.*, **23**, 541.
[430] Smith, H. E., Margon, B. & Conti, P. S. 1973, *Ap. J. (Lett.)*, **179**, L125.
[431] Steiner, J. E. 1977, *Astron. Ap.*, **61**, L35.
[432] Stephenson, C. B. & Sanduleak, N. 1977, *Ap. J. Suppl.*, **33**, 459.
[433] Sterken, C. 1977, *Astron. Ap.*, **57**, 361.
[434] Strittmatter, P. A. 1969, *Ann. Rev. Astron. Ap.*, **7**, 665.
[435] Strittmatter, P. A., Scott, J. *et al.* 1973, *Astron. Ap.*, **25**, 275.
[436] Swank, J., Becker, R. H. *et al.* 1977, *Ap. J. (Lett.)*, **212**, L73.
[437] Tananbaum, H., Gursky, H. *et al.* 1972, *Ap. J. (Lett.)*, **174**, L143.
[438] Taylor, A. R. & Gregory, P. C. 1982, *Ap. J.*, **255**, 210.
[439] Thomas, R. M., Morton, D. C. & Murdin, P. 1979, *MNRAS*, **188**, 19.
[440] Thomas, R. M., Murdin, P. & Morton, D. 1981, *MNRAS*, **195**, 915.
[441] Thorstensen, J., Charles, P. & Bowyer, S. 1978, *Ap. J. (Lett.)*, **220**, L131.
[442] Thorstensen, J. R., Charles, P. *et al.* 1979, *Ap. J. (Lett.)*, **233**, L57.
[443] Thorstensen, J. R., Charles, P. A. & Bowyer, S. 1980, *Ap. J.*, **238**, 964.
[444] Thorstensen, J. R. & Charles, P. A. 1980, *Bull. A.A.S.*, **11**, 721.
[445] Thorstensen, J. R. & Charles, P. A. 1980, *IAU Circular*, No. 3449.
[446] Treves, A., Chiapetti, L. *et al.* 1980, *Ap. J.*, **242**, 1114.
[447] Trimble, V., Rose, W. K. & Weber, J. 1973, *MNRAS*, **162**, 1P.
[448] Tsunemi, H., Matsuoka, M. & Takagishi, K. 1977, *Ap. J. (Lett.)*, **211**, L15.
[449] Tuohy, I. & Cruise, A. M. 1975, *MNRAS*, **171**, 33P.
[450] Ulmer, M. P., Shulman, S. *et al.* 1980, *Ap. J. (Lett.)*, **235**, L159.
[451] Underhill, A. B., Divan, L. *et al.* 1979, *MNRAS*, **189**, 601.
[452] van Beveren, D. 1977, *Astron. Ap.*, **54**, 877.
[453] van den Heuvel, E. P. J. 1975, *Ap. J. (Lett.)*, **198**, L109.
[454] van den Heuvel, E. P. J. 1976, *IAU Symposium*, No. 73, 35.
[455] van den Heuvel, E. P. J. & Heise, J. 1972, *Nature Phys. Sci.*, **239**, 67.
[456] van Genderen, A. M. 1973, *Inf. Bull. Var. Stars*, No. 856.
[457] van Genderen, A. M. 1974, *MNRAS*, **167**, 57P.
[458] van Genderen, A. M. 1977, *Astron. Ap.*, **54**, 307.
[459] van Genderen, A. M. 1977, *Astron. Ap.*, **54**, 683.
[460] van Genderen, A. M. 1977, *Astron. Ap.*, **54**, 733.
[461] van Genderen, A. M. 1977, *Astron. Ap. Suppl.*, **28**, 119.
[462] van Genderen, A. M. 1981, *Astron. Ap.*, **96**, 82.
[463] van Genderen, A. M. & Uiterwaal, G. M. 1976, *Astron. Ap.*, **52**, 139.
[464] van Genderen, A. M. & Windhorst, R. A. 1981, *Astron. Ap.*, **97**, 79.
[465] van Genderen, A. M. & van Groningen, E. 1981, *Astron. Ap.*, **101**, 101.
[466] van Paradijs, J. 1977, *Astron. Ap. Suppl.*, **29**, 339.
[467] van Paradijs, J. 1978, *Nature*, **274**, 650.
[468] van Paradijs, J. 1980, *Astron. Ap.*, **87**, 210.
[469] van Paradijs, J. 1981, *Astron. Ap.*, **101**, 174.
[470] van Paradijs, J. 1981, *Astron. Ap.*, **103**, 140.
[471] van Paradijs, J., Takens, R. J. & Zuiderwijk, E. 1977, *Astron. Ap.*, **57**, 221.
[472] van Paradijs, J. & Zuiderwijk, E. J. 1977, *Astron. Ap.*, **61**, L19.
[473] van Paradijs, J., Zuiderwijk, E. J. *et al.* 1977, *Astron. Ap. Suppl.*, **30**, 195.
[474] van Paradijs, J., Hammerschlag-Hensberge, G. & Zuiderwijk, E. J. 1978, *Astron. Ap. Suppl.*, **31**, 189.
[475] van Paradijs, J., Joss, P. C. *et al.* 1979, *Nature*, **280**, 375.
[476] van Paradijs, J., Verbunt, F. *et al.* 1980, *Ap. J. (Lett.)*, **241**, L161.
[477] van Paradijs, J., Pedersen, H. & Lewin, W. H. G. 1981, *IAU Circular*, No. 3626.
[478] Vidal, N. V. 1974, *PASP*, **86**, 317.

[479] Vidal, N. V., Wickramasinghe, D. T. & Peterson, B. A. 1973, *Ap. J. (Lett)*, **182**, L77.
[480] *Vistas in Astronomy* (ed. P. Beer), Vol. 25, No. 1 (1981).
[481] Walborn, N. R. 1973, *Astron. J.*, **78**, 1067.
[482] Walker, E. N. 1972, *MNRAS*, **160**, 9P.
[483] Walker, E. N. 1980, *IAU Circular*, No. 3517.
[484] Walker, E. N. & Quintanilla, A. R. 1974, *MNRAS*, **169**, 247.
[485] Walker, E. N., Brownlie, G. D. et al. 1976, *Nature*, **263**, 393.
[486] Walker, E. N., Watson, M. G. & Holt, S. S. 1977, *Nature*, **270**, 229.
[487] Walker, E. N. & Quintanilla, A. R. 1978, *MNRAS*, **182**, 315.
[488] Walter, F. M., Bowyer, S. et al. 1982, *Ap. J. (Lett.)*, **253**, L67.
[489] Warner, B. 1976, *IAU Symposium*, No. 73, 85.
[490] Warren, P. R. & Penfold, J. E. 1975, *MNRAS*, **172**, 41P.
[491] Warwick, R. S., Marshall, N. et al. 1981, *MNRAS*, **193**, 865.
[492] Warwick, R. S., Watson, M. G. & Sims, M. R. 1981, *Sp. Sci. Rev.*, **30**, 461.
[493] Watson, M. G. 1976, *MNRAS*, **176**, 19P.
[494] Watson, M. G. & Griffiths, R. E. 1977, *MNRAS*, **178**, 513.
[495] Watson, M. G., Ricketts, M. J. & Griffiths, R. E. 1978, *Ap. J. (Lett.)*, **221**, L69.
[496] Watson, M. G., Warwick, R. J. & Ricketts, M. J. 1981, *MNRAS*, **195**, 197.
[497] Watson, M. G. & Ricketts, M. J. 1978, *MNRAS*, **183**, 35P.
[498] Webster, B. L. & Murdin, P. 1972, *Nature*, **235**, 37.
[499] Westphal, J. A., Sandage, A. & Kristian, J. 1968, *Ap. J.*, **154**, 139.
[500] Whelan, J. 1973, *Ap. J. (Lett.)*, **185**, L127.
[501] Whelan, J. & Wickramasinghe, D. T. 1976, *MNRAS*, **174**, 29.
[502] Whelan, J. A. J., Ward, M. J. et al. 1977, *MNRAS*, **180**, 657.
[503] Whelan, J. A. J., Mayo, S. K. et al. 1977, *MNRAS*, **181**, 259.
[504] White, N. E., Mason, K. O. et al. 1976, *MNRAS*, **176**, 91.
[505] White, N. E., Mason, K. O. et al. 1976, *MNRAS*, **176**, 201.
[506] White, N. E., Mason, K. O. & Sanford, P. W. 1978, *MNRAS*, **184**, 67P.
[507] White, N. E., Charles, P. A. & Thorstensen, J. R. 1980, *MNRAS*, **193**, 731.
[508] White, N. E., Pravdo, S. H. et al. 1980, *Ap. J.*, **239**, 655.
[509] White, N. E. & Swank, J. H. 1982, *Ap. J. (Lett.)*, **253**, L61.
[510] White, N. E., Becker, R. H. et al. 1981, *Ap. J.*, **247**, 994.
[511] White, N. E. & Holt, S. S. 1982, *Ap. J.*, **257**, 318.
[512] Wickramasinghe, D. T. 1975, *MNRAS*, **173**, 21.
[513] Wickramasinghe, D. T., Vidal, N. V. et al. 1974, *Ap. J.*, **188**, 167.
[514] Wickramasinghe, D. T. & Bessell, M. S. 1974, *Nature*, **251**, 25.
[515] Wickramasinghe, D. T. & Whelan, J. 1975, *MNRAS*, **172**, 175.
[516] Williams, R. E. 1980, *Ap. J.*, **235**, 939.
[517] Willis, A. J., Wilson, R. et al. 1980, **237**, 596.
[518] Wilson, R. E. 1979, *Ap. J.*, **234**, 1054.
[519] Wilson, R. E. & Devinney, E. J. 1971, *Ap. J.*, **166**, 605.
[520] Wilson, R. E. & Sofia, S. 1976, *Ap. J.*, **203**, 182.
[521] Wilson, R. E. & Fox, R. K. 1981, *Astron. J.*, **86**, 1259.
[522] Wolff, S. C. & Morrison, N. D. 1974, *Ap. J.*, **187**, 69.
[523] Wright, E. L., Gottlieb, E. W. & Liller, W. 1975, *Ap. J.*, **200**, 171.
[524] Wu, C. C. 1979, *Ap. J.*, **227**, 291.
[525] Wu, C. C., Aalders, J. W. G. et al. 1976, *Astron. Ap.*, **50**, 445.
[526] Wu, C. C., Eaton, J. A. et al. 1982, *PASP*, **94**, 149.
[527] Zuiderwijk, E. J. 1978, Ph.D. thesis, Univ. of Amsterdam.
[528] Zuiderwijk, E. J., van den Heuvel, E. P. J. & Hensberge, G. 1974, *Astron. Ap.*, **35**, 353.
[529] Zuiderwijk, E. J., Hammerschlag-Hensberge, G. et al. 1977, *Astron. Ap. Suppl.*, **27**, 433.
[530] Zuiderwijk, E. J., Hammerschlag-Hensberge, G. et al. 1977, *Astron. Ap.*, **54**, 167.
[531] Skinner, G. K. et al. 1982, *Nature*, **297**, 568; see also *IAU Circular*, No. 3671.
[532] Raymond, J. C. 1982, *Ap. J.*, **258**, 240.

[533] Charles, P. A., Booth, L. *et al.* 1983, *MNRAS*, **202**, 657.
[534] Densham, R. H., Charles, P. A. *et al.* 1982, submitted to *MNRAS*.
[535] Hutchings, J. B., Crampton, D. *et al.* 1982, *PASP*, **94**, 541.
[536] Giess, D. R. & Bolton, C. T. 1982, *Ap. J.*, **260**, 240.
[537] Hutchings, J. B., Cowley, A. P. & Crampton, D. 1983, *PASP*, **95**, 23.
[538] Bruhweiler, F. C., Parsons, S. B. & Wray, J. D. 1982, *Ap. J. (Lett.)*, **256**, L49.
[539] Ponman, T. 1982, *MNRAS*, **200**, 351.
[540] Grindlay, J. E. 1982, private communication.
[541] Hutchings, J. B. 1982, Review paper presented at COSPAR, Ottawa, 17 May–2 June 1982.
[542] van der Klis, M., van Paradijs, J. *et al.* 1982, *MNRAS* (in press).
[543] Kelley, R. L., Rappaport, S. A. *et al.* 1982, preprint MIT-CSR-HEA-82-9.
[544] Kelley, R. L., Jernigan, J. G. *et al.* 1983, *Ap. J.*, **264**, 568.
[545] van Paradijs, J. & van der Woerd, H. 1982, *Astron. Ap.*, **113**, 27.
[546] van Amerongen, S. & van Paradijs, J. 1982, *Inf. Bull. Var. Stars*, No. 2180.
[547] Ohashi, T., Inoue, H. *et al.* 1982, *Ap. J.*, **258**, 254.
[548] Cowley, A. P., Crampton, D. & Hutchings, J. B. 1982, *Ap. J.*, **255**, 596.
[549] Fabian, A. C., Guilbert, P. W. & Rose, R. R. 1982, *MNRAS*, **199**, 1045.
[550] Thorstensen, J. R. & Charles, P. A. 1982, *Ap. J.*, **253**, 756.
[551] Calafat, R., Canal, R. *et al.* 1982, *Astron. Ap.*, **110**, 23.
[552] Watson, M. G., Warwick, R. J. & Corbet, R. H. D. 1982, *MNRAS*, **199**, 919.
[553] Howarth, I. D. 1982, *MNRAS*, **198**, 289.
[554] Motch, C., Ilovaisky, S. A. & Chevalier, C. 1982, *Astron. Ap.*, **109**, L1.
[555] Fabian, A. C., Guilbert, P. W. *et al.* 1982, *Astron. Ap.*, **111**, L9.
[556] van der Klis, M., Hammerschlag-Hensberge, G. *et al.* 1982, *Astron. Ap.*, **106**, 339.
[557] De Loore, C., Burger, M. *et al.* 1981, *Astron. Ap.*, **104**, 150.
[558] Mason, K. O., White, N. E. *et al.* 1976, *MNRAS*, **176**, 193.
[559] Bernacca, P. L. & Bianchi, L. 1981, *Astron. Ap.*, **94**, 345.
[560] Kemp, J. C. & Barbour, M. S. 1983, *Ap. J.*, **264**, 237.
[561] Chevalier, C. & Ilovaisky, S. A. 1982, *Astron. Ap.*, **112**, 68.
[562] Pakull, M., van Amerongen, S. *et al.* 1982, *Astron. Ap.* (submitted).
[563] Densham, R. H. & Charles, P. A. 1982, *MNRAS*, **201**, 171.
[564] Ilovaisky, S. A., Chevalier, C. & Motch, C. 1982, preprint.
[565] Bradt, H. V. & McClintock, J. E. 1983, *Ann. Rev. Astron. Ap.* (to be published); preprint MIT-CSR-HEA-82-1.
[566] Samini, J., Shane, G. H. *et al.* 1979, *Nature*, **278**, 434.

6
ACCRETING MAGNETIC NEUTRON STARS

John G. Kirk
Max-Planck-Institut für Physik und Astrophysik, Institut für Astrophysik

Joachim E. Trümper
*Max-Planck-Institut für Physik und Astrophysik,
Institut für extraterrestrische Physik,
8046 Garching, West Germany*

6.1   Introduction

Neutron stars are of great interest, not only to astronomers, but also to physicists, since they represent the densest form of matter which is available to observation. In addition, they possess the strongest magnetic fields known to exist, exceeding the strongest laboratory field by a factor of one million. In the following, we shall concentrate on this aspect of neutron stars, and on its implications for accreting sources.

Even before the discovery of neutron stars (Hewish *et al.* 1968) it had been conjectured that such objects should be threaded by enormous magnetic fields ($B \sim 10^{12}$ G), on the grounds that flux should be conserved and compressed during the formation phase (Woltjer 1969, Pacini 1967). This suggestion was convincingly confirmed by the detection of the spin-down of radio pulsars, which can only be understood as the result of electromagnetic braking (Goldreich & Julian 1969, Ostriker & Gunn 1969).

The discovery of pulsating X-ray sources and their identification as accreting neutron stars has provided new information about the magnetic fields of these objects. Indeed, the existence of pulsations shows that there must be an effective channelling of material onto a small region of the stellar surface, which immediately gives a lower limit to the surface field of about 10^{10} G (Arons & Lea 1976, 1980). In addition, rough estimates of the stellar magnetic moments may be obtained from an analysis of the spin-up behaviour, leading to values $\sim 10^{30}$ G cm^3, in agreement with the evidence from radio pulsars.

The possibility of direct measurement of a neutron star's magnetic field was opened up by the discovery of sharp spectral lines in the X-ray emission from the pulsating source Her X-1 (Trümper *et al.* 1978). The existence of such features – which must be interpreted as electron-cyclotron resonances – has been confirmed by several experiments. Furthermore, two other pulsating X-ray sources have been found to exhibit similar features. These measurements indicate field strengths in the range $2\text{--}5 \times 10^{12}\,\text{G}$.

Recently, it has been claimed by Mazets *et al.* (1981) that a large fraction of cosmic gamma-ray bursts display cyclotron resonance lines corresponding to similar magnetic field strengths. In addition, emission lines are reported in some burst spectra which occur at the energy of electron–positron annihilation radiation, when account is taken of the gravitational redshift appropriate for a neutron star surface (10–20 per cent). If confirmed, these observations identify neutron stars as the sources of the bursts, and yield direct information about their surface magnetic and gravitational fields.

As well as the magnetic field strength, the observation of cyclotron resonances has another aspect. As in laboratory plasmas, the resonance is a diagnostic tool which can give indications of the physical conditions in the emission region; it thus provides a valuable guideline for the construction of theoretical models.

In the following we shall review the present observational and theoretical status of pulsating X-ray sources and γ-ray-burst sources, emphasising those aspects which relate to the magnetic fields of these objects.

6.2 Pulsating X-ray sources
6.2.1 Timing measurements

Measurements of the time of arrival of the pulses from pulsating X-ray sources are relatively easy to perform, since they do not require good spectral or angular resolution. As a consequence, a large amount of experimental data is available, covering a time interval of about 10 yr. During this period, similar observations of radio pulsars have been made, and it is interesting to contrast the results obtained for these two very different classes of object, both of which are manifestations of rotating magnetic neutron stars (see also Chapter 11, this book).

The quantity of relevance here is the rate of change of the pulse period – the spin-up or spin-down rate of the neutron star. All radio pulsars have a small positive period derivative – corresponding to spin-down on a timescale of about 10^6 yr. Pulsating X-ray sources, on the other hand, have a relatively large period derivative which is usually negative – corresponding to spin-up on a timescale which is in some cases as short as 50 yr. Furthermore, this quantity fluctuates on a timescale of days, sometimes reversing sign. In one source (Vela X-1) the change to spin-down appears to be an effect with a relatively long timescale (Nagase *et al.* 1981, see also Chapters 1 and 11, this book). Contrary to this, radio pulsars have extremely stable period derivatives with only a few relatively small jumps or 'glitches'.

From these data it is possible to make a reasonable estimate of the magnetic moment of a radio pulsar (Gold 1968, Goldreich & Julian 1969, Ostriker & Gunn 1969). In principle, this should also be possible for pulsating X-ray sources. However, the radiation mechanism of these sources is powered by accretion, which makes a calculation of the rate of change of angular momentum much more complicated than it is in the case of radio pulsars, where the energy losses can be estimated using the formula for magnetic dipole radiation. Nevertheless, the rate at which material falls onto the neutron star – the accretion rate – can be ascertained from the X-ray luminosity with reasonable accuracy (factor 3). In order to fall, this material, which originates on a companion star, must lose its angular momentum. If accretion is powered by Roche-lobe overflow of the companion, as is probably the case in most pulsating X-ray sources (Chapter 11, this book), a disc will form around the neutron star. The interaction of the disc with the stellar magnetic field then determines the rate at which angular momentum is transferred to the star, and, therefore, the spin-up or spin-down rate.

The calculations which can be found in the literature (Lamb, Pethick & Pines 1973, Ghosh & Lamb 1979, Anzer & Börner 1980, 1983) deal with special cases in which the magnetic dipole is either parallel to the rotation axis of the star, or perpendicular to it. The results express a relationship between the period P (in seconds) the period derivative $\dot{P}$ (in seconds per year) and the X-ray luminosity. In the case of a magnetic dipole axis parallel to the rotation axis, Lamb, Pethick & Pines (1973) find

$$\frac{\dot{P}}{P} = -6 \times 10^{-5} \zeta^{1/7} M^{-3/7} R_6^{6/7} I_{45}^{-1} \mu_{30}^{2/7} L_{37}^{6/7} P.$$

Here M is the stellar mass in units of the solar mass, R_6 is the radius in units of 10^6 cm, I_{45} is the moment of inertia in units of 10^{45} g cm^2, μ_{30} is the magnetic dipole moment in units of 10^{30} G cm^3 and L_{37} is the X-ray luminosity in units of 10^{37} erg s^{-1}. The quantity ζ expresses some uncertainty about the accretion flow, but is expected to be of order unity. Subsequent considerations (Ghosh & Lamb 1979) have indicated modifications for short-period pulsars which will not be discussed here. (See Chapter 11, this book, for a thorough discussion.) This relation is displayed in Figure 1.6 of Chapter 1. The result is a straight line of unit gradient, with an intercept which can be fitted to the data points to derive a value for the quantity

$$X = -\ln(6 \times 10^{-5} \zeta^{1/7} M^{-3/7} R_6^{6/7} I_{45}^{-1} \mu_{30}^{2/7}).$$

The best value for X is 11.1 ± 1.3 (one standard deviation).

In the case of a magnetic dipole axis perpendicular to the rotation axis, Anzer & Börner (1980) arrive at the formula

$$\frac{\dot{P}}{P} = -7.9 \times 10^{-5} \alpha M^{2/3} I_{45}^{-1} L_{37} P^{4/3}.$$

Here α is the quantity which expresses uncertainty about the details of the accretion flow in this model. It is the only term in the equation which depends on the magnetic moment of the star. For sources in which almost equal amounts of angular momentum are given to the star and removed from it by the flow (e.g. Her X-1), α is small (~ 0.05). See Chapter 11, Section 11.3.10.2 for an alternative explanation for Her X-1. Figure 6.1 shows this relation. The source Vela X-1 has been omitted from the fit because of its spin-down behaviour (Nagase *et al.* 1981). In this case, the best value for $Y = -\ln(7.9 \times 10^{-5} \alpha M^{2/3} I_{45}^{-1})$ is 12.1 ± 1.1.

Although Figure 1.6 and Figure 6.1 provide only a rough interpretation of the data, it should be noted that both X and Y have the order of magnitude to be expected for accreting neutron stars. These quantities are reasonably sensitive to the stellar radius ($X \propto \ln R^{-8/7}$ for fixed μ, and $Y \propto \ln R^{-2}$), so that the possibility

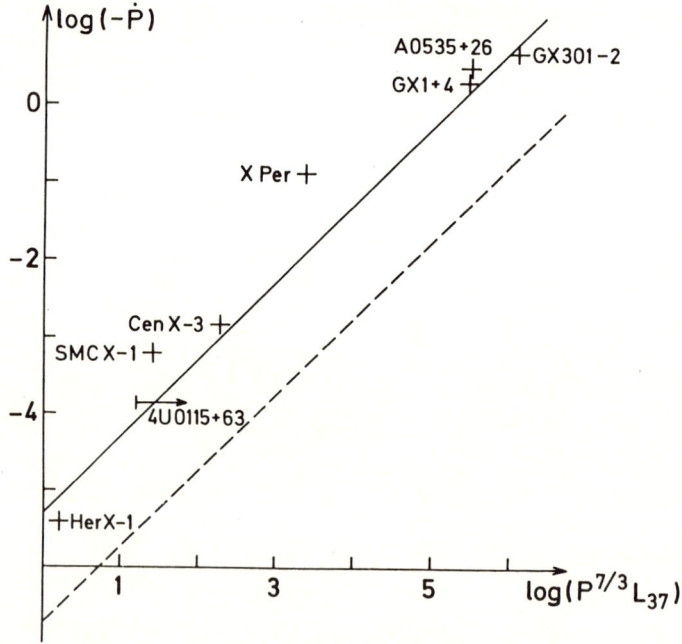

Fig. 6.1. The theoretical expressions for the spin-up of an accreting magnetized neutron star (solid line) and a white dwarf (dashed line) according to Anzer & Börner (1980).

of X-ray pulsars being white-dwarf stars can be ruled out (the lines predicted for white dwarfs are also shown in Figures 1.6 and 6.1).

For a determination of the magnetic moment of neutron stars, however, the situation requires further clarification. Although a value for μ_{30} between 8×10^{-5} and 0.8 can be deduced from the parameter X (in each case one standard deviation above and below the best value, with $\zeta = 1, M = 1, I_{45} = 1$ and $R_6 = 1$), the interpretation offered by Anzer & Börner seems equally plausible from an empirical point of view. However, the complexity of the magnetic field–disc interaction in this case is such that the magnetic moment cannot be deduced from the value of the parameter α. In summary, the best which can be achieved by timing measurements is an order of magnitude estimate of the magnetic dipole moment.

6.2.2 Spectral observations

The first spectroscopic measurement of a neutron star's magnetic field was achieved by the MPE Garching-Tübingen group in a balloon experiment on May 3, 1976 (Trümper et al. 1977, 1978). The observation was of Her X-1, which shows a strong spectral break at ~ 25 keV beyond which the spectrum falls off steeply, making hard X-ray observations very difficult. The data taken during the balloon flight showed (1) that the source continues to pulse at energies above 70 keV and (2) that a strong spectral peak is present around 58 keV as well as a possible second one at ~ 110 keV.

It was immediately clear that it would be difficult to explain a line feature in either emission at 58 keV or absorption at 35 keV by atomic or nuclear processes: In normal magnetic fields the typical energy of a K-shell electron transition is several keV. Only with elements of high atomic number can one achieve sufficiently energetic transitions e.g. the Lyman α line of Pt^{77+} has an energy of about 58 keV when corrected for a gravitational redshift of a few per cent. However, elements with atomic number much greater than that of iron are produced only in extremely small quantities – even in a supernova explosion. The 58 keV line feature in Figure 6.2 accounts for more than 1 per cent of the total X-ray luminosity, and could not be produced by a trace element. In very strong magnetic fields the resonance lines of hydrogen-like ions are shifted towards higher energies (Burdyuzha & Pavlov-Verevkin 1980, Wunner, Ruder & Herold 1981). However, to shift the 7.3 keV resonance line of Fe XXVI up to 58 keV would require a magnetic field strength $\sim 5.10^{13}$ G, which is not impossible, but seems unlikely. In addition it would be difficult to explain the observed line intensity. In the case of a nuclear transition a typical energy is a few MeV, and it is only the relatively rare, heavy nuclei (e.g. Am^{241}) which have excited states lying low enough to produce a resonance at 58 keV. Once again, this explanation fails because of the line intensity.

Thus it was suggested that the feature is produced by the electron-cyclotron resonance in the quasi-homogeneous field at the radiating polar cap of the neutron star. An energy of 58 keV corresponds to an electron-cyclotron resonance in a magnetic field of 5×10^{12} G, which is, of course, precisely the kind of value expected at the surface of a neutron star. If, on the other hand, the feature is interpreted as an absorption line centred at 40 keV, the corresponding magnetic field is 3×10^{12} G. It is tempting to extend this explanation as a cyclotron resonance to the possible feature at ~110 keV, since it lies at the position of the second harmonic of the cyclotron frequency. However, such an identification is in doubt because subsequent observations by the MPE-Tübingen group and the UCSD-MIT group have set upper limits to the flux at ~110 keV which are a factor of 10 below the flux seen in 1976 (Gruber *et al.* 1980, Voges *et al.* 1978, 1982).

On the other hand, the existence of the 58 keV feature has been confirmed by a second balloon observation of the MPE-Tübingen group in September 1977

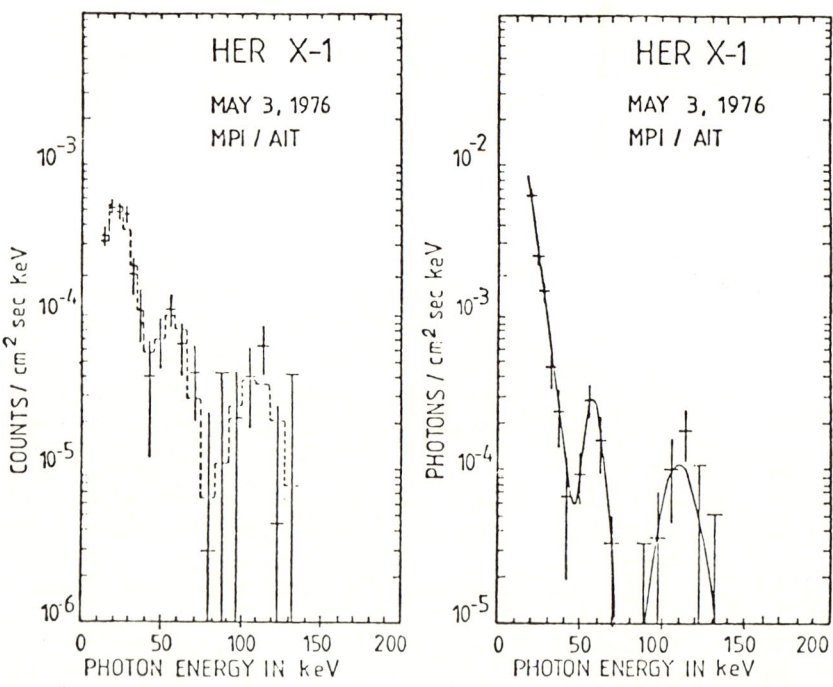

Fig. 6.2. Hard X-ray spectrum of Her X-1 obtained during the pulse phase of the 1.24 s pulsation. The left diagram shows a raw count-rate spectrum. The other diagram shows a deconvoluted spectrum, assuming spectral lines at 58 and 110 keV.

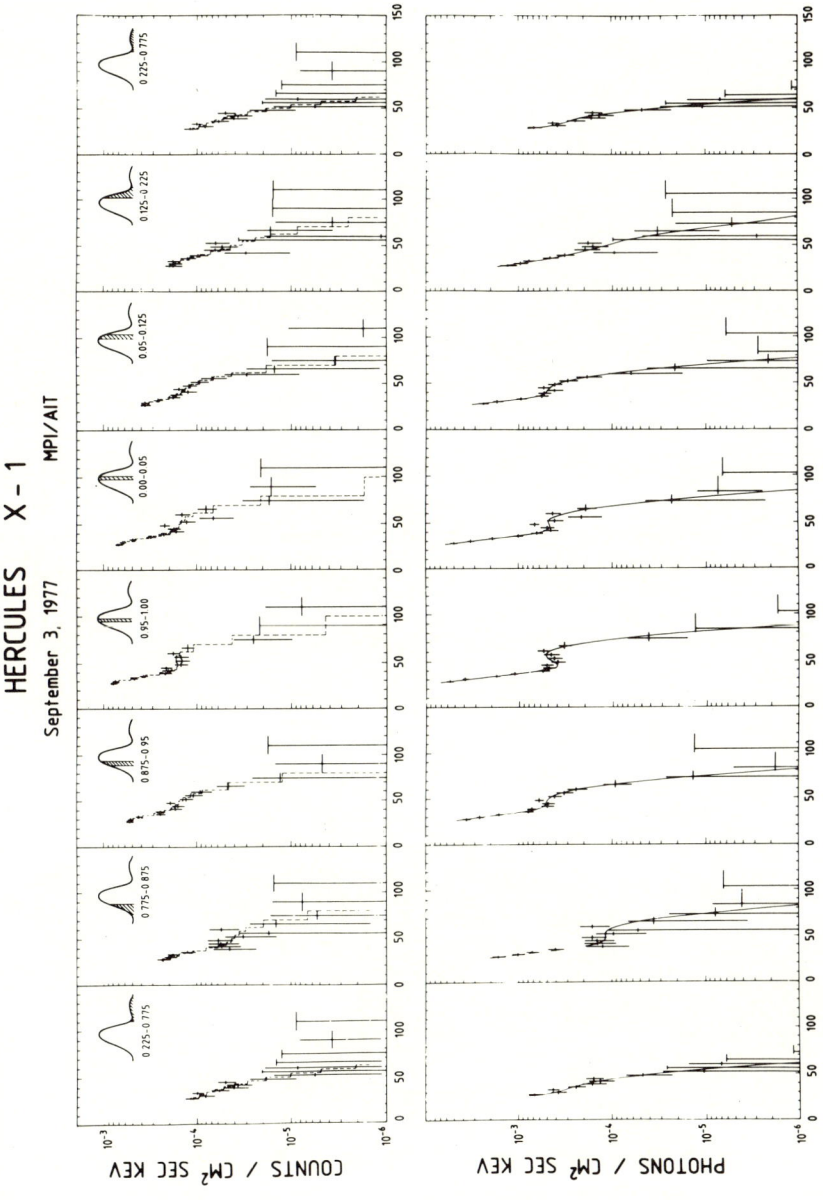

Fig. 6.3. The spectrum of Her X-1 as a function of pulse phase observed in September 1977 by the MPI-AIT balloon experiment. Each spectrum corresponds to the pulse phase interval which is indicated in the insert. Upper spectra are raw data (count rates), lower curves are deconvoluted spectra (Voges et al. 1982).

(Voges *et al.* 1978, 1982) and from satellite data by the UCSD–MIT group (Gruber *et al.* 1980). Since then the feature has also been observed by several other groups (Evans, Quinby & Engel 1980, Pravdo, Bussard & White 1979*b*, Ubertini *et al.* 1981). The improved signal-to-noise ratio of the recent measurements has made it possible to analyse the 58 keV feature in considerable detail. Figure 6.3 shows the spectra taken during different phase intervals of the 1.24 s pulsation (Voges *et al.* 1982). It is obvious that the line feature exists throughout the pulse. A detailed analysis shows that the shape of the spectrum remains constant within the error limits, and this is true in particular for the line energy and the relative line intensity. On the other hand, the HEAO-1 data suggest a small drift of the line energy during the pulse (Gruber *et al.* 1980). One additional piece of information to emerge concerns the line width. Whereas the initial observation in May 1976 had insufficient resolution in frequency to give a reasonable value, the September 1977 flight, and observations by the HEAO-1 group found a width of about 21 keV under the emission-line hypothesis, and about 11 keV under the absorption-line hypothesis (Gruber *et al.* 1980). Recent balloon observations by Tueller *et al.* (1981) also rule out the possibility of a narrow line.

Since the discovery of the Her X-1 cyclotron line there has been considerable effort directed at finding such features in the spectra of other X-ray pulsars, and so far two more candidate sources have been observed. Figure 6.4 displays the spectrum of the source 4U 0115+63 showing an absorption line which corresponds to a magnetic field $\sim 2.10^{12}$ G (Wheaton *et al.* 1979). Figure 6.5 shows the spectrum of 4U 1627−67 with a rather broad feature of similar frequency (Pravdo *et al.* 1979*a*). Thus of 17 pulsating sources, three display lines which may be interpreted as cyclotron resonances. Even a modest dispersion in the values of the surface magnetic field of the remaining 14 pulsars would be sufficient to move the resonance into a frequency range either too high to be observed with present-day detectors, or too low to enable one to exclude the possibility of confusion with an atomic line.

6.2.3 Theory

The energy corresponding to the cyclotron line in the spectrum of Her X-1, is about $\frac{1}{10}$ of the electron rest-mass energy, and appears to be greater than the average thermal energy possessed by a particle in the emission region. This is in contrast with the usual situation in both laboratory plasmas and astrophysical plasmas, and requires that a quantum mechanical description be employed in model building.

In fact the strength of the magnetic field brings about a situation in which the translational degrees of freedom of the electron are to some extent 'frozen-out'. The energy of an electron is a combination of the unquantized kinetic energy of motion along the field lines, and the discrete energy associated with the orbit in the plane perpendicular to the field (Landau levels). Non-relativistically, the spectrum of the discrete levels is that of the harmonic oscillator. An electron of energy less than $\hbar\Omega$ (where Ω is the cyclotron frequency) has only one remaining degree of freedom – translation along the field. The relativistic spectrum was first obtained by Rabi (1928). The eigenvalues of energy are

$$E_{j,s} = mc^2 [1 + (p_z^2/mc)^2 + (2j + s + 1)\hbar\Omega/mc^2]^{1/2},$$

so that regular spacing of the levels is a good approximation only when $E_{j,s}/mc^2 - 1 \ll 1$. For Her X-1 this quantity is about $\frac{1}{10}$ at the cyclotron line, and about $\frac{1}{5}$ at the second harmonic. Further small corrections to the spectrum arise from the theory of quantum electrodynamics, e.g. the level degeneracy is lifted

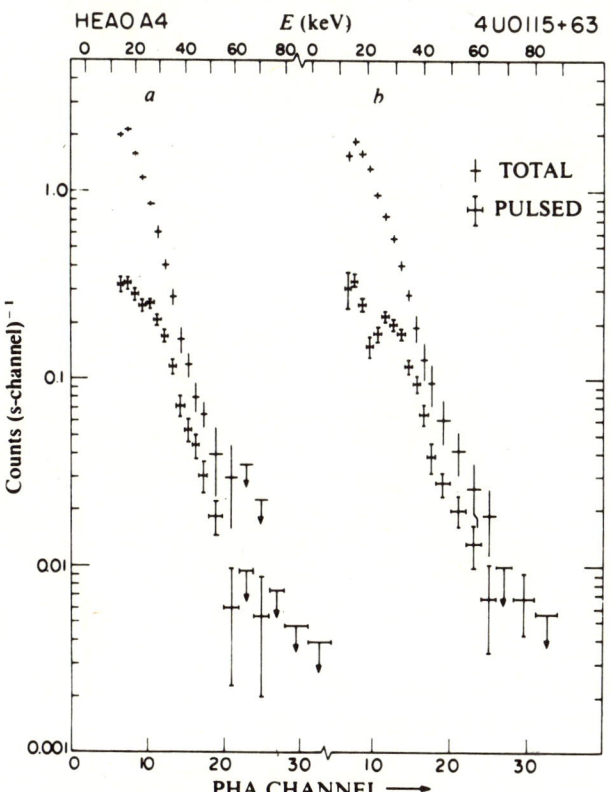

Fig. 6.4. Pulse-height spectra of 4U 0115+63 measured by the HEAO-1-A4 experiment. (a) and (b) represent data from different detectors (LED 1 and 2 respectively). The 20 keV feature shows most clearly in the LED 2 data (from Wheaton et al. 1979).

Fig. 6.5. The spectrum of 4U 1626−67 as a function of pulse phase. Each spectrum corresponds to a successive 0.1 pulse phase starting in the upper left column and continuing down first the left and then the right column. The first spectrum begins at phase 0 which corresponds to JD 2 443 596.5 (Pravdo *et al.* 1979*a*).

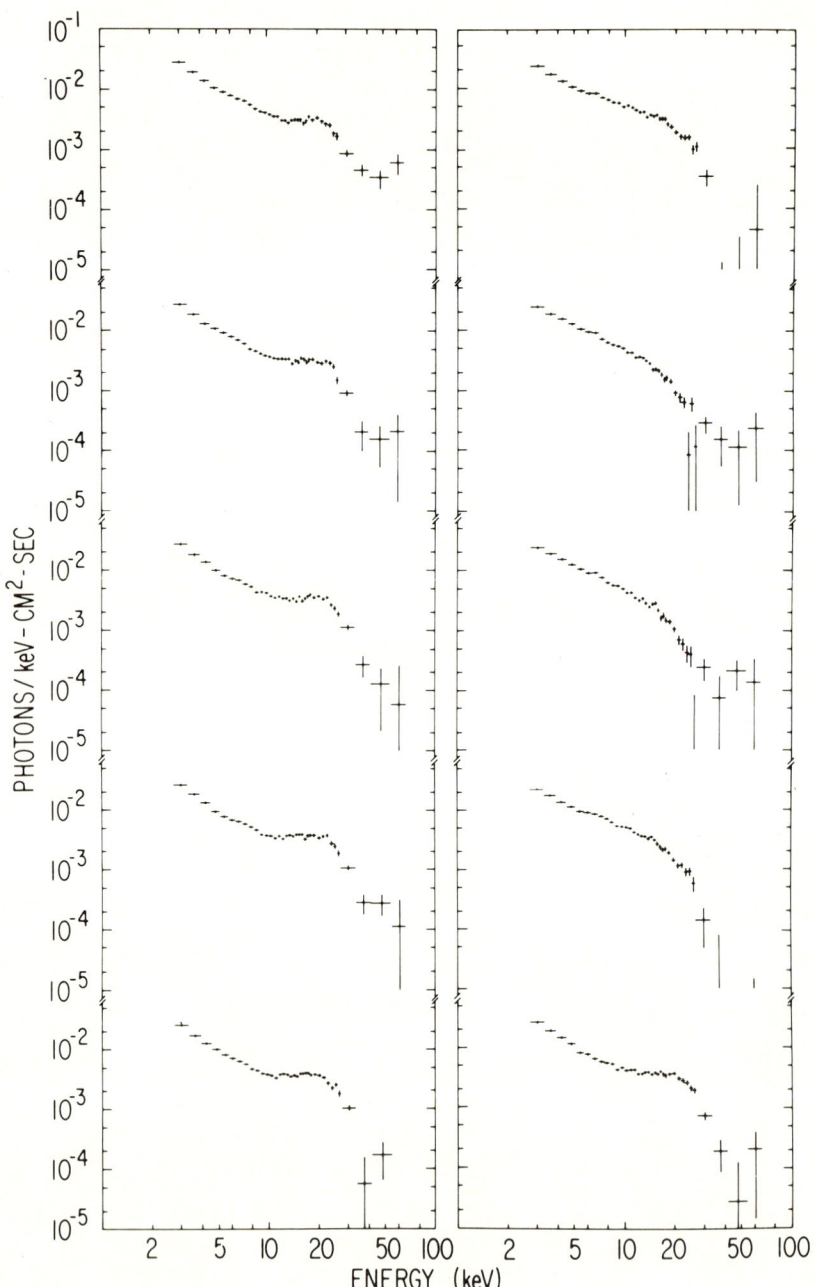

when a precise value of the magnetic moment of the electron is used. However, these effects are too small to be of astrophysical interest.

The spontaneous transition of an electron from the Landau level with $n=1$ to that with $n=0$ is an extremely rapid process in a strong field. The lifetime of the $n=1$ state is

$$\tau \text{ (seconds)} = 2.4 \times 10^{-16} \times B_{12}^{-2}$$

which is much shorter than estimates of the collisional excitation time (the collisional relaxation time given by the Spitzer (1962) formula is of order 10^{-10}s), so that the population of the excited Landau levels may well not correspond to thermal equilibrium. The possibility of observing this Landau transition from a hot (~ 10 keV) plasma on a magnetic neutron star was first pointed out by Gnedin & Sunyaev (1974).

Models of strongly magnetized, accreting neutron stars can be loosely divided into two classes. In the first class an understanding of the features of the accretion flow itself is the goal. To this end it has in every case been found necessary to introduce some simplifying assumptions concerning the geometry of the so-called 'accretion funnel'. An early model of this type was proposed for spherically symmetric accretion by Zel'dovich & Shakura (1969). The deceleration of the flow was assumed to occur either by means of collisions, or by way of a saturated two-stream instability. The influence of radiation pressure on the inflowing material in a cylindrical column was discussed by Davidson (1973). This model aimed at an explanation of the more luminous accreting neutron stars, where one may assume that the gas pressure in the accretion column is always small compared to the radiation pressure. Basko & Sunyaev (1975a) extended this idea and, in a one-dimensional investigation, demonstrated the possibility of exceeding the critical Eddington luminosity. This problem in two dimensions has been investigated recently (Wang & Frank 1981). Further development of models of low-luminosity sources, in which radiation pressure can be neglected, was undertaken by Langer, McCray & Baan (1980) and Langer & Rappaport (1982). In the latter paper the existence of an adiabatic, collisionless shock front is postulated, as had been suggested for spherically symmetric flows by Shapiro & Salpeter (1975). The second class of model attempts to treat the radiation processes in a more consistent manner, but ignores any details of the accretion flow by specifying the density and temperature profiles in the atmosphere. This approach has the attraction that some of the novel effects which are introduced into the micro-physics of the accretion plasma can be investigated in detail. Models of this class are Mészáros (1978), Yahel (1979, 1980), Nagel (1980), Mészáros, Nagel & Ventura (1980), Bonazzola, Heyvaerts & Puget (1979), Wassermann & Salpeter (1980), Kirk & Mészáros (1980) and Nagel (1981a, b).

Thus, the full problem of modelling an X-ray pulsar including a consistent treatment of the radiation transfer and the magnetohydrodynamics has as yet not been attacked. There are, indeed, excellent grounds for this reticence. The cross-sections for both bremsstrahlung and Compton scattering become strongly anisotropic in a strong magnetic field (Canuto, Lodenquai & Ruderman 1971, Gnedin & Sunyaev 1974, Börner & Mészáros 1979, Herold 1979, Pavlov & Panov 1976) and are further complicated by the effects of vacuum polarisation (Mészáros & Ventura 1980, Gnedin, Pavlov & Shibanov 1978). This necessitates a modified approach to the radiative transfer problem (Mészáros, Nagel & Ventura 1980) even when one considers only those frequencies well below the cyclotron resonance. The transfer of cyclotron line photons poses further problems. The structure of the normal modes of propagation of electromagnetic waves in this frequency range displays interesting features which can be attributed to the interplay of vacuum polarisation and thermal effects (Kirk 1980, Herold, Ruder & Wunner 1981). The way in which these modes propagate through a highly magnetized plasma, in which the most common process is a resonant scattering, has been considered by Bonazzola, Heyvaerts & Puget (1979), Nagel (1980), Wassermann & Salpeter (1980), Melrose & Zheleznyakov (1981) and Melrose (1981).

Although the two different approaches to the problem mentioned above have been the subject of considerable activity, no universally accepted solution has been found in either case. The roots of this difficulty lie in the subtleties of the matter–radiation interaction in an optically thick 'accretion plasma'. Fortunately, it now seems that most of these problems may be circumvented if one restricts one's attention to those sources whose luminosity is considerably less than the Eddington value, and which accrete over a relatively large area on the surface.

These conditions are likely to apply in many pulsating X-ray sources: for an accretion-column diameter of a few kilometers, one requires that the luminosity should be less than 10^{37} erg s^{-1} if the flow is homogeneous. If, on the other hand, matter is accreted in clumps or filaments, this requirement is relaxed even further. Thus, Vela X-1 with a luminosity of 10^{36} erg s^{-1} is certainly in this class. Her X-1 (10^{37} erg s^{-1}) may be included also, depending on the actual values of mass, radius, column width and 'clumpiness'. For such sources there is insufficient radiation pressure to decelerate the flow and either a collisionless or a collisional shock must occur before freely falling material is brought to rest at the stellar surface. Although the presence of a collisionless shock cannot be completely excluded, there are theoretical difficulties concerning its realisation: the strong magnetic field tends to stabilize the accreting plasma and render the physics 'one-dimensional'. Investigations of this situation have indicated that collisionless shocks do not form (McKee 1970, Alme & Wilson 1973), which implies that collisions are required to decelerate the flow. To permit these, however, the density across such a shock front must rise significantly – considerably more than the rise obtained across an adiabatic shock front, and one arrives at the conclusion that a substantial fraction of the accretion luminosity must be released in the shock itself. To all intents and purposes this type of shock front rests on the stellar surface, since the material behind it is both slowly moving and dense. A model of the emission region of such a low-luminosity source amounts to a description of the collisional shock. One question of paramount importance is the thickness of this heated layer, or, equivalently, the depth in the atmosphere of the star to which a freely falling proton can penetrate.

The first calculations of this quantity (Basko & Sunyaev 1975b, Pavlov & Yakovlev, 1976) were performed in the limit in which the electrons of the atmosphere are cold, and the protons of the infalling matter have a large gyro radius. They demonstrated that the presence of a strong magnetic field has a marked influence on the deceleration, through its effect on small-angle Coulomb-scattering events. Basko & Sunyaev (1975b) suggested that nuclear collisions could provide a more efficient deceleration, in which case an atmospheric depth of about $50 \,\mathrm{g\,cm^{-2}}$ would be required to stop the accretion flow. However, calculations of small-angle Coulomb scattering in an accretion plasma, in which the proton gyro-radius is comparable in size to the Debye screening length (Kirk & Galloway 1981, 1982), have shown that the stopping distance is significantly reduced. Results of some of these calculations are shown in Figures 6.6 and 6.7. For a standard neutron star (free-fall velocity 0.5 c, Figure 6.6) a penetration depth of $\sim 20\,\mathrm{g\,cm^{-2}}$ is found. This corresponds to a Thomson optical depth of ~ 8. For a low-mass neutron star (0.5 $M_\odot$, $R \sim 20$ km, Figure 6.7) this layer may become optically thin for Thomson scattering.

The radiative heat transfer in such a hot strongly magnetized plasma sheet, allowing for Comptonisation and redistribution of photons in the cyclotron line, has been treated recently by Nagel (1981a). Although these calculations contain several simplifications, such as the assumption of a homogeneous slab with respect to density and temperature, and the neglect of vacuum polarisation and relativistic effects, the results may be relevant to the low-luminosity case discussed here. Figure 6.8 shows the energy spectra for ordinary and extraordinary photons for slabs of different optical thickness τ with a density ρ of $10^{-2}\,\mathrm{g\,cm^{-3}}$ and a temperature of 1.1×10^8 K. As expected, the spectra show an emission line for low optical depths, which becomes self-reversed at intermediate depths. In this case the extraordinary photons are trapped in the resonance and escape in the wings (see also Wassermann & Salpeter 1980). The emission peak seen in the ordinary photons is due to the scattering of trapped extraordinary photons into ordinary

Fig. 6.6. The stopping length of a proton for plasma temperatures 5, 10, 20 and 30 keV and densities (a) 10^{21} cm^{-3}, (b) 10^{23} cm^{-3}, (c) 10^{25} cm^{-3} and (d) 10^{26} cm^{-3}. In each case the initial infall velocity is $c/2$, and the energy is normalised to the kinetic energy of a single proton.

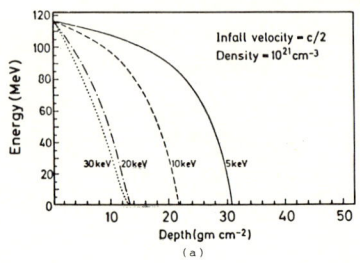
(a)

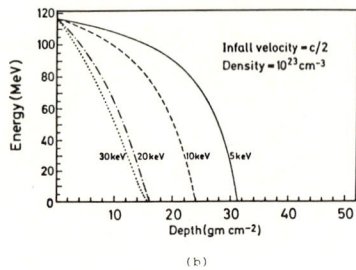
(b)

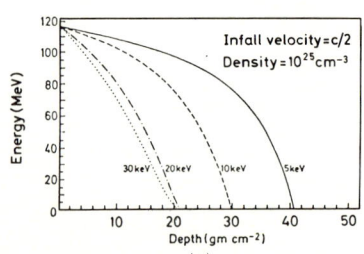
(c)

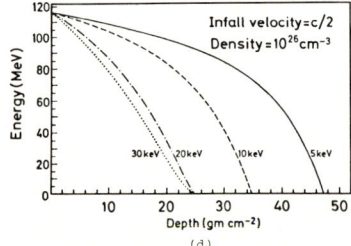
(d)

Fig. 6.7. Same as Figure 6.6(b) but with an infall velocity of $c/4$.

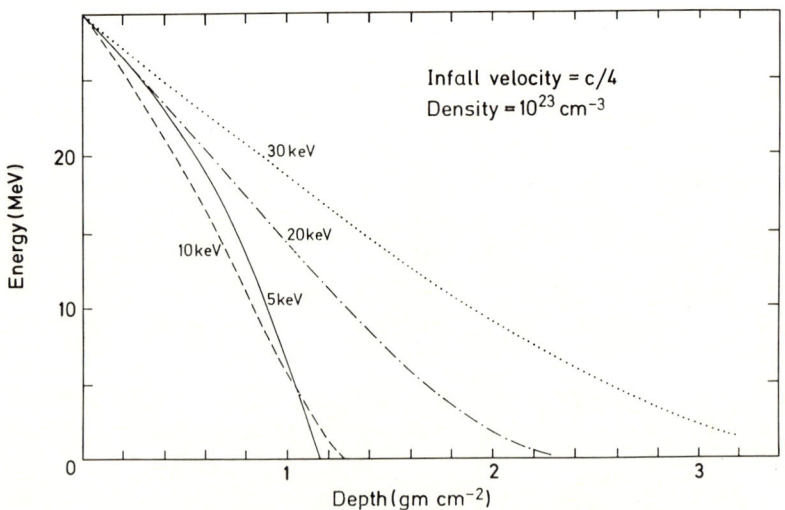

ones, which then escape. This flux of ordinary photons is so strong that it fills the self-reversal dip of the extraordinary photons. At larger optical depths the resonance appears in absorption. At the same time, the luminosity increases with increasing optical depth of the emitting layer and reaches 10^{36} erg s^{-1} for a radiating area of 1 km^2. Larger luminosities may be obtained if the density and/or the temperature of the radiating slab are increased. Figure 6.9 shows the spectrum expected for $\rho = 10$ g cm^{-3} and $\tau = 8$. The luminosity in this case would be 10^{37} erg s^{-1} for $A = 3$ km^2, sufficient to explain Her X-1.

Figure 6.10 shows the angular distribution of radiation at various frequencies which emerges from a homogeneous slab of density 0.1 g cm^{-3}, temperature 1.1×10^8 K and an optical depth with respect to Thomson scattering $\tau \approx 8$ (Nagel 1981b). In the radiative transfer treatment which lies behind these curves,

Fig. 6.8. Photon spectra from slabs of hot strongly magnetized plasma. Solid lines: ordinary photons; dashed lines: extraordinary photons. The uppermost curve represents the Wien spectrum. The plasma parameters are $\rho = 10^{-2}$ g cm^{-3}, $kT = 10$ keV, $\hbar\Omega = 50$ keV. The luminosities, for an area of 1 km^2, are, in order of increasing Thomson optical depths: 8×10^{30}, 6×10^{31}, 6×10^{32}, 2×10^{34}, 4×10^{35} and 10^{36} erg s^{-1} (Nagel 1981a, b).

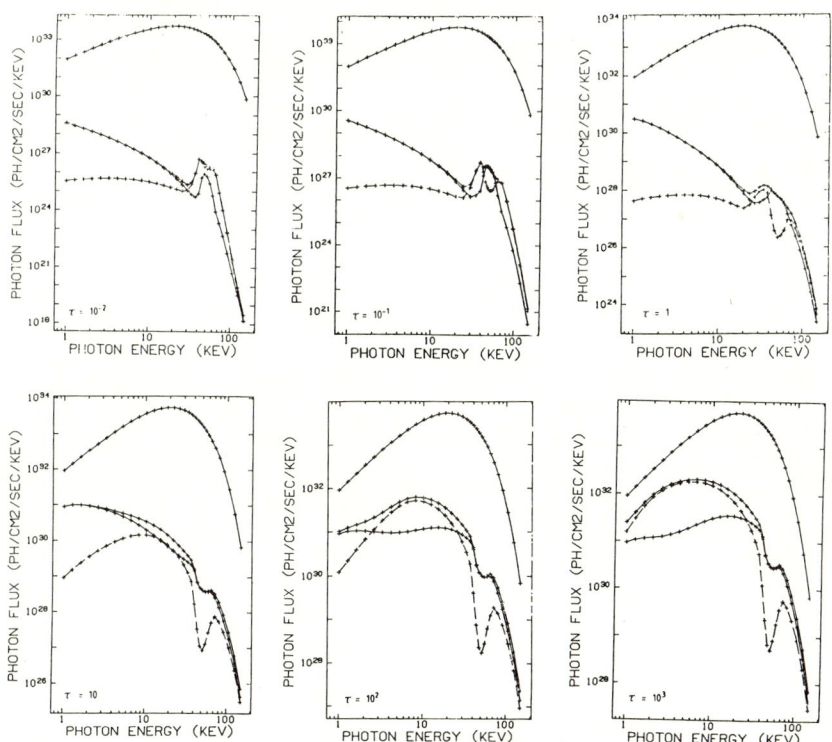

the assumption of coherent transfer was made, but the full anisotropy of the scattering and absorption cross-sections was included. The coherent transfer assumption should, however, be accurate for low frequencies (<40 keV), since then the frequency change suffered by a photon on traversing the slab is small. The results show a minimum in the emitted flux around the magnetic field direction, for frequencies well below the cyclotron frequency. At frequencies approaching the cyclotron frequency this 'hollow cone' disappears. The minimum is due to the fact that the photon–electron cross-sections go to zero for $\theta \to 0$ and $\omega \to 0$. This result had already been discussed qualitatively by Basko & Sunyaev (1975a) in an attempt to explain pulse profiles.

The light curves expected for different orientations of the magnetic and rotational axes with respect to the line of sight are also shown in Figure 6.10. For inclination angles of 45° and 60° the results resemble the observed light curves of Her X-1, which show a double main peak in the range 2–20 keV and single

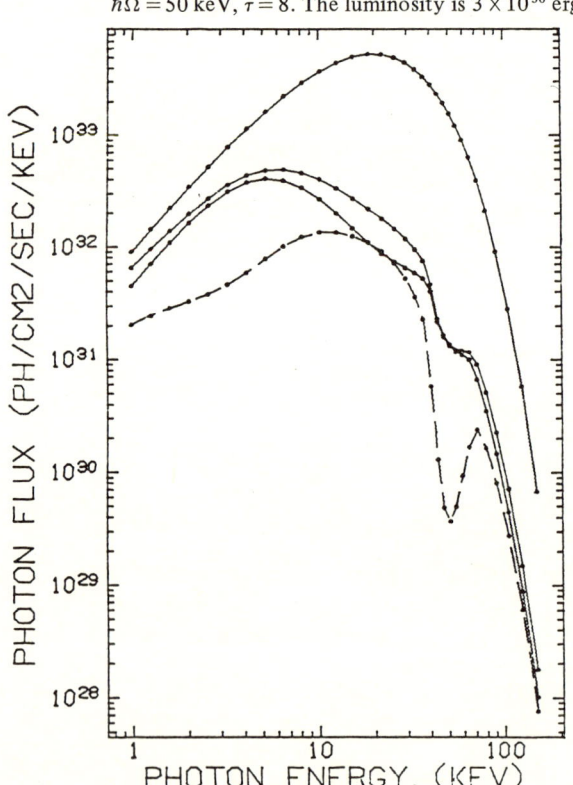

Fig. 6.9. Same as Figure 6.8, for plasma parameters $\rho = 10$ g cm^{-3}, $kT = 10$ keV, $\hbar\Omega = 50$ keV, $\tau = 8$. The luminosity is 3×10^{36} erg s^{-1} (for an area of 1 km^2).

Accreting magnetic neutron stars 277

peaks at $E > 20$ keV. In addition, there is a small interpulse, as in Her X-1. Of course, the computed peaks are symmetric, and cannot be expected to reproduce exactly the Her X-1 pulses, which indicate some asymmetry in the structure of the accretion column.

Assuming that such a model of the heated layer can be assembled consistently, the question arises as to the effect of free-falling material on the outgoing radiation. The cross-section for resonant scattering at the local cyclotron frequency is, of course, much larger than that for ordinary Thomson scattering, let alone free-free absorption. The accreting plasma will thus be effective in scattering frequencies close to the local gyro-frequency. However, for a field anchored in the star itself, the local gyro-frequency in the accretion flow is always less than its value at the surface, because of the Doppler shift and the radial decrease of the

Fig. 6.10. (a) Anisotropy of the radiation from a slab of strongly magnetized plasma, at photon energies 1, 2.5, 5, 10, 25 and 75 keV (from top to bottom). (b) Geometry. (c) The pulse shapes deduced from the beam pattern in (a). The numbers at the top give the two inclination angles i_1 and i_2. Higher curves correspond to higher energies.

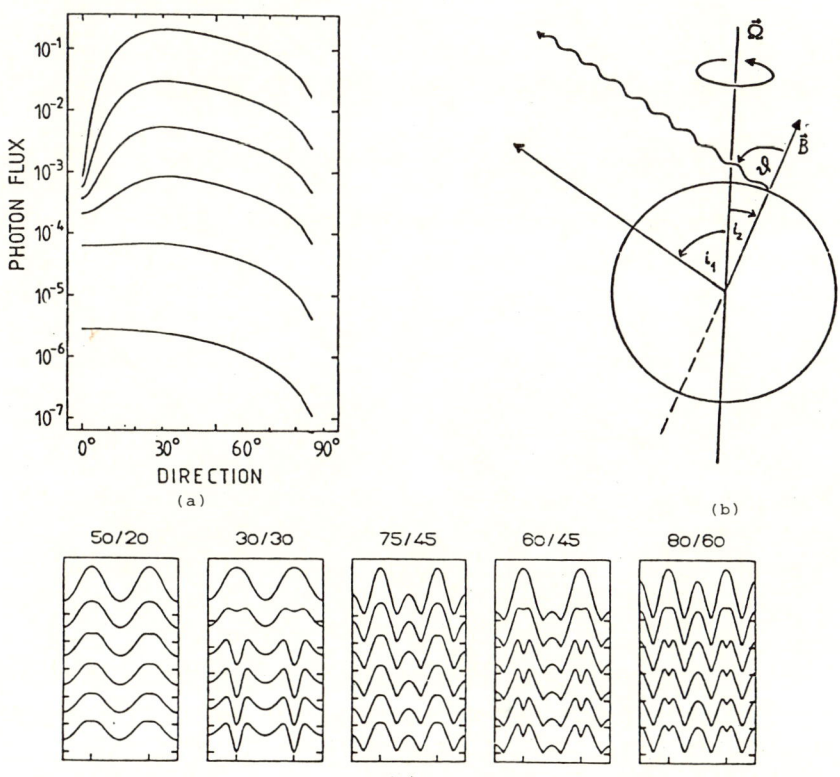

magnetic field strength. Therefore, the flux scattered out of the pulse should contain only frequencies less than that of the cyclotron resonance line (Elsner & Lamb 1976, Ventura 1981). This will have two effects: first, the centre of the hollow-cone beam seen in Figure 6.10 will be made deeper and, consequently, wider. Secondly, that fraction of the total flux from the source which is pulsed should be smaller for low frequencies, increasing towards unity at the cyclotron resonance. This effect may indeed be observed in Her X-1, where the pulsed fraction increases at energies approaching the cyclotron energy (Voges *et al.* 1982). It is also noteworthy that Vela X-1 shows a dramatic increase of the pulsed fraction at energies approaching 100 keV (Staubert *et al.* 1980). Vela X-1 is a low-luminosity source to which our considerations should apply. One can therefore speculate that its surface magnetic field may be of the order 10^{13} G.

The state of the art of modelling accreting magnetized neutron stars leaves much to be desired, especially in the case of a source radiating close to the Eddington limit. Nevertheless, there are several encouraging signs that the ideas about lower-luminosity sources – which comprise, in fact, most X-ray pulsars – are crystallising into a model capable of explaining in a consistent manner several different aspects of the observations.

6.3 Gamma-ray bursts
6.3.1 Observations of gamma-ray bursts

The origin of γ-ray bursts has been a mystery ever since they were first detected by the Vela satellite (Klebesadel, Strong & Olson 1973). Indeed, for a long time the number of theories exceeded the number of events, and one was able to say with some degree of certainty that not all the theories could be correct. Since then, many groups have observed γ-ray bursts, and, especially as a result of the Soviet Konus experiments on the Venera 11 and 12 spacecrafts, the highly speculative theoretical scene seems to have improved (Mazets *et al.* 1981).

These experiments have a spectral coverage which extends down to 20 keV and combine good time resolution with reasonably good spectral resolution, which enables time-resolved spectra to be taken. Several of these spectra for different events, are shown in Figure 6.11. It is possible to identify absorption features between 20 keV and 70 keV, as well as an emission line in this range. A total of about 30 'line events' have been detected, out of 150 bursts. In some of these cases the spectrum evolved during the event in the sense that the spectral features disappeared during the later phases. An event of this type has also been reported by the GSFC group (Dennis *et al.* 1982). In addition to this, and no less important, emission features at 400–460 keV have been observed in 11 of 150 events detected in the Konus experiments.

Accreting magnetic neutron stars

The same line of argument as that used in the case of the Her X-1 line, suggests that the lower energy features should be connected with the electron-cyclotron resonance. From this it follows immediately that γ-ray bursts originate on neutron stars with a surface magnetic field of $2\text{-}7 \times 10^{12}$ G. The higher-energy emission lines are then readily interpreted as electron–positron annihilation lines. In the rest frame, the line feature from a low-energy e^+e^- annihilation event lies at 511 keV. If this were to occur near the surface of a neutron star, the gravitational redshift would move the line by about 10–20 per cent for a distant

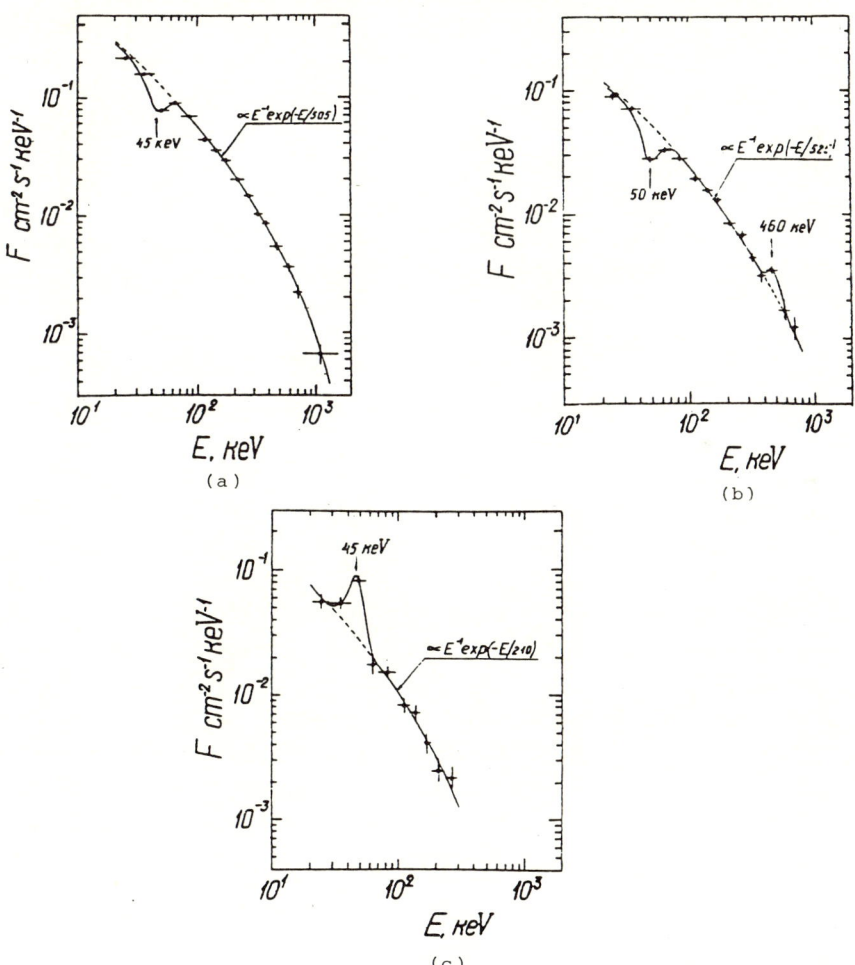

Fig. 6.11. Compilation of γ-ray burst spectra measured with the Konus experiment aboard the Venera 11 and 12 spacecrafts (Mazets *et al.* 1980): (*a*) 1979, 7 March burst. (*b*) 1979, 26 May burst. (*c*) 1979, 22 June burst.

observer. Values of 410-500 keV as in Figure 6.11 are thus in agreement with this interpretation.

Fenimore *et al.* (1982) have pointed out several instrumental effects which could simulate spectral features like those observed in the Konus experiments. However, in our opinion the case in favour of these experiments is very convincing for several reasons: (1) The features appear at different energies in different bursts, (2) they disappear in the later parts of the bursts in several cases, (3) they are observed from two independent spacecraft and (4) the majority of bursts show smooth spectra without line features. On the other hand, spectra measured recently with the γ-ray spectrometer aboard the solar-maximum mission satellite are of the power-law type at high energies, in contrast to the Konus spectra, which are well fitted by an optically thin thermal bremsstrahlung law. Furthermore, no candidates for pair-annihilation lines have been found in any of the events (Share *et al.* 1982). In view of the poor event statistics there is no contradiction with the Konus results. Nevertheless, it appears important to have an independent confirmation of the line features. In addition, it would be desirable to see raw data (count-rate) spectra from the Konus experiments to complement the published deconvoluted spectra.

If confirmed by future investigations, these γ-ray lines in bursts, are exceedingly important; they allow for the first time a simultaneous determination of the magnetic field and the gravitational field of a neutron star.

6.3.2 Theory

Attempts to find a physical explanation of the γ-ray bursts have centred around two ideas: that of a thermonuclear explosion on the surface of a neutron star and that of the sudden accretion of material onto such a star. The possibility of energy release in starquakes has also been discussed in relation to the unique event of 5 March 1979 (Ramaty *et al.* 1980).

In the thermonuclear explosion models, the sequence of events depends strongly on the temperature of the star, and on its steady accretion rate. Thus Woosley & Wallace (1981) and Fryxell & Woosley (1981) consider a relatively high-temperature layer of hydrogen at the magnetic pole of a neutron star. The temperature is sustained by the burning of hydrogen in the CNO cycle. Underneath the hydrogen, a layer of helium accumulates until a critical density is achieved at its base. Explosive burning of the helium then occurs, with an ignition process which is either a detonation wave, or a more slowly travelling deflagration front, according to the physical conditions in the layer. A strong magnetic field is required in this model, not only to focus accretion onto the pole of the star and prevent the subsequent spreading-out of accumulated material, but also to provide the possibility of producing γ-ray bursts rather than X-ray bursts

(Wallace, Woosley & Weaver 1981, Lewin & Joss 1981, see also Chapter 2, this book). The accretion rate required by this model is $10^{-13} M_\odot$ yr^{-1}. It is interesting to compare this figure with the accretion rate for a neutron star of 1 $M_\odot$ in the interstellar medium (Bondi 1952).

$$\dot{M}(M_\odot \text{ yr}^{-1}) \approx 10^{-17} \frac{n}{V_{100}^3}.$$

Here n is the interstellar number density of hydrogen, and V_{100} is either the speed (in units 100 km s^{-1}) of the neutron star through the medium or the sound speed in the ionized surroundings, if this is larger. Since the sound speed is of order 30 km s^{-1} (Shvartsman 1971), one obtains the constraint

$$\dot{M} \leqslant 4 \times 10^{-16} n$$

Thus only an extremely dense region of the interstellar medium ($n > 250$) or a companion star can provide sufficient accretion for the model.

A similar accretion rate is considered by Hameury, Bonazzola, Heyvaerts & Ventura (1981). However, the star is in this case unmagnetized, so that material accretes over the entire surface. If, on the other hand, one were to introduce the assumption that the accretion flow is channelled onto the polar cap, then it is possible that this model could function by accretion from the interstellar medium. The temperatures obtained in the surface layers are somewhat lower, resulting in the prediction of a hydrogen flash initiated by electron capture on protons, which preceeds the main helium explosion by 10^2-10^3 s. The low-temperature regime has also been discussed by Ergma & Tutukov (1980), who find it possible to accumulate enough helium for an explosion even at very low accretion rates. This result is in contradiction with the findings of Hameury et al. Such calculations, however, involve computing reaction rates in the pycnonuclear regime, which seems to involve some uncertainties (Dewitt, Graboske & Cooper 1973, Itoh, Totsuji & Ichimaru 1977).

In addition to thermonuclear models, several suggestions involving sudden accretion have been discussed in the literature (Harwit & Salpeter 1973, Newman & Cox 1980, Colgate & Petschek 1980, Howard, Wilson & Barton 1982). Because this process is more efficient in energy conversion, only $\frac{1}{10}$ as much material is involved as in the case of a thermonuclear explosion. Woosley & Wallace (1981) have pointed out that this should have an effect on the cooling time which raises the possibility of distinguishing between thermonuclear and accretion models by observing the soft X-ray emission associated with the post-burst phase.

One feature which all models share is the inability to account for the optically thin bremsstrahlung spectrum observed in the Konus experiments. The difficulties posed by this spectrum can be formulated by means of a very simple argument: The flux measured is given in terms of the distance to the source D (in parsecs), the volume of the source V (in cm^3), the temperature $T_9 \times 10^9$ K and the number density $n_{22} \times 10^{22}$ cm^{-3} by the formula

$$F \text{ (erg cm}^{-2}\text{ s}^{-1}) = 5 \times 10^{-16} V n_{22}^2 T_9^{1/2} D^{-2},$$

assuming bremsstrahlung from an unmagnetized optically thin plasma. In order that this spectrum be unaffected by Compton scattering, the column density of matter along the line of sight d in the source must satisfy

$$n_{22} d < 1.5 \times 10^2, \quad (d \text{ in cm})$$

if the electrons are relativistic (this condition is only slightly weaker for non-relativistic temperatures – see Bussard & Lamb, 1982). Combining these expressions gives

$$D \text{ (parsec)} < 3 \times 10^{-6} F^{-1/2} T_9^{1/4} V^{1/2} d^{-1}.$$

If estimates of the number of neutron stars in the Galaxy (Ostriker, Rees & Silk 1970) are not vastly in error, and if γ-ray bursts originate on such objects, then $D > 10$ pc, which implies that the emission region must have an unusual geometry:

$$V^{1/3}/d > 30.$$

A more realistic distance estimate makes this requirement more severe. Such a restriction poses a problem for all models. In fact, it may indicate that the spectrum, whilst having the optically thin bremsstrahlung form, is nevertheless produced by some other process in the highly magnetized plasma of the emitting region (Bussard & Lamb 1982).

A further difficulty in interpretation of the spectrum concerns the absorption features observed by Mazets *et al.* (1981). If these are interpreted as cyclotron resonances, the temperature of the absorbing layer can be estimated from the line width to be about 20 keV. This is very small compared with the temperatures seen in the continuum, and therefore one has to have an extreme temperature stratification in the source. Furthermore, only one narrow line is seen in most cases. If this is due to the first harmonic cyclotron resonance the absence of the second harmonic absorption line sets very severe limitations to the column density of the absorbing layer (Trümper 1982), which must then lie between 2×10^{-4} and 2×10^{-3} g cm^{-2}.

Although theoretical speculation has been to some extent focussed onto neutron stars by the observations of the Leningrad group, there remains much uncertainty. In the near future, however, one can look forward to further important observations. Not only will the number of γ-ray-burst events continue to rise, but simultaneous X-ray observations should provide valuable information. Increased accuracy in position determination may lead to the identification of companion stars,* and observations in the UV band may detect continuous emission from the neutron star itself (Kirk & Stoneham 1982).

* An optical flash which occurred in 1928 at the site of the γ-burst of 19 November 1978 has recently been discovered by Schaefer (1981)

Acknowledgements

We are grateful to U. Anzer, G. Börner, P. Kafka and F. Meyer for many discussions.

References

Alme, M. L. & Wilson, J. R. 1973, *Ap. J.*, **186**, 1015.
Anzer, U. & Börner, G. 1980, *Astron. and Ap.*, **83**, 133.
Anzer, U. & Börner, G. 1983, *Astron. Ap.*, in press.
Arons, J. & Lea, S. M. 1976, *Ap. J.*, **207**, 914.
Arons, J. & Lea, S. M. 1980, *Ap. J.*, **235**, 1016.
Basko, M. M. & Sunyaev, R. A. 1975a, *Astron. Ap.*, **42**, 311.
Basko, M. M. & Sunyaev, R. A. 1975b, *Sov. Phys. JETP*, **41**, 52.
Basko, M. M. & Sunyaev, R. A. 1976, *Mon. Not. Roy. Astron. Soc.*, **175**, 395.
Bonazzola, S., Heyvaerts, J. & Puget, J. G. 1979, *Astron. Ap.*, **78**, 53.
Bondi, H. 1952, *Mon. Not. Roy. Astron. Soc.*, **112**, 195.
Börner, G. & Mészáros, P. 1979, *Plasma Phys.*, **21**, 357.
Burdyuzha, V. V. & Pavlov-Verevkin, V. B. 1980, Preprint No. 505, Space Research Institute, Moscow.
Bussard, R. W. & Lamb, F. K. 1982, *In: Gamma Ray Transients and Related Astrophysical Phenomena*, eds. R. E. Lingenfelter, H. S. Hudson & D. M. Worrall, American Institute of Physics, New York.
Canuto, V., Lodenquai, J. & Ruderman, M. A. 1971, *Phys. Rev. D*, **3**, 2303.
Colgate, S. A. & Petschek, A. G. 1980, *Ap. J.*, **236**, L115.
Davidson, K. 1973, *Nature Phys. Sci.*, **246**, 1.
Dennis, B. R., Frost, K. J., Kiplinger, A. L., Orwij, L. E., Desai, U. & Cline, T. L. 1982, *In: Gamma Ray Transients and Related Astrophysical Phenomena*, eds. R. E. Lingenfelter, H. S. Hudson & D. M. Worral, American Institute of Physics, New York.
DeWitt, H. E., Graboske, H. C. & Cooper, M. S. 1973, *Ap. J.*, **181**, 439.
Elsner, R. F. & Lamb, F. K. 1976, *Nature*, **262**, 356.
Ergma, E. V. & Tutukov, A. V. 1980, *Astron. Ap.*, **84**, 123.
Evans, A. J., Quenby, J. J. & Engel, A. R. 1980, *Astron. Ap. Supp.*, **41**, 13.
Fenimore, E. E., Laros, J. G., Klebesadel, R. W., Stockdale, R. E. & Kane, S. 1982, *In: Gamma Ray Transients and Related Astrophysical Phenomena*, eds. R. E. Lingenfelter, H. S. Hudson & D. M. Worrall, American Institute of Physics, New York.
Fryxell, B. A. & Woosley, S. E. 1981, *Lick Observatory Bulletin*.
Ghosh, P. & Lamb, F. K. 1979, *Ap. J.*, **234**, 296.
Gnedin, Y. N. & Sunyaev, R. A. 1974, *Astron. Ap.*, **36**, 379.
Gnedin, Y. N., Pavlov, G. G. & Shibanov, Y. A. 1978, *Sov. Astron. Lett.*, vol. 4, no. 3, p. 117.
Gold, T. 1968, *Nature*, **218**, 731.
Goldreich, P. & Julian, W. H. 1969, *Ap. J.*, **157**, 869.
Gruber, D. E., Matteson, J. L., Nolan, P. L., Knight, F. K., Buty, W. A., Rothschild, R. E., Peterson, L. E., Hoffmann, J. A., Scheepmaker, A., Wheaton, W. A., Primini, F. A., Levine, A. M. & Lewin, W. H. G. 1980, *Ap. J.*, **240**, L127.
Hameury, J. M., Bonazzola, S., Heyvaerts, J. & Ventura, J. 1981, submitted to *Astron. Ap.*
Harwit, M. & Salpeter, E. E. 1973, *Ap. J.*, **186**, L37.
Herold, H. 1979, *Phys. Rev. D*, **19**, 2868.
Herold, H., Ruder, H. & Wunner, G. 1981, *Plasma Phys.*, **23**, 775.
Hewish, A., Bell, S. J., Pilkington, J. D., Scott, P. F. & Collins, R. A. 1968, *Nature*, **217**, 709.
Howard, W. M., Wilson, J. R. & Barton, R. T. 1982, *In: Gamma Ray Transients and Related Astrophysical Phenomena*, eds. R. E. Lingenfelter, H. S. Hudson & D. M. Worrall, American Institute of Physics, New York.

Itoh, N., Totsuji, H. & Ichimaru, S. 1977, *Ap. J.*, **218**, 477.
Kirk, J. G. 1980, *Plasma Phys.*, **22**, 639.
Kirk, J. G. & Galloway, D. J. 1981, *Mon. Not. Roy. Astr. Soc.*, **195**, 45P.
Kirk, J. G. & Galloway, D. J. 1982, *Plasma Phys.*, **24**, 339 and 1025 (erratum).
Kirk, J. G. & Mészáros, P. 1980, *Ap. J.*, **241**, 1153.
Kirk, J. G. & Stoneham, R. J. 1982, *Mon. Not. Roy. Astron. Soc.*, **201**, 1183.
Klebesadel, R. W., Strong, I. B. & Olson, R. A. 1973, *Ap. J.*, **182**, L85.
Lamb, F. K., Pethick, C. J. & Pines, D. 1973, *Ap. J.*, **184**, 271.
Langer, S. H., McCray, R. & Baan, W. A. 1980, *Ap. J.*, **235**, 731.
Langer, S. H. & Rappaport, S. 1982, JILA-University of Colorado preprint.
Lewin, W. H. G. & Joss, P. C. 1981, *Space Sc. Rev.*, **28**, 3.
McKee, C. F. 1970, *Phys. Rev. Lett.*, **24**, L990.
Mazets, E. P., Golenetskii, S. V., Ilyisiskii, V. N., Guryan, Y. A., Aptekar, R. L., Panov, V. N., Sokolov, I. A., Sokolova, Z. Y. & Kheritonova, T. V. 1981, preprint, A. F. Ioffe Institute.
Melrose, D. B. 1981, *Astron. Ap.*, **101**, 284.
Melrose, D. B. & Zehleznyakov, V. V. 1981, *Astron. Ap.*, **95**, 86.
Mészáros, P. 1978, *Astron. Ap.*, **63**, L19.
Mészáros, P., Nagel, W. & Ventura, J. 1980, *Ap. J.*, **238**, 1066.
Mészáros, P. & Ventura, J. 1979, *Phys. Rev. D*, **19**, 3565.
Nagase, F., Hayakawa, S., Kunieda, H., Makino, F., Masai, K., Tawara, Y., Inoue, H., Koyama, K., Makishima, K., Matsuoka, M., Murakami, T., Oda, M., Ogawara, Ohashi, T., Shibazaki, N., Tanaka, Y., Tanaka, Y., Kondo, I., Miyamoto, S., Tsunemi, H. & Yamashita, K. 1981, *Nature*, **290**, 572.
Nagel, W. 1980, *Ap. J.*, **236**, 904.
Nagel, W. 1981a, *Ap. J.*, **251**, 278 (1981).
Nagel, W. 1981b, *Ap. J.*, **251**, 288 (1981).
Newman, M. J. & Cox, A. N. 1980, *Ap. J.*, **242**, 319.
Ostriker, J. P. & Gunn, J. E. 1969, *Ap. J.*, **157**, 1395.
Ostriker, J. P., Rees, M. J. & Silk, J. 1970, *Ap. Lett.*, **6**, 179.
Pacini, F. 1967, *Nature*, **216**, 567.
Pavlov, G. G. & Panov, A. N. 1976, *Sov. Phys. JETP*, **44**, 300.
Pavlov, G. G., Shibanov, Y. A. & Gnedin, Y. N. 1980, *Sov. Phys. JETP (Lett.)*, **30**, 125.
Pavlov, G. G. & Yakovlev, D. G. 1976, *Sov. Phys. JETP*, **43**, 389.
Pravdo, S. H., White, N. E., Boldt, E. A., Holt, S. S., Serlemitsos, P. J., Swank, J. H. & Szymkowiak, A. E. 1979a, *Ap. J.*, **231**, 912.
Pravdo, S. H., Bussard, R. W. & White, N. E. 1979b, *Mon. Not. Roy. Astron. Soc.*, **188**, 5.
Rabi, I. I. 1928, *Z. Physik*, **49**, 509.
Rappaport, S. & Joss, P. C. 1977, *Nature*, **266**, 683.
Ramaty, R., Bonazzola, S., Cline, T. L., Kazanas, D. & Mészáros, P. 1980, *Nature*, **287**, 122.
Schaefer, B. E. 1981, *Nature*, **294**, 722.
Shapiro, S. I. & Salpeter, E. E. 1975, *Ap. J.*, **198**, 671.
Share, G. H., Kurfess, J. D., Dee, S., Chupp, E. L., Ryan, J. M., Forrest, D. J., Lanigan, J., Rieger, E., Kanbach, G. & Reppin, C. 1982, *In: Gamma Ray Transients and Related Astrophysical Phenomena*, eds. R. E. Lingenfelter, H. S. Hudson & D. M. Worrall, American Institute of Physics, New York.
Shvartsman, V. F. 1971, *Sov. Astron. AJ*, **14**, 662.
Spitzer, L. 1962, *Physics of fully ionized gases*, Wiley, London.
Staubert, R., Kendziorra, E., Pietsch, W., Reppin, C., Trümper, J. & Voges, W. 1980, *Ap. J.*, **239**, 1010.
Trümper, J., Pietsch, W., Reppin, C., Secco, B., Kendziorra, E. & Staubert, R. 1977, *Annals New York Acad. Sci.*, **302**, 538.
Trümper, J. 1982. *In: Gamma Ray Transients and related Astrophysical Phenomena*, eds. L. E. Lingenfelter, H. S. Hudson & D. M. Worrall, American Institute of Physics, New York.

Trümper, J., Pietsch, W., Reppin, C., Voges, W., Staubert, R. & Kendziorra, E. 1978, *Ap. J.*, **219**, 105.
Tueller, J., Chine, T., Paciesas, W., Teegarden, B., Bodet, D., Durouchoux, P., Hameury, J. & Haymes, R. 1981, to appear in *Proc. 17th Int. Cosmic Ray Conf.* Paris.
Ubertini, P., Bozzano, A., La Padule, C. D., Polcero, V. F., Vialetto, G. & Manchanda, R. K. 1981, *15th ESLAB Symposium: X-ray Astronomy*, Amsterdam.
Ventura, J. 1981, *IAU Symposium No. 95*, eds. W. Sieber & R. Wielebinski, D. Reidel, Dordrecht, p. 263.
Voges, W., Pietsch, W., Reppin, C., Trümper, J., Kendziorra, E. & Staubert, R. 1978, *Proc. 21st Cospar Meeting*, Innsbruck (X-ray Astronomy, eds. W. A. Baity & L. E. Peterson, Pergamon Press, Oxford, p. 485).
Voges, W., Trümper, J., Pietsch, W., Reppin, C., Staubert, R. & Kendziorra, E. 1982, *MPI für extraterrestrische Physik*, preprint, Munich.
Wallace, R. K., Woosley, S. E. & Weaver, T. A. 1981, *Lick Observatory Bulletin*.
Wang, Y. M. & Frank, J. 1981, *Astron. Ap.*, **93**, 255.
Wheaton, W. A., Doty, J. P., Primini, F. A., Cooke, B. A., Dobson, C. A., Goldman, A., Hecht, M., Hoffman, J. A., Howe, S. K., Scheepmaker, A., Tsiang, E. Y., Lewin, W. H. G., Matteson, J. L., Gruber, D. E., Baity, W. A., Rothschild, R., Knight, F. K., Nolan, P. & Peterson, L. E. 1979, *Nature*, **28**, pp. 240-3.
Wassermann, I. & Salpeter, E. E. 1980, *Ap. J.*, **241**, 1107.
Woltjer, L. 1969, *Ap. J.*, **140**, 1309.
Woosley, S. E. & Wallace, R. K. 1981, *Lick Observatory Bulletin*.
Wunner, G., Ruder, H., Herold, H. 1981, *Ap. J.*, **247**, 374.
Yahel, R. Z. 1979, *Ap. J.*, **229**, L73.
Yahel, R. Z. 1980, *Ap. J.*, **236**, 911.
Zel'dovich, Y. B. & Shakura, N. I. 1969, *Sov. Astron. AJ.*, **13**, 175.

7

SS 433

Bruce Margon
Astronomy Department, University of Washington, Seattle, Washington, USA

7.1 Spectroscopy and the basic kinematic model

Petterson (1981) has remarked that SS 433 is the star that made its catalogue famous. While this statement is something of an injustice to the objective-prism spectroscopists whose patient work helped lead to the discovery of this and many other unusual objects, it does nicely illustrate the surprise with which this most unusual 14th magnitude star emerged into the scrutiny of the astronomical community. Discovered as a strong source of $H\alpha$ emission on objective-prism plates taken 20 years ago (Stephenson & Sanduleak 1977), the object resurfaced in 1978 when three independent groups called attention to its unusual properties. Seaquist *et al.* (1979) and Ryle *et al.* (1978) noted that it is an intense source of radio emission located near the center of the extended, radio-emitting supernova remnant W50, while Clark & Murdin (1978) suggested it as the optical counterpart both of this radio source and also a previously-known variable X-ray source, A1909+04 (Seward *et al.* 1976).

My own involvement with SS 433 began shortly thereafter, when a series of spectra taken by myself and colleagues showed intense, broad, optical emission lines whose profiles and, most surprisingly, wavelengths, changed drastically from night to night (Margon *et al.* 1979b). One can see from Figure 7.1, amongst the very first spectra that we obtained, that this is hardly a subtle effect! Although we initially resisted the most simple interpretation of such strong features as Doppler-shifted Balmer and He I lines (e.g., Margon 1980), further observations of spectra with multiple lines all sharing the same redshift, and a different set of (also multiple) features with identical blueshift, rapidly made all other interpretations of the 'moving' spectral features untenable (Margon *et al.* 1979a; Liebert *et al.* 1979). Figure 7.2 provides a particularly vivid example of these data. The amplitudes of the Doppler shifts of these emission lines are unprecedented, up to 50 000 km s^{-1} of redshift, and 30 000 km s^{-1} of blueshift, but yet another surprise emerged from this first year of observations: the modulation of the Doppler shifts is cyclical, with period 164 d (Margon *et al.* 1979*a*).

The earliest theoretical attempts at interpreting these phenomena turned out to be remarkably close to the still-current models. Fabian & Rees (1979) pointed out that narrowly-collimated, colinear beams ejected in opposite directions from a central object could explain some facets of the data. Milgrom (1979*a*), without knowing that Margon (1979) had detected the 164 d spectroscopic period (which was announced by Margon [1979] prior to appearance of his manuscript), suggested on the basis of the few then published data points, that the velocity variations of the emission might be periodic, possibly caused by a cyclical rotation

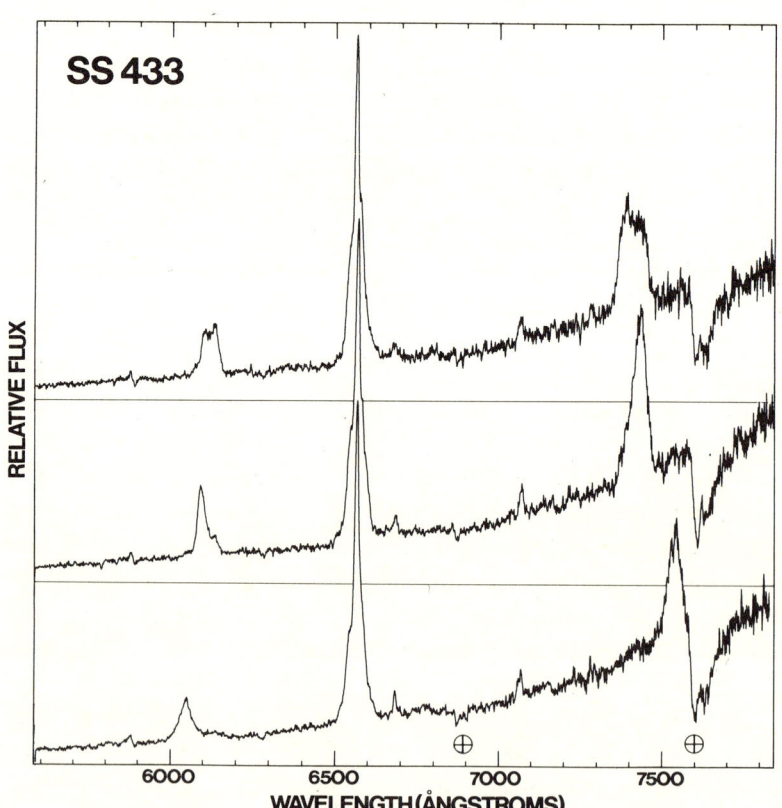

Fig. 7.1. Spectra of SS 433 obtained on three out of four consecutive nights in October 1978, using the Lick 0.9 m reflector, with resolution approximately 10 Å. The dramatic changes in the profiles and wavelengths of the Doppler-shifted Hα emission features near 6100 and 7400 Å are readily apparent. The telluric A and B absorption bands are also indicated, and weak stellar emission at He I λ5876, 6678, 7065 can also be seen (adapted from Margon *et al.* (1979*b*)).

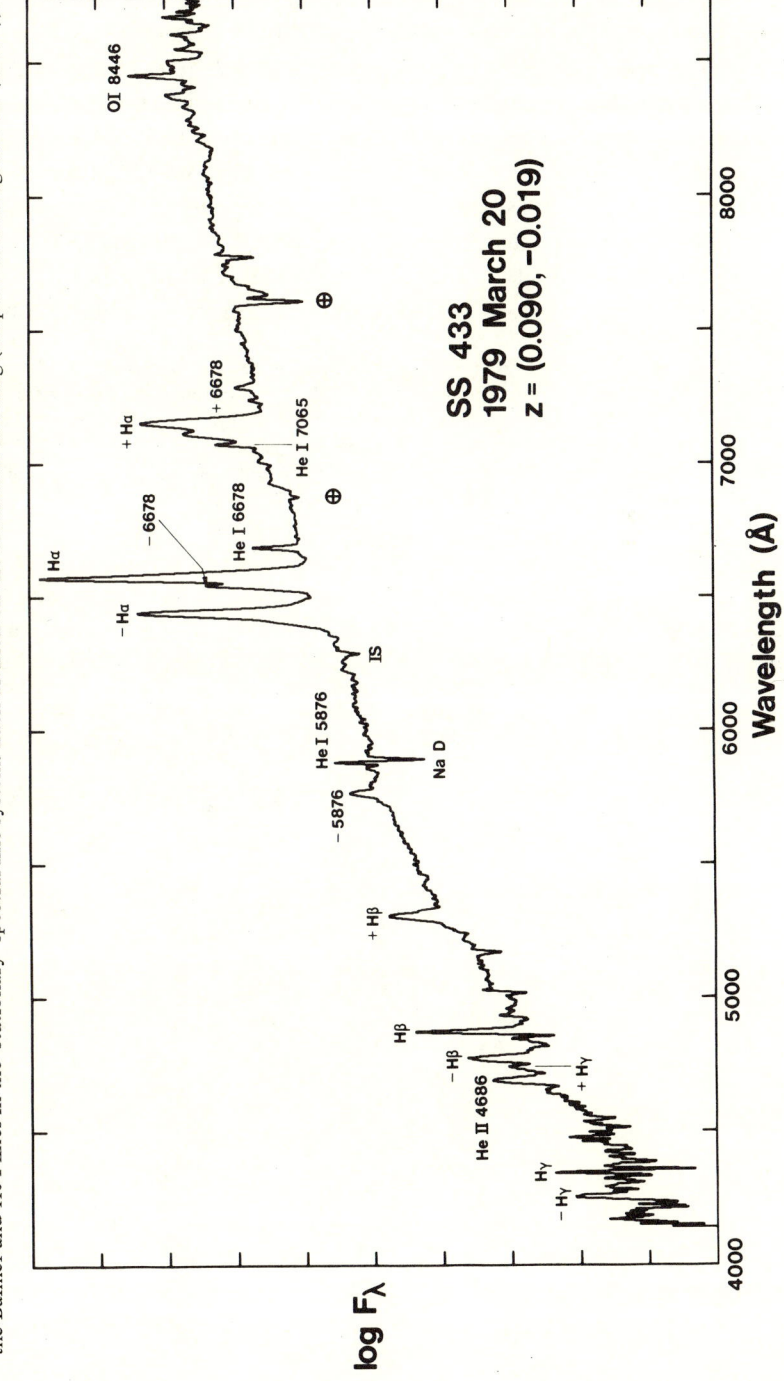

Fig. 7.2. The spectrum of SS 433 on 20 March 1979, when the pattern of multiple red- and blueshifted emission lines is particularly obvious. Emission lines prefixed with a 'plus' are redshifted features, and the 'minus' prefix denotes blueshifted features. This spectrum was obtained with the Lick 3 m Shane reflector. Note the very strong stellar absorptions of Fe II λ5169 and O I λ7773, as well as P-Cygni absorption components to the Balmer and He I lines in the 'stationary' spectral line system. Each division on the ordinate is 0.83 mag (adapted from Margon et al. (1979a)).

of the beam around an axis inclined to the line of sight. Abell & Margon (1979), who had the benefit of an extended series of spectroscopic observations by ten different University of California observers, were able to confirm and considerably elaborate Milgrom's suggestion, and solve for the values of the free parameters of what has come to be called the 'kinematic model' of SS 433.

The current status of the kinematic model is summarized in Figure 7.3 (see also Margon, Grandi & Downes 1980). Here we see data extending over almost three complete years, from June 1978 through April 1981, gathered chiefly by the author and University of California colleagues. There are almost 500 individual values of redshifts and blueshifts of the SS 433 emission lines in this data

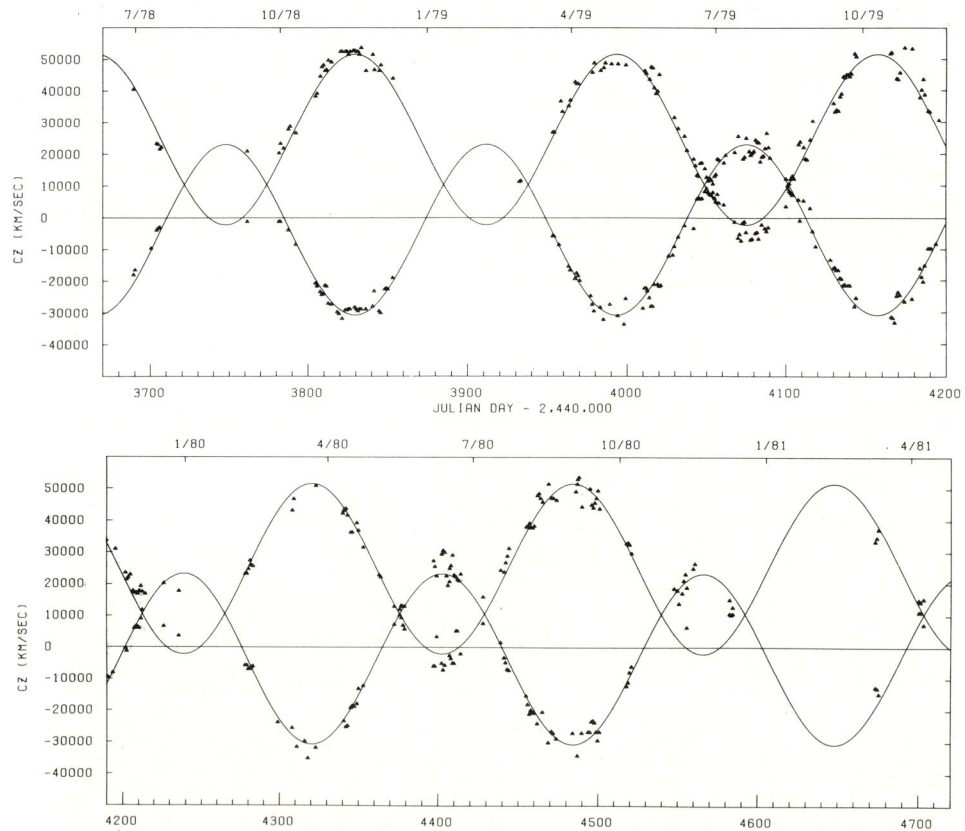

Fig. 7.3. The values of red- and blueshifts of SS 433 over an almost 3 yr period; displayed are 500 separate Doppler-shift values obtained on 300 separate nights. Most of the data have been obtained by the author and University of California colleagues, although a few points are from the literature (cited in Margon *et al.* 1980). The solid line is the best-fit to the simple kinematic model (Abell & Margon 1979), with free parameter values given in Table 7.1.

base, gathered on 292 separate nights. A simple nonlinear least-squares fit to the kinematic model yields free parameter values shown in Table 7.1, and the best-fit model with these parameters is also displayed in the figure. Some additional spectroscopic observations have been presented by Mammano *et al.* (1980), Murdin *et al.* (1980), and Ciatti *et al.* (1981), and are in agreement with the more extended data discussed here.

It is interesting to note from the table that although there are currently 400 per cent more data extending over 300 per cent of the timebase of the earliest kinematic solution (Abell & Margon 1979), the estimates of the values of the free parameters are basically identical. The systemic ejection velocity is 0.26 c, and the two inclination angles in the system (the rotation axis to the plane of the sky, and the ejection axis to the rotation axis) are 79° and 20°. The optical data alone cannot distinguish which angle is which, and indeed the ambiguity of the model goes much deeper than this: a wide variety of different physical models (infalling jets, expelled jets, disks illuminated by rotating beams) are potentially compatible with the spectroscopic data. As we shall see shortly, more recent radio observations have neatly resolved these ambiguities, and give us confidence not only that the kinematic model is correct, but that the ejected twin-jet picture is the proper physical description of the system.

It is gratifying that such a simple model fits the enormous velocity variation (80 000 km s^{-1}) so nicely. On the other hand, it is clear from Figure 7.3 that, at least in a statistical sense, the kinematic model is a terrible overall fit to the entire data base. The accuracy of an individual Doppler-shift determination is typically better than 500 km s^{-1}, and is limited not by the resolution of the observations, but rather by ambiguity in centroiding the broad, irregular spectral features. Yet the root-mean-square residual of the least-squares fit to the data of Table 7.1 is 3000 km s^{-1}, far in excess of this worst-case measurement error. Figure 7.4 is

Table 7.1. *Kinematic parameters of SS 433 as of April 1981*†

$P = 163.51$ d ± 0.13
$t_0 =$ JD 2 443 558.59 ± 0.51
$v/c = 0.260 \pm 0.002$
$\theta = 19.77° \pm 0.15$
$i = 79.00° \pm 0.15$
$va = (8.63 \pm 0.06) \times 10^{-2}$
$vb = (4.67 \pm 0.04) \times 10^{-2}$
$\gamma = 1.035\ 60 \pm 0.000\ 45$
$\alpha = 0.3409$

† Uncertainties quoted are 1σ errors; parameter notation as in Margon, Grandi & Downes (1980).

a histogram of the difference of each observation from the best-fit model prediction at that time, and clearly illustrates that there must be additional physical effects yet to be described which grossly influence the radial velocities, at a level ~10 per cent of the sinusoidal variation. We find that a stochastic jitter of the beam direction of very modest amplitude nicely describes this excess scatter. The beam-opening angle is inferred to be of order a few degrees, from the ratio of the moving line width to the beam systemic velocity. A jitter in the beam-pointing direction of this same angular scale will reproduce the observed scatter amplitude in the radial velocities quite well. To reverse the problem, it would be quite surprising if the mechanism which 'aims' this beam of velocity 0.26 c also has accuracy greatly in excess of the beam width; in retrospect, deviations of the observed amplitude are perhaps expected. Thus this hypothesized jitter in the pointing direction is a consistent, although not unique, explanation of the large observed deviations from the most simple kinematic model.

It is straightforward to argue (e.g., Abell & Margon 1979; Begelman *et al.* 1980) that because the inferred kinetic energy in the ejected beam model grossly exceeds the rotational energy stored in a compact star, the 164 d clock in SS 433 is unlikely to be simple rotation. Most authors (e.g., Katz 1980) have therefore considered precession as the most likely mechanism for rotating the jet axis;

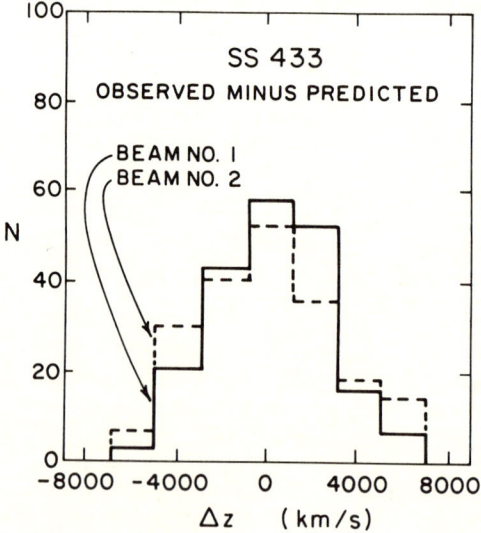

Fig. 7.4. The redshift difference between the best-fit kinematic model and the actual observations for each of the two beams. The characteristic half-width of these histograms, 8000 km s^{-1}, is 10 per cent of the full amplitude of the basic 164 d variation, and far exceeds the line measurement errors, indicating the presence of second-order kinematic effects (adapted from Margon (1981*a*)).

then of course a mechanism to drive this precession is needed. One is thus led naturally to suspect that SS 433 is a member of a close-binary system. One arrives at the same inference by noting that the total energy dissipated by SS 433 ($\sim 10^{39}$ erg s^{-1}) is not dissimilar from the radiated X-ray luminosity of the more luminous X-ray binary systems; for some reason SS 433 has chosen to convert its accretional energy to kinetic form rather than radiation (Shklovskii 1979). Firm evidence for this companion emerged when Crampton, Cowley & Hutchings (1980) discovered a 13 d small amplitude spectroscopic variation in the 'stationary' Balmer and helium emission lines, i.e. those lines at their laboratory wavelengths in the spectrum. If interpreted as orbital motion of the companion star, the velocity curve suggests a mass ratio near unity, and thus in the picture where the compact star is a neutron star, would imply a main-sequence star of spectral type near F as the companion. Such an object would be too subluminous to detect directly in the spectrum, but for reasonable parameters one expects photometric eclipses of the luminous disk (in this picture responsible for the stationary emission lines and most of the continuum) by the dark companion (e.g., Whitmire & Matese 1981). Recently, however, an orbital interpretation of the radial-velocity variations of the 'stationary' Balmer lines has been put in doubt since Crampton & Hutchings (1981) have found that the He II 4686 line shows a much larger radial-velocity amplitude and a 90° phase difference with the Balmer lines. The 195 km s^{-1} velocity amplitude of the He II 4686 line would suggest a companion mass $\geqslant 12\,M_\odot$, i.e. a high-mass-binary system, similar to the massive X-ray binaries.

7.2 Photometry and polarimetry

In a reversal of the normal order of observations of peculiar stars, spectroscopic studies of SS 433 preceded photometric work on the system. However, extensive photometric studies of this variable object, also called V1343 Aquilae, are now becoming available (e.g. Kemp *et al.* 1980; Cherepashchuk 1981; Mazeh, Leibowitz & Lahav 1981). There is a general agreement that the V-band magnitude of the object varies on timescales of days and months by up to 1 mag, but that any color variations are considerably less extreme. This amplitude of variability seems too large to be entirely due to the 'moving' spectral lines passing in and out of the photometric bandpass (although a non-negligible portion of the observed variability must be due to this cause), and therefore the evidence seems firm that the continuum does indeed vary. However, the situation regarding periodicities in this variation is at the moment more uncertain. All three of the above groups present evidence for a periodic variation with timescale almost equal to, but significantly different than, one-half of the 13 d spectroscopic period. Whether this discrepancy is profound is not yet certain. In addition the

Soviet and Israeli groups have presented evidence for a 164 d photometric variation, from both historical and modern plate material. If this can be confirmed it will be especially interesting as it will present the best evidence that the 164 d period cannot be changing on a short timescale.

One may conservatively conclude from the current photometry that deep eclipses with a 13 d period are absent. We might then be confronted with an incompatibility with the 13 d spectroscopic data, but there are a variety of resolutions available. In terms of a low-mass model the system may have a high and inverted mass ratio (Whitmire & Matese 1981) like WZ Sge, or the binary orbital plane may be misaligned with the symmetry plane of the precession (making the true orbital inclination of the system indeterminate and not necessarily equal to the inclination angle determined from the kinematic model). On the other hand, in a high-mass model, an interpretation of the photometric variations in terms of eclipses has been suggested by Cherepaschuk (1980; see also van den Heuvel 1981). Regardless of the luminosity of the primary, it is uncertain if this star is visible in the SS 433 optical spectrum, as has been suggested by Murdin, Clark & Martin (1980), van den Heuvel, Ostriker & Petterson (1980), and Hut & van den Heuvel (1981), simply because the observed spectrum does not resemble *any* type of a star. For example, accompanying the broad Balmer and He I emission lines are extremely strong Fe II $\lambda 5169$ and O I $\lambda 7773$ absorption lines, and the Fe II strengths are modulated with the 164 d period. This suggests that the entire 'stationary' spectrum originates from the accretion disk.

Recent observations of the linear polarization of SS 433 suggest interesting behavior. The object is polarized a few per cent (e.g., Liebert *et al.* 1979; Michalsky *et al.* 1980), but until recently it has been unclear that this is not all interstellar in origin. However, the polarization proves to be variable in amplitude (McLean & Tapia 1980), and this variation may be synchronized with the 164 d period (Tapia & McLean 1981). This may imply that a significant component of the optical continuum is non-thermal in origin, consistent with the above conclusion that the spectrum is not that of the primary star.

7.3 X-ray and radio observations

X-ray observations of SS 433 from the *Einstein Observatory* (Seward *et al.* 1980) have provided us with important additional clues to the nature of SS 433. Approximately 90 per cent of the X-ray flux from the region is contained in an unresolved point source coincident with the optical position of SS 433, but, perhaps more interestingly, the remaining 10 per cent originates from two collimated, colinear jets that originate at SS 433 and extend for 0.5°

quite accurately aligned with the major axis of the radio emission from W50 (Geldzahler, Pauls & Salter 1980). The X-ray data are displayed in Figure 7.5. These data provide unique confirmation of the arguments of Begelman *et al.* (1980) that the evolution of W50 has been profoundly influenced by the ejected beams; the radio object in fact may not be a supernova remnant at all. Any remaining doubts concerning the association of SS 433 and W50 are also removed by these data, an important point as most of the remaining radio sources postulated by Ryle *et al.* (1978) to be analogs of SS 433 have been shown to be chance background superpositions on supernova remnants (Margon, Downes & Gunn 1981 and references therein). Finally, the length of the observed X-ray jets and a knowledge of the ejection speed from the kinematic data give us a firm lower limit of 10^3 yr on the lifetime of SS 433 in its current stage. This number is uniquely difficult to estimate by any other technique, especially as we lack any analogous objects, and places stringent restrictions, certainly less than 1 per cent, on any possible time derivative of the 164 d period. It would be very interesting to learn of any 13 or 164 d modulation of the X-ray flux from the point source, but these data may be beyond the capability of post-*Einstein* experiments.

Perhaps the most exciting recent developments in observations of SS 433 have emerged from radio observations. The source shows extended structure, with

Fig. 7.5. An X-ray image of SS 433, obtained by the *Einstein Observatory* Imaging Proportional Counter, in the 0.5–4 keV range. The 10 per cent of the flux located in the collimated, opposing jets is readily apparent, and this axis accurately matches that of the extended radio source W50 (adapted from Seward *et al.* (1980)).

major axis of this extension aligned with the long axis of W50 and the X-ray 'jets', and this structure persists over a wide range of wavelengths and spatial scales. On scales of a few arcsec, recent studies include Spencer (1979), Hjellming & Johnston (1981a, b), and Gilmore et al. (1981); structure at the 0.1" has been explored by Niell, Lockhart & Preston (1981); and at the 10 mas level by Schilizzi et al. (1979, 1981). Most of this work has been done at 2 and 6 cm, although the extension is detected to frequencies as low as 408 MHz. All observers with long time baselines (months) find that the spatial structure changes with time, in a manner which gives us some confidence that we are directly observing the effects of the 164 d precession of ejected jets on the surrounding ambient medium. Figure 7.6 shows an example of this time-variable structure, from the work of Hjellming & Johnston (1981a). Although these maps at first glance may appear to reveal highly-complex structure and chaotic changes therein, it is amazing that a quite simple, quantitative interpretation of these data is available. The locus of the precessing jets in three dimensions can be algebraically 'unfolded' and mapped onto the two-dimensional celestial sphere, using the jet ephemeris from optical observations of the moving lines (e.g., Table 7.1). The resulting

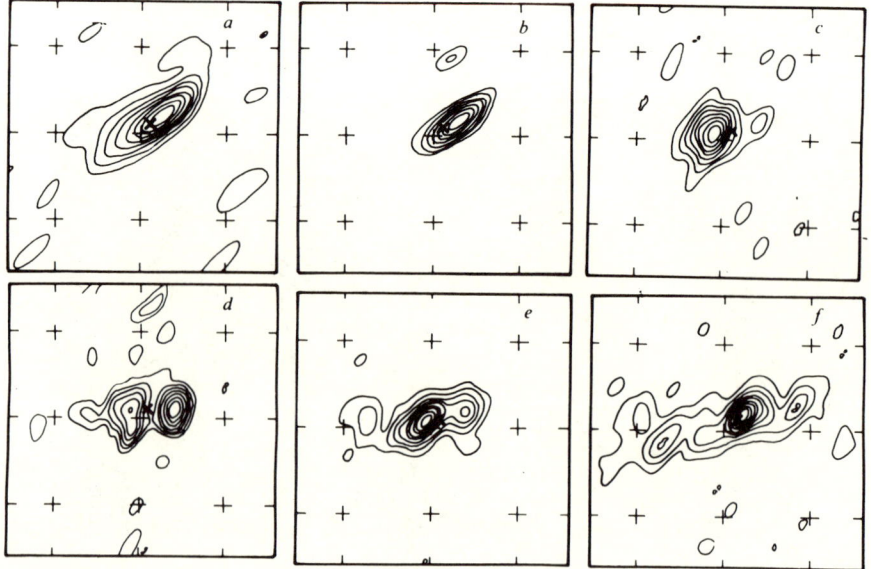

Fig. 7.6. Maps of SS 433 obtained with the Very Large Array at 4885 MHz over a 9 mth period. The intensity contours are at 10 per cent levels starting with 90 per cent, and the grid of crosses denote 2 arcsec intervals. Maps (a) through (f) were made on 8–9 September 1979; 16 September, 7 December 1979; March, 5 April, and 20 June 1980 respectively. Spatial changes during these one and one-half precession periods of the jet are obvious (adapted from Hjellming & Johnston (1981a)).

pattern is a corkscrew-like shape that agrees astonishingly well with the major features of the radio maps.

One can reverse this procedure (e.g., Hjellming & Johnston 1981*b*) and derive values for the kinematic parameters from the radio observations, entirely independent of the optical spectroscopy. One can say at the moment that these values so derived are entirely consistent with the optical observations; as the time baseline of the radio observations increases, there is no obvious reason why the radio kinematic parameters cannot equal the optical data in quality. Indeed, the radio-emitting regions may be free of the small-scale jitter which we know profoundly effects the optical radial-velocity data and, potentially, may therefore yield more accurate kinematic solutions.

These radio data are already vitally important to our picture of SS 433 for several reasons. The most basic is that they quite unambiguously point to the ejected twin-jet model as the correct interpretation of the kinematic data, an interpretation that is consistent with but not demanded by the optical data. In the case of the radio data, we are actually observing the locus of the ejected blobs as they coast ballistically away from the central source. Thus more exotic models for SS 433 involving, for example, rings and disks around massive black holes are almost surely ruled out. A second, uniquely important consequence of the radio observations is that they resolve certain ambiguities from the optical kinematic solution, and go further to calculate parameters that are unavailable from the spectroscopy. The ambiguity between the two inclination angles is removed in the radio kinematic solution; only the 80°-angle for the precession plane can be made to fit the data. Furthermore, the radio-emitting volume is large enough that we know light travel times are important, while conversely we know the optical emitting region is small enough that light travel-delays cannot be observed. The VLA maps show significant flux several arcsec away from the core, corresponding to 10^{17} cm at the source, while the optical-emitting region must be equal to or less than 1 lt d (10^{15} cm), or the motion of the moving lines would be badly smeared. Therefore light travel-delays must be included in the radio kinematic solution to obtain consistent results. This adds the third dimension to an observation of proper motion of the coasting blobs, and therefore allows the distance to SS 433 to be explicitly calculated. The resulting initial estimates of 5.1 kpc (Hjellming & Johnston 1981*a*) agree well with far more indirect and uncertain estimates from optical observations, which earlier had set lower limits to the distance of 3.5 kpc (Margon *et al.* 1979*b*). Finally, one can even infer that the sense of the beam rotation is clockwise (left-handed) from the VLA data.

7.4 Outstanding problems

The combination of the optical and radio observations of SS 433 described above leave us with an accurate measure of the kinematics of the system, and some confidence that we understand what these kinematic equations are describing: the ejection of matter through two collimated, opposing jets. Unfortunately the specific physical processes associated with this ejection are considerably more poorly understood than the kinematics. Some general physical arguments regarding conditions in the jets have been given by Begelman *et al.* (1980) and Davidson & McCray (1980), and need not be repeated here. Reviews listing much of the current theoretical work are also given by Margon (1981*a, b*), Shaham (1980*a*, 1981) and Petterson (1981). An attempt to enumerate the theoretical difficulties might proceed as follows:

(1) The data leave no obvious signature of the source of the beam's kinetic energy although, as discussed above, accretion certainly seems the most attractive choice, by analogy to the X-ray binaries.

(2) The details of the acceleration mechanism have yet to be worked out, although Milgrom (1979*b*) has pointed out that line-locking due to radiative absorptions in the sub-Lyman continuum would yield exactly the observed $v = 0.26\,c$.

(3) The collimation mechanism has yet to be worked out in detail, although presumably the accretion disk plays an important role (e.g., Katz 1980).

(4) The nature of the 164 d clock has yet to be defined. As discussed above, most workers opt for precession, but whether or not general relativistic effects are important, and what body drives the precession, is not at all clear (cf. Martin & Rees 1979; Begelman *et al.* 1980; Fabian 1980; Shaham 1980*b*; Sarazin, Begelman & Hatchett 1980; DeCampli 1980; Hut & van den Heuvel 1981).

(5) The physical conditions in the optically radiating volume are as yet obscure. Simple recombination seems appealing to explain the moving line spectrum, but selective mechanisms have also been considered. Does the continuum originate from the accretion disk or the companion star?

(6) What is the nature of the compact star? Many models have little preference for a 1 $M_\odot$ neutron star over a 1 $M_\odot$ black hole, although the analogy to X-ray binaries plus conservatism probably favors the neutron star.

(7) What is the nature of the companion star? As mentioned earlier, the simplest interpretation of the 13 d spectroscopic curve, a mass ratio unity system, has problems, and more massive companions may have to be considered.

(8) The evolutionary status of SS 433 is grossly uncertain. The lifetime of the phenomenon is probably quite short, $\lesssim 10^4$ yr just due to the large mass-loss rate, but whether this behavior is a precursor to some, most, or all traditional X-ray binaries (see Chapter 8, this book) is certainly not obvious. Observation of even one additional such object would greatly help to clarify this point, but at the moment such an analog eludes us, and there is no reason to believe that any of

the recently-suggested candidates are in fact similar to SS 433 (Margon, Downes & Gunn 1981).

Although the length of the above list of questions may seem disheartening, there can be little argument that our understanding of SS 433 has progressed enormously in the three years that the object has been under study. We have gone from a star that is 'coming and going at the same time', as described by the popular press, to an enormous set of microscopic observations, at both optical and radio wavelengths, of this most unusual astronomical phenomenon.

Acknowledgements

The author's work on SS 433 is an ongoing collaboration with S. A. Grandi and R. A. Downes of UCLA, and S. F. Anderson of the University of Washington. Spectra generously provided by the University of California astronomers enumerated in Margon *et al.* (1980) made possible much of our understanding of the kinematic model. Financial support for this work has been provided by the U.S. National Science Foundation, under grant AST 77-27745, and the Alfred P. Sloan Foundation.

Note added in proof

Several new observations have become available since the first draft of this manuscript. A 6.28 d period has become apparent in the radial velocity of the 'moving' spectral lines, of amplitude approximately 10 per cent of the 164 d period (Katz *et al.*, *Ap. J.*, **260**, 780, 1982; Mammano *et al.*, *Astron. Ap.*, in press, 1983). A 6.06 d variation previously suggested by Newsom & Collins (*A. J.*, **86**, 1250, 1981) is probably the same phenomenon, with the different period due to those authors' unique model for the system. Katz *et al.* have interpreted the periodicity as a nodding motion of the accretion disk, caused by the periodic gravitational torque of the binary companion on the disk, as the 13 d orbital period is not negligible compared with the precession period. As this interpretation invokes only Newtonian mechanics, it is quite model-independent, and thus represents interesting evidence both for the existence of the extended accretion disk, and the binary nature of the system. In addition, evidence is growing that the 164 d period is unstable, with period derivatives of both signs now detected (Anderson *et al.*, *Ap. J.*, **273**, in press, 1983). This detection of a second period derivative negates the very remarkable suggestion (Collins & Newsom, *IAU Circ.*, 3547, 1980) that the system lifetime will be only a few tens of years, due to a monotonic period change. The second period derivative is of yet uncertain analytic form, implying that it will be difficult to precisely predict values for Doppler shifts of SS 433 more than a few months in advance. Finally, photometry over a four year period by Anderson, Margon & Grandi (*Ap. J.*, **269**, 607, 1983) has removed the possible discrepancy between the spectroscopic and photometric orbital periods, with detections of (double-peaked) variations at 13.090 ± 0.007 and 6.545 ± 0.007 d, as well as strongly confirming the existence of photometric analogs to the precession variations, with periods of 162 ± 1 and 81 ± 1 d. The recent photometric and nodding motion data both favor a massive ($\sim 14\,M_\odot$) companion star.

References

Abell, G. O. & Margon, B. 1979, *Nature*, **279**, 701.
Begelman, M. C., Sarazin, C. L., Hatchett, S. P., McKee, C. F. & Arons, J. 1980, *Ap. J.*, **238**, 722.
Cherepashchuk, A. M. 1981, *MNRAS*, **194**, 761.
Ciatti, F., Mammano, A. & Vittone, A. 1981, *Astron. Ap.*, **94**, 251.
Clark, D. H. & Murdin, P. 1978, *Nature*, **276**, 44.
Crampton, D., Cowley, A. P. & Hutchings, J. B. 1980, *Ap. J. (Lett.)*, **235**, L131.
Crampton, D. & Hutchings, J. B. 1981, *Ap. J.*, **251**, 604.
Davidson, K. & McCray, R. 1980, *Ap. J.*, **241**, 1082.
DeCampli, W. M. 1980, *Ap. J.*, **242**, 306.
Fabian, A. C. & Rees, M. J. 1979, *MNRAS*, **187**, 13P.
Fabian, A. C. 1980, *MNRAS*, **192**, 11P.
Geldzahler, B. J., Pauls, T. & Salter, C. J. 1980, *Astron. Ap.*, **84**, 237.
Gilmore, W. S., Seaquist, E. R., Stocke, J. T. & Crane, P. C. 1981, *Astron. J.*, **86**, 864.
Hjellming, R. M. & Johnston, K. J. 1981a, *Nature*, **290**, 100.
Hjellming, R. M. & Johnston, K. J. 1981b, *Ap. J. (Lett.)*, **246**, L141.
Hut, P. & van den Heuvel, E. P. J. 1981, *Astron. Ap.*, **94**, 327.
Katz, J. I. 1980, *Ap. J. (Lett.)*, **236**, L127.
Kemp, J. C., Barbour, M. S., Arbabi, M., Leibowitz, E. M. & Mazeh, T. 1980, *Ap. J. (Lett.)*, **238**, L133.
Liebert, J., Angel, J. R. P., Hege, E. K., Martin, P. G. & Blair, W. P. 1979, *Nature*, **279**, 384.
Mammano, A., Ciatti, F. & Vittone, A. 1980, *Astron. Ap.*, **85**, 14.
Margon, B. 1979, *IAU Circular*, No. 3345.
Margon, B. 1980, *In:* 'Proc. Ninth Texas Symposium on Relativistic Ap.', *Ann. N.Y. Acad. Sci.*, **336**, 550.
Margon, B. 1981a, *In:* 'Proc. Tenth Texas Symposium on Relativistic Ap.', *Ann. N.Y. Acad. Sci.*, **375**, 403.
Margon, B. 1981b, *In: Galactic X-ray Sources*, eds. P. W. Sanford, P. Laskarides & J. Salton (Chichester: J. Wiley), p. 417.
Margon, B., Downes, R. A. & Gunn, J. E. 1981, *Ap. J. (Lett.)*, **249**, L1.
Margon, B., Ford, H. C., Grandi, S. A. & Stone, R. P. S. 1979a, *Ap. J. (Lett.)*, **233**, L63.
Margon, B., Ford, H. C., Katz, J. I., Kwitter, K. B., Ulrich, R. K., Stone, R. P. S. & Klemola, A. 1979b, *Ap. J.*, **230**, L41.
Margon, B., Grandi, S. A. & Downes, R. A. 1980, *Ap. J.*, **241**, 306.
Martin, P. G. & Rees, M. J. 1979, *MNRAS*, **189**, 19P.
Mazeh, T., Leibowitz, E. M. & Lahav, O. 1981, *Ap. Lett.*, **22**, 185.
McClean, I. S. & Tapia, S. 1980, *Nature*, **287**, 703.
Michalsky, J. J., Stokes, G. M., Szkody, P. & Larson, N. R. 1980, *Publ. ASP*, **92**, 654.
Milgrom, M. 1979a, *Astron. Ap.*, **76**, L3.
Milgrom, M. 1979b, *Astron. Ap.*, **78**, L9.
Murdin, P., Clark, D. H. & Martin, P. G. 1980, *MNRAS*, **193**, 135.

Niell, A. E., Lockhart, T. G. & Preston, R. A. 1981, *Ap. J.*, **250**, 248.
Petterson, J. A. 1981, *Adv. Space Res.*, **1**, 13, 49.
Ryle, M., Caswell, J. L., Hine, G. & Shakeshaft, J. 1978, *Nature*, **276**, 571.
Sarazin, C. L., Begelman, M. C. & Hatchett, S. P. 1980, *Ap. J. (Lett.)*, **238**, L129.
Schilizzi, R. T., Norman, C. A., van Breugel, W. & Hummel, E. 1979, *Astron. Ap.*, **79**, L26.
Schilizzi, R. T., Miley, G. K., Romney, J. D. & Spencer, R. E. 1981, *Nature*, **290**, 318.
Seaquist, E. R., Garrison, R. F., Gregory, P. C., Taylor, A. R. & Crane, P. C. 1979, *Astron. J.*, **84**, 1037.
Seward, F. D., Page, C. G., Turner, M. J. L. & Pounds, K. A. 1976, *MNRAS*, **175**, 39P.
Seward, F., Grindlay, J., Seaquist, E. & Gilmore, W. 1980, *Nature*, **287**, 806.
Shaham, J. 1980a, *Com. on Ap.*, **9**, 1.
Shaham, J. 1980b, *Ap. Lett.*, **20**, 115.
Shaham, J. 1981, *Vistas in Astronomy*, **25**, 217.
Shklovskii, I. S. 1979, *Pis'ma Astron. Zh.*, **5**, 644.
Spencer, R. E. 1979, *Nature*, **282**, 483.
Stephenson, C. B. & Sanduleak, N. 1977, *Ap. J. Suppl.*, **33**, 459.
Tapia, S. & McClean, I. S. 1981, in preparation.
van den Heuvel, E. P. J., Ostriker, J. P. & Petterson, J. A. 1980, *Astron. Ap.*, **81**, L7.
van den Heuvel, E. P. J. 1981, *Vistas in Astronomy*, **25**, 95.
Whitmire, D. P. & Matese, J. J. 1981, *MNRAS*, **194**, 293.

8

FORMATION AND EVOLUTION OF X-RAY BINARIES

Edward P. J. van den Heuvel

Astronomical Institute, University of Amsterdam, Roetersstraat 15, 1018 WB Amsterdam, The Netherlands

8.1   Introduction and summary

The existence of X-ray binaries and binary radio pulsars shows that binary systems can, under certain circumstances, survive the supernova (SN) explosion of one of the components. In the case of massive X-ray binaries this survival is a clear consequence of the large-scale *mass transfer* which precedes the supernova explosion of the initially more-massive component of the system, such that at the moment of the explosion this star has become the less massive of the two. Explosive mass ejection from the less-massive component will in general not lead to disruption of the system (Blaauw 1961; Boersma 1961) unless the effects of impact and ablation are very large (see Wheeler, Lecar & McKee 1975). In the case of low-mass X-ray binaries it is much more difficult to see why the systems were not disrupted. Here, however, one can show semi-empirically that - in contrast to the case of the massive X-binaries - the formation of such a system is an extremely rare event and therefore must require a very exceptional evolutionary history (see Section 8.2.4).

In this chapter we explore how the various types of binaries that contain compact objects may have formed. In Section 8.2 we first summarize the observed characteristics of the various types of these binaries (for details about the observations of X-ray binaries we refer to the reviews in this book by Rappaport & Joss, Lewin & Joss, and van Paradijs). In Sections 8.3 through 8.5 we review the evolution of close binaries - with mass transfer and mass loss - and discuss the various possible final states of the systems. In Section 8.6 we discuss the possible formation mechanisms of globular cluster sources and bulge sources.

8.2   Observational data on compact objects in binary systems
8.2.1   X-ray binaries

The binary X-ray sources that contain compact objects can roughly be divided into two groups, the massive ones ($M_s \gtrsim 10 \, M_\odot$) and low-mass ones

($M_s \leq 2 M_\odot$), of which each can be subdivided further into several subclasses (M_s is the mass of the non-degenerate companion star), as follows.

(i) *Massive X-ray binaries.* Two broad groups can be distinguished (see Table 8.1), which differ in a number of physical characteristics as outlined in Figure 8.1. The 'standard' systems such as Cen X-3, SMC X-1 are strong sources, characterized by the occurrence of regular X-ray eclipses and by double-wave ellipsoidal light variations produced by tidally-deformed giants or supergiant companion stars that (nearly) fill their critical potential lobes. With one exception (4U 1223−62) their binary periods are between 1.4 and 10 d. The optical luminosities and spectral types of the companions indicate original main-sequence masses of $\gtrsim 20 M_\odot$, corresponding to O-type progenitors.

On the other hand, in the Be-X-ray binaries, first recognized as a group by Maraschi, Treves & van den Heuvel (1976), the optical companions are rapidly-rotating B-emission stars belonging to the main sequence (luminosity Class III-V). At present 11 such systems are known. The Be stars are deep inside their Roche lobes as is indicated by the generally long binary periods ($\gtrsim 15$ d) and by the absence of X-ray eclipses and of regular ellipsoidal light variations. According to their luminosities and spectral types the companion stars have masses in the range 10-20 $M_\odot$ (spectral types O9-B3, III, IV, V). The X-ray emission from the

Table 8.1. *The standard 'massive' X-ray binaries (top) and some examples of Be-X-ray binaries (bottom)*

Source	Optical counterpart	Spectral type	Pulse period (s)	Orbital period (d)	Eccentricity
LMC X−4	Sk−Ph	O7 III−V	13.5	1.408 d	$e = 0.00$
Cen X-3	Krz's star	O6.5 II−III	4.84	2.087	$e = 0.00$
4U 1700−37	HD 153919	O6.5f	−	3.412	−
SMC X-1	Sk 160	B0Ib	0.714	3.89	$e = 0.00$
4U 1538−52	#12	B0Iab	529	3.73	$e = 0.00$
Cyg X-1	HD 226868	O9.7Iab	−	5.60	−
4U 0900−40	HD 77581	B0.5Ib	283	8.965	$e = 0.09$
GX 301−2	Wra 977	B1.5Ia	696	41.4	$e = 0.47$
4U 0115+63	Johns' star	Be	3.61	24.3 d	$e = 0.35$ transient
4U 0352+30	X Per	O9.5 (III−V)e	835	580 d	Very weak steady
A 0535+26	HD 245770	B0Ve	104	≥ 18	transient
4U 1145−61	Hen 715	B1Vne	292	≥ 3.5	Highly variable
4U 1258−61	MMV star	B2Vne	272	≥ 13	Highly variable

Be X-ray systems tends to be extremely variable, ranging from complete absence to large transient outbursts. The X-ray outbursts are most probably related to the irregular optical outbursts generally observed in Be stars, which indicate sudden bursts of mass ejection, presumably due to rotation-driven instability in the equatorial regions of these stars. For a further discussion of the characteristics of the two groups of massive X-ray binaries we refer to Chapter 1, this book, and Rappaport & van den Heuvel (1982).

Fig. 8.1. Schematic of a 'standard' massive X-ray binary vs. a Be–X-ray binary system.

STANDARD MASSIVE X-RAY BINARY

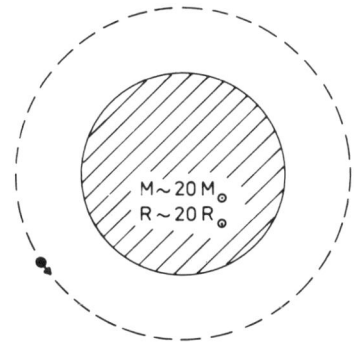

- Companion evolved to fill critical potential lobe
- $2^d \lesssim P_{orb} \lesssim 10^d$
- Circular Orbit
- Eclipses likely
- "Steady" X-ray emission
- Mass transfer(10^{-10}–$10^{-8}\,M_\odot\,yr^{-1}$) by Roche lobe overflow, stellar wind or both

$M \sim 20\,M_\odot$
$R \sim 20\,R_\odot$

Be-STAR X-RAY BINARY

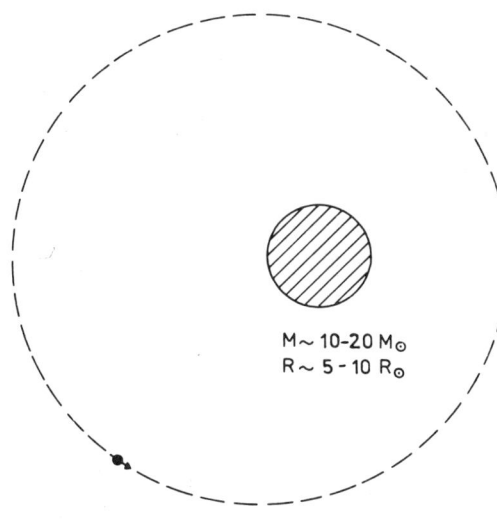

- Companion unevolved, underfills critical potential lobe
- $P_{orb} \gtrsim 20^d$
- Eccentric orbit
- Eclipses rare
- Transient X-ray emission
- Mass transfer(10^{-10}–$10^{-13}\,M_\odot\,yr^{-1}$) by episodic ejection of matter in the orbital plane and/or steady stellar wind

$M \sim 10$–$20\,M_\odot$
$R \sim 5$–$10\,R_\odot$

(iii) *Low-mass X-ray binaries.* Only a few of these systems show direct evidence of binary character. These are listed in Table 5.4 of Chapter 5 (see also Chapter 2, Section 2.4.2.7.3). Two low-mass X-ray binaries contain pulsars (Her X-1 and 4U 1626−67) and have hard X-ray spectra, like the massive X-ray binaries. The others have somewhat softer X-ray spectra and in only a few of them the light of an optical companion star can be detected. In the vast majority of these so-called 'Type II' sources (which make up more than half of all strong galactic sources) no trace of an optical companion is visible, and the optical spectrum is that of a hot accretion disk. Their large X-ray fluxes and the fact that many of them are bursters leave little doubt that these sources are neutron stars; the recent discovery of regular periodicities in some of them and the discovery of red-dwarf companions in some systems during an X-ray low state strongly suggests that the Type II sources are in close binaries with a low-mass ($\leqslant 1$ M$_\odot$) companion star (see Chapters 2 and 5).

There are at least two population types among the low-mass X-ray binaries as illustrated in Figure 8.2, i.e. (1) Her X-1 is a young object, definitely belonging to stellar Population I, since its ~ 2 M$_\odot$ companion cannot be older than $\sim 5 \cdot 10^8$ yr. The location of Her X-1 at 3 kpc from the galactic plane is very unusual for a Population I object and indicates that it has been shot out of the galactic plane with a velocity $\gtrsim 125$ km s^{-1} (Sutantyo 1975a; Bouwman 1980). This is clearly the result of the explosive mass ejection in the supernova that created the neutron star. The same may be true for Cyg X-2. (2) The bulge sources and globular cluster sources belong to the oldest stellar population in the Galaxy (ages $\sim 5 \cdot 10^9 - \gtrsim 10^{10}$ yr). These do not show runaway characteristics since most bulge sources and bursters are within 0.6 kpc from the galactic plane; also, a runaway velocity $\gtrsim 10$ km s^{-1} would have caused the globular-cluster sources to escape from their clusters.

(iii) *Peculiar systems.* The strongly radio-emitting, peculiar X-ray binaries Cyg X-3, Cir X-1 and SS 433 might form a separate category (Table 8.2). They are characterized by occasional, strong radio outbursts with a synchrotron spectrum, and large IR luminosities. The 3.49 s X-ray pulsar 1E 2259+586 may belong to the same category (Fahlman & Gregory 1981). SS 433 (Chapter 7) and 1E 2259+586 are surrounded by large radio-X-ray shells similar in appearance to supernova remnants, and show beam-like structures, extending from the central source out to the shell.

8.2.2 Binary radio pulsars

Three binary radio pulsars are known, listed in Table 8.3. The best-studied one, PSR 1913+16, probably consists of two neutron stars (Taylor & Weisberg 1982; Srinivasan & van den Heuvel 1982). Its large orbital eccentricity

Table 8.2. *Peculiar sources with strong radio and IR emission*

Source	Sp. type	P_{orb}	L_x/L_{opt}	z	Ref.
Cir X-1	O–Be	16.6 d	10^{-2}	In gal. plane	—
SS 433	—	13.1 d	10^{-3}	130 pc	Crampton & Hutchings (1981) Chapter 7, this book
Cyg X-3	—	4.8 h	?	In gal. plane	

Table 8.3. *Binary radio pulsars*

Name	P_{pulse}	P_{orb}	e	Type of companion	Ref.
PSR 0656+64	0.196 s	24 h 41 min	0.00	White dwarf ?	Damahsek *et al.* (1982)
PSR 0820+02	0.865 s	3.2 yr	0.00	$M \sim 0.85 M_\odot$ (?)	Manchester *et al.* (1980)
PSR 1913+16	0.059 s	7 h 45 min	0.62	Compact star	Taylor *et al.* (1976)

Fig. 8.2. Schematic of the two categories of low-mass X-ray binaries: 'Her X-1'-type and 'old-population'-type systems.

YOUNG-POPULATION LOW-MASS X-RAY BINARY (Her X-1-type)

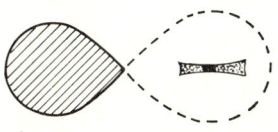

$M \sim M_{neutron\ star}$

- $P_{orb} \sim$ 0.2 to $\sim$ 10 days
- Optical light from companion observable
- Eclipses possible
- Runaway velocity 50-200 km/s
- Age < few billion years

OLD POPULATION LOW-MASS X-RAY BINARY (type II)

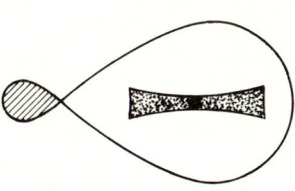

$M \ll M_{neutron\ star}$

- $P_{orb} \sim$ hours
- Optical light dominated by disk
- Eclipses unlikely
- No clear runaways
- Age $\geqslant 5 \times 10^9$ yrs ?

is clear evidence of large mass ejection in the supernova that produced the youngest neutron star in the system. Nothing is known with certainty about the nature of the companions in the other two systems (see Damashek *et al.* 1982).

8.2.3 Do all X-ray binaries contain neutron stars? The case of Cyg X-1

With the sole exception of Cyg X-1, the massive X-ray binaries with known orbits are either pulsars or have optical mass functions low enough to fit a neutron star. This also holds for SS 433 (Crampton & Hutchings 1981). The same is true for the low-mass X-ray binaries and for the bulge and globular cluster sources; for these the burster character gives strong and independent evidence that we are dealing with neutron stars (see Chapter 2, this book). For the case of Cyg X-1 it still remains possible to construct plausible models in which this object is a neutron star — either by considering triple-star models or by constructing low-mass models for the optical star (see van den Heuvel & Ostriker 1973; Bahcall 1978). In any event, since after more than a decade of observing X-ray sources Cyg X-1 remains the only plausible black hole candidate, the formation of a stellar-mass black hole must be an extremely rare event. In Section 8.4.11 a possible explanation for this is suggested.

8.2.4 Rough relative formation rates of massive and low-mass X-ray binaries

The numbers of massive ('Type I') and low-mass ('Type II') X-ray systems in the Galaxy are roughly the same (this also holds in M31, see Chapter 3, this book). However, their expected lifetimes are very different, i.e. $\sim 10^5$ yr and 10^8–10^9 yr respectively (see Chapter 9, this book).

This implies that the galactic *formation rate* of massive X-ray systems is some 10^3–10^4 times larger than the formation rate of low-mass systems. This holds even more for the relative formation rates (i.e. among stars in the same mass range): in view of the shape of the initial mass function, the formation rate of low-mass stars ($M \sim 1$–$2\,M_\odot$) is some 10^2–10^3 times larger than that of massive stars ($M \geqslant 15\,M_\odot$), i.e.: the *fraction* of X-ray binaries formed (per unit time) among massive stars is some 10^5–10^7 times larger than among low-mass stars. If anything, this shows that, while among massive stars the formation of a massive X-ray binary is a relatively normal event, the formation of a low-mass X-ray binary is an *extremely rare event* and, consequently, must require very special conditions.

8.3 Evolution of close binaries and the formation of compact objects
8.3.1 Introduction: the 'scenario' concept

Much progress has been made in the calculation of the evolution of close-binary systems (see the reviews by Paczynski 1971 and Thomas 1977; and in *IAU Symposia*, Nos. 73, 83, 88, 93 and *IAU Colloquia*, Nos. 42, 46, 53). In such calculations it is often assumed that during phases of mass transfer the total mass and total orbital angular momentum of the system are conserved; this is so-called 'conservative' evolution.

It appears, however, that conservative evolution will occur only under certain restricted conditions (see Sections 8.4.3 and 8.5) and that, in general, mass and angular momentum may be lost from the systems during several stages of the evolution.

The precise amounts of mass and angular momentum lost during various evolutionary phases are poorly known, except in some clear cases such as supernova mass ejection. For this reason it is, at present, not possible to calculate precisely all the steps in the evolutionary history of a close binary up to the final stages.

Nevertheless, by combining the results of precise calculations of evolution with mass transfer, with calculations of the advanced evolutionary stages of single stars, and by making reasonable assumptions about the amounts of mass and angular momentum lost during certain critical phases (such as the supernova event), one may construct plausible 'scenarios' for the evolutionary origin of binary X-ray sources. In a good scenario one makes as few *ad hoc* assumptions as possible, and tries to rigorously calculate as many steps in the evolutionary sequence as possible. The final goal is to find a solid physical justification for each of the assumptions (see Kippenhahn & Thomas 1980; Tutukov 1981).

It appears that, for the massive X-ray binaries, where in most cases only a moderate fraction of the total mass and total orbital angular momentum were lost during the evolution, 'quasi-conservative' scenarios can be used which involve relatively few assumptions (see van den Heuvel 1974, 1976, 1977, 1978). On the other hand, for the low-mass X-ray binaries where a very large fraction ($\geqslant 90$ per cent) of the total original mass and orbital angular momentum may have been lost during the evolution, the uncertainties are much larger (the same holds for the origin of their close relatives, the cataclysmic-variable binaries). Still, here also progress has been made, due to the introduction of the concept of common-envelope evolution (Paczynski 1976; Ritter 1976; Ostriker 1976).

8.3.2 Types of close-binary evolution and final states of the primary star

In binaries with periods up to about ten years, the envelope of the primary star may, at some stage of the evolution, overflow a critical surface (Roche lobe or tidal lobe) and be lost to the companion star or from the system. Such binaries, in which the stars interact during some stage of their evolution, we will call 'close'. The way in which the two stars interact depends on the evolutionary state of the core of the primary star at the onset of the mass transfer, on the structure of the envelope of this star at that moment, and also on the mass ratio $q = M_2/M_1$ of the components. The classification, by Kippenhahn & Weigert (1967), in terms of the evolutionary state of the core at the onset of the mass transfer, is particularly useful if one wishes to study the possible final evolutionary state of the primary star, i.e. the kind of remnant that will be left. We will in this section concentrate on this problem. On the other hand, if one wishes to know whether or not the system will be disrupted by the supernova of the primary star, one should know how much mass is captured by the other star, and how the orbital period is affected by the mass transfer. These factors will be discussed in Sections 8.4 and 8.5.

Figure 8.3 illustrates the cases A, B and C of close-binary evolution, as defined by Kippenhahn & Weigert. For example, consider a binary system with a primary mass of 9 $M_\odot$ and a secondary of 4.5 $M_\odot$. If the binary period is less than 0.652 d the primary star already overflows its Roche lobe when it is on the zero age main sequence (ZAMS); in order to fill its Roche lobe before the end of core-hydrogen burning, the orbital period should be less than 1.90 d. Such binaries are said to undergo case A of mass transfer when the primary star fills its Roche lobe. If the binary period is longer than 1.90 d but shorter than 394 d the primary star will fill its Roche lobe after the end of core-hydrogen burning but before core-helium ignition. In this case one speaks of case B. If the orbital period is longer than 394 d but shorter than about 800 d the primary will fill its Roche lobe after helium ignition but before carbon ignition; this is case C. The precise values of the limiting periods for the onset of the cases A, B or C depend mainly on the mass of the primary star, and only slightly on the mass ratio q and the chemical composition. Figure 8.4 displays the critical binary periods for the onset of cases A, B and C as a function of the primary mass, for $q = 0.5$ (and $X = 0.70$, $Z = 0.02$ after Webbink 1979). For $q = 1$ they are larger by a factor of 1.0334.

Since cases B and C evolution occur over a very wide range in binary periods, these are by far the most common types of evolution in nature. The fraction of

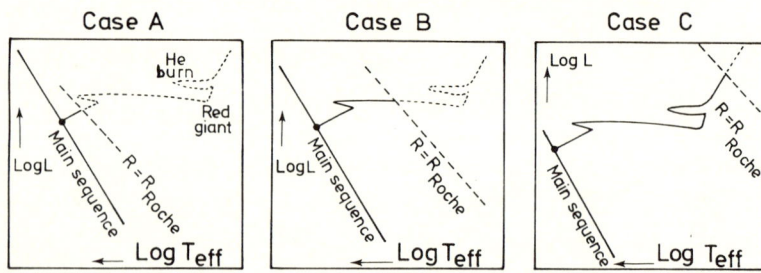

Fig. 8.3. The three basic cases A, B and C of mass exchange in a close-binary system, as defined by Kippenhahn & Weigert (1967), illustrated on the evolutionary track of the primary star in the Hertzsprung-Russell diagram.

the systems that evolve according to case A is relatively small: for $M_1 \lesssim 10\,M_\odot$, less than 10 per cent; only among the massive O-type stars this percentage may be as high as 25 per cent (see Garmany, Conti & Massey 1980).

In cases B and C, the evolution of the primary star is little affected by whether or not the mass transfer is conservative. It appears that here in most cases as soon as the primary star begins to overflow its critical lobe most of the hydrogen-rich envelope, which may contain as much as 80 per cent of the stellar mass, is lost on a thermal (Kelvin–Helmholtz) – or even shorter – timescale, either to the secondary, or partly from the system (see Massevitch, Tutukov & Yungelson 1976; van Beveren 1980, 1982). The Kelvin–Helmholtz timescale is roughly given by (see van den Heuvel 1978):

$$\tau_{KH} = GM_1^2/R_1 L_1 \approx 3 \times 10^7 (M/M_\odot)^{-2} \text{ yr}, \tag{8.1}$$

Fig. 8.4. Types of close-binary evolution as a function of primary mass and orbital period, for binaries with initial mass ratio $q = 0.5$ and $X = 0.70$, $Z = 0.03$ (partly after Webbink 1979). (The orbital periods correspond to binaries in which the primary star just fills its Roche lobe.) The cases A, B and C are defined in Figure 8.3 and in the text. Above the convective boundary line the primary stars have convective envelopes at the onset of the Roche-lobe overflow. This leads to mass transfer on a dynamical timescale (Paczynski & Sienkiewicz, 1972) which presumably causes the formation of a common envelope in which the cores of the two stars will spiral-in on a very short timescale (Meyer & Meyer-Hoffmeister 1979; Webbink 1979).

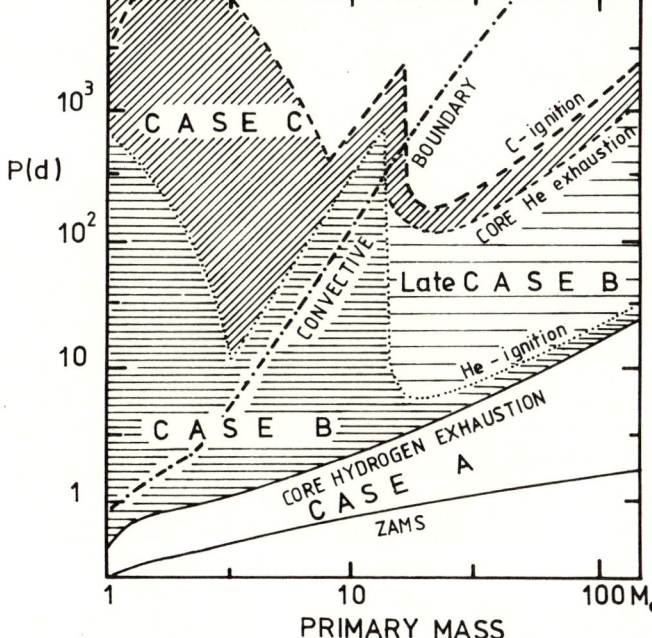

After the transfer, practically only the core of the primary star, consisting of helium (and in case C, of heavier elements) remains. The further evolution of the primary stars can therefore relatively simply be described in terms of the evolution of the helium core. In view of this, and since cases B and C are by far the most common cases of binary evolution in nature, we will, in what follows, only concentrate on these cases (the outcome of 'late' case A evolution, in which the mass transfer starts when the central fractional hydrogen content of the primary star is $\leqslant 0.1$ is the same as in case B, see Massevitch, Tutukov & Yungelson 1976). We will neglect the possible presence of a small ($\lesssim M_\odot$) hydrogen-rich envelope around the helium star, as this does not affect its further evolution. Calculations of advanced evolution of helium stars were, for example, carried out by Arnett (1978), Sugimoto & Nomoto (1980), Miyaji *et al.* (1980) and Delgado & Thomas (1981). The results show that helium stars with masses up to about $3.0\,M_\odot$ may become subgiant or giant stars during some phase in helium-shell burning. In a binary system such stars may therefore lose part (or all) of their envelope.

The expected final evolution of helium stars of various masses, single as well as in binaries, can be roughly summarized as follows (see Table 8.4, and Nomoto 1981):

(1) $M_{He} \leqslant 2\,M_\odot$: the CO core degenerates during helium-shell burning and the outer layers rapidly expand. When C-burning ignites, the entire core is incinerated (carbon deflagration, see Mazurek & Wheeler 1980) and probably explodes as a supernova. Arnett (1979) has suggested that this is a Type I supernova. Probably no remnant is left. In a binary system, this carbon deflagration is avoided since the giant's envelope is transferred to the companion and only a CO white dwarf remains.

(2) $2\,M_\odot \leqslant M_{He} \leqslant 2.8$–$3.0\,M_\odot$: the CO core left by helium burning is not highly degenerate and quietly ignites carbon burning, which leaves behind a degenerate O–Ne–Mg core with a mass in the range 1.2–$1.4\,M_\odot$. Helium-shell burning around this core gradually increases its mass and makes the envelope expand to giant size. In a binary, the envelope overflows the Roche lobe and is lost. A $1.2\,M_\odot$–$1.4\,M_\odot$ O–Ne–Mg white dwarf is left behind.

In a single star, the O–Ne–Mg core grows by helium-shell burning until it approaches the Chandrasekhar limit, and the central density becomes so high that e-captures on ^{24}Mg set in and make the core collapse (Miyaji *et al.* 1980). This e-capture supernova will lead to the formation of a neutron star.

(3) $M_{He} \geqslant 2.8$–$3.0\,M_\odot$: the O–Ne–Mg core left by carbon burning is larger than the Chandrasekhar limit and evolves through all stages of nuclear burning up to the formation of an Fe core. This core finally undergoes a photodisintegration collapse (Fe-photodisintegration supernova), leaving behind a neutron star

(or possibly, a black hole). The limit of about 2.8 $M_\odot$ holds for single stars or wide binaries ($P \gtrsim 30$-100 d). In short-period binaries, where the core has less time to grow by helium-shell burning before the envelope is lost, the limit is increased to about 3.0 $M_\odot$. (The precise values of these limiting masses still depend on the initial composition, on assumptions about convection and possibly also on rotation; see footnote †.)

> † Habets (1983) recently obtained boundary masses for the various types of final evolution of helium stars which differ considerably from the above-given values. Treating convection with a diffusion approximated and implicitly solving for the composition variables, he found that in helium stars with masses as low as 2.3 $M_\odot$ the CO core still grows to the Chandrasekhar mass. According to these results, in a binary, the mass range for leaving an O–Ne–Mg white dwarf would be very narrow, i.e. 2.0–2.3 $M_\odot$.

Table 8.4. *Rough summary of expected final evolution of helium stars* (see Nomoto 1981; see also, however, Habets 1983)

M_{He}	Single	In a binary
$\leqslant 2\,M_\odot$	– growing degenerate CO core – envelope expands to giant size – C-ignition $\Rightarrow$ C-deflagration $\Rightarrow$ no remnant?	– envelope lost during He-shell burning $\Rightarrow$ CO white dwarf
2–2.8(3.0) $M_\odot$	– non degenerate C ignition – formation of growing degenerate O–Ne–Mg core $\Rightarrow$ e-capture SN $\Rightarrow$ neutron star	– envelope lost during He-shell burning $\Rightarrow$ O–Ne–Mg white dwarf
2.8(3.0)–~3.0 $M_\odot$	– CO core larger than Chandrasekhar-mass – envelope expands considerably – evolution to Fe-photodisintegration $\Downarrow$ neutron star	part (or all) of envelope lost, but core evolves as in a single star $\Downarrow$ neutron star
~3.0 $M_\odot$–60 $M_\odot$	same as for 2.8–3.0 $M_\odot$, but no significant envelope expansion	no important mass loss before core collapse
$M > 60\,M_\odot$	core evolves to pair-creation collapse: no remnant?	no remnant?

In binaries, helium stars with $M \lesssim 3.0\,M_\odot$ may still lose a considerable part of their envelope by mass transfer, before the core collapses; such stars with masses just above the lower limit for the photodisintegration (between 2.8 and $3.0\,M_\odot$) may, in binaries, be reduced to almost bare cores of about a Chandrasekhar mass at the moment of the photodisintegration collapse. In such a case, the mass of the ejected supernova shell will be negligible. This is, of course very favourable if one wishes to avoid disruption of the system (see Section 8.5).

(4) $M_{He} \gtrsim 60\,M_\odot$: here the core evolves to a pair-creation collapse and probably no remnant is left. Since a helium star of $60\,M_\odot$ corresponds to a main-sequence mass of about 80–$100\,M_\odot$, and since stars with $M \gtrsim 50\,M_\odot$ are extremely rare, this case is of little practical importance.

8.3.3 Relation between helium-core mass and initial primary mass

Numerical calculations show that the mass of the remnant of the primary star after case B (and late case A) mass transfer can be well represented by

$$M_{1f} = 0.1\,(M_1/M_\odot)^{1.4} \tag{8.2}$$

for $X = 0.602, Z = 0.044$ (Massevitch, Tutukov & Yungelson 1976) and by

$$M_{1f} = 0.073\,(M_1/M_\odot)^{1.42} \tag{8.3}$$

for $X = 0.700, Z = 0.030$ (see van Beveren 1980), where M_i is the initial mass of the primary star. The remnant mass depends hardly at all on the initial binary period and mass ratio, nor does it depend on either the adopted criterion for semi-convection (see Massevitch, Tutukov & Yungelson 1976) or on Z (the results for $X = 0.70, Z = 0.02$ and $X = 0.70, Z = 0.03$ hardly differ). In case C the core still grows considerably by H-shell burning during the core-helium-burning stage: the works of Weaver & Woosley (1980) and of Lamb, Iben & Howard (1976) show that in a $15\,M_\odot$ star the core mass grows from 3 to $4.5\,M_\odot$ during this phase. Hence, for the same primary mass there is a jump in core mass between cases B and C; similarly, for the same core mass there is a jump in initial primary mass.

Taking these considerations into account, we have converted the helium-star masses from the preceding paragraph into initial primary masses. Figure 8.5 schematically depicts the results. For a primary star in a binary, the heavily-dashed, nearly-vertical line with a jump at the He-ignition line is the expected lower-mass limit for terminating the evolution with an Fe-photodisintegration collapse (i.e. $M_{He} \approx 2.8$–$3.0\,M_\odot$). Primary stars to the right of this line directly evolve towards a supernova and leave a neutron star, regardless of whether or not part (or all) of the layers outside the core are lost to the companion.

The solid, nearly-vertical line corresponds to primaries that leave a 2 $M_\odot$ helium star. To the left of this line, the helium cores that remain after case B mass transfer will undergo a second mass-transfer phase (case BB) during helium-shell burning and leave a CO white dwarf. Between this line and the heavily-dashed line helium stars lose their envelopes by case BB mass transfer in most cases after C ignition and leave O–Ne–Mg white dwarfs (see, however, the footnote on page 313).

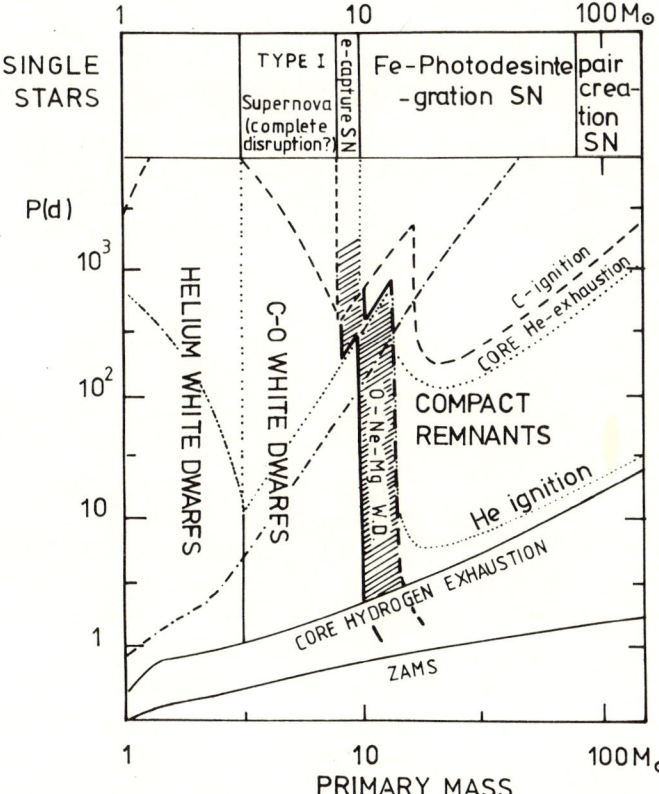

Fig. 8.5. Classification of expected final evolutionary states of primary stars of close binaries as a function of primary mass and initial orbital period (for mass ratio 0.5 and $X = 0.70$). At the top of the figure the expected final evolution of single stars is indicated. In a subsequent phase of reversed mass-transfer CO white dwarfs may be triggered to explode as a Type I supernova (complete disruption, see Nomoto 1981; Arnett 1979); old CO white dwarfs with a mass close to the Chandrasekhar limit, and O–Ne–Mg white dwarfs may be triggered by accretion to implode to a neutron star (see text). (See also the note added above Table 8.4.)

8.3.4 Difference between final evolution of single stars and of primaries of close binaries

At the top of Figure 8.5 are indicated the approximate boundaries of the expected final evolution of single stars (or components of very wide binaries). A 2 $M_\odot$ core is formed at the end of the evolution in stars of about 8 ± 1 $M_\odot$; a 2.8 $M_\odot$ core in stars of about 10 ± 2 $M_\odot$. The main difference between the evolution of single stars and of primaries of close binaries are (for $X = 0.70$, $Z = 0.02$–0.03):

(1) for $4 \leqslant M_1 \leqslant 8\, (\pm 1)$ $M_\odot$ the primaries of close binaries leave CO white dwarfs instead of evolving to C-deflagration.

(2) for $8\,(\pm 1)\,M_\odot \leqslant M_1 \leqslant \begin{matrix} 10\,(\pm 2)\,M_\odot \text{ (case C)} \\ 14\,(\pm 2)\,M_\odot \text{ (case B)} \end{matrix}$

primaries of close binaries leave O–Ne–Mg white dwarfs instead of evolving to e-capture collapse and leaving neutron stars.

(3) for case C systems the lower mass limit for evolving to a photodisintegration collapse is about the same as for single stars, i.e. $\sim 10\,(\pm 2)\,M_\odot$; for case B systems it is about $14\,(\pm 2)\,M_\odot$.

8.4 (Quasi-) conservative evolution: formation of the massive X-ray binaries

8.4.1 Introduction

We consider primary stars which directly evolve to core collapse ($M_{core} \gtrsim 2.8$–$3.0\,M_\odot$). In binaries that evolve conservatively, the remnant of the primary star at the time of its collapse will be the less-massive component of the system. For circular orbits the conservative assumptions imply that the orbital radius a changes according to the equation (see Chapter 9, this book).

$$a/a_0 = [M_1^0 (M - M_1^0)/M_1 (M - M_1)]^2$$
$$= [(1+q)^2 \cdot q_0/(1+q_0)^2 q]^2, \tag{8.4}$$

where M is the total mass of the system, M_1 is the mass of the primary, $q = (M_2/M_1) = (M - M_1)/M_1$, and index zero indicates the initial situation. The change in the orbital period is given by Kepler's third law:

$$P/P_0 = (a/a_0)^{3/2}. \tag{8.5}$$

Since the explosion of the less-massive component is unlikely to disrupt the system, even if the effects of impact and ablation are taken into account (Sutantyo 1974a, 1975a; Wheeler, Lecar & McKee 1975; Fryxell & Arnett 1981; De Cuyper 1981) one expects that the compact stars in conservatively evolving systems will practically always remain bound after the explosion. This must have been the case in the massive X-ray binaries since their orbital periods are so short that extensive mass transfer must have taken place during their evolution (van den Heuvel & Heise 1972; Tutukov & Yungelson 1973). For the most massive systems among them one can, moreover, identify direct progenitors, which are the Wolf–Rayet (WR) binaries (van den Heuvel 1973; Tutukov & Yungelson 1973). As was already suggested by Paczynski (1967) and confirmed by later works (Kippenhahn 1969; Smith 1973; Tutukov & Yungelson 1973) these must be the result of case B mass transfer (and in a few cases also of late case A) in very massive close-binary systems. In later years it has become clear that also stellar-wind mass loss may have played an important part during their formation (see De Loore 1980; van Beveren & Conti, 1980; van Beveren 1982).

The WR stars are (with one exception) always the less-massive components of their systems, but are, at the same time, the most evolved ones, already in the stage of core-helium burning or beyond (Conti 1981). The WN (nitrogen) types have helium-rich atmospheres and show the products of the CNO cycle on their surface. The WC (carbon) types show the products of helium burning on their surface. The companions of WR stars tend to be less evolved O- and early B-type stars which are mostly still in the stage of core-hydrogen burning. Table 8.5 lists some important data of a number of representative WR binaries and also gives the system parameters which would result if the WR star were to explode as a super-

Table 8.5. *Galactic WR+O systems (SB2) from Massey (1982), P_f, e and V_g are the orbital period, eccentricity and runaway velocity of the systems, which would result after a symmetric supernova explosion of the WR star. It is assumed that at the moment of the explosion the WR star has a mass of $10 M_\odot$ (except for HD190918) and that it leaves a $1.4 M_\odot$ compact remnant*

Name	P (d)	$M_{WR} + M_{OB}$ ($M_\odot$)	P_f (d)	e	V_g (km s^{-1})
HDE 320102	8.83 d	11 + 35	14.7 d	0.19	55
HD 90657	8.2	18 + 34.6	13.7	0.24	70
HD 94546	4.9	10 + 28.2	9.3 (8.2)	0.29	91
HD 186943	9.56	13 + 25	19.9	0.33	77
HD 190918	112.8	9 + 35	176.2	0.21	26
V 444 Cyg	4.212	17 + 45	6.2 (5.9)	0.19	77
CX Cep	2.217	11 + 25.6	4.3 (3.7)	0.32	126
HD 311884	6.34	50 + 60	8.5 (8.2)	0.14	57
GP Cep	6.69	10 + 45.5	9.9 (9.6)	0.18	65
CQ Cep	1.6	23 + 19.3	4.3 (3.2)	0.41	155
HD 97152	7.886	20 + 35	13.1	0.24	70
γ^2 Vel	78.5002	20 + 35	130.8	0.24	33
CV Ser	29.707	13 + 25	61.8	0.33	53
HD 152270	8.893	20 + 60	>11.9	0.14	51
HD 63099	27.63	–	>40	–	–
θ Mus	18.34	–	>27	–	–

nova and leave a 1.4 $M_\odot$ neutron star. We assumed that at the time of the explosion all WR stars had $M = 10\,M_\odot$ as a consequence of preceding stellar-wind mass loss (see Section 8.4.7). Clearly, the systems will never be disrupted and will, after the supernova explosion, have binary periods (and masses) in the range of those of the 'standard' massive X-ray binaries (see van den Heuvel 1973) and of some of the Be-X-ray binaries. Therefore, they must be direct progenitors of such systems. Thus, in exploring the evolution of the most massive binaries, one should study how the WR binaries were formed. In Table 8.5 the P_f-values after tidal synchronization and circularization of the orbits are added in parentheses.

8.4.2 Examples of conservative evolution of massive binaries

Figure 8.6 (*a* and *b*) depicts as examples the conservative evolution of two case B systems, with initial component masses of $25 + 10$ and $16 + 9.6\,M_\odot$, adapted from van den Heuvel (1974) and Rappaport & van den Heuvel (1982) respectively. The initial orbital period in both cases is 5.0 d, and the initial composition $X = 0.60, Z = 0.044$.

The system parameters were chosen such that the first system can produce a fairly wide 'standard' massive X-ray binary and the second one a Be-X-ray binary. The following evolutionary phases can be distinguished (parameters for the second system are always in parentheses):

(a)–(b): *Unevolved close binary.*

(b)–(c): *First stage of mass transfer:* 4.71×10^6 yr (6.88×10^6 yr), after the birth of the system the primary star overflows its Roche lobe and transfers its hydrogen-rich envelope of 16.5 $M_\odot$ (12.0 $M_\odot$) to its companion in about 10^4 ($3 \cdot 10^4$) yr.

(c)–(d): *Helium-star (Wolf-Rayet?) binary:* the system consists of an 8.5 $M_\odot$ (4 $M_\odot$) helium star and a 26.5 $M_\odot$ (21.6 $M_\odot$) main-sequence star. The orbital period has changed to 6.84 d (28.1 d). The system of Figure 8.6(*a*) may be a WR binary at this stage (see Table 8.5).

(d) *Supernova explosion:* $5 \cdot 10^5$ yr (1.1×10^6) yr after the mass transfer the helium star has exploded as a photodisintegration supernova; we assumed that it has left a 1.4 $M_\odot$ neutron star. The orbital period has suddenly increased to 11.9 (35.6) d, the orbit has obtained an eccentricity $e = 0.35$ (0.10), and the system has obtained a runaway velocity of 71 (18) km s^{-1} (the amount of matter ripped off by the impact is negligible).

(e) *Persistent X-ray binary* (not depicted in Figure 8.6(*b*)): $3 \cdot 8 \times 10^6$ ($3 \cdot 6 \times 10^6$) yr after the explosion the secondary has left the main sequence and is a blue supergiant. It will within $\sim 10^4$ yr overflow its Roche lobe. Mass accretion from the strong wind of the supergiant or from beginning Roche-lobe overflow turns the neutron star into a strong X-ray source for about 10^4 yr (see Chapter 9, this book).

The conservative evolutionary scheme of Figure 8.6(a, b) can explain the existence of 'standard' massive X-ray binaries with relatively long periods ($\geqslant 9$ d) such as 4U 0900−40 and 4U 1223−62 (see Table 8.1), which both contain an early-type supergiant with a strong wind which presumably powers the X-ray source. On the other hand, the evolutionary scheme of Figure 8.6(b) can also explain the existence of the Be–X-ray binaries, for the following reasons. After the mass-transfer phase, in stage (c), one expects the secondary star to be a very rapid rotator, since the exchanged matter has a large specific angular momentum and will flow in through a disk (Shu & Lubow 1981). Arriving at the stellar equator with a Keplerian velocity it will have spun-up the rotation of this star. As the exchanged amount of matter is of the same order or larger than the original mass of the secondary and because of the compression of the underlying matter, this star will have been spun-up to nearly rotational break-up (see van der Linden, 1982). This spinning-up is a general phenomenon in mass-transfer binaries such as U Cephei and β Lyrae; in the latter case the mass-receiving component is a disk rather than a star (Plavec 1981). The rapid rotation of many Be stars, such as ϕ Per (B 0e, $P = 126$ d) and HR 2142 (B1 Ve, $P = 81$ d) is expected to have arisen in this way. Once an O or B star is a rapid rotator it goes from time to time through emission-line phases by ejecting clouds of gas from its equator (see Section 8.2.1). During such phases the neutron star will become a transient X-ray source.

Since for $P \geqslant 10$-15 d, tidal deceleration plays a negligible role as long as the secondary is on the main-sequence (see De Greve, De Loore & Sutantyo 1975), one expects the system of Figure 8.6(b) to exhibit transient X-ray outbursts during most of the time between phases (d) and (e). Of course, before the neutron star can become an X-ray pulsar, its rotation should have slowed down sufficiently to allow the accreting matter to enter its magnetosphere; the spin-down timescale of the neutron star in the wind of its companion may be relatively short, i.e. 10^5-10^6 yr (see Chapter 11, this book).

The fact that no Be–X-ray binaries with orbital periods $\lesssim 15$ d are observed, then finds a quanitative explanation in the more efficient tidal deceleration of the secondary's rotation in these closer systems (when the rotation of the secondaries in such systems is still rapid – i.e. just after phase (d) – the neutron star still spins too fast to allow accretion). In systems with $P < 10$-15 d the neutron star is therefore expected to become an X-ray source only when the secondary star either becomes a supergiant with a strong wind, or begins to overflow its Roche lobe (see Chapter 9, this book, and Section 8.4.8). It thus appears that conservative scenarios such as those of Figure 8.6(a, b) can explain the existence of the Be–X-ray binaries and of the wider ones ($P \gtrsim 9$ d) among the 'standard' massive X-ray binaries.

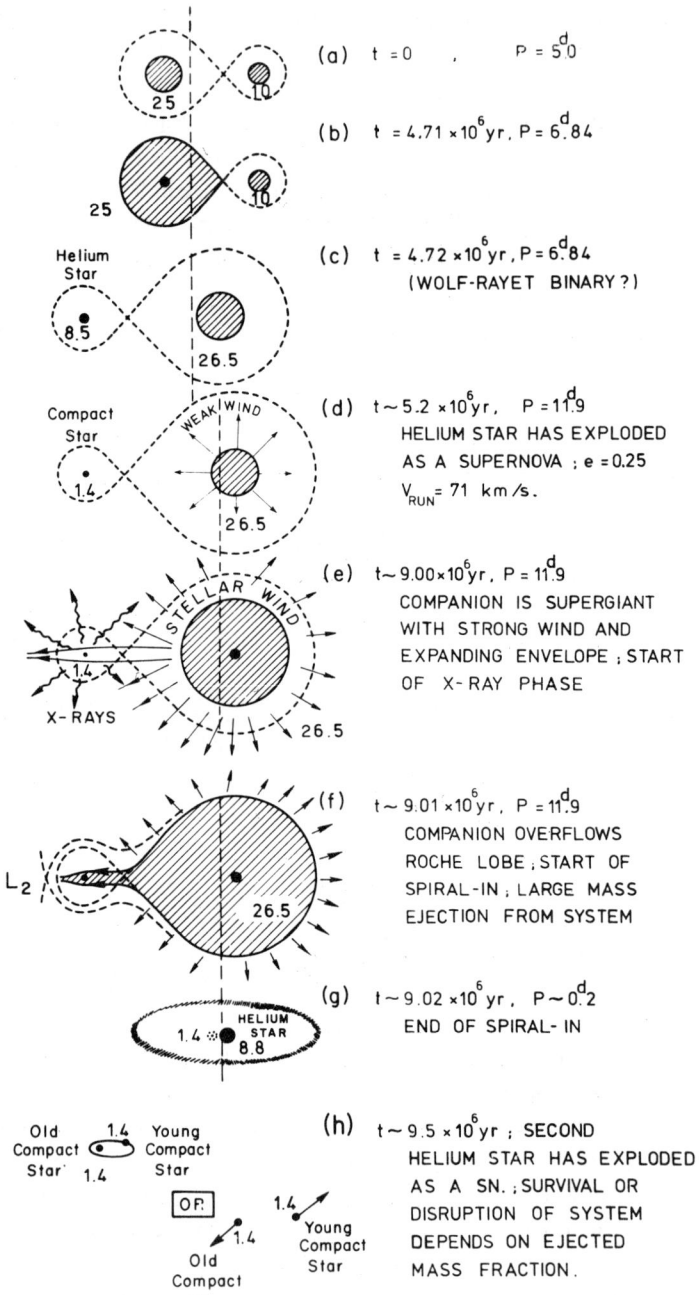

Fig. 8.6 (a) Conservative evolution of a close binary system with initial masses of 25 and 10 M$_\odot$. Each stage is labelled with the approximate age of the system and the orbital period in days. The numbers inside the representations of the stars indicate mass (M$_\odot$). The system becomes a wind-powered 'standard' massive X-ray binary for some 10^4 yr, in stage e (see text). For the subsequent evolutionary stages f, g and h: see Section 8.4.10. (b) Conservative evolutionary scenario for the formation of a Be–X-ray binary out of a close pair of early B stars with masses of 16 M$_\odot$ and 9.6 M$_\odot$. The numbers again indicate mass (M$_\odot$). After the end of the mass transfer the Be star presumably has a circumstellar disk or shell of matter associated with the rapid rotation (induced by the previous accretion of matter with high angular momentum).

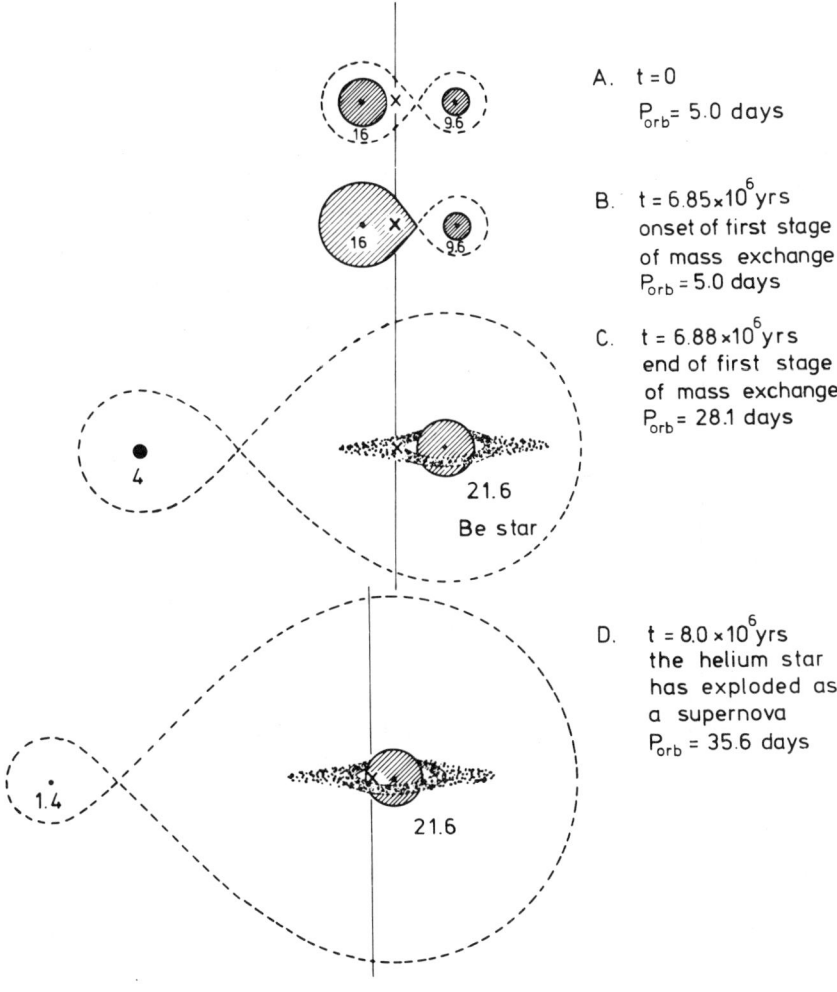

8.4.3 Problems with conservative scenarios for short-period, 'standard' massive X-ray binaries: breakdown of the conservative assumptions

Equations (8.4) and (8.5) show that in order to obtain orbital periods $\lesssim 2\text{-}3$ d (such as those of Cen X-3 and LMC X-4) after conservative mass transfer, one has to start out with a very low initial mass ratio: $q \lesssim 0.3$. This is due to the fact that for $M_1 \sim 20\text{-}30\,M_\odot$ the remaining helium core of the primary star contains only some 25-35 per cent of its original mass (see Equations (8.2) and (8.3)) However, such a scenario is problematic since a short-period system with a low q may grow into deep contact during the exchange for several reasons, i.e.:

(1) Kippenhahn & Meyer Hofmeister (1977) and Flannery & Ulrich (1977) have pointed out that the secondary star needs time – roughly its own thermal timescale – to accept the matter which is transferred to it. Their numerical calculations show that in massive binaries for $q \leqslant 0.5\text{-}0.6$ the mass transfer proceeds so rapidly that the radius of the secondary will swell up greatly due to accretion, and this star may begin to overflow its Roche lobe, such that a contact configuration is formed surrounded by a common envelope.

(2) The formation of a contact system is further facilitated by a low value of $q\,(\leqslant 0.3)$, since conservative evolution then predicts the system to shrink very much during the mass exchange (see Equations (8.4) and (8.5)). The minimum separation is reached when the components have equal masses; for $q \leqslant 0.3$ the separation has decreased at that moment by more than a factor of 2.

Shu & Lubow (1981) have argued that point (1) may not be as serious at the numerical calculations suggest. Their argument is that the calculated large swelling-up of the secondary due to accretion might be a computational artifact and would not occur if a more realistic – i.e. shock – outer boundary condition were used in the computations, just as for accreting protostars (see Larson 1972). In the latter case, the stellar radius will not increase by more than about a factor of 2 over its thermal equilibrium radius. Although this somewhat relaxes the lower limit on q mentioned under point (1), still, because of point (2), for short-period systems with a low q, the formation of a contact configuration cannot be avoided. For example, for $M = 25\,M_\odot$, $P = 5.0$ d, one finds that, if the secondary swells up by a factor of 2 (over its thermal equilibrium radius) during accretion, all systems with $q < 0.4$ will still grow into contact at minimum separation. So, only for $q \geqslant 0.4$ may such systems be expected to evolve conservatively. Therefore, the scenario of Figure 8.6(a) is probably just allowed. For $q = 0.3$, with the same condition on the radius of the secondary, one finds that only systems with $P \geqslant 9$ d can avoid contact. The post-supernova orbital periods of these systems are in both cases $\geqslant 10$ d. Hence, with conservative evolution it appears to be impossible to reach post-supernova orbital periods $\leqslant 10$ d, *even if the initial mass*

Formation and evolution of X-ray binaries 323

ratio q is very low and the initial period is very short (see Kippenhahn & Thomas 1980).

8.4.4 Effects of moderate angular momentum losses during a common-envelope phase: formation of short-period 'standard' massive X-ray binaries

When a common envelope forms, one expects that mass loss may start from the outer Lagrangian points L_2 or L_3 (if corotation is maintained out to that distance). The mass lost in this way always has a specific angular momentum that is considerably larger than the average orbital momentum of the components (see Kuiper 1941; Nariai 1975). Therefore, such mass loss *will always cause the system to shrink rapidly*. For example, in systems with roughly equal-mass components the specific angular momentum of matter at L_2, L_3 is some four times the specific orbital angular momentum. In that case the rate of loss of (orbital) angular momentum J is given by

$$\frac{dJ}{dM} = 4(J/M), \quad \text{such that } J\,(:)\,M^4, \tag{8.6}$$

where M is the total mass of the system. As an example, consider the evolution of a system of $30 + 10\,M_\odot$ with $P_{\text{orb}} \cong 5$ d, in which it is assumed that $5\,M_\odot$ of the exchanged $20\,M_\odot$ is lost during the mass transfer in the way described by Equation (8.6). One then finds that the post-mass-transfer system has $10 + 25\,M_\odot$ and a period of 2.27 d. If we assume that the $10\,M_\odot$ helium star loses another $3\,M_\odot$ by wind – as it is a Wolf–Rayet star (see Section 8.4.7) before it explodes and leaves a $1.4\,M_\odot$ remnant – one finds that the post-supernova period and eccentricity are $P = 3.5$ d and $e = 0.2$ respectively. (Effects of impact and ablation have been neglected and may still slightly change these values, see Sutantyo 1975a; De Cuyper 1981, 1982.) The resulting system closely resembles SMC X-1 and 4U 1538−52. From the above, we conclude that:

(i) Systems with $q \lesssim 0.4$ will evolve through a common-envelope phase during which mass and angular momentum will be lost; this is the only way to explain the short orbital periods of some of the massive X-ray binaries and WR binaries, such as Cen X-3, LMC X-4, SMC X-1, CX Cephei ($P = 1.6$ d) and CQ Cephei ($P = 2.14$ d). The, for a WR star, anomalously high mass ratio (1.12) of the latter system can certainly not be understood without mass loss from the system.

(ii) Systems with $q \gtrsim 0.4$ may evolve more-or-less conservatively and will then produce post-supernova binaries with periods $\gtrsim 9$–10 d. In order to know what fraction of the unevolved binaries will evolve into short-period and long-period massive X-ray binaries we will now consider the distribution of mass ratios and periods of unevolved systems.

8.4.5 The distribution of mass ratios and periods of unevolved early-type close binaries

Figures 8.7 and 8.8 depict the distribution of mass ratios q and periods P of short-period O-type spectroscopic binaries, as obtained by Garmany, Conti & Massey (1980); the sample is expected to be complete for $P \lesssim 30$ d. Lucy &

Fig. 8.7. Mass ratio distribution of unevolved massive close binaries; short-period systems after Garmany *et al.* (1980; as interpreted by De Loore 1981). The overall distribution of short- and long-period systems together, is after Abt & Levy (1978).

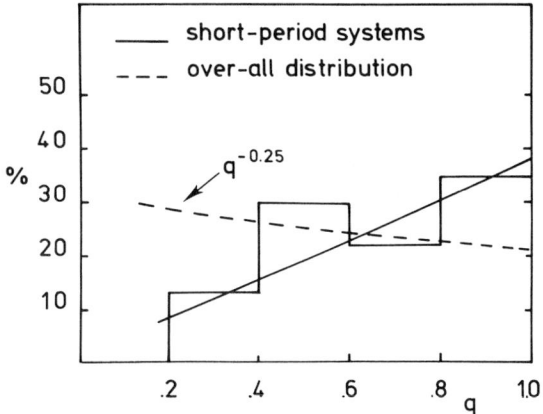

Fig. 8.8. Solid line and histogram: distribution of orbital periods of unevolved O- and B-type close binaries ($P \leq 30$ d) after Garmany *et al.* (1980). Dashed and dash-dotted lines: resulting expected post-mass-transfer distributions for very massive and moderately-massive systems, as explained in the text. After the supernova explosion of the primary star the post-mass-transfer periods are increased by ~60 per cent for the very massive and ~40 per cent for the moderately massive systems.

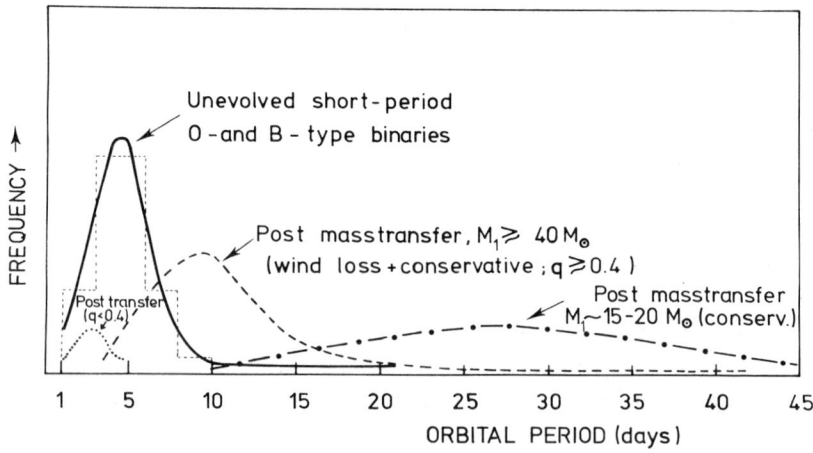

Ricco (1979) and Kraicheva et al. (1979) independently showed that short-period binaries of other spectral types have a similar q-distribution, also peaking at $q \sim 0.8$-0.9. If one includes binaries with $P > 100$ d one seems to find that the distribution is double peaked with one peak at $q \approx 0.8$ and the other at $q \approx 0.3$ (Trimble 1974; Kraicheva et al. 1979). Longer-period systems ($P \gtrsim 100$ d) therefore seem to have a q-distribution that peaks at low q-values (because of strong selection effects against low q-values, the real peak might be below $q = 0.3$). The general trend of a mild increase towards the lower q-values is confirmed by the investigations of Abt & Levy (1978) who found that, for spectroscopic binaries of spectral type B-G with all kinds of periods, the frequency distribution of q varies roughly as $q^{-0.25}$ (the dashed curve in Figure 8.7). Numerical studies by Lucy (1981) and Bodenheimer (1981) suggest that the peak near $q \approx 0.8$-0.9 for short-period systems ($P \lesssim 30$ d) is due to the fragmentation process of a rotating, collapsing cloud. As to the period distribution of unevolved systems: Figure 8.8 shows that most O-type systems have P between 1 and 10 d with a peak in the distribution between 2 and 6 d. For B-type spectroscopic binaries van Albada & Blaauw's (1968) work shows that the short-period systems have a similar P-distribution (see also Mantegazza et al. 1980), while the semi-major axes a of the wider systems ($a \gtrsim 10$ AU) are distributed according to a function

$$f(a) = \text{const}/a \tag{8.7}$$

(i.e. $F(\log a) = $ const (van Albada (1968)). Abt & Levy (1978) and Kraicheva et al. (1979) found that roughly this distribution also holds for $a \sim 1$-10 AU.

8.4.6 Expected period distributions after the mass transfer

We consider only systems with $P \lesssim 30$ d. We will assume $X = 0.70$, $Z = 0.03$ and consider conservative ($q \gtrsim 0.4$) and non-conservative ($q < 0.4$) systems separately.

8.4.6.1 Conservative systems: dependence of final period on initial primary mass

One observes from Equations (8.2) and (8.3) that the fraction p of the mass of the primary star left after the mass transfer increases from 0.2 for $M_1 = 10$ M$_\odot$ (spectral type $\sim$B3V) to $\geqslant 0.4$ for $M_1 \gtrsim 60$ M$_\odot$ (spectral type about O4V). This has important implications for the orbital periods resulting after the mass transfer, as can be seen from Equations (8.4) and (8.5).

One observes, e.g. for $q = 0.6$, that, for $M_1 = 16$ M$_\odot$ ($p = 0.234$), the period increases by a factor of 6.6, whereas for $M_1 = 60$ M$_\odot$ ($p = 0.41$), it increases by only a factor of 1.88 (for $X = 0.60, Z = 0.044$, the increases are by a factor of 3.6 and 1.23 respectively, i.e. again different by a factor of 3).

Since O- and B-type systems have $\langle q \rangle \approx 0.6$-$0.7$ and roughly-similar distributions of periods – one thus expects with conservative evolution that the post-mass-transfer orbital periods of B-type close binaries are a factor of 3 or more larger than those of early O-type close binaries. With about a factor of 7 increase in P for the early B-type systems and a factor of 2 increase for the O-type systems the P-distribution of the mass-transfer remnants of early B-type spectroscopic binaries is expected to peak somewhere between 14 and 24 d, whereas that of the early O-type systems is expected to peak between 4 and 12 d. (After the supernova explosion of the primary the periods will be increased by another 30-70 per cent.)

We suggest that this difference is one of the reasons why the orbital periods of the Be–X-ray binaries (as well as other Be spectroscopic binaries) tend to be much longer than those of the 'standard' massive X-ray binaries, and their progenitors, the Wolf–Rayet binaries (see also, however, Section 8.4.8). This difference in final period remains present also if one takes into account that the most massive systems may undergo strong stellar-wind mass loss prior to the mass transfer, as we will show in the next section.

8.4.6.2 Non-conservatively evolving systems

Figure 8.7 shows that only ~14 per cent of the unevolved short-period O- and B-type spectroscopic binaries have $q < 0.4$. These systems are expected to lose mass and angular momentum and may produce very short-period WR binaries and X-ray binaries (see Section 8.4.4). We thus conclude that the majority of the systems will evolve more-or-less conservatively and, therefore, after the supernova will produce systems with $P \gtrsim 5$–6 d. Figure 8.8 depicts the above results schematically.

8.4.7 Effects of stellar-wind mass loss on the evolution of very massive close binaries; the origin of Wolf–Rayet binaries

Observations show that stars more massive than about 15-20 $M_\odot$ develop strong stellar winds ($\dot{M}_w \gtrsim 10^{-6} M_\odot \, \text{yr}^{-1}$) during some stage of their evolution. In stars more massive than about 30 $M_\odot$ this does already occur during the later phases of core-hydrogen burning, when the stars become Of stars with mass-loss rates $10^{-6} M_\odot \, \text{yr}^{-1}$ ($M \sim 30 M_\odot$) to $3.10^{-5} M_\odot \, \text{yr}^{-1}$ ($M \sim 60 M_\odot$) (see for example the reviews by De Jager 1980; Cassinelli 1979 and, especially, Lamers 1981). Using the equations for the mean mass-loss rates as a function of M, L and R (R and L are radius and luminosity) as given by Lamers (1981), one finds that stars of 60 $M_\odot$, 50 $M_\odot$, 30 $M_\odot$, 20 $M_\odot$ and 16 $M_\odot$ lose 14 $M_\odot$, 10.8 $M_\odot$, 4.2 $M_\odot$, 1.7 $M_\odot$ and 1.0 $M_\odot$ up till core-hydrogen exhaustion.

(The wind mass loss is presumably due to radiation–pressure-driven pulsational instability of the stellar interior which sets in when the helium abundance Y in the core increases, see van den Heuvel 1979; Maeder 1981; Noels & Masereel

1982.) Since, during such gradual mass loss the core mass and stellar luminosity do hardly change (see Chapter 1, Figure 1.13, this book), the mass fraction p contained in the helium core increases from 0.30 to 0.37 in a 30 M$_\odot$ star and from 0.41 to 0.53 in a 60 M$_\odot$ star (for $X = 0.7$). Also the q of a binary will change due to this process. For example, starting with $q = 0.6, M_1 = 60$ M$_\odot$ (and using Lamers's equation for $\dot{M}_w$) the values of p and q at the end of core-hydrogen burning will be 0.53 and 0.67 respectively, which together result in $P_f/P_o^m = 1.37$, where P_o^m is the period at the onset of the mass transfer and P_f is the period after the transfer.

On the other hand, for a system which starts with $M_1 = 16$ M$_\odot$, $q = 0.6$, p and q will hardly have changed during core-hydrogen burning and the period increase remains $P_f/P_o^m = 6.6$. For the system with $M_1 = 60$ M$_\odot$, P_o^m will differ from the period P_o at birth, as a consequence of the angular momentum carried away in the wind. For a simple isotropic wind mass loss at very high velocities ($v_w \gg V_{orb}$) one obtains after Huang (1963; see Hadjidemetriou 1967).

$$P_o^m/P_o = (M_i/M_m)^2, \quad (a_m/a_o = (M_i/M_m)), \tag{8.8}$$

where P and a are the orbital period and semi-major axis respectively, and indices i and m denote the situation at the beginning and end of the wind mass-loss stage (notice that this expression differs slightly from the one used by Massey 1981). Inserting $M_i = 60 + 36 = 96$ M$_\odot$, $M_m = 46 + 31 = 77$ M$_\odot$, $q_i = 0.6$, $q_m = 0.67$, one obtains for the above massive system: $P_o^m/P_o = 1.55$. Together with $P_f/P_o^m = 1.37$ this yields after the mass exchange: $P_f/P_o \approx 2.13$ (for $X = 0.60$ one similarly obtains $P_f/P_o = 1.55$). So, after stellar-wind mass loss and mass transfer, the period of very massive systems is on the average expected to increase by at most about a factor of 2.

The latter expectation is consistent with the finding by Massey (1981) that WR binaries and short-period ($P \lesssim 30$ d) O-type close binaries have roughly-similar distributions of orbital periods, their mean periods differing by not more than a factor of 2. The *average* mass of a WR star is, according to Massey, some 20 M$_\odot$ (the range is from $\gtrsim 5$ M$_\odot$ to 45 M$_\odot$) which for $X \approx 0.70$ corresponds to a typical progenitor mass of about 50 M$_\odot$. (It should, however, be noted that the average WR-star mass might be estimated too high by a factor of 2 as it depends on the adopted masses of the O-type companion stars, which are uncertain by about this factor.)

The average mass ratio of 0.4 of WR binaries can probably not be obtained with purely *conservative* evolution which predicts mean post-mass-transfer mass ratio around 0.3 (for an average initial mass ratio $\langle q \rangle = 0.7$ and $M_1 \approx 30$–60 M$_\odot$) as was noticed by De Greve et al. (1978). Wind mass loss at the above-mentioned rates brings q closer to 0.4. Therefore, wind mass loss prior to the mass transfer is important for understanding the present system parameters of WR binaries.

8.4.8. Modes of mass transfer in persistent massive X-ray binaries; reasons for the dominance of short-period systems: a selection effect

As argued in Sections 8.4.4. and 8.4.6 and observable in Table 8.5, one expects the vast majority of the systems after the first supernova explosion to have $P \gtrsim 5$ d. Nevertheless, five out of the eight known 'standard' massive systems have $P \lesssim 5$ d. We believe that this is due to a selection effect produced by the much larger X-ray lifetimes of short-period systems than of long-period ones, for the following reasons. There are two ways in which a massive post-supernova system can become a strong, persistent X-ray source (see Ziolkowski 1977, and Chapter 9, this book) i.e.:

(1) *By beginning atmosphere Roche-lobe overflow*, if the optical star is still in the core-hydrogen burning stage when it begins to overflow its Roche lobe. This is only possible if $P \lesssim 4$-5 d. The source can then be powered by a moderate mass-transfer rate ($\sim 10^{-9}$-10^{-8} M$_\odot$ yr^{-1}) for some $\sim 10^5$ yr, after which mass transfer on a thermal timescale ensues and the source will be quenched.

(2) *If the optical star has a strong wind* ($\dot{M}_w \gtrsim 10^{-6}$ M$_\odot$ yr^{-1}),

(a) for $M_{opt} \lesssim 30$ M$_\odot$ the wind becomes sufficiently strong only when the star has terminated core-hydrogen burning and is a blue supergiant. This requires $P \gtrsim 5$d. The expected lifetime of such systems is $\lesssim 2 \cdot 10^4$ yr as the companion is in a stage of rapid envelope expansion when it approaches its Roche lobe. 4U 0900−40 ($P = 8.96$ d) and GX 301−2 (4U 1223−62) are probably systems of this type;

(b) for $M \gtrsim 30$ M$_\odot$ the wind already becomes strong during later phases of core-hydrogen burning, when the star becomes an Of star, such that a persistent wind-powered source may exist for several times 10^6 yr. The orbital period in this case should be short ($\lesssim 5$ d) as high accretion rates can only be reached in the low-velocity portion of the wind, close to the star (because of the V^{-4} dependence of the cross-section for wind accretion). We suggest that 4U 1700−37 (HD 153919) and 4U 1538−52 are systems of this type.

Thus, systems with $P \lesssim 4$-5 d have lifetimes as a strong, persistent source one to two orders of magnitude longer than the X-ray lifetimes of systems with $P \gtrsim 5$ d. Hence, to make a representative comparison between the five short-period 'standard' systems and the three 'standard' systems with $P \gtrsim 5$ d, the latter number should be multiplied by a factor of between 10 and 100. This indeed shows that the bulk of the post-supernova systems must have $P > 5$ d and that the short-period systems are in reality only a small fraction (<10 per cent) of the total.

8.4.9 Tidal synchronization and tidal instability

If the orbital period is sufficiently short, tidal forces will be strong enough to circularize and synchronize the orbit within a few million years after the supernova explosion. The perfectly-circular orbits of Cen X-3, SMC X-1 and Her X-1 show that in these systems this must have happened. The most important parameter for this process is the internal viscosity in the star (Sutantyo 1974b; Hut 1980, 1982). Zahn (1977) has argued that radiative damping of the dynamical tide can indeed produce the required short circularization timescale in the massive X-ray binaries (see also Savonije & Papaloizou 1982). Darwin (1908) and Counselman (1973) have shown that if, after tidal circularization and synchronization, the ratio of the orbital and rotational angular momenta

$$J_{orb}/J_{rot} \leq 3, \tag{8.9}$$

then the system is tidally unstable and the compact star will inevitably spiral down into the massive star. In LMC X-4, 4U 1700−37 and Cen X-3 this condition is probably fulfilled as one can see as follows.

The rotational angular momentum is concentrated in the optical star and can be expressed as

$$J_{rot} = k^2 M_1 \omega R_1^2, \tag{8.10}$$

where k is the radius of gyration of the star and M_1, R_1 are its mass and radius. The orbital angular momentum is

$$J_{orb} = \omega M_1 M_2 a^2/(M_1 + M_2), \tag{8.11}$$

where a is the orbital radius. Substitution of Equations (8.10) and (8.11) into Equation (3.9) yields

$$k^2 > M_2 a^2/3R_1^2(M_1 + M_2). \tag{8.12}$$

The observed best values of M_1, R_1, a (after Rappaport & Joss, Chapter 1, this book) and the resulting lower limits to k^2 for the five shortest-period systems are listed in Table 8.6. For stars with M_1 between 15 and 30 $M_\odot$ the theoretically calculated values of k^2 decrease from about 0.075 to 0.030 between the zero-age main sequence and core-hydrogen exhaustion (see De Greve, De Loore & Sutantyo 1975). Since the optical stars in LMC X-4, Cen X-3 and 4U 1700−37 are still in an early stage of core-hydrogen burning (see Chapter 1, Figure 1.13, this book), the systems of LMC X-4 and 4U 1700−37 are almost certainly tidally unstable and Cen X-3 presumably is as well.

Table 8.6. *Lower limits to* k^2 *of the optical component required for tidal instability in five short-period massive X-ray binaries*

System	P (d)	$a/R_\odot$	$R_{opt}/R_\odot$	$M_{opt}/M_\odot$	Unstable if $k^2 \geq$
Cen X-3	2.087 d	18.3	12.2	19	≥0.050
LMC X-4	1.402	16.9	10.5	32	≥0.031
SMC X-1	3.89	27.4	16.5	17	≥0.057
4U 1538−52	3.73	26	16.0	16	≥0.069
4U 1700−37	3.41	19.9	24	≥30	≥0.023

The fact that the orbital period of Cen X-3 is observed to be decreasing on a timescale of $4 \cdot 10^5$ yr (Kelley 1981, van der Klis 1983) might be a manifestation of this instability (see also Chevalier 1975). For SMC X-1 and 4U 1538−52 the case is less clear since the stars are more evolved and may have k^2 just smaller than their critical values.

The fact that the three systems with the shortest orbital periods are all likely to be tidally unstable has important consequences (a) for their evolutionary history, and (b) for their future evolution. As to (a): a decrease of the orbital period of Cen X-3 on a timescale of $4 \cdot 10^5$ yr implies that less than $4 \cdot 10^5$ yr ago the orbital period of this system may have been > 4 d. The same holds for LMC X-4 and 4U 1700−37. This relaxes the problems of the evolutionary origin of the standard systems with the shortest orbital periods, since all of them may thus have started out with $P_{orb} \gtrsim 3$-4 d just after the supernova explosion. Notice also that a system which was tidally stable at birth (just after the supernova) may become tidally unstable later on as a consequence of the evolutionary increase in radius of the optical star, which causes its moment of inertia $k^2 M_1 R_1^2$ to increase (despite the fact that k^2 decreases during the evolution).

8.4.10 Final evolution of massive X-ray binaries – formation of binary radio pulsars and runaway radio pulsars

Also in systems which are tidally stable such as the longer-period ones, the neutron star will at a certain moment in the evolution be engulfed by matter from the envelope of the massive companion. In systems with $P \gtrsim 4$-5 d the envelope of the companion is in the rapid post-main-sequence expansion stage when it begins to overflow the Roche lobe and within a few thousand years the mass-transfer rate grows to $\sim 10^{-3}$ $M_\odot$ yr^{-1} (see Savonije 1978). Presumably the matter will flow in towards the compact star through a thick accretion disk. But, arriving near the disk centre at maximum only some 10^{-7} $M_\odot$ yr^{-1} can be accepted by the neutron star (the Eddington limit – with possible corrections for neutrino emission) such that the bulk of the inflowing matter must be expelled. Because of the super-Eddington luminosity of the inner parts of the disk, this expulsion is presumably driven by radiation pressure and, for geometrical reasons, most probably takes place in directions perpendicular to the disk (Shakura & Sunyaev 1973; Abramowicz & Piran 1980).

In this way, beams such as those of SS 433 and 1E 2259 + 586 might originate (Katz 1980; van den Heuvel, Ostriker & Petterson 1980; Hut & van den Heuvel 1981). Although the precise mechanism for the beam formation is still far from clear, an evolution as outlined above seems inevitable for all massive X-ray binaries with $P \gtrsim 4$-5 d, including the B-emission ones. As the expelled matter has large specific angular momentum (i.e. the orbital angular momentum of the compact star) the orbit will shrink rapidly; the spiralling-in is expected to terminate when only the evolved core of the massive star is left: at that moment the orbital period is a few hours (van den Heuvel & De Loore 1973). The resulting system may resemble Cyg X-3, a peculiar source with a very high X-ray luminosity that in several respects - notably its synchrotron radio outbursts - resembles SS 433 and Cir X-1. (A similar type of spiral-in evolution is also expected if the neutron star is literally engulfed by the envelope of its companion, see Taam, Bodenheimer & Ostriker 1978; Taam 1979; Delgado 1979.) When, finally, the helium star explodes as a supernova, the system may become unbound if more than half of the total mass of the system is ejected in the explosion. In this case two runaway radio pulsars are formed with space velocities of several hundreds of km s^{-1}. If less than half of the mass of the system is ejected (i.e. if $M_{He} \lesssim 4.2\,M_\odot$ for an assumed neutron-star mass of $1.4\,M_\odot$) the system will remain bound and consists of two neutron stars in a very eccentric short-period orbit. These types of final evolution are outlined in Figure 8.6(a) (f, g, h). (If the helium core has a mass $\lesssim 2.8\,M_\odot$ the final system may consist of a white dwarf and a neutron star in a circular orbit.) Notice that in the case of the binary radio pulsar PSR 1913+16 the observed radio pulsar must be the oldest of the two neutron stars in the system, which was spun-up back to a very short rotation period (0.059 s) during the phase of strong accretion; this is the only way to explain its peculiar combination of short pulse period and low surface-magnetic field strength (Smarr & Blandford 1976; Srinivasan & van den Heuvel 1982; Damashek *et al.* 1982).

8.4.11 Maximum mass of the exploding star - possible reason for the absence of black hole remnants

The upper mass limit M_L for instability of a star against radiation pressure depends on the mean molecular mass μ of the stellar material as

$$M_L = M_c/\mu^2,$$

where M_c is a constant (Eddington 1926). Since no hydrogen-rich ($\mu \approx \frac{1}{2}$) stars more massive than about 80-100 $M_\odot$ are known in the galaxy we estimate $M_c \approx 20\text{-}25\,M_\odot$. Adopting this value, one finds for helium-rich stars ($\mu = \frac{4}{3}$), and for stars consisting of elements heavier than helium ($\mu \approx 2$), that M_L has a value of $\sim 11\text{-}14\,M_\odot$, and $\sim 5\text{-}6\,M_\odot$ respectively. (Precise calculations for He stars by Noels & Masereel (1982) yield $M_L \sim 16\,M_\odot$.) Thus, since stars more massive than $\sim 40\,M_\odot$ develop helium cores more massive than $\sim 14\,M_\odot$, such stars will gradually become unstable to radiation pressure during core-hydrogen burning. To regain stability they have to shed their excess mass. This may well be the reason why all very massive stars gradually develop very strong winds and lose their envelopes (van den Heuvel 1979; Maeder 1981; Noels & Masereel 1982). The same mechanism holds in WR stars: helium-burning helium stars with masses $\gtrsim 7$–$9\,M_\odot$ develop carbon–oxygen cores more massive than ~ 5–$6\,M_\odot$ and become pulsationally unstable and shed their excess mass. This can explain the strong winds of $\sim 3.10^{-5}\,M_\odot\,\mathrm{yr}^{-1}$ of WR stars.

Consequently, only He stars less massive than about 7–9 $M_\odot$ can evolve without much mass loss, and the more massive ones are expected to be reduced to $\lesssim 7$–$9\,M_\odot$ by strong wind mass loss. Thus, at the moment of the supernova explosion all WR stars are expected to have $M \lesssim 7$–$9\,M_\odot$. Since CO stars with masses $\lesssim 5$–$6\,M_\odot$ develop collapsing cores of just one Chandrasekhar mass they are expected to leave neutron stars (see Arnett 1978). In this way the instability of massive stars against radiation pressure might very well prevent the formation of stellar-mass black holes. Notice that the above reasoning is independent of whether or not the star is single or in a binary. *We thus expect that stellar mass black holes are rather unlikely to be produced by stellar evolution.*

8.5 The formation of young low-mass X-ray binaries and young runaway pulsars

8.5.1 Evolution of massive systems with a low initial mass ratio ($\lesssim 0.3$)

As argued in Section 8.4.3 these systems will evolve through a common-envelope stage. In this stage, one expects the secondary to spiral down towards the core of the primary star on a short timescale, in a way similar to that outlined for the compact star in Figure 8.6(a) (f. g. h). During this spiral-in the secondary hardly collects any matter from the envelope of its companion. If $M_\mathrm{core} \lesssim 2.8\,M_\odot$ the outcome of such evolution is presumably an ultrashort-period system consisting of a white dwarf and a normal dwarf star, i.e. resembling a cataclysmic-variable (CV) binary (Paczynski 1976; Ritter 1976; Meyer & Meyer-Hofmeister 1979; see Chapter 4, this book). If $M_\mathrm{core} \gtrsim 2.8\,M_\odot$ the core of the primary will explode as a supernova and leave a neutron-star remnant. As the post-spiral-in system is expected to be very close ($P_\mathrm{orb} \sim 0.5$ d) the effects of impact of the supernova shell and the resulting ablation of the secondary will be very large and in most cases are expected to disrupt the system. In case of disruption the released young neutron star and its low-mass companion both have runaway velocities $\geqslant 100\,\mathrm{km\,s}^{-1}$ (van den Heuvel 1981a, b).

Post-spiral-in systems that were not disrupted will have a high eccentricity and a high space velocity (Sutantyo 1975a). As the separation at periastron has not changed due to the explosion, tidal forces will be very strong and will rapidly circularize the orbit. The result will be a low-mass X-ray binary with a circular orbit and a high space velocity ($> 100\,\mathrm{km\,s}^{-1}$). This scenario might be appropriate for the Her X-1 system (see Figure 8.2) and may apply to Cyg X-2 and

Sco X-1 as well. Cyg X-2 could have evolved from a Her X-1 system by subsequent mass transfer (van den Heuvel 1981a, b).

Since spectroscopic binaries with long orbital periods tend to have low mass ratios (see Section 8.4.5), such a system may either produce a single runaway radio pulsar or a runaway low-mass X-ray binary, such as Her X-1. In view of the high disruption probability of post-spiral-in systems the Her X-1 systems are expected to be very rare (van den Heuvel 1981a, b). Taking into account that over $\frac{2}{3}$ of all early-type stars are found in spectroscopic binaries with P less than a few decades, the above results, together with those of Section 8.4.10, imply that over 70 per cent of all neutron stars are expected to be runaway stars with $v > 100$ km s^{-1} as a consequence of a preceding spiral-in history followed by a supernova.

This may account for the large space velocities ($\langle v \rangle \approx 200$ km s^{-1}) of most radio pulsars (Lyne 1981; see van den Heuvel 1981a, b).

8.6 Possible formation mechanisms of Type II sources
8.6.1 Globular cluster sources

As pointed out by Gursky (1973) and Katz (1975) the incidence of strong X-ray sources among globular cluster stars is some 10^2–10^3 times higher than in the galaxy as a whole (11 strong and 2 weaker sources in ~130 observed globular clusters, with a combined mass of ~$10^7 M_\odot$, vs. some 10^2 strong ($>10^{36}$ erg s^{-1}) sources in the entire galaxy with a mass of ~$1.3 \times 10^{11} M_\odot$. A similar discrepancy holds in M31 where over 19 globular cluster sources with $L_x \geq 10^{37}$ erg s^{-1} are found among 270 observed globular clusters (within a 27.5' circle around the centre) while only 90 sources $\geq 10^{37}$ erg s^{-1} are found in that entire galaxy (see also Chapter 3, this book).

The excessively large incidence of X-ray sources in globular clusters indicates that the conditions for the formation of neutron-star binaries are much more favourable in globular clusters than anywhere else in the galaxy. The same appears to hold for binary formation by close encounters, followed by capture: in view of the very high star densities (~10^5 stars pc^{-3}) and low relative velocities (≤ 10 km s^{-1}) in globular cluster cores, the probability for binary formation by capture is much higher there than anywhere else in a galaxy. (Due to the low relative velocities, little energy has to be dissipated in the first encounter, in order to achieve capture.)

For these reasons it seems most likely that the globular cluster sources were formed in the cluster core by capture of a normal star by a compact remnant, either a neutron star or black hole, as first suggested by Clark (1975). As specific mechanisms for such capture formations, Sutantyo (1975b) considered direct red-giant–neutron star collisions in the cluster core, and Fabian et al. (1975)

considered neutron-star–main-sequence star tidal collisions. We now know that the X-ray sources in globular clusters are indeed neutron-star binaries (see Chapter 2, this book). One expects that during the first 10^8 yr of the life of a globular cluster many thousands of neutron stars will have formed (see below) of which a fraction may still be present in the clusters. Because of their higher-than-average mass such stars will have sunk to the cluster cores where the star density and the collision probability are highest.

A rough estimate of the number of neutron stars per cluster required to explain the observed incidence of X-ray sources can be made as follows. The calculations by Hills & Day (1976) for some 60 globular clusters (see also Heggie 1977; Press & Teukolsky 1977; Lightman & Shapiro 1978; see especially also Hanes & Madore 1980) show that, on the average, 3.3 per cent of all main-sequence stars in the cores of globular clusters underwent a collision in 10^{10} yr. The same is expected to hold for the neutron stars in the clusters. To explain the presence of a steady-state population of 13 globular cluster sources, this average collision frequency requires that some 30 neutron stars are present per cluster if the lifetime on an X-ray source is 10^9 yr, or some 300 neutron stars if the lifetime is 10^8 yr. (An X-ray lifetime of $\sim 10^8$–10^9 yr is expected if the companion is a star of $\sim 0.4\,M_\odot$ and the mass transfer rate is 10^{-10}–$10^{-9}\,M_\odot\,\mathrm{yr}^{-1}$, as inferred from the X-ray luminosity of 10^{36}–$10^{37}\,\mathrm{erg\,s}^{-1}$.) Since the above-mentioned 3.3 per cent is already an average over clusters with a variety of core densities, these numbers are not significantly altered if one takes into account that the X-ray sources are predominantly found in the clusters with the highest central densities (which correlate with a high metallicity and a high optical luminosity), where the collision probability per star may be an order of magnitude larger than average (Hills & Day 1976). In the case of a red-giant–neutron-star collision (about 20 per cent of the systems may have formed in this way, van der Woerd 1983) the companion to the neutron star will be a hydrogen-shell burning star with a degenerate helium core. Calculations by Savonije, Rappaport & Webbink (1982) show that internal evolution of such stars may drive a mass-transfer rate as high as $10^{-8}\,M_\odot\,\mathrm{yr}$ to the neutron star. This might be the reason why some of the globular cluster X-ray sources have very high X-ray luminosities (10^{37}–$10^{38}\,\mathrm{erg\,s}^{-1}$), which are very hard to obtain with a low-mass main-sequence companion (see Chapter 9, this book).

On the basis of this result one would expect two groups of globular cluster X-ray sources to exist, i.e. a low-luminosity group ($L_x \leqslant 10^{36}\,\mathrm{erg\,s}^{-1}$) in which the companion is a main-sequence star, and a high-luminosity group ($L_x \sim 10^{37}$–$10^{38}\,\mathrm{erg\,s}^{-1}$) in which the companion is a hydrogen-shell burning sub-giant.

The requirement of 30–300 neutron stars per globular cluster is a very modest one since, from the extrapolation of the cluster mass distribution beyond the lower mass limit M_L for neutron-star formation (by using Salpeter's IMF), one expects that in the early days of a typical globular cluster some 4000 (for $M_L = 8 M_\odot$) to 12 000 (for $M_L = 4 M_\odot$) neutron stars were formed (see Sutantyo 1975b). In view of the uncertainties in the shape of the IMF these numbers may be uncertain by as much as a factor of 4. Hence, even if less than 30 per cent (and possibly $\leqslant 3$ per cent) of the originally formed neutron stars have remained in a cluster, the present high incidence of X-ray sources can be explained.

Notice that if all originally formed neutron stars had remained in the clusters, the incidence of X-ray sources should be about an order of magnitude larger than is observed, and in each centrally concentrated cluster several sources should be present.

Thus, the observed incidence of sources in globular clusters indicates that the majority (~ 90 per cent) of the neutron stars that were originally formed in globular clusters have escaped (van den Heuvel 1980, 1981c).

This inference is in good agreement with the present-day observations of radio pulsars which show that the vast majority of the radio pulsars at birth receive runaway velocities of ~ 200 km s^{-1} (Lyne 1981). Since the escape velocity from a globular cluster core is less than 10-20 km s^{-1}, only the neutron stars in the low-velocity tail of the distribution of runaway velocities are expected to have remained in the clusters. Since the large runaway velocities of most neutron stars may well be due to the disruption of close-binary systems (see Section 8.5), the neutron stars with low runaway velocities are presumably products of the evolution of single stars, or very wide binaries. Thus, we are faced with the conclusion that in globular clusters those neutron stars that started out their evolution in binary systems have become single runaways that have left the clusters, whereas those that started out as a single star may have remained in the clusters and, in a number of cases, have later on captured themselves a companion.

Since white dwarfs are abundant in globular clusters, also many white-dwarf binaries are expected to have formed by capture in globular cluster cores (Hut & Verbunt 1983). Indeed, cataclysmic variables appear to be anomalously abundant in globular cluster cores (Webbink 1981). An alternative scenario for the formation of neutron-star binaries in globular clusters is, therefore, accretion-induced collapse of a white dwarf in a cataclysmic-variable binary (see below). The rapid burster MXB 1730−335 is a good candidate for formation by such a process, as the likely presence of a magnetic field in its neutron star suggests that this star may not be much older than about 10^8 yr (Taam & van den Heuvel 1983).

8.6.2 Formation of bulge sources and other Type II sources
8.6.2.1 Possibility of formation by capture

The relative velocities of the stars in the galactic bulge are at least an order of magnitude higher (i.e. $\geqslant 100$ km s^{-1}) than in globular clusters. Consequently, formation by capture seems most unlikely here since, in order to form a bound system, the kinetic energy to be dissipated in the first encounter is over two orders of magnitude larger than in globular clusters. This is larger than the entire binding energy of a dwarf star. Hence, a 'capturing' encounter in the bulge would at the same time lead to the complete disruption of the dwarf companion (see Finzi 1978 for details).

The following alternative 'capture' mechanisms for the formation of the low-mass Type II X-ray binaries have been suggested:

(1) They have escaped from globular clusters. Although exchange collisions may impart a recoil velocity to the newly formed neutron-star binary, these velocities are always small since a dwarf star of a few tenths of a solar mass is exchanged for a relatively massive neutron star (~ 1.5 $M_\odot$); the resulting system velocity deviates little from the original velocity of the neutron star (see Hills 1976). Therefore, escape from the cluster is unlikely.

(2) They were formed in globular clusters that have subsequently evaporated. Since globular clusters are very loosely bound systems, the formation of a very 'hard' central binary may be sufficient to evaporate the cluster (Heggie 1977). Long & van Speybroeck (Chapter 3, this book) have pointed out that the 19 sources within 400 pc (2 arcmin) from the centre of M31 resemble the globular cluster sources in being more than twice as bright as the other bulge sources in M31, and may have formed in globular clusters that have subsequently merged with the bulge. Since the X-ray globular clusters are more strongly concentrated towards the M31 centre than the general globular cluster population, this seems a viable possibility for the sources within 400 pc from the centre.

However, the spatial distribution of globular clusters from the innermost few hundred parsecs out through the bulge, follows the same $1/r^3$ law as all the other bulge population stars. This indicates that beyond a few hundred parsecs from the centre hardly any globular clusters can have merged with the bulge.

The other formation mechanism suggested for the bulge sources beyond 400 pc is:

8.6.2.2 Origin from a cataclysmic-variable-type close binary

Here the white dwarf is driven over the Chandrasekhar limit by accretion, and undergoes electron-capture collapse. Computations by Canal & Schatzman (1976) and by Canal, Isern & Labay (1980) suggest that carbon-oxygen (CO) white dwarfs with a mass very close to the Chandrasekhar limit may indeed be induced to collapse in this way. Sugimoto & Nomoto (1980), Miyaji *et al.* (1980) and Nomoto (1981) have shown that the same is true for O–Ne–Mg white dwarfs in close binaries. The implosion and the possibly induced mass ejection will in general not disrupt the system, although it may become detached during a considerable length of time following the explosion (van den Heuvel 1977, 1981*a*, *b*).

Figure 8.9 depicts this evolutionary scenario for the bulge X-ray sources, in which it is assumed that a 0.2 $M_\odot$ supernova shell is explosively ejected. A few billion years after the implosion the system will have shrunk sufficiently – due to gravitational radiation losses – to resume mass transfer by Roche-lobe overflow. It then becomes a low-mass X-ray binary of the bulge type. As the neutron star is already old by the time it becomes an X-ray source, it may have lost most of its magnetic field, and the source may become an X-ray burster as well. Consequently, the difference between these sources and the globular cluster X-ray sources is not expected to be large, since both types of objects are expected to consist of an old neutron star, together with a red (or, in some cases, a degenerate) dwarf. The main difference may be in the metal abundance. Since a lower metal abundance seems to correlate with a greater mean X-ray luminosity (Long & van Speybroeck, Chapter 3, this book), this might explain why the globular cluster sources in M31 are more than a factor of 2 brighter than the bulge sources between 2′ and 5′ from the centre.

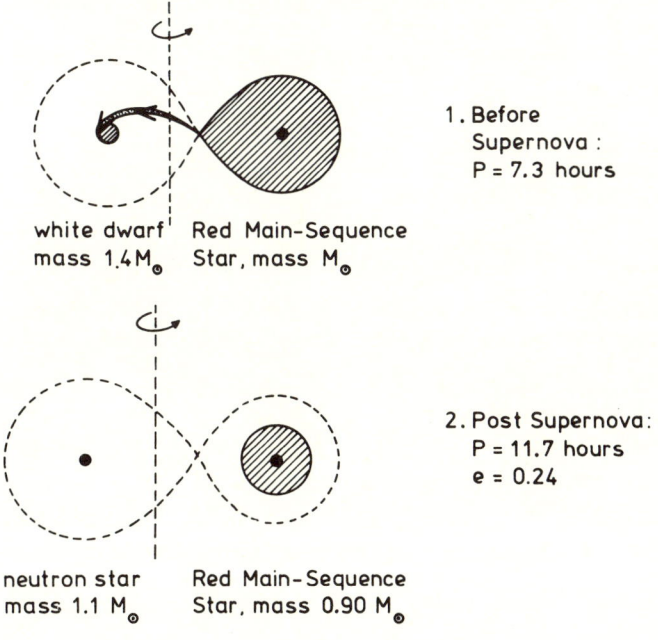

Fig. 9.9. Possible way to produce a low-mass neutron-star binary by accretion-induced implosion of a massive white dwarf in a cataclysmic binary. It is assumed that a 0.2 $M_\odot$ supernova shell is ejected with $v = 10^4 \text{km s}^{-1}$. As a result of the impact and mass loss, the system is detached for several billion years after the supernova. If less mass is ejected and the companion has a mass $\leq 0.1 M_\odot$, the duration of the detached phase may become as short as 10^8 yr (Taam & van den Heuvel 1983).

The scenario is a rare one since, in order to implode to a neutron star, the white dwarf must have been born with a mass that differs less than 0.01 $M_\odot$ from the Chandrasekhar limit.

Vader *et al.* (1982) have argued that this scenario might be the only one that can explain the observed distributions of novae and bulge X-ray sources in M31. The scenario is expected to be a rare one since, in order to implode to a neutron star, the white dwarf as well as the accretion rate must fulfill very special conditions (Canal, Isern & Labay 1980; Sugimoto & Nomoto 1980). For CO white dwarfs the conditions become more favourable after several billion years as a consequence of cooling and gravitational settling of oxygen in the stellar core (Labay, Canal & Isern 1983). This might explain why the low-mass X-ray binaries tend to belong to an old stellar population.

References

Abramowicz, M. A. & Piran, T. 1980, *Ap. J.*, **241**, L7.
Abt, H. A. & Levy, S. G. 1978, *Ap. J. Suppl.*, **36**, 241.
Arnett, W. D. 1978, *In: Physics and Astrophysics of Neutron Stars and Black Holes* (R. Giacconi & R. Ruffini, eds.) North Holland, Amsterdam, p. 356.
Arnett, W. D. 1979, *Ap. J.*, **230**, L37.
Bahcall, J. N. 1978, *Ann. Rev. Astron. Ap.*, **16**, 241.
Blaauw, A. 1961, *Bull. Astron. Inst. Neth.*, **15**, 265.
Blaauw, A. & van Albada, T. S. 1974 (unpublished preprint).
Bodenheimer, P. 1981, *In: Fundamental Problems in the Theory of Stellar Evolution* (D. Sugimoto *et al.*, eds.), Reidel, Dordrecht, pp. 5-26.
Boersma, J. 1961, *Bull. Astron. Inst. Neth.*, **15**, 291.
Bouwman, P. 1980, *Ap. Space Sci.*, **68**, 525.
Canal, R., Isern, J. & Labay, J. 1980, *Ap. J.*, **241**, L33.
Canal, R. & Schatzman, E. 1976, *Astron. Ap.*, **46**, 229.
Cassinelli, J. P. 1979, *Ann. Rev. Astron. Ap.*, **17**, 275.
Chevalier, R. 1975, *Ap. J.*, **199**, 189.
Chevalier, R. 1981, *Ap. J.*, **246**, 267.
Clark, G. W. 1975, *Ap. J.*, **199**, L143.
Conti, P. S. 1981. Paper presented at *IAU Symp.*, 99: *Wolf-Rayet Stars* (C. De Loore & A. Willis, eds.), Reidel, Dordrecht, **3**, 22.
Counselman, C. C. 1973, *Ap. J.*, **180**, 307.
Crampton, D. & Hutchings, J. B. 1981, *Ap. J.*, **251**, 604.
Damashek, M., Backus, P. R., Taylor, J. H. & Burkhardt, R. K. 1982, *Ap. J.*, **253**, L57.
Darwin, G. H. 1908, *Scientific Papers*, vol. 2, Cambridge University Press.
De Cuyper, J. P. 1981, *In: Pulsars* (W. Sieber & R. Wielebinski, eds.), Reidel, Dordrecht, pp. 399-401.
De Cuyper, J. P. 1982, *In: Binary and Multiple Stars as Tracers of Stellar Evolution* (Z. Kopal & J. Rahe, eds.), Reidel, Dordrecht (in press).
De Greve, J. P., De Loore, C. & Sutantyo, W. 1975, *Ap. Space Sci.*, **38**, 301.
De Greve, J. P., De Loore, C. & van Dessel, E. 1978, *Ap. Space Sci.*, **53**, 105.
De Jager, C. 1980, *The Brightest Stars*, Reidel, Dordrecht.
Delgado, A. 1979, *In: Mass Loss and Evolution of O-type Stars* (P. S. Conti & C. De Loore, eds.), Reidel, Dordrecht, p. 145.
Delgado, A. & Thomas, H. C. 1981, *Astron. Ap.*, **96**, 142.
De Loore, C. 1981, *In: Effects of Mass Loss on Stellar Evolution* (C. Chiossi & R. Stalio, eds.), Reidel, Dordrecht, pp. 405-30.
De Loore, C. 1980, *Space Sc. Rev.*, **26**, 113.

Eddington, A. S. 1926, *The Internal Constitution of the Stars*, Cambridge University Press, pp. 16ff.
Fabian, A. C., Pringle, J. E. & Rees, M. 1975, *Month. Not. Roy. Astron. Soc.*, **1972**, 15P.
Fahlman, G. G. & Gregory, P. C. 1981, *Nature*, **293**, 202.
Finzi, A. 1978, *Astron. Ap.*, **62**, 149.
Flannery, B. P. & Ulrich, R. K. 1977, *Ap. J.*, **212**, 533.
Fryxell, B. A. & Arnett, W. D. 1981, *Ap. J.*, **243**, 994.
Garmany, C. D., Conti, P. S. & Massey, P. 1980, *Ap. J.*, **242**, 1063.
Gursky, H. 1973. Lecture presented at NATO Advanced Study Institute on 'Physics of compact objects', Cambridge, England.
Hadjidemetriou, J. 1967, *Adv. in Astron. Ap.*, **5**, 131.
Habets, G. 1983, *Astron. Ap.* (to be published).
Hanes, D. & Madore, B. 1980, *Globular Clusters*, Cambridge University Press.
Heggie, D. C. 1977, *Comments on Ap. and Space Phys.*, 7, 43.
Hills, J. G. 1976, *Month. Not. Roy. Astron. Soc.*, **175**, 1P.
Hills, J. G. & Day, C. A. 1976, *Ap. Lett.*, **17**, 87.
Huang, Su-Shu 1963, *Ap. J.*, **138**, 471.
Hut, P. 1980, *Astron. Ap.*, **92**, 167.
Hut, P. 1982, *Astron. Ap.*, **110**, 37.
Hut, P. & van den Heuvel, E. P. J. 1981, *Astron. Ap.*, **94**, 327.
Hut, P. & Verbunt, F. 1983, *Nature*, **301**, 587-9.
Katz, J. I. 1975, *Nature*, **253**, 698.
Katz, J. I. 1980, *Ap. J.*, **236**, L127.
Kelley, R. 1981 (private communication).
Kippenhahn, R. 1969, *Astron. Ap.* 3, 83.
Kippenhahn, R. & Meyer-Hofmeister, E. 1977, *Astron. Ap.*, **54**, 539.
Kippenhahn, R. & Thomas, H. C. 1980, *Annals N.Y. Acad. Sci.*, **336**, 579.
Kippenhahn, R. & Weigert, A. 1967, *Zeits. f. Ap.*, **65**, 251.
Kraicheva, Z. T., Popova, E. I., Tutukov, A. V. & Yungelson, L. R. 1979, *Sov. Astron.*, **23**, 290.
Kuiper, G. P. 1941, *Ap. J.*, **93**, 133.
Labay, J., Canal, R. & Isern, J. 1983, *Astron. Ap.*, **117**, L1-L4.
Lamb, S. A., Iben, I. & Howard, W. M. 1976, *Ap. J.*, **207**, 209.
Lamers, H. J. 1981, *Ap. J.*, **245**, 593.
Larson, R. B. 1972, *Monthly Not. Roy. Astron. Soc.*, **157**, 121.
Lightman, A. P. & Shapiro, S. L. 1978, *Rev. Mod. Phys.*, **50**, 437-81.
Lucy, L. B. 1981, *In: Fundamental Problems in the Theory of Stellar Evolution* (D. Sugimoto et al., eds.), Reidel, Dordrecht, p. 75.
Lucy, L. B. & Ricco, E. 1979, *Astron. J.*, **84**, 401.
Lyne, A. G. 1981, *In: Pulsars* W. Sieber & R. Wielebinski, eds.), Reidel, Dordrecht, pp. 423-36.
Maeder, A. 1981, *Astron. Ap.*, **99**, 97.
Montegazza, L., Paolicchi, P., Farinella, P. & Luzny, F. 1980, *In: Close Binary Stars: Observations and Interpretation* (M. Plavec et al., eds.), Reidel, Dordrecht, pp. 23-6.
Maraschi, L., Treves, A. & van den Heuvel, E. P. J. 1976, *Nature*, **259**, 292.
Massevitch, A. G., Tutukov, A. V. & Yungelson, L. R. 1976, *Ap. Space Sci.*, **40**, 115.
Massey, P. 1981, *Ap. J.*, **246**, 153.
Massey, P. 1982, *In: Wolf-Rayet Stars* C. De Loore & A. Willis, eds.), Reidel, Dordrecht, pp. 251-63.
Mazurek, T. J. & Wheeler, J. C. 1980, *Fundamentals of Cosmic Physics*, 5, 193.
Meyer, F. & Meyer-Hofmeister, E. 1979, *Astron. Ap.*, **78**, 167.
Miyaji, S., Nomoto, K., Yokoi, K. & Sugimoto, D. 1980, *Publ. Astron. Soc. Japan*, **32**, 303.
Nariai, K. 1975, *Astron. Ap.*, **43**, 309.
Noels, A. & Masereel, C. 1982, *Astron. Ap.*, **105**, 293.

Nomoto, K. 1982, *Ap. J.*, **253**, 798.
Nomoto, K. 1982, *In: Fundamental Problems in the Theory of Stellar Evolution* D. Sugimoto *et al.*, eds.), Reidel, Dordrecht, p. 295.
Ostriker, J. P. 1976, *In: Structure and Evolution of Close Binary Systems* (P. Eggleton *et al.*, eds.), Reidel, Dordrecht, p. 206.
Paczynski, B. 1967, *Acta Astron.*, **17**, 355.
Paczynski, B. 1971, *Ann. Rev. Astron. Ap.*, **9**, 183.
Paczynski, B. 1976, *In: Structure and Evolution of Close Binary Systems* (P. Eggleton *et al.*, eds.), Reidel, Dordrecht, p. 75.
Paczynski, B. & Sienkiewicz, R. 1972, *Acta Astron.*, **22**, 73.
Plavec, M. J. 1981, *In: Effects of Mass Loss on Stellar Evolution* (C. Chiosi & R. Stalio, eds.), Reidel, Dordrecht, pp. 431-56.
Press, W. H. & Teukolsky, S. A. 1977, *Ap. J.*, **213**, 183.
Rappaport, S. & van den Heuvel, E. P. J. 1982, *In: Be-stars* (M. Jaschek & H. G. Groth, eds.), Reidel, Dordrecht, pp. 327-46.
Refsdal, S., Roth, M. L. & Weigert, A. 1974, *Astron. Ap.*, **36**, 113.
Ritter, H. 1976, *Mon. Not. Roy. Astron. Soc.*, **175**, 279.
Salpeter, E. E. 1959, *Ap. J.*, **129**, 608.
Savonije, G. J. 1978, *Astron. Ap.*, **62**, 317.
Savonije, G. J. & Papaloizou, J. 1983, *MNRAS*, **203**, (in press).
Savonije, G. J., Rappaport, S. & Webbink, R. 1983, *Ap. J.* (in press).
Shakura, N. I. & Sunyaev, R. A. 1973, *Astron. Ap.*, **24**, 337.
Shu, F. H. & Lubow, S. H. 1981, *Ann. Rev. Astron. Ap.*, **19**, 277.
Smarr, L. L. & Blandford, R. 1976, *Ap. J.*, **207**, 574.
Smith, L. F. 1973, *In: Wolf-Rayet and High Temperature Stars* (M. K. V. Bappu & J. Sahade, eds.), Reidel, Dordrecht, pp. 15-35.
Srinivasan, G. & van den Heuvel, E. P. J. 1982, *Astron. Ap.*, **108**, 143-7.
Sugimoto, D. & Nomoto, K. 1980, *Space Science Rev.*, **25**, 155.
Sutantyo, W. 1974a, *Astron. Ap.*, **31**, 339.
Sutantyo, W. 1974b, *Astron. Ap.*, **35**, 251.
Sutantyo, W. 1975a, *Astron. Ap.*, **41**, 47.
Sutantyo, W. 1975b, *Astron. Ap.*, **44**, 227.
Taam, R. E. 1980, *Ap. Lett.*, **20**, 29.
Taam, R. E., Bodenheimer, P. & Ostriker, J. P. 1978, *Ap. J.*, **222**, 269.
Taam, R. E. & van den Heuvel, E. P. J. 1983, *Astron. Ap.* (in press).
Taylor, J. H. & Weisberg, J. M. 1982, *Ap. J.*, **253**, 908.
Thomas, H. C. 1977, *Ann. Rev. Astron. Ap.*, **15**, 127.
Trimble, V. 1974, *Astron. J.*, **79**, 967.
Tutukov, A. V. & Yungelson, L. R. 1973, *Nautsnie Inform.*, **27**, 58.
Tutukov, A. V. 1981, *In: Fundamental Problems in the Theory of Stellar Evolution* (D. Sugimoto *et al.*, eds.), Reidel, Dordrecht, p. 137.
Vader, J. P., van den Heuvel, E. P. J., Lewin, W. H. & Takens, R. J. 1982, *Astron. Ap.*, **113**, 328-35.
van Albada, T. S. 1968, *Bull. Astron. Inst. Neth.*, **20**, 47.
van Beveren, D. 1980, Ph.D. Thesis, Vrije Univ., Brussels.
van Beveren, D. 1982, *Astron. Ap.*, **105**, 260.
van Beveren, D. & Conti, P. S. 1980, *Astron. Ap.*, **88**, 230.
van den Heuvel, E. P. J. 1973, *Nature Phys. Sci.*, **242**, 71.
van den Heuvel, E. P. J. 1974, *Proc. 16th Solvay Conf. on Physics*, University of Brussels Press, p. 119.
van den Heuvel, E. P. J. 1976, *In: Structure and Evolution of Close Binary Systems* (P. Eggleton *et al.*, eds.), Reidel, Dordrecht, p. 35.
van den Heuvel, E. P. J. 1977, *Ann. N.Y. Acad. Sci.*, **302**, 14.
van den Heuvel, E. P. J. 1978, *In: Physics and Astrophysics of Neutron Stars and Black Holes* (R. Giacconi & R. Ruffini, eds.), North Holland, Amsterdam, p. 828.
van den Heuvel, E. P. J. 1979, *In: Mass Loss and Evolution of O-type Stars* (P. S. Conti & C. De Loore, eds.), Reidel, Dordrecht, pp. 491-8.

van den Heuvel, E. P. J. 1980, *In: X-ray Astronomy* (R. Giacconi & G. Setti, eds.), Reidel, Dordrecht, pp. 115-27.
van den Heuvel, E. P. J. 1981a, *In: Fundamental Problems in the Theory of Stellar Evolution* (D. Sugimoto, D. Q. Lamb & D. N. Schramm, eds.), Reidel, Dordrecht, pp. 155-75.
van den Heuvel, E. P. J. 1981b, *In: Pulsars* (W. Sieber & R. Wielebinski, eds.), Reidel, Dordrecht, pp. 379-96.
van den Heuvel, E. P. J. 1981c, *Space Sci. Rev.*, **30**, 623-42.
van den Heuvel, E. P. J. & Heise, J. 1972, *Nature Phys. Sci.*, **239**, 67.
van den Heuvel, E. P. J. & De Loore, C. 1973, *Astron. Ap.*, **25**, 387.
van den Heuvel, E. P. J. & Ostriker, J. P. 1973, *Nature Phys. Sci.*, **245**, 99.
van den Heuvel, E. P. J., Ostriker, J. P. & Petterson, J. A. 1980, *Astron. Ap.*, **81**, L7.
van der Klis, M. 1983, Ph.D. Thesis, Univ. of Amsterdam.
van der Linden, Th. J. 1982, Ph.D. Thesis, Univ. of Amsterdam.
van der Woerd, H. 1983, *Astron. Ap.* (in preparation).
Weaver, J. A. & Woosley, S. E. 1980, *Proc. 9th Texas Symp. on Relativ. Astrophys., Ann. N.Y. Acad. Sci.*, **336**, 335.
Webbink, R. F. 1979, *In. IAU Colloq.*, No. 53, *White Dwarfs and Variable Degenerate Stars* (H. M. van Horn & V. Weidemann, eds.), Rochester University Press, p. 426.
Wheeler, J. C., McKee, C. F. & Lecar, M. 1974, *Ap. J.*, **192**, L71.
Wheeler, J. C., Lecar, M. & McKee, C. F. 1975, *Ap. J.*, **200**, 145.
Zahn, J. P. 1977, *Astron. Ap.*, **57**, 383.
Ziolkowski, J. 1977, *Ann. N.Y. Acad. Sci.*, **302**, 47-54.

9

EVOLUTION AND MASS TRANSFER IN X-RAY BINARIES

Gert Jan Savonije
Astronomical Institute, University of Amsterdam

9.1 Introduction

The about-one-hundred bright X-ray sources ($L_x \gtrsim 10^{35.5}$ erg s^{-1}) found in our galaxy can be divided into two broad classes, commonly referred to as Type I and Type II X-ray sources. There are some 30 Type I sources which characteristically have hard X-ray spectra ($kT \gtrsim 9$ keV) and often show pulsations (see also Chapters 1 and 8, this book). The optical counterparts of these X-ray sources are generally massive ($M \gtrsim 15$ M$_\odot$) and luminous early-type stars that belong to the extreme Population I stars with an average height above the galactic plane of $|z| \lesssim 150$ pc (see Chapter 5, this book).

The Type II X-ray sources, on the other hand, have softer spectra (3 keV $< kT < 7$ keV) and do not show regular pulsations. These X-ray sources have undetected or very faint optical counterparts which show in general no stellar absorption features. Type II X-ray sources are strongly concentrated towards the galactic centre and are probably related to the old disk population of low-mass stars in the galactic bulge (see Chapter 2, this book).

The commonly accepted model for the bright galactic X-ray sources of both types consists of a compact star (neutron star or black hole) which accretes matter from a close companion star. When this matter falls down the deep gravitational well of the compact star it gains an enormous amount of kinetic energy which – when dissipated at or close to the surface of the compact star – is liberated as strong X-ray emission. For the binary character of Type I sources, also commonly referred to as massive X-ray binaries, there is an overwhelming observational confirmation since these sources often show eclipses and in many cases show readily measurable doppler orbits and regular photometric variations. The observational support for the binary character of Type II X-ray sources is increasing (e.g. see Chapters 2 and 5, this book). Since the only plausible environment in which a compact star will be able to accrete matter at a rate higher than

$\approx 10^{-11}$ $M_\odot$ yr^{-1} (as implied by the observed X-ray flux of galactic bulge sources scaled to a distance of 10 kpc) seems the close neighbourhood of another, mass-spilling star, we will assume here that all Type II X-ray sources are indeed binary stars.

In the following sections we will discuss the character of the mass transfer towards the compact companion of these binary systems. The details of this mass transfer depend strongly on the evolutionary history and present evolutionary state of the donor star, which are unfortunately rather poorly understood, especially for the Type II X-ray sources (see Chapter 8, this book).

In Section 9.2 we start with a brief review of the main principles of mass transfer, especially by Roche-lobe overflow, in close-binary systems. We will apply these results in Section 9.3 to both Type I and Type II X-ray sources in an attempt to explain the observed X-ray luminosities of these sources.

9.2 The main principles of mass transfer in binary systems
9.2.1 Mass transfer and orbital evolution

If M_1 is the mass of the mass-losing star, M_2 that of its companion and $M = M_1 + M_2$ the total mass of the system, the total orbital angular momentum of the binary system (with a circular orbit) can be expressed as:

$$J_{\text{orb}} = \frac{M_1 M_2}{M} \Omega a^2, \tag{9.1}$$

where a is the orbital separation, Ω is the orbital angular velocity given by $\Omega^2 = GM/a^3$, and G the constant of gravity. Suppose that a fraction α of the transferred mass leaves the system, then one can express the rate of change of the orbital separation as:

$$\frac{\dot{a}}{a} = -2 \left\{ 1 + (\alpha - 1)\frac{M_1}{M_2} - \frac{1}{2}\alpha\left(\frac{M_1}{M}\right) \right\} \frac{\dot{M}_1}{M_1} + 2\frac{\dot{J}_{\text{orb}}}{J_{\text{orb}}}. \tag{9.2}$$

For conservative evolutions (i.e. no mass loss from the system), ($\dot{J}_{\text{orb}} = 0$; $\alpha = 0$) this becomes:

$$\frac{\dot{a}}{a} = -2\left(1 - \frac{M_1}{M_2}\right)\frac{\dot{M}_1}{M_1}, \tag{9.2a}$$

which shows that the mass transfer ($\dot{M}_1 < 0$) gives rise to an expanding orbit if $M_1/M_2 < 1$ and to a decaying orbit if $M_1/M_2 > 1$. If $\alpha > 0$, which occurs, for example, when the mass-losing star ejects a stellar wind, the orbital evolution depends very much on the specific angular momentum $\alpha^{-1}(\dot{J}_{\text{orb}}/\dot{M}_1)$ of the outflowing matter. Because this quantity is poorly known, the effect of a stellar wind makes the evolution of the binary system uncertain. The binary evolution may be non-conservative, see above, even when $\alpha = 0$. Tidal interaction, for example, will exchange angular momentum between the orbit and the rotation of the binary components. But, except for rather close binaries, this is in general not very important for the orbital evolution because $J_{orb} \gg J_{spin}$. Stars in tight binary systems with very short orbital periods revolve so rapidly that they emit gravitational waves which carry away orbital angular momentum at rate (e.g. Landau & Lifshitz, 1951):

$$\frac{\dot{J}_{orb}}{J_{orb}} = -\frac{32}{5}\frac{G^3}{C^5} M_1 M_2 M a^{-4} \text{ s}^{-1}, \qquad (9.3)$$

which can influence the orbital evolution noticeably. For a sufficiently small orbital separation the angular momentum losses given by Equation (9.3) dominate Equation (9.2) and will make $\dot{a}/a < 0$, even when mass is transferred to a more-massive companion. Very narrow binaries are therefore 'captured' by gravitational radiation and will become even narrower (Section 9.3.2.1).

9.2.2 Critical radius for Roche-lobe overflow (synchronous contact star)

The primary star in an X-ray binary system is distorted by the gravitational field of its compact companion, which produces two tidal bulges at its surface. These tides are stationary only if the primary rotates synchronously with the orbital revolution of the compact companion. One can then derive the net force on each gas element in the primary from a potential (e.g. Kopal 1959, Kruszewski 1966). An infinite family of concentric equipotential surfaces can be constructed, which become more and more distorted from spherical symmetry when going outwards from the centre of the star. The particular equipotential surface which passes through the inner Lagrangian point L_1, located between the two stars, is called the critical lobe or Roche lobe. If the secondary has expanded enough to fill this critical equipotential surface the unbalanced gas pressure at L_1 (where the net gravity vanishes) pushes the gas out, after which it falls down towards the companion as a relatively narrow jet of gas (e.g. Lubbow & Shu 1975). This type of mass transfer to the companion is called Roche-lobe overflow (e.g. Plavec 1968).

It is usual to measure the size of the Roche lobe in terms of the critical radius R_c which is defined as the radius of the sphere with the same volume as the Roche lobe. Approximating the gravitational field in the binary by that of two pointmasses (most of the mass is in the relatively undisturbed inner regions of the stars) one can calculate R_c relatively easily (e.g. Kopal 1959) and compare it with the radius of the one-dimensional (spherically symmetric) stellar model as calculated by a stellar evolution code (e.g. Kippenhahn et al. 1967).

Paczynski (1971) has derived a simple approximate expression for the critical radius around the primary component (with mass M_1):

$$\frac{R_c}{a} = 0.38 + 0.2\,{}^{10}\!\log\!\left(\frac{M_1}{M_2}\right) \quad \text{for } 0.3 \lesssim \frac{M_1}{M_2} \lesssim 20, \tag{9.4}$$

$$\frac{R_c}{a} = 0.462 \left(\frac{M_1}{M}\right)^{1/3} \quad \text{for } \frac{M_1}{M_2} \lesssim 0.8, \tag{9.4a}$$

where a is the orbital separation of the binary and M_1 and M_2 are the mass of the contact star and its (compact) companion respectively. We have seen that matter will be transferred to the companion if the star begins to exceed its Roche lobe, i.e. $R > R_c$. The mass-transfer rate to the companion can be approximated by (e.g. Jedrezejec 1969, Webbink 1975, Savonije 1978, 1979):

$$\dot{M} \approx \int_{\psi_c}^{\psi_s} \rho c_s \frac{dA}{d\psi}\, d\psi, \tag{9.5}$$

where $\rho(\psi)$ and $c_s(\psi)$ are the density and sound speed in the vicinity of L_1 evaluated as functions of ψ. The gravitational potential ψ (corrected for the centrifugal force) assumes the values ψ_c and ψ_s at the Roche lobe and the stellar surface respectively. A is the cross-sectional area of the flow near L_1 and can be calculated from a series expansion of ψ near the gravitational saddlepoint (Savonije 1978),

$$\frac{dA}{d\psi} = -2\pi(1-\phi)^{-1/2}\phi\Omega^{-2},$$

where ϕ is a dimensionless function of the binary mass ratio and Ω is the orbital angular velocity. It is convenient to translate $\Delta\psi = \psi_s - \psi_c$ into a radius excess $\Delta R = R - R_c$ by the approximate relation:

$$\Delta\psi \approx -\left(\frac{GM_1}{RR_c}\right)\Delta R.$$

The mass-transfer rate that results from Equation (9.5) depends on

(1) how the stellar radius R reacts on the mass loss and
(2) how the critical radius R_c changes when the binary mass ratio is altered by the mass transfer.

These two effects will be discussed in the next section.

9.2.3 The mass-transfer rate during Roche-lobe overflow

Let us first discuss how the critical radius reacts on the mass transfer. According to Equations (9.4) (R_c/a) always decreases during mass transfer from M_1 to M_2. Yet, the critical radius itself may increase if the binary mass ratio M_1/M_2 is small enough (Equation (9.2a)). But for a mass ratio $M_1/M_2 > 1$, the critical radius must necessarily shrink during mass transfer. In such a situation mass transfer is always rapid (see below) and takes place on a thermal timescale or even shorter, depending on whether the outer layers of the contact star are radiative or convective (cf. Paczynski 1970). Let us discuss both cases in turn.

9.2.3.1 The contact star's envelope is radiative

If the envelope of the contact star is radiative the temperature gradient in the envelope must be subadiabatic, so that the specific entropy of the stellar material increases outwards. When the star loses some mass to its companion it will react first by expanding on a (short) dynamical timescale to recover hydrostatic equilibrium. After this adiabatic expansion (with constant entropy) the star will be slightly smaller sized than it was before the mass loss. But the star is not yet in thermal equilibrium. In order to attain that the welled-up and expanded material must first absorb some energy from the radiation flux to restore the equilibrium entropy gradient. During this thermal relaxation process the contact star will tend to re-expand to almost its initial size before the beginning of the mass loss. But, depending on the amount of mass transferred and the mass ratio of the binary system, the Roche lobe may have shrunk enough to inhibit this. If the contact star is more massive than its companion it will certainly exceed its then shrunken Roche lobe before it has relaxed thermally. This results in further mass transfer, so that thermal equilibrium will be unattainable. The Roche lobe will shrink ever more, so that a thermal runaway process develops, during which the mass-transfer rate increases steeply. The large amount of energy absorbed by the envelope makes the contact star in the end far less luminous than a corresponding star in thermal equilibrium would be. The rapid mass transfer will continue until the contact star has lost so much mass that the mass ratio of the binary is reversed, after which the Roche lobe starts to expand when more mass is transferred. One would expect (e.g. Paczynski 1971), as is indeed corroborated by detailed calculations, that the peak mass-transfer rate during such a thermal runaway is of order:

$$\dot{M}_1 \approx \frac{M_1}{\tau_{KH}} \approx 3 \times 10^{-8} \frac{RL}{M_1} \ M_\odot \, yr^{-1}, \tag{9.6}$$

where τ_{KH} is the contact star's Kelvin-Helmholtz timescale and R, L and M_1 the contact star's radius, luminosity and mass in solar units respectively.

The mass transfer will be far more peaceful if the contact star is less massive than its companion because the Roche lobe will then expand during mass transfer. The contact star must therefore expand beyond its original size in order to continue the mass transfer and has time to relax to thermal equilibrium. The mass transfer then takes place on the natural expansion timescale of the contact star (and stops if the star shrinks during its further evolution).

A main-sequence star expands on a long, nuclear timescale and can thus drive a rather mild mass transfer if the companion is the more-massive component. But, if the contact star is in the rapid core contraction phase just after exhaustion of hydrogen in the core, the envelope expands very rapidly, so that the star will drive a strong mass transfer, even to a more-massive companion.

9.2.3.2 The contact star's envelope is convective

Suppose that the contact star has a deep convective envelope with a super-adiabatic temperature gradient. The entropy of the envelope matter will then decrease outwards. Suppose further that the binary mass ratio M_1/M_2 is large enough that the Roche lobe remains constant or shrinks during the mass transfer. After the transfer of some matter the contact star expands adiabatically to restore hydrostatic equilibrium (entropy remains constant), whereby the upwelling material shows an entropy excess with respect to the equilibrium configuration. The star will therefore tend to shrink when it evolves back to thermal equilibrium (if the situation would allow the star to do so). The mass transfer depends thus critically on the size of the star just after the adiabatic expansion. If this size is larger than the Roche lobe the mass transfer will continue on the very fast adiabatic expansion timescale and result in a runaway that is even more violent than for a radiative envelope (Paczynski, Ziolkowski & Zytkow 1969). Extremely high mass-transfer rates are then expected. But, these very high mass fluxes will affect the convective envelope in a complicated and as yet unpredictable way, so that no reliable estimates of $\dot{M}$ can be made.

The mass transfer will slow down if the binary mass ratio M_1/M_2 has become sufficiently small for the Roche lobe to expand enough to compensate for the mass-loss-induced growth of the stellar radius. The mass transfer will then again take place on the evolutionary expansion timescale of the star.

9.2.4 Some comments on non-synchronous rotation and non-conservative evolution

Above we have introduced the usual idealised model of a circularised binary in which the contact star rotates in synchronism with the orbital revolution of the companion and transfers mass conservatively, i.e. without spilling mass and angular momentum from the binary system. Although these assumptions seem in general rather unrealistic we may hope that the results obtained give us at least a qualitative framework from which we can derive an idea of how a real mass-transferring binary evolves.

Let us briefly mention some of the difficulties that are encountered if one tries to consider more realistic binary models.

If the rotation of the contact star is not synchronised with the orbital revolution of the compact companion, the latter star will induce a tidal wave running along the surface of the contact star. The amplitude of this so-called 'dynamical tide' may be substantially smaller or larger than the amplitude of the equilibrium tide that would be raised up if the binary were synchronised. The latter can be the case when the relative orbital period of the companion ($P_{orb} = 2\pi/|\Omega - \omega|$, where ω and Ω are the angular velocity of the star's rotation and the orbital revolution respectively) happens to be close to the period of one of the normal modes of the contact star. Natural modes with relevant periods (~ days) are the g-modes (Cowling 1943) or r-modes (Papaloizou & Pringle 1978). Strong synchronising torques can result if one of those modes is close to resonance with the orbital revolution of the companion (Zahn 1977, Savonije & Papaloizou 1983). Such a resonant mode may give rise to overflow in an otherwise non-mass-transferring binary system, or may increase the mass-transfer rate substantially over a period of time. As the contact star evolves it may pass through a number of resonances.

A second complication in non-synchronised or non-circularised binary systems is related to the fact that one can no longer apply Equation (9.4) for the radius of the critical lobe. When the tidal wave is non-stationary in the reference frame corotating with the binary system, one can no longer express the effective force (due to gravity and inertial effects) as the gradient of a (Roche) potential. Thus, as seems to have been first noted by Limber (1963), equipotential (e.g. Roche-) lobes have in that case little meaning. Pratt & Strittmatter (1976) evaluated an expression for the critical radius of a non-synchronous binary component. Their expression is derived on the basis of a modified Roche potential and is only valid (approximately) if the star rotates almost synchronously ($\omega/\Omega \approx 1$). Lubow (1979) studied this case ($\omega/\Omega \approx 1$) in some detail and derived a general form for the velocity field in the deformed star. He also estimated the angle between isobars and isodensity surfaces, etc.

One may conclude that mass transfer in non-synchronous and non-circular binary systems is a complicated and, as yet, unsolved problem. Particle-trajectory calculations such as those performed by Kopal (1959) and Kruszewski (1966) give some idea of how this type of mass transfer may occur, but only to a limited extent, since the interaction between the gas elements – which seems crucial – is not included in these calculations. An appreciable fraction of the overflowing matter may fall back onto the non-synchronous star, especially when this star is much more massive than its companion, as in massive X-ray binary systems (Petterson 1978). Under these circumstances it seems impossible to define a 'critical lobe' for mass transfer. Only detailed 3-dimensional hydrodynamical calculations (which can handle the shocks that will arise when the ejected matter falls back) can give us insight into the mass-transfer process.

Another difficult and unsolved problem is how much mass and angular momentum is lost from the system during mass transfer. Stellar-wind losses can be substantial for massive stars and evolved giants. One can estimate the effect of the loss of mass and *orbital* angular momentum by such a stellar wind. Suppose that the star with mass M_1 loses mass isotropically at a rate $\dot{M}_1$ and that nothing of this wind material is accreted by the companion. The specific orbital angular momentum of the wind material is (circular orbit) $j_1 = (M_2/M)^2 a^2 \Omega$, so that the orbital angular momentum losses by the stellar wind are:

$$\frac{\dot{J}_{\text{orb}}}{J_{\text{orb}}} = \left(\frac{M_2}{M}\right)\frac{\dot{M}_1}{M_1}.$$

Substituting this into Equation (9.2) and putting $\alpha = 1$ yields

$$\frac{\dot{a}}{a} = -\frac{\dot{M}_1}{M}. \tag{9.7}$$

This would mean that the stellar-wind ejection tends *to expand* the orbit of the binary (independent of whether M_1 or M_2 is the most massive star). It would thus tend to stabilise simultaneously occurring Roche-lobe overflow (cf. Basko et al. 1977). However, the stellar-wind material also carries away rotational angular momentum, so that the rotation of the mass-losing star is slowed down. But tidal interaction will tend to spin the star up to a synchronous state. During this process angular momentum is taken out of the orbit and transferred to the contact star's rotation to compensate for the stellar-wind losses, thereby tending to *decrease* the orbital separation. Because one does not know how much spin angular momentum is carried off by the wind and how effective the tidal interaction in the binary is, the net effect of a stellar wind is very uncertain. Depending on the precise balance between the widening and narrowing effect of stellar-wind losses, the orbit may actually shrink or expand.

The rotational braking effect of stellar-wind losses may be particularly strong if the outflowing wind material is magnetically forced to corotate with the star up to an appreciable distance from the photosphere (cf. Mestel 1968). There is observational evidence (e.g. Middelkoop 1981) that this can be an important effect in stars with cool, convective envelopes. It is thought that these stars also tend to undergo an effective tidal interaction when situated in a close-binary system (e.g. Zahn 1977). Eggleton (1976) and, more recently, Verbunt & Zwaan (1981) have stressed the possible importance of this effect for the mass-transfer rates in low-mass, close-binary systems, like cataclysmic variables and Type II X-ray binaries. When the mass-transfer rate during Roche-lobe overflow becomes so high that it cannot all be accreted by the companion, substantial mass loss from a binary system may occur. If the companion star is compact, the radiation pressure from the liberated accretion energy may limit the accretion rate and blow part of the infalling gas out of the system. For spherically symmetrically infalling fully-ionised gas the maximum accretion rate onto a neutron star of mass M_x ($M_\odot$) is given by (cf. Eddington 1926, Lightman *et al.* 1978),

$$\dot{M}_{Edd} = 1.5 \times 10^{-8} M_x \; (M_\odot \, yr^{-1}). \tag{9.8}$$

For non-spherical infall the maximum accretion rate may be higher, but probably not by more than a factor of about 5 (e.g. Rees 1974, Thorne 1974). When the accreting star is a non-degenerate star, it will swell up if the infall time is less than its thermal relaxation time (assuming that its envelope has a sub-adiabatic temperature gradient), as first suggested by Benson (1970). When the accreting star swells up beyond its Roche lobe, a common-envelope binary is formed, from which mass may be lost through the outer Lagrangian points (cf. Flannery & Ulrich 1977, Kippenhahn & Meyer-Hofmeister 1977).

All these effects are, however, very poorly understood and no reliable estimates for the resulting mass and angular momentum losses from the binary system exist at present.

9.3 Application to X-ray binaries
9.3.1 Mass transfer in Type I X-ray binaries

The massive early-type stars found in Type I X-ray binaries are known to emit fairly strong stellar winds, like all massive stars. Estimates for the wind losses vary from roughly $10^{-8} \, M_\odot \, yr^{-1}$ for unevolved main-sequence stars up to $10^{-5} \, M_\odot \, yr^{-1}$ or even more for some exotic-type stars like Of stars or Wolf-Rayet stars (cf. Conti & Garmany 1980, Lamers 1981). These stellar winds from massive, hot stars are generally thought to be driven by radiation pressure on the resonance lines of highly ionised atoms (cf. Lucy & Salomon 1970 and Cassinelli 1979 and references therein). Because of the high (hypersonic) velocity of this wind material – with typical terminal velocities of about three times the escape velocity from the stellar surface (cf. Abott 1978) – only a small fraction f can be accreted by the orbiting compact companion (cf. Bondi & Hoyle 1944, Davidson & Ostriker 1973), where f is of order:

$$f \approx \left(\frac{GM_x}{av_w^2}\right)^2,$$

where M_x is the mass of the compact companion, a the orbital separation and v_w the speed of the wind material with respect to the compact star. The accretion rate onto the compact star depends thus very sensitively on the wind speed v_w close to the X-ray source, which is in general not very well known. Because most of the accretion energy is liberated as X-rays we can estimate the X-ray luminosity by (e.g. McCray 1977):

$$L_x \approx \left(\frac{GM_x}{R}\right) \dot{M} \approx 10^{20} \dot{M} \text{ erg s}^{-1}, \qquad (9.9)$$

where the neutron star is assumed to have a mass of $1.4 M_\odot$ and an approximate radius $R \approx 10^6$ cm. Hence, accretion rates $\dot{M}$ of about 10^{18} g s^{-1} are required to power a bright Type I X-ray source with $L_x \approx 10^{38}$ erg s^{-1}.

We do not know whether all Type I X-ray sources can be powered by the stellar winds of their massive companions. It has been argued (cf. Lamers *et al.* 1976; Petterson, 1978; Savonije 1978) that the very bright sources, like Cen X-3 and SMC X-1, have accretion rates which cannot be explained by accretion from the stellar wind of their companion. They suggest that these very bright sources are powered by Roche-lobe overflow. The companion of many bright Type I X-ray sources do indeed seem to be close to or filling their Roche lobes (Avni 1977; Chapters 1 and 5, this book).

Fully developed Roche-lobe overflow from a massive star, say from a $20 M_\odot$ main-sequence star with $L \approx 10^4 L_\odot$ and $R \approx 10 R_\odot$, onto a $1.4 M_\odot$ neutron star is rather violent. Equation (9.6) predicts a peak mass-transfer rate of about $10^{-3} M_\odot$ yr^{-1} for such a star. This is much higher than the critical Eddington rate of a neutron star (Equation (9.8)), so that one expects that most X-ray photons will be absorbed by the material which piles up around the X-ray source. Roche-lobe overflow by a massive star can thus presumably produce an X-ray source only during the period – roughly equal to the thermal timescale of the star – before the mass transfer has reached its peak value. This initial phase of Roche-lobe overflow can be described as follows. According to Equation (9.5) we have approximately:

$$\dot{M}_1 \approx -\frac{2\pi}{RR_c}\frac{\phi}{(1-\phi)^{1/2}}\left(\frac{GM_1}{\Omega^2}\right)\int_{R_c}^{\infty} \rho c_s \, dr.$$

Initially only the atmospheric layers of the star can flow towards the companion as the star's photospheric radius R is still smaller than the Roche-lobe radius R_c. For our purpose a simple isothermal representation of the atmosphere is adequate, i.e. we may approximate the density distribution by $\rho(r) = \rho_{ph} \exp[-(r-R)/H]$, where R and H are the stellar radius and scale height at the photosphere. Integrating the above equation yields:

$$\dot{M}_1 \approx -2\pi \frac{\phi}{(1-\phi)^{1/2}} \frac{M_1}{M}\left(\frac{a^3 H}{RR_c}\right) \rho_{ph} c_s \exp\left(\frac{R-R_c}{H}\right), \qquad (9.10)$$

where M is the total mass of the binary, a the orbital separation and c_s the sound speed in the (isothermal) atmosphere of the contact star. By inserting into

Equation (9.10) numerical values typical for a massive Type I X-ray binary system one finds that the atmospheric mass-transfer rate ($\sim 10^{-8}$ M$_\odot$ yr^{-1}) can be sufficiently high to power a bright X-ray source. And yet, the change of the mass ratio caused by the beginning mass transfer is too insignificant to influence the size of the Roche lobe noticeably. The massive star itself is also hardly perturbed by the initially relatively weak mass transfer, so that it continues its evolutionary expansion at almost the same pace. The argument of the exponential function in Equation (9.10) may therefore be approximated by:

$$(R - R_c)/H \approx (R_o - R_c)/H + (\dot{R}/H)\tau,$$

where $\dot{R}$ is the evolutionary expansion rate of the massive star and where τ is the time elapsed since the mass-transfer rate reached a value of say 10^{-10} M$_\odot$ yr^{-1}, by which the X-ray source starts to become moderately bright. The exponential rise of the beginning mass transfer is then given by:

$$\dot{M}_1 \approx A \exp\left(\frac{\tau}{\tau_0}\right), \qquad (9.11)$$

where $A \approx 10^{-10}$ M$_\odot$ yr^{-1} and $\tau_0 = (H/\dot{R})$ is the relevant characteristic expansion time of the massive star. The X-ray lifetime of a source powered by Roche-lobe overflow is then roughly $\tau_x \approx 6\tau_0$ which is the time τ it takes to reach a super-Eddington mass-transfer rate of $10^{-7.5}$ M$_\odot$ yr^{-1}. The characteristic time τ_0 depends on the evolutionary state of the contact star. During the core hydrogen-burning phase the massive stars speed up their expansion rate, so that τ_0 and τ_x diminish slowly. For a typical 20 M$_\odot$ star in the middle of its core hydrogen-burning phase, $\tau_0 \approx 5 \times 10^3$ yr. Because τ_0 decreases with the main-sequence age of the star, the X-ray lifetime $\tau_x \approx 6\tau_0$ of a Roche-lobe-overflow powered X-ray source is a function of the orbital period of the binary system in which it is situated. This is so because the massive star must be more evolved to reach contact in a wider binary system. Figure 9.1 shows the expected X-ray lifetime τ_x as a function of the orbital period of the X-ray binary system for three different masses of the contact star, namely 16 M$_\odot$, 22 M$_\odot$ and 32 M$_\odot$. The 16 M$_\odot$ and 22 M$_\odot$ star were evolved at constant mass before reaching contact, while the 32 M$_\odot$ star was evolved under the assumption that it loses mass by a strong stellar wind (as is indeed observed for such massive stars). At the end of its main-sequence life this star has a substantially reduced mass of only 22 M$_\odot$ left (cf. Savonije 1979).

As first remarked by Ziolkowski (1977) and De Loore, De Greve & van Beveren (1978) the strong stellar-wind losses will make the star brighter and larger than a star of the same mass evolved at constant mass. This can be observed in Figure 9.1 as well: the 22 M$_\odot$ remnant can come into contact during its main-sequence life in a much wider binary system ($P_{orb} \lesssim 4.5$ d) than the

22 $M_\odot$ star evolved at constant mass ($P_{orb} \lesssim 2.8$ d). The minima for τ_x that occur in Figure 9.1 are caused by the structural changes of a star near the end of the core hydrogen-burning phase. When the core hydrogen content has been depleted to about $X_x \approx 0.1$ the stellar-expansion rate reaches a maximum, after which the expansion slows down until the massive star actually starts to shrink. The duration of the transient phase of slowed down expansion is of order 10^5 yr (3×10^5 yr for a 20 $M_\odot$ star). If the massive star happens to reach contact during this slow expansion phase a fairly long-lived X-ray source will be generated. This corresponds to the peaks in τ_x at the maximum orbital period attainable for the core hydrogen-burning contact stars.

After the main-sequence phase the hydrogen-exhausted core begins to contract rapidly whereby the outer layers above the hydrogen-burning shell source start to re-expand very fast. During this post-main-sequence phase, τ_0, the characteristic expansion time, becomes shorter than 50 yr (20 $M_\odot$ star) and one does not expect that Roche-lobe overflow can generate a relatively long-lived X-ray source under such conditions.

Wide X-ray binaries, like Vela X-1 and 4U 1223−62 (see Table 9.1), are therefore more likely to be powered by a stellar-wind type of mass transfer.

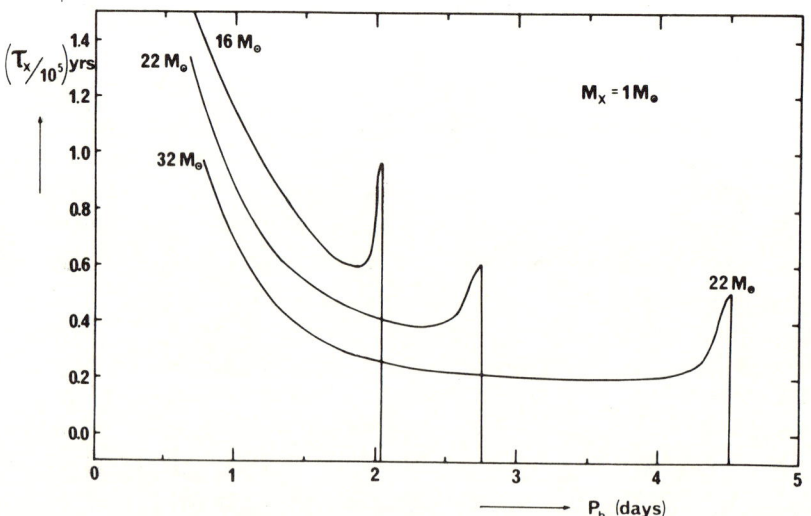

Fig. 9.1. Estimated X-ray lifetimes for beginning Roche-lobe overflow in massive X-ray binaries systems, as a function of their orbital period P_{orb}. The massive stars are in the core hydrogen-burning phase and the rotation of the outer layers is assumed to be synchronised with the 1 $M_\odot$ neutron-star companion. The 16 $M_\odot$ and 22 $M_\odot$ stars are assumed to evolve at constant mass before coming into contact, while the 32 $M_\odot$ is assumed to undergo a strong stellar-wind mass loss ($\dot{M} = 10^{-6} M_\odot \text{ yr}^{-1}$). This reduces its mass to about 22 $M_\odot$ at the end of its core hydrogen-burning phase.

The possibility of Roche-lobe overflow as the dominant mode of mass transfer in a massive X-ray binary depends thus strongly on the evolutionary state of the massive component. Unfortunately, these stars seem to obey a rather anomalous mass-luminosity relation, which makes it difficult to compare their estimated position in the HR diagram with calculated evolutionary tracks of massive stars. As remarked above, it has been suggested (cf. Ziolkowski 1977, De Loore *et al.* 1978) that the 'over-luminosity' (for a given mass) of these stars is caused by heavy mass loss by a stellar wind during their previous evolution. However, calculations with a strong upper limit for the mass-loss rate for a radiation-pressure-driven stellar wind showed that this explanation is not always satisfactory (Savonije 1980). It may well be that the observationally inferred over-luminosity is at least partly spurious and due to incorrect mass and/or luminosity estimates (see Chapter 1, this book).

Most of the Type I X-ray sources show X-ray pulsations which reflect the rotation of the (magnetised) neutron star. Unlike wind-fed sources (cf. Shapiro & Lightman 1976), X-ray sources powered by Roche-lobe overflow are likely to be surrounded by a Keplerian accretion disk. This is because the matter in the flow from L_1 has substantial angular momentum with respect to the neutron star and cannot fall in radially. The accreted matter tends therefore to spin the neutron star up (e.g. Pringle & Rees 1972; Rappaport & Joss 1977; Mason 1977; Lamb 1977) to a short period of a few seconds. Sources powered by Roche-lobe overflow are thus expected to show rapid X-ray pulsations (see Table 9.1), whereas wind-fed pulsars are expected to be slow rotators (see Chapter 11, this book, for a review on this subject). If the X-ray pulsars are able to maintain a spin rate near their 'equilibrium rate' corresponding to the instantaneous value of the mass-transfer rate, the spin-up must take place on

Table 9.1. *Characteristics of some Type I X-ray binaries*

Source	Orbital period (d)	X-ray pulse period (s)	L_x erg s^{-1}	Spectrum of companion	Probable mode of mass transfer
SMC X-1	3.81	0.71	6×10^{38}	B0 I	RLO
Cen X-3	2.09	4.84	4×10^{37}	O6.5 II–III	RLO
LMC X-4	1.40	13.51	2×10^{38}	O8 (IV–V)e	RLO
4U 1700−37	3.41	–	3×10^{36}	O6.5 f	Wind
Vela X-1	8.97	283	1.4×10^{36}	B0.5 Iab	Wind
4U 1538−52	3.73	529	4×10^{36}	B0 Ib	Wind (?)
4U 1223−62	40.8 (?)	699	1×10^{37}	B1.5 Ia	Wind

a timescale (cf. Savonije 1978 and Chapter 11, this book):

$$-\frac{P}{\dot{P}} \approx \frac{7}{3}\tau_0 \approx \frac{1}{3}\tau_x, \qquad (9.12)$$

where P is the pulse period of the X-ray pulsar. The observed spin-up times of SMC X-1 and Cen X-3, two candidates for mass transfer by Roche-lobe overflow, are ≈ 1300 yr and ≈ 3600 yr respectively. If these two X-ray pulsars do indeed spin near their equilibrium rate (which is perhaps not true) then these spin-up rates imply a rather short X-ray lifetime. We refer to Chapter 11, Section 11.3.10.2, this book, for a further discussion.

9.3.2 Mass transfer in Type II X-ray binaries

In contrast to the Type I X-ray sources for which the details of the mass-transfer process and binary evolution have still to be found out, but for which exists a coherent general picture, the Type II X-ray binaries form a mysterious group (about 60 in our galaxy). The evolutionary history of these X-ray binaries is rather obscure (see Chapter 8, this book), so that we can only speculate about the mass-transfer mechanism(s) responsible for their existence. We can therefore do no better than enumerate a number of possible binary configurations and likely mass-transfer mechanisms. We will limit ourselves to Roche-lobe overflow since mass transfer by a stellar wind seems not very important for the binary systems considered. The main observational facts to be explained by these models are that many Type II binary systems show X-ray luminosities in the range 10^{36}–10^{37} erg s^{-1}, although the distance estimates to many individual sources are rather uncertain. A fraction (12) of the sources with 10^{35} erg s$^{-1} \lesssim L_x \lesssim 10^{37}$ erg s^{-1} are situated near the centre of globular clusters whereas there are some 10 very bright sources with $L_x > 10^{38}$ erg s^{-1} in the vicinity of the galactic centre.

9.3.2.1 Roche-lobe overflow from a main-sequence dwarf

A sub-set of the Type II sources is formed by the transient X-ray sources, like Aql X-1 and Cen X-4. The origin of the X-ray outbursts in these systems is not known but may be related to the quasi-regular outbursts shown by dwarf-nova systems (e.g. Robinson 1976). During the quiescent phase of the Type II X-ray transients one can, for some systems, observe the absorption spectrum of the companion, which appears to be a low-mass main-sequence star (of type G-K, see Chapter 5, this book) which presumably undergoes (time-dependent ?) Roche-lobe overflow.

Assuming a K-dwarf companion of $\approx 0.7\,M_\odot$ and a neutron star mass of $1.4\,M_\odot$ we find from Equation (9.4a) that its radius must be $R \approx 0.32a$, where a is the orbital separation. We know, on the other hand, that main-sequence dwarfs in the range $0.3 \leq M/M_\odot \leq 0.9$ obey the approximate mass–radius relation (Savonije, unpublished):

$$R/R_\odot = 0.86(0.94 - 0.18 \ln X)(M/M_\odot)^{0.9}, \qquad (9.13)$$

accurate to about 5 per cent for $0.3 \leq X \leq 0.7$ ($Z = 0.02$) and where X, the core hydrogen fraction, describes the evolution away from the zero-age main-sequence. We conclude from Equation (9.13) that a $0.7\,M_\odot$ K-dwarf, with say $X = 0.6$, has a radius of about $0.64\,R_\odot$. Assuming that the dwarf fills its Roche lobe we deduce for the orbital separation $a \approx 2\,R_\odot$ and an orbital period $P_{\rm orb} \approx 5.4\,{\rm h}$. From Equation (9.3) we can derive a typical timescale for the evolution caused by the emission of gravitational radiation:

$$\tau_\gamma \approx \frac{-J_{\rm orb}}{\dot{J}_{\rm orb}} = 1.6 \times 10^8 \left(\frac{M_1 M_2 M}{M_\odot^3}\right)^{-1} \left(\frac{a}{R_\odot}\right)^4 \,{\rm yr}. \qquad (9.14)$$

For our binary system $\tau_\gamma \approx 1.5 \times 10^9$ yr which is more than an order of magnitude shorter than the nuclear evolution timescale ($\tau_{\rm nuc} \approx 5 \times 10^{10}$ yr) of the K-dwarf. Without angular momentum losses this latter timescale would be relevant to the mass transfer ($M_1 < M_2$, see Section 9.3). With gravitational radiation the mass transfer is, however, entirely driven by the induced *shrinking* of the Roche lobe (the τ_γ^{-1} term dominates in Equation 9.2).

Evolutionary calculations of binary systems with gravitational radiation losses have been performed by many authors, e.g. Paczynski (1967), Faulkner (1971), Chau & Lauterborn (1977), Taam, Flannery & Faulkner (1980), Rappaport, Joss & Webbink (1981), Paczynski & Sienkiewicz (1981) and references therein. Since the thermal relaxation timescale of the K-dwarf ($\tau_{\rm KH} \approx GM^2/RL \approx 5 \times 10^7$ yr) is much smaller than the evolution time τ_γ, the star is not driven out of thermal equilibrium by the mass transfer. We may therefore simply apply the equilibrium mass–radius relation (9.13) with X kept constant ($\tau_\gamma \ll \tau_{\rm nuc}$) during the mass transfer. Let us derive an expression for the resulting mass transfer rate by substituting Equation (9.14) into Equation (9.2) and putting $\alpha = 0$:

$$\frac{\dot{a}}{a} = 2\left(\frac{M_1}{M_2} - 1\right)\frac{\dot{M}_1}{M_1} - 2\tau_\gamma^{-1}. \qquad (9.15)$$

By writing the mass–radius relation (9.13) in a more general way as

$$\frac{R}{R_\odot} = \alpha \left(\frac{M}{M_\odot}\right)^\beta, \qquad (9.16)$$

we can express the condition of Roche-lobe overflow $\dot{R}_1 \approx \dot{R}_c$ (where R_c is given by Equation (9.4a)) as:

$$\frac{\dot{a}}{a} = (\beta - \tfrac{1}{3})\frac{\dot{M}_1}{M_1}. \tag{9.15a}$$

Eliminating $\dot{a}/a$ by combining Equations (9.15) and (9.15a) yields:

$$\frac{\dot{M}_1}{M_1} = -\left(\frac{5}{6} + \frac{1}{2}\beta - \frac{M_1}{M_2}\right)^{-1} \tau_\gamma^{-1}.$$

Inserting Equation (9.14) for τ_γ, in which we eliminate a by using the condition $R_1 = R_c$ in the form:

$$\frac{a}{R_\odot} = \frac{\alpha}{0.462}\left(\frac{M_1}{M_\odot}\right)^{\beta - 1/3}\left(\frac{M}{M_\odot}\right)^{1/3},$$

we obtain the mass transfer rate in $(M_\odot \text{ yr}^{-1})$:

$$\dot{M}_1 = -3.7 \times 10^{-11} \alpha^{-4}\left(\frac{5}{6} + \frac{\beta}{2} - \frac{M_1}{M_2}\right)^{-1}\left(\frac{M_1}{M_\odot}\right)^{(10/3) - 4\beta}\left(\frac{M_2}{M_\odot}\right)\left(\frac{M}{M_\odot}\right)^{-1/3}. \tag{9.17}$$

If we approximate the mass–radius relation by $\alpha = \beta = 1$ and introduce $\mu = M_1/M$, we find (Li et al. 1980):

$$\dot{M} = -1.0 \times 10^{-10}(1 - \mu)^2(4 - 7\mu)^{-1}\mu^{-2/3} \, M_\odot \text{ yr}^{-1}. \tag{9.18}$$

Equation (9.18) shows that the gravitational radiation-driven mass transfer from main-sequence dwarfs is *always of order* $\dot{M} \approx 10^{-10} \, M_\odot \text{ yr}^{-1}$ (to within a factor of about 2) for all realistic values of μ. Such a moderate transfer rate can only power a less bright, steady Type II source, with $L_x \approx 10^{36}$ erg s^{-1} (Equation (9.9)). We lack enough detailed observations (besides that, the individual distance estimates are poor) to tell whether such an accretion rate is sufficient to explain the *time-averaged* X-ray emission of the transient Type II sources. It could be that the outbursts are due to a sudden discharge of the accumulated matter in the accretion disc. But anyhow, gravitational radiation dominated mass transfer from a G- or K- dwarf cannot explain the brighter Type II sources.

Eggleton (1976) and Verbunt & Zwaan (1981) have suggested that rotational braking by a magnetic stellar wind may be important in close, low-mass binary systems. Verbunt & Zwaan extrapolated the braking law observed for single G-stars (Skumanich 1972, Smith 1979) $dV_e/dt = -fV_e^3$, where f is a 'constant' and V_e the star's equatorial velocity, to the low-mass dwarfs in close-binary systems and concluded that such braking can dominate the binary evolution and speed up the mass transfer to $\dot{M} \approx$ few times $10^{-9} \, M_\odot \text{ yr}^{-1}$. This figure, however, is rather uncertain, because it involves an extrapolation of the braking law for single G-stars to mass-losing binary stars rotating at ≈ 30 per cent of their break-up speed. Nevertheless, the proposed magnetic braking may indeed be important and the idea certainly needs further study.

9.3.2.2 Roche-lobe overflow from a low-mass degenerate contact star

We have seen above that the gravitational, radiation-driven evolution of the red dwarf–neutron star binaries proceeds on a timescale τ_γ, fast compared to the nuclear evolution time τ_{nuc}, so that the nuclear evolution of the red dwarf is effectively frozen. During the mass transfer these stars therefore move down on a track parallel to the ZAMS (Taam et al. 1980). But, the thermal timescale of the dwarf, given by $\tau_{\text{KH}} \approx 2 \times 10^8 \, (M/M_\odot)^{-1.2}$ yr, increases by an order of magnitude when its mass decreases from $0.7 \, M_\odot$ to about $0.1 \, M_\odot$. Thus when the dwarf has arrived at the lower-end of the main sequence $\tau_{\text{KH}} \approx \tau_\gamma$. It is then no longer allowed to use the simple thermal equilibrium mass–radius relation (9.13). At about the same time the hydrogen fusion stops as the central temperature of the contact star becomes sub-critical and the dwarf becomes a strongly degenerate configuration. The star then obeys a different mass–radius relation, approximately given by (Paczynski 1967):

$$\frac{R}{R_\odot} = 0.013 \, (1 + X)^{5/3} \left(\frac{M}{M_\odot}\right)^{-1/3}. \tag{9.19}$$

This relation shows the well-known property of a degenerate star, namely that it expands when its mass decreases. This implies (e.g. Equation (9.15a)) that the whole binary system must expand because otherwise the degenerate contact star would start to grossly exceed its Roche lobe. The thus resulting mass transfer (to a more-massive companion) would certainly dominate the τ_γ^{-1} term in Equation (9.15) and increase the orbital separation. Thus the binary system will pass through a minimum orbital period, which turns out to be about 80 min (cf. Paczynski & Sienkiewicz 1981, Rappaport et al. 1982). After that we may use the thermal equilibrium relation (9.19) to calculate the mass-transfer rate. Assuming again $X = 0.6$ and substituting $\alpha = 0.028$ and $\beta = -1/3$ from Equation (9.19) into Equation (9.17) we obtain:

$$\dot{M}_1 = -2 \times 10^{-4} (1 - \mu)^2 (2 - 5\mu)^{-1} \mu^{14/3} \left(\frac{M}{M_\odot}\right)^{16/3} M_\odot \, \text{yr}^{-1}, \tag{9.20}$$

where again $\mu = M_1/(M_1 + M_2)$. For a low-mass degenerate dwarf (with, say $M_1 = 0.08 \, M_\odot$) orbiting around a $1.4 \, M_\odot$ neutron star μ is small, so that we have approximately $\dot{M} \approx -10^{-4} \mu^{14/3} (M/M_\odot)^{16/3} \, M_\odot \, \text{yr}^{-1}$, i.e., mass-transfer rates below $10^{-9} \, M_\odot \, \text{yr}^{-1}$ result.

Figure 9.2 shows how the mass-transfer rate from a $0.08 \, M_\odot$ degenerate hydrogen-rich dwarf evolves in time. The mass-transfer rate would be much higher if the binary had followed a different evolution by which the contact star manages to exhaust its central hydrogen before becoming degenerate.

The radius of a degenerate helium star ($X = 0$) is about half as small as that of a hydrogen-rich degenerate dwarf of the same mass (Equation 9.19), so that the orbital separation must also be about half as small during the contact phase. For the degenerate helium star the gravitational radiation losses are therefore about 16 times higher, with a correspondingly faster mass transfer during the contact phase. Roche-lobe overflow from a helium white dwarf of $\approx 0.2\,M_\odot$ onto a $1.4\,M_\odot$ neutron star can reach mass-transfer rates of $\approx 10^{-8}\,M_\odot\,yr^{-1}$, sufficient to power a very bright X-ray source. However, from Figure 9.2 it can be observed that the mass-transfer rate decays rapidly, so that one would expect many more weak X-ray sources than bright X-ray sources. This seems not to have been observed by the Einstein Observatory (see Chapters 1 and 3, this book), although a concentration of many weak X-ray sources may have escaped identification as distinct point

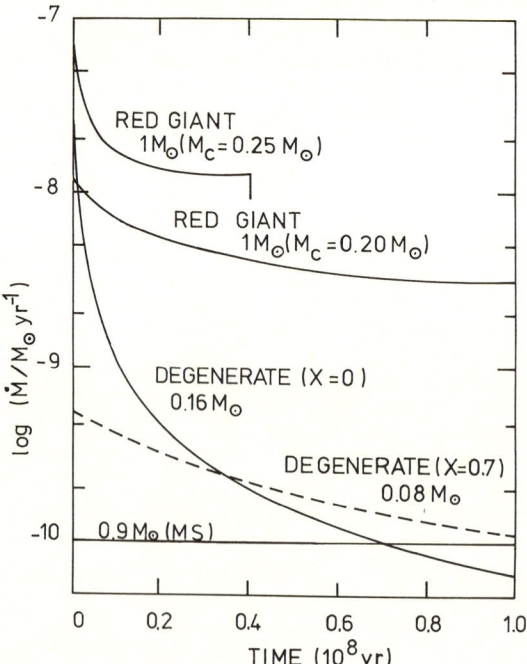

Fig. 9.2. The evolution in time of the mass-transfer rate during Roche-lobe overflow towards a $1.4\,M_\odot$ neutron star from different types of contact stars, as indicated in the diagram. For a $0.9\,M_\odot$ main-sequence star the mass transfer is rather slow and changes on a timescale much longer than considered in the figure ($\tau_\gamma \approx 10^{10}$ yr). The mass-transfer rate from a low-mass degenerate dwarf (with either $X = 0.7$ or $X = 0$) is faster but decreases sharply in time because of the rapidly diminishing effect of gravitational radiation losses when the orbit expands (after Rappaport, Joss & Webbink, 1982). In contrast to the previous cases, mass transfer from a $\approx 1\,M_\odot$ red giant is not driven by gravitational radiation, but by nuclear evolution. The mass-transfer rate is rather high and levels off to an almost constant value (after Webbink, Rappaport & Savonije, 1982).

sources. The reason why the mass transfer decays rapidly is that a degenerate star, and thus the binary system as well (Equation (9.15a), with $\beta < \frac{1}{3}$), expands during the mass transfer phase. The gravitational radiation loss and the mass-transfer rate thus decay rapidly.

In the next two sections we will discuss binary systems in which high, but rather steady mass-transfer rates can be expected to occur. The contact stars (helium stars and red giants) are rather bright stars compared to the dwarf stars considered up to now.

9.3.2.3 Roche-lobe overflow from a non-degenerate helium star

From calculations by Paczynski (1971) we derive an approximate mass-radius relation for pure helium stars ($Z = 0.03$) in thermal equilibrium:

$$\frac{R}{R_\odot} \approx 0.20 \left(\frac{M}{M_\odot}\right)^{0.86}, \qquad (9.21)$$

valid for $0.5 < M/M_\odot < 1.5$. These stars thus have small radii implying strong gravitational radiation losses, but they do not expand when losing mass. Substituting $\alpha = 0.20$ and $\beta = 0.86$ into Equation (9.17) for the mass-transfer rate, we obtain:

$$\dot{M} = -2.3 \times 10^{-8} \left(1.26 - \frac{M_1}{M_2}\right)^{-1} \left(\frac{M_1}{M_\odot}\right)^{-0.107} \left(\frac{M_2}{M_\odot}\right) \left(\frac{M}{M_\odot}\right)^{-1/3} M_\odot \, \text{yr}^{-1}. \qquad (9.22)$$

This expression shows that very high mass-transfer rates are to be expected from Roche-lobe overflow by pure helium stars. Table 9.2 shows how $\dot{M}$ develops in time for a binary consisting of a helium star of 0.7 $M_\odot$ and a companion of 1 $M_\odot$. It appears that helium stars can give rise to very high, almost constant mass-transfer rates of $\geq 10^{-8} M_\odot \text{yr}^{-1}$ for a period of roughly 2×10^7 yr. During this phase of the evolution the thermal timescale τ_{KH} of the contact star is much smaller than the actual evolutionary timescale τ_γ (roughly 10^6 and 10^7 yr

Table 9.2 *Evolution of a binary consisting of a Roche-lobe overflowing helium star (initial mass 0.7 $M_\odot$ and a 1 $M_\odot$ companion)*

M_1 ($M_\odot$)	M_2 ($M_\odot$)	$\dot{M}$ ($M_\odot$ yr^{-1})	P_{orb} (min)
0.7	1.0	3.6×10^{-8}	36.0
0.6	1.1	3.1×10^{-8}	32.0
0.5	1.2	3.0×10^{-8}	27.6
0.4	1.3	2.9×10^{-8}	23.2
0.3	1.4	2.9×10^{-8}	18.5

Note the short orbital periods and steady, high mass-transfer rate.

respectively), so that the helium star remains close to thermal equilibrium. The nuclear timescale τ_{nuc}, on the other hand, is an order of magnitude larger than τ_γ. Equations (9.21) and (9.22) thus remain approximately valid. But at about the time that the helium star's mass has shrunk to 0.3 $M_\odot$, the thermal timescale and the evolution timescale τ_γ become comparable and the helium star begins to depart appreciably from thermal equilibrium due to the fast mass transfer. More detailed numerical calculations with a full evolution code (Savonije, unpublished) show that the luminosity of the helium star drops rapidly below 1 $L_\odot$, to very small values, as ever more of the nuclear energy liberated is absorbed by the envelope (Section 9.2.3.1). After some time the central temperature drops below the critical value for core helium burning and the contact star settles down as a degenerate helium dwarf of about 0.25 $M_\odot$. Just like the degenerate hydrogen-rich dwarfs (previous section), the exponent β in the mass–radius relation then drops below $\frac{1}{3}$ and the binary system must re-expand in order to keep the contact star inside its Roche lobe. For helium stars the corresponding minimum orbital period turns out to be about 10 min. The precise values of these minimum orbital periods seem, however, rather uncertain as a result of the poorly understood equation of state and opacity of stellar matter in these regions of the Hertzsprung-Russell diagram.

As described in the previous section, the mass transfer from a 0.2 $M_\odot$ degenerate helium star decays below 10^{-9} $M_\odot$ yr^{-1} in about 10^7 yr (Figure 9.2). Starting with an initially non-degenerate helium star thus gives a considerably longer phase of strong mass transfer and helps to alleviate the problem that one observes a relatively large ratio of strong to weak X-ray sources in the galactic bulge. But it remains to be seen whether nature can indeed produce systems which consist of a neutron star in a close orbit around a helium star, and that preferentially near the galactic centre.

9.3.2.4 Roche-lobe overflow from a red giant

It is speculated that the progenitor systems of cataclysmic variables and Type II X-ray binary systems have evolved through a common-envelope phase with a subsequent spiral-in, after which the binary seems to end up with typical orbital periods of about 0.5 d (see Chapter 8, this book). This period is about the limiting period (for stars of about 1 $M_\odot$) below which gravitational radiation losses dominate the evolution and lead to very narrow binaries (Taam *et al.* 1980). Hence, when there is some spread in the orbital periods after spiral-in, then some systems may be forced to become very narrow and follow the evolution as briefly sketched in Sections 9.3.2.1 and 9.3.2.2. The wider binary systems, however, may escape this fate and manage to follow the 'normal' evolution and evolve into very wide binaries with a red-giant contact star (e.g. Kippenhahn *et al.* 1967).

A similar system may be formed by capture of a red giant by a neutron star in regions of high star densities, as in globular clusters (cf. Sutantyo 1975; Hills & & Day 1976).

The structure and evolution of such a red giant with a degenerate helium core appears to depend very sensitively on the mass contained in this core and to be almost independent of the envelope mass (e.g. Refsdal & Weigert 1970). As the hydrogen-burning shell source eats itself through the remaining hydrogen-rich envelope, the core mass increases steadily as the nuclear ash (helium) settles down onto the degenerate helium core. This growth of the core mass is related to the giant's luminosity L (which is almost completely generated in the shell source) by

$$\dot{M}_c = 1.37 \times 10^{-11} \left(\frac{L}{L_\odot}\right) M_\odot \text{ yr}^{-1}, \qquad (9.23)$$

where the constant 1.37×10^{-11} follows from the amount of energy liberated per gram envelope matter ($X = 0.7$) burnt into helium ($\approx 4.5 \times 10^{18}$ erg). The gravitational luminosity is negligible, because $R/\dot{R} \gg \tau_{KH}$. The growing core mass results in a continuous expansion and luminosity increase corresponding to the giant's ascent of the giant branch (e.g. Cox & Giuli 1968). For a metal content of $Z = 0.02$ one can derive the following approximate relations between the giant's luminosity L, radius R and degenerate core mass M_c expressed as $y = \ln(M_c/M_\odot)$ (Webbink et al. 1983):

$$\ln\left(\frac{R}{R_\odot}\right) = 4.96 - 4.89y - 7.16y^2 - 1.71y^3$$
$$\ln\left(\frac{L}{L_\odot}\right) = 7.89 - 5.87y - 9.47y^2 - 2.13y^3. \qquad (9.24)$$

These three relations describe the giant's evolution in the H–R diagram fairly well and can be used to obtain the mass-transfer rate by Roche-lobe overflow when the giant is situated in a binary (by requiring $\dot{R} = \dot{R}_c$):

$$\dot{M}_1 = -\frac{(4.89 + 14.32y + 5.13y^2)}{2(M_1/M_2 - 1) + 1/3}\left(\frac{M_c}{M_1}\right)^{-1}\dot{M}_c. \qquad (9.25)$$

Figure 9.2 shows the resulting mass transfer as calculated by Webbink et al. (1983). It can be seen that $\dot{M}$ depends very sensitively on the mass of the degenerate core M_c. Since this core mass depends on the unknown evolutionary history of the giant, one has some freedom to vary its value. For $M_c \approx 0.25 \, M_\odot$ the mass transfer is very fast ($\dot{M} > 10^{-8} \, M_\odot \text{ yr}^{-1}$), so that after some 4×10^7 yr almost the entire hydrogen-rich envelope will be lost and only the (then) $\approx 0.3 \, M_\odot$ degenerate helium core remains, after which the mass transfer stops abruptly. During the mass-transfer phase the orbital period increases from about 20 d to well over a 100 d, while the giant's luminosity increases from about 50 $L_\odot$ to some 180 $L_\odot$. A value of $M_c \approx 0.2$ $M_\odot$ results in a steady value of $\dot{M} \approx 3 \times 10^{-9}$ $M_\odot$ yr^{-1} which can be maintained much longer ($\approx 2 \times 10^8$ yr).

If indeed some fraction of the Type II sources are situated in wide binary systems with a giant contact star (the orbital period of most Type II sources is unknown) - one could explain the observed relatively high number of bright galactic centre sources in terms of the above model. It is possible that the extremely bright X-ray sources as seen in M31 by the Einstein Observatory (see Chapter 3, this book) are also powered by mass transfer from an ≈ 1 $M_\odot$ red giant. Less bright X-ray sources with orbital periods of ≈ 10 d (like Cyg X-2) may contain a (sub-) giant contact star with a slightly smaller helium core.

This ends our enumeration of possible binary configuration for low-mass X-ray binary systems. The last two possibilities produced steady and high mass-transfer rates ($\geqslant 10^{-8}$ $M_\odot$ yr^{-1}), sufficient to power the brightest galactic X-ray sources. But it remains to be seen whether these models are consistent with more detailed observations of these systems. Both the 0.7 $M_\odot$ helium star and the 1 $M_\odot$ red-giant contact star are rather bright stars (but with widely different intrinsic effective temperatures of $\approx 40\,000$ K and ≈ 3000 K respectively). Unless the accretion disc could screen off the X-ray source from the companion star (Milgrom 1978), the contact star would be expected to show a strong periodic heating effect caused by the impinging X-rays on the side facing the X-ray source. This heating could severely modify the contact star's spectral type.

Acknowledgements

The author thanks Drs E. van den Heuvel, H. Henrichs, M. van der Klis, W. H. G. Lewin, J. van Paradijs and S. A. Rappaport for stimulating discussions and Dr P. Eggleton for providing a ZAMS starting model for the helium-star evolution calculations.

References

Abbott, D. C. 1978, *Ap. J.*, **225**, 893.
Avni, Y. 1977, *In: Highlights of Astronomy*, Vol. 4, Part 1, ed. H. Müller (Dordrecht, Reidel), p. 137.
Basko, M. M., Hatchett, S., McCray, R. & Sunyaev, R. A. (1977) *Ap. J.*, **215**, 276.
Benson, R. S. 1970, Ph.D. Thesis, Univ. of California, Berkeley.
Bondi, H. & Hoyle, F. 1944, *Mon. Not. Roy. Astron. Soc.*, **104**, 273.
Bradt, H. V., Doxsey, R. E. & Jernigan, J. G. 1979, *In: X-ray Astronomy*, eds. W. A. Baity & L. E. Peterson (Oxford, Pergamon), p. 3.
Cassinelli, J. P. 1979, *Ann. Rev. Astron. Ap.*, **17**, 275.
Castor, J. I., Abbott, D. C. & Klein, R. I. 1975, *Ap. J.*, **195**, 157.
Chau, W. Y. & Lauterborn, D. 1977, *Ap. J.*, **214**, 540.
Conti, P. S. & Garmany, C. D. 1980, *Ap. J.*, **238**, 190.

Cowling, T. G. 1941, *Mon. Not. Roy. Astron. Soc.*, **101**, 367.
Cox, J. P. & Giuli, R. T. 1968, *Principles of Stellar Structure* (New York, Gordon and Breach, Science Publ.).
Davidson, K. & Ostriker, J. P. 1973, *Ap. J.*, **179**, 583.
De Loore, C., De Greve, J. P. & van Beveren, D. 1978, *Astron. Ap.*, **67**, 373.
Eddington, A. S. 1926, *In: The Internal Constitution of the Stars* (reprinted) 1959 (New York, Dover Publ. Comp.).
Eggleton, P. P. 1976, *IAU Symposium*, **73**, 209 (eds. P. Eggleton *et al.*).
Faulkner, J. 1971, *Ap. J.*, **170**, L99.
Flannery, B. P. & Ulrich, R. K. 1977, *Ap. J.*, **212**, 533.
Heuvel, van den, E. P. J. 1978, *In: Proc. of the Int. School of Physics*, 'Enrico Fermi', Course 65, p. 828 (eds. R. Giaconni & R. Ruffini).
Hills, J. G. & Day, C. A. 1976, *Ap. Letts.*, **17**, 87.
Jedrezejec, E. 1969, MS Thesis, Warsaw Univ.
Kippenhahn, R., Kohl, K. & Weigert, A. 1976, *Z. f. Ap.*, **66**, 58.
Kippenhahn, R., Weigert, A. & Hofmeister, E. 1967, *In: Methods in Comp. Phys.*, Vol. 7, eds. B. Alder *et al.* (New York, Academic Press), p. 129.
Kippenhahn, R. & Meyer-Hofmeister, E. 1977, *Astron. Ap.*, **54**, 539.
Kopal, Z. 1959, *Close Binary Systems* (New York, John Wiley).
Kruszewski, A. 1966, *Acta Astron.*, **13**, 106.
Lamers, H. J., van den Heuvel, E. P. J. & Petterson, J. A. 1976, *Astron. Ap.*, **49**, 327.
Lamers, H. J. G. L. M. & Snow, T. P. 1978, *Ap. J.*, **219**, 504.
Lamers, H. J. 1981, *Ap. J.*, **245**, 593.
Lamb, F. K. 1977, *In: Proc. 8th Texas Symp. on Relativ. Astrophys.* (*Ann. N.Y. Acad. of Sci.*, **302**, 481).
Landau, L. & Lifshitz, E. 1958, *Theory of the Fields* (Oxford, Pergamon Press).
Li, F. K., Joss, P. C., McClintock, J. E., Rappaport, S. & Wright, E. L. 1980, *Ap. J.*, **240**, 628.
Lightman, A. P., Shapiro, S. L. & Rees, M. J. 1978, *In: Proc. of the Int. School of Physics*, 'Enrico Fermi', Course 65, p. 786 (eds. R. Giacconi & R. Ruffini).
Limber, D. N. 1963, *Ap. J.*, **138**, 1112.
Lubow, S. H. & Shu, F. H. 1975, *Ap. J.*, **198**, 383.
Lubow, S. H. 1979, *Ap. J.*, **229**, 1008.
Lucy, L. B., Solomon, P. M. 1970, *Ap. J.*, **159**, 879.
Mason, K. O. 1977, *Mon. Not. Roy. Astron. Soc.*, **178**, 81P.
McCray, R. 1977, *Highlights of Astron.*, vol. 4, part 1, p. 155.
Mestel, L. 1968, *Mon. Not. Roy. Astron. Soc.*, **138**, 359.
Middelkoop, F. 1981, *Astron. Ap.*, **101**, 295.
Milgrom, M. 1978, *Astron. Ap.*, **67** (L25).
Paczynski, B. 1967, *Acta Astron.*, **17**, 287.
Paczynski, B., Ziolkowski, J., Zytkow, A. 1969, *In: Mass Loss from Stars*, ed. M. Hack (Dordrecht, Reidel), p. 237.
Paczynski, B. 1971, *Acta Astron*, **21**, 1.
Paczynski, B. 1970, *In: Mass Loss and Evolution of Close Binaries*, eds. K. Gyldenkerne & R. West, Kopenhagen Univ. Publ. Funds, p. 142.
Paczynski, B. 1971, *Ann. Rev. Astron. Ap.*, **9**, 183.
Paczynski, B. & Sienkiewicz, R. 1981, *Ap. J.*, **248**, L27.
Papaloizou, J. & Pringle, J. 1978, *Mon. Not. Roy. Astron. Soc.*, **182**, 423.
Petterson, J. A. 1978, *Ap. J.*, **224**, 625.
Plavec, M. 1968, *Adv. in Astron. Ap.*, **6**, 201.
Pratt, J. P. & Strittmatter, P. A. 1976, *Ap. J.*, **204**, L29.
Pringle, J. E. & Rees, M. J. 1972, *Astron. Ap. J.*, **21**, 1.
Rappaport, S. & Joss, P. C. 1977, *Nature*, **266**, 683.
Rappaport, S., Joss, P. C. & Webbink, R. 1982, *Ap. J.*, **254**, 616.
Rappaport, S., Savonije, G. J. & Webbink, R. 1982, in preparation.

Rees, M. J. 1974, *Proc. 16th Solvay Conf. Brussels*, Univ. of Brussels.
Refsdal, S. & Weigert, A. 1971, *Astron. Ap.*, **13**, 367.
Robinson, E. L. 1976, *Ann. Rev. Astron. Ap.*, **14**, 119.
Savonije, G. J. 1978, *Astron. Ap.*, **62**, 317.
Savonije, G. J. 1979, *Astron. Ap.*, **71**, 352.
Savonije, G. J. 1980, *Astron. Ap.*, **81**, 25.
Savonije, G. J. & Papaloizou, J. C. 1983, *MNRAS*, **203**.
Shaprio, S. L. & Lightman, A. P. 1976, *Ap. J.*, **204**, 555.
Skumanich, A. 1972, *Ap. J.*, **171**, 565.
Smith, M. A. 1979, *P.A.S.P.*, **91**, 737.
Sutantyo, W. 1975, *Astron. Ap.*, **35**, 251.
Taam, R. E., Flannery, B. P. & Faulkner, J. 1980, *Ap. J.*, **239**, 1017.
Thorne, K. S. 1974, *Ap. J.*, **191**, 507.
Verbunt, F. & Zwaan, C. 1981, *Astron. Ap.*, **100**, L7.
Webbink, R., 1975, Ph.D. Thesis, University of Cambridge.
Webbink, R., Rappaport, S. & Savonije, G. J. 1983, *Ap. J.* (in press).
Zahn, J. P. 1977, *Astron. Ap.*, **57**, 383.
Ziolkowski, J. 1977, *Proc. 8th Texas Symposium on Relativ. Astrophys.* (Ann. N.Y. Acad. Sc., **302**, 47).

10

ACCRETION DISKS IN CLOSE-BINARY SYSTEMS

Jacobus A. Petterson
Physics Department, New Mexico Institute of Mining & Technology

10.1 The occurrence of accretion disks
10.1.1 Introduction

This review restricts itself to disks in stellar binaries containing a mass-losing star and a compact object, i.e., a white dwarf, a neutron star, or a black hole. Accretion disks may occur in other types of systems, e.g. in quasars or in radio galaxies, but it is always hazardous to extrapolate results described here to models for these other systems.

The pre-history of binary-disk modelling includes publications by Prendergast & Burbidge (1968), Shakura (1972), and Pringle & Rees (1972). History may be said to start with the appearance of a monumental paper by Shakura & Sunyaev (1973), which marked the beginning of attempts to find a detailed self-consistent model, resolving both the hydrodynamic structure and the radiative properties of disks. The paper reappeared in edited form as part of an early review (Novikov & Thorne, 1973). Since then, the literature on disks has grown enormously but, in spite of that, many of the major uncertainties in the models of 1973 are still with us at this time.

The present review emphasizes results about disks and disk models which are now relatively secure. It also points out some remaining major issues. More encyclopaedic reviews are available in the literature; a fairly complete and recent one is Pringle (1981).

10.1.2 Formation of disks in binaries

Consider a binary system in which one of the two stars is a compact object, the other, its 'companion', a star which loses mass. The flow of mass between the stars can take place in different patterns, which depend on the binary parameters of the system (including such variables as the rotation rate of the companion, and the nature and magnetic properties of the compact object). However, they depend in an important way also on the process by which mass is expelled from the donor star (see also Chapter 9, Section 9.3 and Chapter 11, Section 11.3, this book):

(1) The most common expulsion mechanism is expansion of the companion beyond its critical equipotential lobe, which leads to overflow in a gas stream from the inner critical point of the binary toward the compact object (see Figure 10.1 and Chapter 9, Section 9.2). For companion stars co-rotating with the binary system, the critical lobe and inner critical point are respectively called 'Roche lobe' and 'inner Lagrangian point' (L_1).

(2) Massive stars may expel significant amounts of gas steadily in a *stellar wind*, or irregularly in 'blobs'. These forms of expulsion may well be active *in addition to* overflow. Indeed, any combination of them is conceivable for stars whose mass is not too low. If it is below $\sim 2\,M_\odot$, stellar winds drive out too little mass to be of interest here (see e.g. van den Heuvel, 1975 and Chapter 8, Section 8.4, this book).

(3) Gas outflow may be further complicated by radiation from the accreting object which heats the facing side of the companion. This radiation can stimulate an *extra* outflow from the companion's surface in the vicinity of, but not precisely at, the inner critical point. It has been estimated (Basko, Hatchett & Sunyaev, 1977) that this extra outflow is not sufficiently strong to close a feedback loop to the radiation source, which would have made the source 'self-fueling' and capable of operating in the absence of other gas expelling mechanisms; see Davidson & Ostriker (1973) on 'self-exciting' winds.

After leaving the companion star, gas may start forming a disk in the system. To do so, it must at least satisfy the following criteria:

(a) Its specific potential energy ϵ_p with respect to the compact object must exceed twice its specific kinetic energy $\epsilon_k = \frac{1}{2} v_{rel}^2$, where v_{rel} is its velocity relative to the object. This allows gravitational capture by the receptor star. (Other capturing mechanisms, e.g. magnetic ones, do not in general lead to the formation of an accretion disk.)

(b) The specific angular momentum J of the captured part of the gas must be large enough with respect to the compact object to prevent direct (quasi-radial) infall. This allows the gas to go into a Keplerian orbit, from where it can slowly drift inwards as a result of viscosity.

(a') The success of the gravitational capture process largely depends on a low enough value for v_{rel}, which is determined by the outflow velocity from the donor star and the binary geometry. A good estimate for v_{rel} is hard to obtain. Even in the two extreme (and simplest) cases, namely mass loss by *pure* overflow and by *pure* stellar wind, analytical estimates are rough and still depend on parameters which can rarely be estimated accurately in real systems (e.g. the rotation rate of the companion). This suggests that the predictive power of these estimates for intermediate cases, and for partially illumination-driven outflows is very small. A few results follow for the simplest cases:

Gas leaving the companion star by flowing over the critical lobe near L_1 has a velocity relative to the stellar surface of the same order as the local sound velocity v_s, since L_1 acts as a nozzle for the outgoing fluid. This velocity is very small compared to orbital velocities in disk and binary. When the rotation of the mass-losing star does not deviate too drastically from co-rotation, v_{rel} will be small enough to ensure capture by the object (see e.g. Shu & Lubow, 1981). When the deviation is large, capture becomes strongly dependent on the mass ratio M_1/M_2 (M_1 is the mass of the compact object, M_2 that of the companion). For instance, capture by the object of a gas stream from the inner critical point of a *non*-rotating companion requires $M_1/M_2 \gtrsim 0.1$ (Petterson, 1978), a ratio probably *not* obtained in every X-ray binary. This estimate is rough, but it should caution the reader against automatically assuming capture whenever the mass loss occurs by overflow. Uncaptured, outflowing gas may either leave the system or fall back onto the companion (see also Chapter 9, Section 9.2, this book).

Gas expelled from the companion in a stellar wind receives an initial velocity with a wide spread in directions and a magnitude at least comparable to orbital speeds in the system. Capture is expected only for that part of the gas which passes by the compact object within the *'accretion radius'* r_A, which is of order

$$r_A \approx \frac{2M_1}{v_{rel}^2} \tag{10.1}$$

(Davidson & Ostriker, 1973). The theoretical value of v_{rel} is strongly dependent on the particular wind model used, and of course, on the binary parameters. Estimates vary widely. However, it is agreed that v_{rel} must be supersonic in this case, so that capture is always difficult and involves the passage of the accreting material through a shockfront. Speculations about the details of this process could fill a book. In typical situations, the compact object captures only a small fraction $\pi r_A^2/4\pi a^2$ (of order 10^{-3}, where a is the binary separation) of the total mass loss in the wind. The rest leaves the system.

(b') The minimum specific angular momentum preventing direct infall is of order

$$J_{min} \approx (GM_1 r_{min})^{1/2}, \tag{10.2}$$

where G is the gravitational constant, and r_{min} the radius of the innermost stable orbit. For black holes this radius lies between GM_1/c^2 (maximally rotating hole) and $6GM_1/c^2$ (non-rotating hole), and one finds

$$J_{min} \approx \frac{2GM_1}{c}, \tag{10.3}$$

assuming that the rotations of black hole and gas are in the same sense. For non-magnetic neutron stars or white dwarfs, r_{min} can be set equal to the radius of the object, which in the former case leads to a value of J_{min} similar to (10.3), and in the latter to a value that is roughly a factor of 30 higher. When the object possesses a strong magnetic field, r_{min} may be taken of order r_M, the radius of the magnetospheric boundary. Within this boundary the field dominates the flow (see Chapter 11, this book). For the strongest fields known to occur at the surface of compact stars ($\sim 10^{12}$ Gauss for neutron stars, and $\sim 10^7$ Gauss for white dwarfs), r_M is approximately $10^2 R_1$. Fields of that strength on white dwarfs may dominate the flow all the way out to the surface of the companion star. This would prevent the formation of a disk; a case exemplified by the Am Her type objects (see Chapter 4, Section 4.6, this book).

The part of the gas, expelled by the companion, which satisfies both the energy and momentum criteria, goes into orbit around the compact object. When this has happened, the viscosity in the gas will start to minimize the energy in the flow without (of course) being able to change its total angular momentum. First, the random velocities in the 'vertical' direction (parallel to the average angular momentum vector of the gas at that radius) are dissipated until thermal values are reached. The disk becomes thin. This process is relatively fast because particles at the same radius can communicate effectively.

Several other *simultaneous* processes are also taking place, each one mainly the result of the viscous interactions between disk 'rings' at different radii:

* Orbital ellipticities are reduced until the orbit with lowest energy for the given angular momentum is reached, which is circular unless it passes close to the companion.
* If rings at different radii lie in different planes, these planes are adjusted to one another as much as possible. This leads to a flat (planar) disk, unless there is a precessional effect which drives the disk towards a twisted shape (see Section 10.4.1).
* Mass is transported radially. The total energy is lowered by moving the bulk of it inward to smaller radii, while moving a small part outward to conserve the total J: The orbiting material tends to *spread out* radially.

The outward expansion of the disk continues until, just before the Roche lobe is reached, tidal interaction with the companion can remove the excess angular momentum from the disk edge, and feed it back into the binary orbital motion (Papaloizou & Pringle, 1977). In steady state, new mass is constantly added to the outer part of the disk, which influences its shape and radius to some extent. Figure 10.1 summarizes the disk formation process in the case of overflow.

10.1.3 Observational evidence for disks

The cataclysmic variables (CVs), which are mass-transferring binaries containing a white dwarf, are the largest group of systems for which evidence is available for the presence of a disk. Their optical emission is generally dominated by light from a source centered on the white dwarf and extending to near the Roche lobe. The disk-like proportion of this source is easily measured in eclipsing systems, but in others this is obtained from the splitting of emission lines. This splitting corresponds to a Doppler shift appropriate for twice the orbital velocity at a radius near the disk edge. In a few cases much more subtle effects can be noticed which confirm the disk hypothesis. For instance, in DQ Her a wavelength-dependent phase-shift phenomenon not only allows a determination of the range of orbital velocities in the disk, but also establishes that the disk material rotates in the same sense as the central white dwarf (Chanan, Nelson & Margon, 1978). Extensive (but not very recent) reviews of the details of observational evidence for disks in CVs are contained in Warner (1976) and Robinson (1976). See also Chapter 4, Section 4.5, this book.

Amongst the *X-ray binaries*, the presence of disks is probably common but not universal. For most systems the evidence is not very direct. The best case is Her X-1, for which many intricate details are known about time variations

Fig. 10.1. The formation of an accretion disk by overflow. (*a*) initial ring; (*b*) spreading due to viscosity; (*c*) full size steady state disk; (*d*) side view (cross-section).

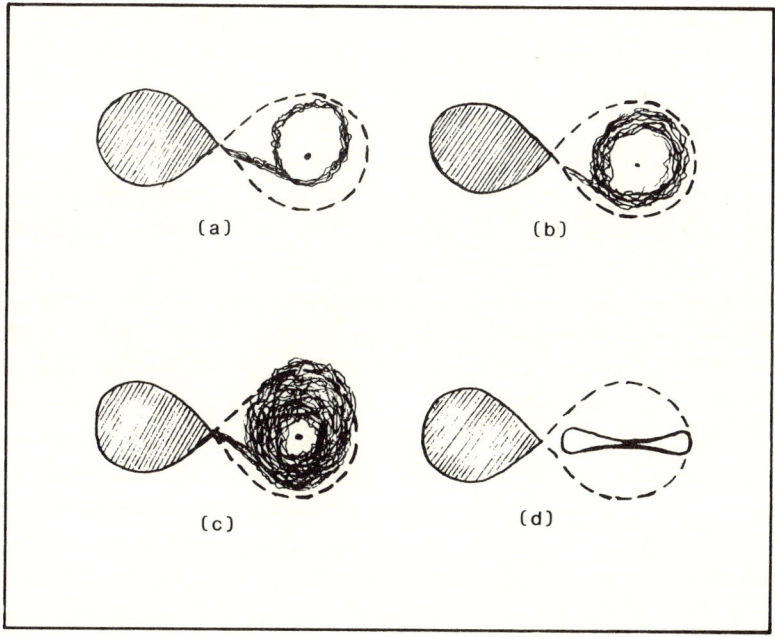

in the optical and X-ray emission. These are well accommodated by models (Katz, 1973; Gerend & Boynton, 1976; Petterson, 1977) containing a large disk (again of 'Roche-lobe' size), whose total angular momentum vector is not parallel to that of the binary system. The fact that the X-ray source is clearly seen to illuminate and heat the companion star uninterruptedly, while the directly observed X-rays switch periodically on and off, is explained in these models by obscuration of the X-rays due to a precession of the 'misaligned' disk. In several other X-ray binaries the presence of a disk is supported by an order of magnitude agreement between the observed spin-up rate for the rotation of the compact object and theoretical predictions based on disk accretion (see Chapter 11, this book), and sometimes by the presence of 'extra' optical light in the lightcurve (see e.g. van Paradijs & Zuiderwijk, 1977, on SMC X-1, and Chapter 5). On the presence of disks in low-mass systems, see Chapter 5, Section 5.2.3, this book. Finally, the occurrence and properties of optical bursts occurring in X-ray burst sources may be explained by the presence of a disk which reflects and slightly smears the X-ray bursts. (Pedersen, 1981; Chapter 2, Section 2.4.2.7, this book).

10.1.4 Applicability of disk models

It is reassuring to note that disk models around compact objects, no matter how uncertain they still are, can at least explain the *most* general characteristics of bright X-ray sources and CVs. Compact objects form deep gravitational wells which can extract a substantial part of the *total* energy of accreting material, and turn it into radiation. Accreting neutron stars and black holes can convert up to 10 per cent of that energy; white dwarfs less by a factor of 1000. For modest mass-transfer rates of $\dot{M} \approx 10^{17}$ g s^{-1}, one expects luminosities of 10^{37} erg s^{-1} and 10^{34} erg s^{-1} respectively. If emitted thermally from an area a few times the size of the object, this radiation must predominantly fall in the X-ray (1 keV) and EUV (10 eV) radiation bands in the two cases. Higher characteristic temperatures occur when the object has a strong magnetic field and funnels all accreting matter onto a small area near its two poles, or when for some reason the emission is (partially) non-thermal. The small size of the region from which almost all the radiation is emitted, even in the absence of a magnetic field, allows large flux variations on very short timescales (seconds for white dwarfs, and milliseconds for black holes and neutron stars). Thus the characteristic property of X-ray binaries and CVs to emit a large part of their excess energy in the form of rapidly fluctuating X-ray or UV radiation is accommodated naturally by disk models around compact objects. If the geometry were not disk-like but 'spherical' some of the above arguments could also hold, but we would not be able to look at the compact object directly.

10.2 The structure of a disk model
10.2.1 Hydrodynamic description

If a binary system contains a disk-like gas flow, transferring at least 10^{16} g s^{-1} ($\sim 10^{-10}$ M$_\odot$ yr^{-1}) between the two stars, the mean free path λ of the atoms and ions in the gas is very much smaller than the size of the binary system. For particle densities $n \gtrsim 10^{14}$ cm^{-3}, and a total cross-section $\sigma_{\text{tot}} \gtrsim 10^{-18}$ cm^2 (Coulomb cross-section for $T < 10^7$ K), one finds $(n\sigma_{\text{tot}})^{-1} \lesssim 10^4$ cm, while the separation between the stars is typically larger than 10^{10} cm. It is therefore appropriate to describe the motion of the gas by *hydrodynamics*, not by particle mechanics (Burbidge & Prendergast, 1968).

Without *viscosity*, the gas would settle in nearly-Keplerian orbits and remain there. It is the viscosity which makes the gas move inward. Thus, to describe the flow, the hydrodynamic equations for a viscous (compressible) fluid must be used. They are:

(a) *Mass conservation law*:

$$\frac{\partial \rho}{\partial t} + \nabla \cdot (\rho \mathbf{v}) = 0 \tag{10.4}$$

or 'continuity equation', where $\mathbf{v}$ is the flow velocity.

(b) *Momentum conservation law*:

$$\rho \left[\frac{\partial \mathbf{v}}{\partial t} + (\mathbf{v} \cdot \nabla)\mathbf{v} \right] = \rho(\mathbf{F} - \nabla \Phi) - \nabla p - \nabla \cdot \mathbf{t} \tag{10.5}$$

or 'Euler's equation', expressing the change in momentum of a fluid element of unit volume (left-hand side) in terms of the forces acting on it. $\mathbf{F}$ takes external forces into account, and also corrects Equation (10.5) in case a non-inertial frame is used. For a frame rotating around the center of mass (CM) of the binary, with the same angular velocity Ω as the two stars (a 'co-rotating frame'), $\mathbf{F}$ takes the form

$$\mathbf{F} = -2\mathbf{\Omega} \times \mathbf{v} - \mathbf{\Omega} \times [\mathbf{\Omega} \times (\mathbf{r} - \mathbf{r}_{\text{CM}})]. \tag{10.6}$$

The first term in (10.6) represents the Coriolis force, the second the centrifugal force. The gravitational potential Φ is given by

$$\Phi = -\frac{M_1}{|\mathbf{r} - \mathbf{r}_1|} - \frac{M_2}{|\mathbf{r} - \mathbf{r}_2|}, \tag{10.7}$$

where M_1 and M_2 are the masses of, respectively, the compact object and the mass-losing star, and $\mathbf{r}_1$ and $\mathbf{r}_2$ the distances of their centers of mass to an arbitrary origin (from here on taken to coincide with the compact object; thus $|\mathbf{r} - \mathbf{r}_1| = r$). The symbols p and $\mathbf{t}$ denote pressure and viscous stress in the gas.

(c) *Energy conservation law*:

$$\rho\left[\frac{\partial \epsilon}{\partial t} + \mathbf{v} \cdot \nabla \epsilon\right] = -p\nabla \cdot \mathbf{v} - (\mathbf{t} \cdot \nabla) \cdot \mathbf{v} - \nabla \cdot \mathbf{q}, \tag{10.8}$$

expressing the change of internal energy ϵ of a moving fluid element of unit volume (left-hand side) in terms of the work done by pressure, the heat generated by viscous stress, and the loss due to cooling. The flux of energy $\mathbf{q}$ includes radiative, convective, and conductive losses.

These five conservation laws ((b) contains three equations) are insufficient to determine the velocity field $\mathbf{v}$ as well as the emitted radiation spectrum: auxiliary equations linking $\epsilon, p, \rho, \mathbf{t}$, and $\mathbf{q}$ are needed.

10.2.2 Two-dimensional approximation

A first step towards determining the field $\mathbf{v}$ can be made by assuming that the flow takes place in a very thin layer near the binary plane, neglecting the perpendicular velocity component and the entire vertical structure. This reduces the problem to two dimensions. Since the equation of vertical balance between pressure and gravity is ignored in this approach, pressures and temperatures cannot be calculated, so that the emitted radiation spectrum cannot be determined. Nevertheless, an approximate picture of the flow pattern in the disk plane can be found.

The two-dimensional problem is still formidable because of the boundary conditions, requiring that (in the simplest case) new gas is steadily added to the flow in a stream from the mass-losing star. This stream collides with the disk gas at supersonic relative velocities, and creates a shock in the interaction region (see Figure 10.2). With simple assumptions about the value of p and $t_{\phi r}$ (the only relevant stress tensor component in the flow plane, as shown in Section 10.2.3), the problem can be solved numerically (Prendergast & Taam, 1974; Lin & Pringle, 1976). The results show that:

* The disk is *very large*: it can even extend beyond the Roche lobe. This is possible because the lowest outlet of the equipotentials outside the Roche lobe lies at the outer Lagrangian point L_2. The potential at L_2 is higher than at L_1 (Figure 10.2).
* Away from the outer boundary, the flow occurs in approximately *circular orbits* around the compact object.
* Near the outer boundary *non-circularities* occur, caused by the tidal effect from the companion star and the impact of the stream of new gas into the disk.

A three-dimensional numerical simulation of the flow, simultaneously determining its structure and outcoming radiation, is not attainable at this time. However, rough, analytical, three-dimensional models can be constructed,

Accretion disks in close-binary systems

starting from the above results: the gas moves in approximately circular orbits around the compact object in the orbital plane of the binary. Instead of neglecting the vertical structure entirely, one may work – as a next step – with vertically-averaged disk quantities. Stationary models of this type which can resolve the most basic features of the radial structure of the disk will here be called 'basic disk models'.

10.2.3 The five hydrodynamic equations for a basic stationary axisymmetric disk

The hydrodynamic equations, Equations (10.4), (10.5), and (10.8), can be simplified by making the following additional assumptions:

(1) *The disk is axisymmetric.* This implies that the gravitational influence of the mass-losing star is neglected. Thus, it is convenient to use an inertial frame ($\Omega = 0$), and a cylindrical coordinate system, r, ϕ, z with $z = 0$ the

Fig. 10.2. Density contours of steady disk. Equipotential surfaces of the binary system are dotted, and the Lagrangian points L_1, L_2 shown (adapted from Prendergast & Taam 1974).

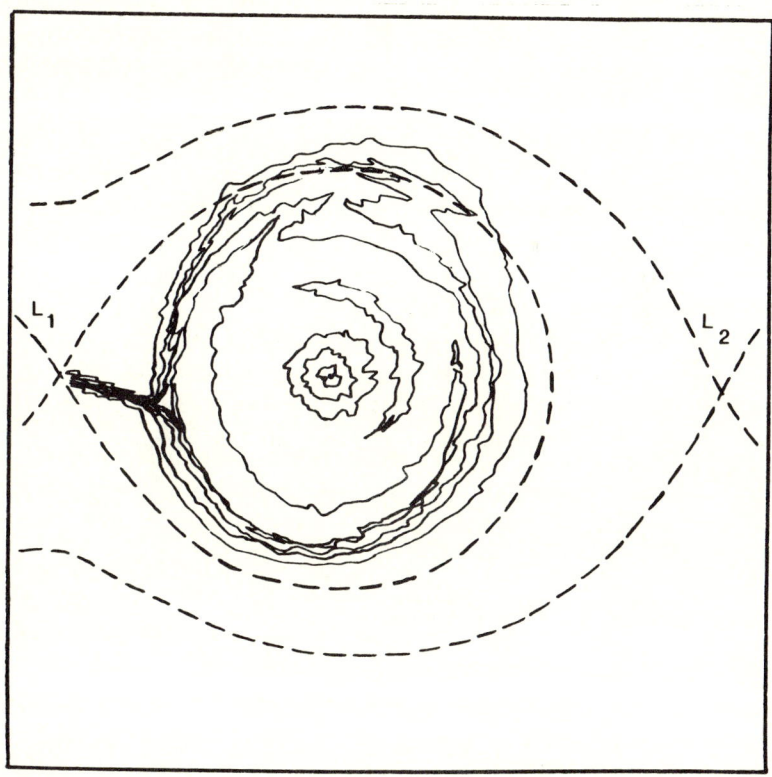

central plane of the disk, taken to be the binary orbital plane. The disk is assumed symmetric about this plane.

(2) *The disk is thin.* Its half-thickness h satisfies the inequality $h/r \ll 1$ at every radius r. This implies that pressure forces in the disk are much smaller than gravitational forces (pressure thickens the disk): $p \ll \rho v^2$. It also implies that the vertical velocity component v_z, which changes the thickness of a fluid element as it moves inward (hence $v_z \sim (h/r) v_r$), must be much smaller than the radial component v_r.

(3) *Viscosity influences the flow only slightly.* None of the stress-tensor components exceeds the pressure substantially: $t_{ij} \lesssim p$. An assumption is made here about the viscosity mechanism, which is further discussed in Section 10.3.3. Combining this with the previous assumption gives $t_{ij} \ll \rho v^2$. Since viscosity causes the radial velocity, this velocity component must be small: $v_r \ll v_\phi$.

(4) *The disk is stationary.* All variables and parameters are time-independent, including the mass flux $\dot{M}$ in the gas stream.

A lowest-order approximation of the hydrodynamic equations based on the above assumptions, which include $v_z \ll v_r \ll v_\phi$, can now be made. Vertical integration of the results removes all remaining z-derivatives, and gives the equations:

(a) *mass conservation*:

$$\frac{\partial}{\partial r}(r\rho h v_r) = 0 \tag{10.9}$$

or after radial integration

$$r\rho h v_r = -\frac{\dot{M}}{4\pi}, \tag{A}$$

where the constant of integration is chosen such that $\dot{M}$ is the total accretion rate onto the compact object. This equation expresses the fact that the rate of mass inflow across a cylinder of radius r must equal $\dot{M}$ at every r. Vertical winds which carry off mass are assumed negligible. The half-thickness h of the disk is defined as

$$h = \frac{1}{\rho}\int_0^\infty \rho(z)\,dz, \tag{10.10}$$

(b1) *momentum conservation (r)*:

$$v_\phi^2 = \frac{GM_1}{r}; \tag{B}$$

the rotation is Keplerian.

(b2) *momentum conservation (ϕ):*

$$r\rho v_r \frac{\partial}{\partial r}(rv_\phi) = -\frac{\partial}{\partial r}(r^2 T_{\phi r}),\tag{10.11}$$

where

$$T_{\phi r} \equiv \int_0^\infty t_{\phi r}(z)\,dz.\tag{10.12}$$

Using (A), one can radially integrate (10.11), and obtain

$$\frac{\dot M}{4\pi} r v_\phi = r^2 T_{\phi r},\tag{C}$$

neglecting the integration constant. The significance of this neglect is discussed in Section 10.2.4.

(b3) *momentum conservation (z):*

$$p \sim \rho h^2 \frac{GM_1}{r^3},\tag{10.13}$$

or vertical pressure balance. This can, with the help of (B), be written as

$$\frac{p}{\rho} \sim \left(\frac{h}{r} v_\phi\right)^2.\tag{D}$$

(c) *energy conservation:*

$$Q^- = -T_{\phi r} r \frac{\partial}{\partial r}\left(\frac{v_\phi}{r}\right),\tag{10.14}$$

where

$$Q^- = \int_0^\infty \frac{\partial q}{\partial z}\,dz = q(\infty)$$

is the flux lost from the disk surface, per unit area. Substitution of (B) and (C) gives

$$Q^- = \frac{3}{8\pi}\dot M \frac{GM_1}{r^3},\tag{E}$$

showing that for a stationary disk, the emitted flux Q^- at a certain radius r is determined completely by M_1 and $\dot M$. The key to the absence of any dependence on the heat transport and viscosity mechanisms, is the assumption of a stationary state.

10.2.4 Modifications near the inner and outer-disk edges

The compact object may exchange angular momentum (and energy) with the disk via its inner edge. The interaction occurs mainly by direct contact in a narrow boundary layer on the star's surface if $r_i \approx R_1$ (Pringle & Savonije, 1979), or magnetically in an extended area containing the magnetospheric boundary if $r_i \approx r_M$ (Gosh & Lamb, 1978; also Chapter 11, this book). For black holes it occurs in an area of very-strongly-curved geometry by means of the frame-dragging effect. In each one of these cases the interaction itself is complicated, but its net effect on the disk equations, Equations (A)–(E), *outside of the interaction region* can be expressed in a simple way.

Integrate (10.11) *without* neglecting the integration constant:

$$\frac{\dot{M}}{4\pi} r v_\phi = r^2 T_{\phi r} + \frac{\dot{J}}{4\pi}. \tag{10.15}$$

The constant ($\dot{J}$) represents the net rate at which angular momentum is lost from the disk at the inner radius r_i through the interaction with the object. For instance, if the torque $T_{\phi r}$ vanishes at r_i, then $\dot{J} = \dot{M} r v_\phi$. In this case the object receives the entire angular momentum $(r v_\phi)_i$ per unit mass, which the accreting material has at radius r_i. For an arbitrary torque at r_i one may use

$$\dot{J} = \beta_L \dot{M}(r v_\phi)_i, \tag{10.16}$$

where β_L is a dimensionless parameter of order 1. Inside the interaction region, β_L depends on r, outside it does not. Where ever β_L may be considered constant, Equation (C) becomes

$$\dot{M} r v_\phi \left[1 - \beta_L \sqrt{\left(\frac{r_i}{r}\right)}\right] = 4\pi r^2 T_{\phi r}, \tag{C'}$$

and (E) becomes

$$Q^- = \frac{3}{8\pi} \dot{M} \frac{GM_1}{r^3} \left[1 - \beta_L \sqrt{\left(\frac{r_i}{r}\right)}\right]. \tag{E'}$$

Equations (A), (B), and (D) remain unchanged. The modifications of (C) and (E) are clearly unimportant at disk radii $r \gg r_i$: it is an inner-edge effect.

Another inner-edge effect consists of *general relativistic corrections* to the disk equations. These will become of interest as soon as any reliable, lower-order approximation exists for the structure and radiated flux of the innermost part of the disk. That point is not in sight. In anticipation of it, Novikov & Thorne (1973) have derived the corrections to the disk equations (A)–(E), Cunningham (1975) those to the properties of the emitted photons.

A general relativistic effect that can in principle exert its influence out to 10^2–10^3 times the radius of a neutron star or black hole has been described by

Bardeen & Petterson (1975): if the compact object is rapidly rotating around an axis which is not perpendicular to the binary plane (for whatever reason), the Lense-Thirring effect (or 'dragging of inertial frames') due to the object's rotation, forces the inner part of the disk to lie in the equatorial plane of the object. For a 1 $M_\odot$ black hole rotating at ~10 per cent of its maximum frequency or a neutron star rotating with a period of ~1 s, the disk makes its gradual transition from binary to equatorial plane at radii of order 10^8 cm.

Near the *outer edge*, the energy equation, Equation (E), needs several important modifications. Newly incoming disk material delivers a considerable part of its kinetic energy to the matter in the disk rim by ramming into it. This adds substantially to the energy that is locally produced by the viscous stresses. Another, and probably even more important extra source of energy for the outer-disk edge, is from reabsorbed radiation emitted near the compact object. At the outermost boundary of a disk around a strong X-ray source with luminosity $\gtrsim 10^{36}$ erg s^{-1}, this reabsorbed radiation can so dominate the local energy balance that (E) is not even approximately correct anymore.

Similar comments can be made about the applicability of the momentum equations near the outer edge. The orbital motion of the gas becomes non-Keplerian due to the influence of the companion, affecting both (B) and (C). The momentum equation, Equation (C), must be modified also because angular momentum is removed from the outer edge by tidal interaction. The tidal motions are dissipated by viscosity, and thus form yet another extra source of energy to be included in (E).

10.2.5 How to build a basic disk model

The five hydrodynamic equations in the algebraic (non-differential) form, (A)–(E), are the starting point for the construction of almost all stationary axisymmetric disk models. They do not describe the disk edges very well, but they may form a good lowest-order approximation at intermediate radii. For disks in X-ray binaries, which may extend over five decades of radii (10^6–10^{11} cm), this allows the modelling of a large part of the disk. However, disks in CVs maximally extend over only about two radial decades ($10^{8.5}$–$10^{10.5}$ cm) and often less if the white dwarf possesses a substantial magnetic field. In such cases the disk is little more than an inner and an outer edge together. It is in fact hardly a disk.

The set of hydrodynamic equations is not algebraically complete. Thus, to obtain a solution even at intermediate radii, several auxiliary equations need to be added. A sufficient set of these consists of an *equation of state* for the gas, a *viscosity law*, and an equation that specifies the *cooling process* transporting heat from the disk interior to the surface. Make a choice for these three equations.

The entire set of eight can then be solved: This gives the temperature T, and the velocity field $\mathbf{v}$ of the gas, both 'averaged' over one- (vertical) half of the disk. It also gives the half-thickness h and the emitted photon spectrum. The result is a 'basic disk model'.

10.3 Basic stationary axisymmetric models
10.3.1 General comments on viscosity

Uncertainties in all accretion disk models are not only due to our ignorance about the viscosity mechanism, but also to uncertainties related to the equations of state and cooling, the vertical disk structure, and various boundary effects, among other things.

It is, of course, only *an assumption* that the mechanism causing radial inflow in disks can be described as a viscosity (i.e. as the product of a viscosity coefficient and a velocity gradient). However, if this assumption is made, the observational restrictions on the viscosity mechanism are much tighter than is often suspected. The component of the viscous stress tensor responsible for radial inflow in axisymmetric disk models is

$$t_{\phi r} \approx -\eta \left(\frac{\partial v_\phi}{\partial r} - \frac{v_\phi}{r} \right), \tag{10.17}$$

where η is the dynamic viscosity coefficient. Insertion of the Keplerian value for v_ϕ, and substitution of (10.17) into (C) gives the remarkably simple relation

$$\dot{M} \approx 3\pi h \eta. \tag{10.18}$$

This shows that in a stationary axisymmetric disk the value of $h\eta$ is uniquely determined by the accretion rate $\dot{M}$. At any radius in the disk where the viscosity mechanism does not provide a high enough value for η, matter will pile up and increase $h\eta$ until (10.18) is satisfied.

Equation (10.18) allows a direct estimate of η. Since $h \ll r$ implies $h \lesssim 10^{10}$ cm, one finds at a typical accretion rate of $\dot{M} \sim 10^{17}$ g s^{-1}, that $\eta \gtrsim 10^6$ g cm^{-1} s^{-1}. This value is orders of magnitude higher than the expected particle viscosity (Rossi & Olbert (1970) give $\eta < 10^2$ g cm^{-1} s^{-1} for a completely ionized hydrogen plasma at $T < 10^7$ K). Hence, the viscous mechanism in disks must be much more effective than particle viscosity. Only a few physical processes present themselves as candidates.

Turbulence, an efficient angular momentum transporter in fluids, can mimic the effects of a large viscosity in certain situations, e.g. in pipes, or between two rotating cylinders. Consideration of this mechanism as a possible viscosity source in accretion disks may *a-priori* seem entirely out of place, because these disks are textbook examples of flows which satisfy Rayleigh's stability criterium ($\partial J/\partial r > 0$). Furthermore, finding turbulence-feeding instabilities outside the main flow in axisymmetric disk models has failed consistently. Nevertheless, many authors are reluctant to give it up. Explicitly or implicitly, they prefer to question the validity of a criterion unverified under circumstances anywhere near the ones in accretion disks. Here, the flow is strongly shearing and supersonic, and has a very high Reynolds number

$$R \approx \frac{v_\phi r \rho}{\eta} \approx \frac{p}{t_{\phi r}} \left(\frac{r}{h}\right)^2 \gg 1, \qquad (10.19)$$

as follows from (D) and (10.17).

It is not even clear that turbulence in an accretion disk, if it *could* occur there, would have the effects of a viscosity. It may simply reduce angular momentum differences in the flow by particle exchanges between different orbits (as it would in the most naive model). This would drive near-Keplerian flows toward a uniform angular momentum distribution $J = r v_\phi =$ constant. (A viscosity would drive it toward uniform rotation $\omega = v_\phi/r =$ constant.) The result would be that instead of spreading the disk material over a wider range of radii, it would force the matter into a narrow ring at the radius where the Keplerian value for J equals the average J-value of all the gas in the disk. Such a mechanism would obviously not provide a good cause for radial inflow. More complicated, non-isotropic forms of turbulence don't need to have these utterly disqualifying properties. Still, turbulence seems a problematic candidate for the main viscous mechanism in disks.

The principal remaining candidate, capable of providing viscosity at the required high level, is the action of small scale magnetic fields in the gas. This tends to reduce velocity gradients, so it drives Keplerian flows towards uniform rotation, as desired for a viscosity. While entering the disk with the transferred matter, the field may have been very small. Once it is in the disk, the Keplerian shear amplifies it at a rate

$$\frac{\partial \mathbf{B}}{\partial t} = -\nabla \times (\mathbf{v} \times \mathbf{B}), \qquad (10.20)$$

which does not change B_r, but stretches the B_ϕ component as

$$\frac{\partial B_\phi}{\partial t} = r \frac{\partial \omega}{\partial r} B_r. \qquad (10.21)$$

The viscous stress tensor component $t_{\phi r}$ due to these magnetic fields has a value of order

$$t_{\phi r} \approx \frac{B_r B_\phi}{4\pi}. \qquad (10.22)$$

Since the shearing motions of the gas tend to string the fields out in the

ϕ-direction, $t_{\phi r}$ is expected to be smaller than the magnetic pressure $B^2/8\pi$. If this pressure exceeds the internal disk pressure, the fields develop the tendency to bulge out of the disk, reconnect, and leave, like coronal loops in the sun. Thus, (10.22) can be expressed in the form

$$t_{\phi r} \approx \alpha p, \qquad (10.23)$$

with $\alpha \lesssim 1$.

This value of the dimensionless parameter α depends on how effectively the B-field, increasing steadily due to shearing, can be limited by other processes, e.g. magnetic reconnection. Lightman & Eardley (1974) and Ichimaru (1977) have estimated theoretical values for α in the range $10^{-2} - 1$. On the other hand, Galeev, Rosner & Vaiana (1979) argue, although Coroniti (1981) disputes, that magnetic reconnection does not occur sufficiently fast to balance the increase of B inside the disk. Its value is then only limited by the escape process of field loops (Parker instability). Under those circumstances α could not be much less than unity.

The problem of determining α is reminiscent of the debate, a little less than a decade ago, about how large the radius of a mass-losing star in an X-ray binary is, compared to its Roche lobe. One side argued that it could not be smaller than the Roche lobe or the star would not lose mass. The other side, that it could not be larger because this lobe forms the theoretical limit, and whatever goes beyond would be transferred. An observation of Her X-1 by Middleditch & Nelson (1976) settled the matter in a convenient way for all. It established that the star *precisely* filled the Roche lobe so accurately that the pointed part of the lobe at L_1 forced the star to be pointed there too.

Quite analogous to the radius of the star, the value of α in *some* disks is now also well determined by observations, although not with so high an accuracy. The decay timescale of outbursts in dwarf-novae systems (a few days) and in hard transient X-ray sources (a few weeks), if interpreted as due to the dissipation of a structure of accreting matter with disk-like properties, leads to values for the viscosity parameter α of order 1 (Bath & Pringle, 1981; see Section 10.4.2 for time-dependent disk properties). Similar values have been obtained from studies of the precession of disks in Her X-1, and SS 433 (Merritt & Petterson, 1980; Katz, 1980). An uncertainty in this result for α of one order of magnitude must still be admitted, but no author should get away anymore with claims that α could have any arbitrary value not significantly exceeding unity in any realistic accretion disk. Of course, the possibility remains open that its value differs substantially from one system to another so that the observational limits cannot be applied uniformly.

10.3.2 The Shakura–Sunyaev (SS) model

A simple choice for the three auxiliary equations is the following:

(1) The ordinary *equation of state* for a mixture of radiation and ideal gas

$$p = \tfrac{1}{3} a T^4 + \rho \frac{k}{\mu m_p} T. \tag{10.24}$$

(2) A *cooling equation* based on radiative transport occurring mainly in the z-direction through an optically-thick disk (optical depth $\tau > \kappa \rho h \gg 1$)

$$q_z = -\frac{1}{\kappa \rho} \frac{\partial}{\partial z} \left(\frac{a}{3} T^4 \right). \tag{10.25}$$

Vertical integration over one half of the disk gives

$$Q^- = \frac{caT^4}{\kappa \rho h}, \tag{10.26}$$

where the values of T, ρ, and the opacity κ are all taken in the central ($z=0$) disk plane. The surface term from the integration, proportional to the surface temperature T_s to the fourth power, has been neglected in (10.26) under the assumption that $T_s < T$. The main contributions to κ are expected to be Thomson and free-free opacity. Thus,

$$\kappa \approx (0.4 + 0.6 \times 10^{23} \rho T^{-7/2}) \text{cm}^2 \text{ g}^{-1}, \tag{10.27}$$

with ρ and T respectively expressed in g cm^{-3} and °K. The vertical distributions of density and energy production have some influence on the numerical factor c in (10.26), but very little on the resulting disk structure.

(3) A *viscosity law* of the form (10.23), based on viscosity due to small scale magnetic fields.

This choice for the auxiliary equations was first made by Shakura & Sunyaev (1973), although they did not restrict (10.23) to viscosity from magnetic fields (turbulence allows a similar estimate $t_{\phi r} \approx \alpha p$). Together with the hydrodynamic laws (A)–(E), these three equations form a complete set from which the radial disk structure and emitted radiation Q^- can be calculated by purely algebraic manipulations. The result is often called the α-disk model. (See also Novikov & Thorne (1973) for a detailed description.) Its most important features are:

* *The disk is concave.*

It thickens outward as $h \sim r^\gamma$, with $\gamma > 1$, except perhaps very near the compact object. At radii $r > 10^8$ cm, $h \sim r^{9/8}$. This property is seen to be realistic for disks in the cataclysmic variables DQ Her and UX UMa. In these systems, rapidly oscillating emission from the vicinity of the white dwarf, observed after having been reflected by the disk, can be used to reconstruct the shape of the reflecting surface, and indicates concaveness (Chanan, Nelson & Margon, 1978; Chester, 1979; Petterson, 1980).

* *The disk is extremely thin.*
Even at the outer edge ($r \approx 10^{11.5}$ cm), where the disk is thickest, h is at most a few per cent of the radius r. This property is probably not realistic for many disks, particularly those in X-ray binaries and bursters. Pedersen *et al.* (1982) show that the solid angle spanned by a disk, as seen from the compact object in the burster MXB 1636-53, must be an order of magnitude larger than the Shakura-Sunyaev model predicts if optical bursts from this object are to be explained by absorption and reradiation of X-ray bursts by the disk. In most other X-ray binaries, reabsorption of the steady flux of X-rays from the central object by the concave outer parts of the disk probably leads to a considerable thickening, an effect which the basic SS model does not take into account (see Chapter 5, Section 5.2.3, this book, for observational evidence of the 'thickening').

* *The disk radially extends to the Roche lobe.*
This is not a property of the SS model specifically, but results from the disk's need to remove its excess angular momentum, as discussed in Section 10.1.2. Observational support for large disks in CVs was mentioned before. In the eclipsing X-ray binary Her X-1 a measurement of the disk radius can be made straightforwardly from the width and depth of the secondary optical minimum in the lightcurve: The disk is of noble size (Gerend & Boynton, 1976).

* *The mass of the disk is low.*
For $M_1 \sim 1\, M_\odot$ and $\dot{M} \sim 10^{17}$ g s^{-1} one finds a total mass of $\sim 10^{-9}\, M_\odot$. This value becomes larger for increasing M_1, $\dot{M}$, and α^{-1}. Self-gravity of the disk may be completely neglected.

* *The disk is optically thick.*
This criterion is met easily at all radii if $\dot{M}$ and M_1 have typical values. It implies that the disk can radiate its energy thermally (by blackbody radiation).

* *The disk has a $\nu^{1/3}$ spectrum.*
If all radiation from the disk is indeed thermally emitted, intergration of the flux over the entire range of radii gives a $\nu^{1/3}$ frequency dependence over a large frequency interval (Pringle, 1981). Observational support for this disk feature is strikingly present in the spectra of the cataclysmic variables EX Hya and VW Hya (Bath, Pringle & Whelan, 1980; see Figure 10.3 see also Chapter 4, Figure 4.1(*a*), this book). The bulk of the energy radiated from the disk and the compact object comes out in X-rays around a neutron star or black hole, and in UV around a white dwarf, as anticipated in Section 10.1.4. Some X-rays are also produced in many CVs (see Chapter 4, Section 4.4 and Table 4.2, this book), and some exceedingly hard X-rays (>100 keV) are emitted by CYG X-1, the

system that contains the black hole candidate. Both phenomena probably involve non-thermal emission, and can therefore not be accommodated in the simple Shakura–Sunyaev optically thick disk model.

10.3.3 Other models

A different choice for the three auxiliary equations gives a different basic model. The possible variations are almost unlimited. Let us mention a few alternative choices for the equations of state and cooling. The viscosity law has been discussed at length above.

* *Equation of state.*
Accreting gas is almost exclusively cooled through its electrons, while its ions gain most of the infall energy. Thus, an effective coupling between the two types of particles is essential to keep them at the same temperature. If the coupling is assumed to take place by Coulomb interactions only, it may fail in the immediate vicinity of black holes and neutron stars. The result would be a two-temperature gas (Shapiro, Eardley & Lightman, 1976) with an equation of state of the form

$$p = \rho \frac{k}{\mu m_p} (T_i + T_e), \tag{10.28}$$

to be supplemented by a relation coupling the two temperatures. This equation of state leads to a disk model with much higher temperatures (T_i) than the SS model. It may be applicable to CYG X-1, which emits some exceedingly hard X-rays, as mentioned above.

Fig. 10.3. Dependence of disk radiation intensity on wavelength (adapted from Bath *et al.* 1980).

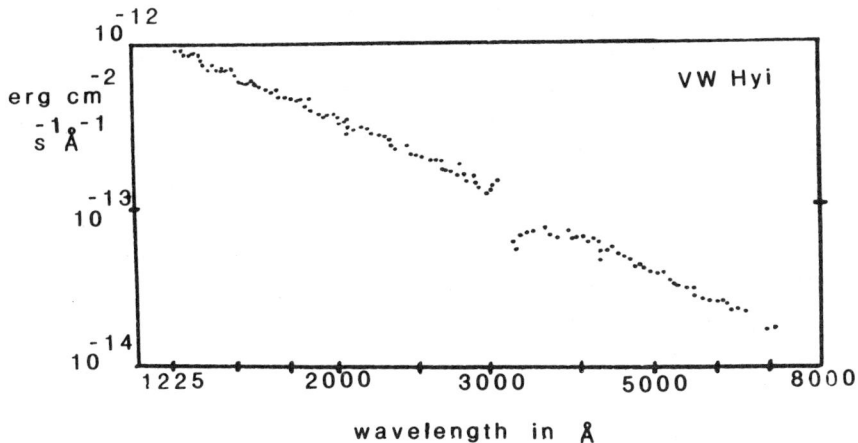

* *Cooling equation.*

Convective cooling has been explored by a number of authors (Liang, 1977; Bisnovatyi-Kogan & Blinnikov, 1977; Shakura, Sunyaev & Zilitinkevitch, 1978; and Taylor, 1980; see also the references in Chapter 4, Section 8, this book), all from a slightly different perspective. It appears that self-consistent disk models based on convective instead of radiative cooling are possible. The problem with them is that they do not predict such strikingly characteristic features that observations could easily distinguish them.

Inverse Compton cooling may be the appropriate mechanism in extremely hot parts of the disk. Shapiro, Eardley & Lightman (1976) used it in their two-temperature disk model which was tailored to Cyg X-1.

10.4 Additional features of disk models
10.4.1 Twists

Up to here it has been assumed that the binary plane was the symmetry plane for the disk flow. This is the simplest assumption, but not the only possible one. If somehow the orbiting matter came to lie in an inclined plane, the tidal force of the companion acting on it would acquire a vertical component. This would cause the orbits to precess with an angular velocity ω, which is strongly dependent on radius ($\omega \sim -r^{3/2}$). The negative sign of ω indicates that this precession is retrograde, i.e. in the opposite sense of the binary motion, if the orbital rotation of the disk matter itself is prograde, as is certainly expected.

Accretion disks do not behave as solid bodies, so rings of matter at different radii will attempt to precess at different rates, each in accordance with its own value of ω. On the other hand, disk rings are coupled by viscosity, which opposes the 'winding-up' resulting from differential precession. If viscosity is successful in stopping the winding up, the result is a *twisted disk* with stationary shape, precessing with a single period. This period is some weighted average over the precession periods of the individual rings. How far the disk gets wound-up depends on the strength of the viscosity. It is just like the distortion of an elastic body: a small viscosity allows a large twist, and a large viscosity only a small one. Any tilted disk must necessarily get twisted to *some* extent.

To describe a twist, characterize the orientation of each ring of disk matter by the two Eulerian angles β and γ respectively, denoting the inclination angle of the ring with respect to the orbital plane, and the azimuthal angle of its line of nodes. The twist equations can now be obtained by minimizing the *total* energy contained in the viscous shear of the deformed disk.

For small twists ($r\beta', r\gamma' \ll 1$), the potential energy density in the deformation is

$$\epsilon_{\text{twist}} \approx \tfrac{1}{2}\eta\omega\sigma^2, \tag{10.29}$$

where $\eta\omega$ plays the role of an elasticity constant, and the strain

$$\sigma \approx r(\beta' \cos\phi + \gamma' \sin\beta \sin\phi) \tag{10.30}$$

is a measure of the local deformation of the disk. Primes denote radial derivatives. The total potential energy in the twist is obtained by integrating ϵ_{twist} over the entire disk

$$E_{\text{twist}} \approx \int \epsilon_{\text{twist}} r \, dr \, d\phi \, dz. \tag{10.31}$$

The ϕ-integration is straightforward, and the z-integration can be performed in an approximate way by multiplying the integrand by $2\,h(r)$. Equation (10.18) shows that the product $h\eta$ does not depend on r and may be placed outside the integral. Thus, one is left with

$$E_{\text{twist}} \approx 2\pi h\eta \sqrt{(GM_1)} \int r^{3/2} [(\beta')^2 + (\gamma')^2 \sin^2\beta] \, dr. \tag{10.32}$$

Determine β and γ by assuming that the equilibrium shape for the twist is the one that minimizes E_{twist}, subject to the particular set of boundary conditions for β and γ. Independent variation of β and γ gives the two Euler-Lagrange equations

$$\begin{aligned}\frac{\partial}{\partial r}\left(\frac{\partial L}{\partial \beta'}\right) &= \frac{\partial L}{\partial \beta} \\ \frac{\partial}{\partial r}\left(\frac{\partial L}{\partial \gamma'}\right) &= \frac{\partial L}{\partial \gamma},\end{aligned} \tag{10.33}$$

where L is defined by $E_{\text{twist}} = \int L \, dr$. Substitute L into (10.33) and find two coupled non-linear differential equations in β and γ: *the twist equations* (Petterson, 1977, 1978a). Hatchett, Begelman & Sarazin (1981) discovered that these two equations can be written as a single *linear complex* differential equation for the variable $w = -i\beta \exp(i\gamma)$:

$$w'' + c\frac{w'}{r} = 0. \tag{10.34}$$

The constant c, which is 3 in the above case, may come out somewhat differently if η is allowed to have different values in the horizontal and vertical directions. Equation (10.34) is homogeneous, but 'twist-driving forces', such as the ones due to the companion star, can easily be included if the above derivation is performed by the method of Lagrange multipliers. The resulting equilibrium disk shape is obtained by integration of (10.34).

Extensive discussions of the observational evidence supporting the presence of a tilted, and perhaps significantly twisted, precessing accretion disk in the system Her X-1 can be found in Deeter, Gerend, Crosa, & Boynton (1976), and Petterson (1977). Other systems (e.g. SS 433, and LMC X-4) may show the same phenomenon (van den Heuvel *et al.*, 1979; Lang *et al.*, 1981), but these are less well documented.

10.4.2 Time dependencies

Assume that the flow around a compact object satisfies all the conditions for a basic accretion disk model formulated in Section 10.2.3, with the exception of stationarity. The vertically-integrated continuity equation, Equation (10.9), then becomes

$$\frac{\partial \Sigma}{\partial t} + \frac{1}{r}\frac{\partial}{\partial r}(r\Sigma v_r) = 0, \tag{10.35}$$

where the disk surface density Σ is defined as $2\rho h$. Neither *this* equation, nor the vertically integrated ϕ-momentum equation, Equation (10.11), can now be radially integrated, in the way it was done in Section 10.2.1. The other three hydrodynamic equations (B, D, and E) remain as they were if the variations take place on a timescale no shorter than that of the radial drift.

Since time-dependency only enters the disk equations via the term $\partial \Sigma/\partial t$ in (10.35), one may use the complete set (including the auxiliary equations) *without* (10.35) to obtain expressions for all disk variables in terms of r and Σ. The continuity equation, Equation (10.35), can be solved to find how Σ evolves in time, which then immediately shows how all other variables evolve. To do this, write (10.35) in the form

$$\frac{\partial \Sigma}{\partial t} = -\frac{1}{r}\frac{\partial}{\partial r}\left[\frac{4}{v_\phi}\frac{\partial}{\partial r}(r^2 T_{\phi r})\right] \tag{10.36}$$

with the help of (B) and (10.11), and express $T_{\phi r}$ in terms of r and Σ. The result depends on the particular form of the auxiliary equations used for $T_{\phi r}$, and will generally have the form of a non-linear diffusion-type equation for Σ. In analogy to the diffusion coefficient μ in the linear diffusion equation

$$\frac{\partial \Sigma}{\partial t} = \mu \frac{\partial^2 \Sigma}{\partial r^2}, \tag{10.37}$$

one can define an 'effective diffusion coefficient' in the non-linear equation, Equation (10.36), which must be positive for a stable 'smooth' diffusion (Lightman, 1974 a, b). Lightman & Eardley (1974) have demonstrated that this condition is not necessarily satisfied in every disk model. In particular, μ is negative in radiation-pressure dominated parts of the SS disk model (LE instability). Further discussions of disk instabilities, including thermal ones, can be found in Pringle, Rees & Pacholczyk (1973), Shakura & Sunyaev

(1976), Pringle (1976), Giora & Shaviv (1977, 1981), and Piran (1978).

Variations in the mass-transfer rate, radially diffusing through the disk, lead to variations in the luminosity of the compact object, which are not only time-delayed but also somewhat dispersed. Figure 10.4 demonstrates the process. Both the delaying and the dispersing effect increase, the longer it takes the gas to spiral inward. This is closely analogous to the way a twisted disk develops in time if it has not yet reached an equilibrium shape. In both cases, perturbations diffuse due to viscosity, so the two are expected to have the same timescale. Their similarity is illustrated by the fact that Figure 10.4 can also be interpreted as describing the process of a few tilted disk rings aligning themselves to the rest of the disk (see Merritt *et al.*, 1980).

The equation for time-dependent twists (Petterson, 1978*b*; Hatchett *et al.*, 1981)

$$\frac{\partial w}{\partial t} = -\tfrac{1}{3} r v_r \left(w'' + c \frac{w'}{r} \right) + w i \omega \tag{10.38}$$

shows its diffusion character distinctly. The precession rate $\omega(r)$ of a disk ring at radius r due to external forces is included in (10.38) to show how the rings would behave if there were no viscosity. The effective diffusion coefficient μ_{twist} has the value $-\tfrac{1}{3} r v_r$ if the viscosity coefficient η can be considered a scalar. For *inward* radial mass flow in a disk $v_r < 0$, so that $\mu > 0$; as required for a stable diffusion.

Fig. 10.4. Time development (t_1, t_2, t_3, t_4) of δ-function-type perturbation in stationary axisymmetric accretion disk. Ordinate may equally well be interpreted as Σ (surface density) or as β (ring inclination angle).

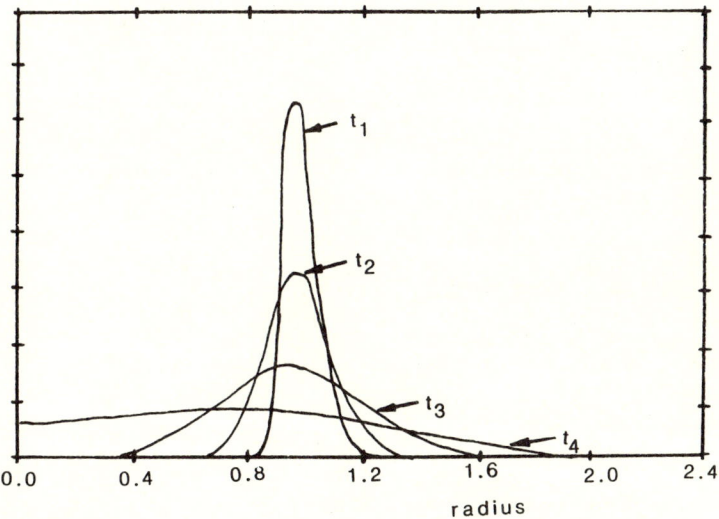

10.4.3 Conclusions

When the radial structure of a disk is solved, as it is in the basic models, one can return to the vertical pressure balance (10.13), and attempt a lowest-order approximation to the vertical disk structure for the particular three auxiliary equations assumed. The outcome depends on r, of course. Before a vertical solution can be obtained, however, one *extra* assumption still has to be made about the vertical T-structure (e.g. that it is polytropic) to make the set of equations complete.

The problem can be extended beyond the disk surface, and made as complicated as desired by including a disk *atmosphere* and/or *corona* (Icke, 1976; Liang & Price, 1977; Packzynski, 1978; Liang & Thompson, 1979), an outflowing disk '*wind*' (Berger & Bardeen, 1978; Piran, 1977), and even a *jet* squirting out along the z-axis (Shakura & Sunyaev, 1973; Bisnovatyi-Kogan & Blinnikov, 1977). For the evidence of winds or jets from CV disks see Chapter 4, Section 4.5. The latter may turn out to be natural in disks with a mass accretion rate exceeding the capacity of the compact object to accrete. The binary SS 433 may show this phenomenon, with the extra twist that the disk in this system is probably also 'misaligned' and precessing (Petterson, 1980; van den Heuvel *et al.*, 1980; Chapter 7, this book).

Acknowledgements

The author is greatly endebted to Dr F. Verbunt for discussions and extensive help with the present article. He also provided Figure 10.1.

References

Bardeen, J. M. & Petterson, J. A. 1975, *Ap. J. (Lett.)*, **195**, L65.
Basko, M. M., Hatchett, S. & Sunyaev, R. A. 1977, *Ap. J.*, **215**, 276.
Bath, G. T. & Pringle, J. E. 1981, *MNRAS*, **194**, 967.
Bath, G. T., Pringle, J. E. & Whelan, J. A. J. 1980, *MNRAS*, **190**, 185.
Berger, B. K. & Bardeen, J. M. 1978, *Ap. J.*, **221**, 105.
Bisnovatyi-Kogan, G. S. & Blinnikov, S. I. 1977, *A. and A.*, **59**, 111.
Boynton, P. E. 1978, In: *Physics and Astrophysics of Neutron Stars and Black Holes*, Varenna, Eds. R. G. Giacconi & R. Ruffini, North-Holland, Amsterdam.
Chanan, G. A., Nelson, J. E. & Margon, B. 1978, *Ap. J.*, **226**, 963.
Chester, T. J. 1979, *Ap. J.*, **230**, 167.
Coroniti, F. V. 1981, *Ap. J.*, **244**, 587.
Cunningham, C. T. 1975, *Ap. J.*, **202**, 788.
Davidson, K. & Ostriker, J. P. 1973, *Ap. J.*, **179**, 585.
Deeter, J. E., Crosa, L., Gerend, D. & Boynton, P. E. 1976, *Ap. J.*, **206**, 861.
Eardley, D. M. & Lightman, A. P. 1975, *Ap. J.*, **200**, 187.
Galeev, A. A., Rosner, R. & Vaiana, G. S. 1979, *Ap. J.*, **229**, 318.
Gerend, D. & Boynton, P. E. 1976, *Ap. J.*, **209**, 562.
Gosh, P. & Lamb, F. K. 1978, *Ap. J. (Lett.)*, **223**, L83.
Ichimaru, S. 1977, *Ap. J.*, **214**, 840.
Icke, V. 1976, In: 'Structure and evolution of close binary systems', *IAU Symposium*, 73, eds. P. Eggleton, S. Mitton & J. Whelan, P. 267.
Katz, J. I. 1973, *Nature Phys. Sci.*, **246**, 87.

Katz, J. I. 1980, *App. Lett.*, **20**, 135.
Lang, F. L., Levine, A. N., Bautz, M., Hauskins, S., Howe, S., Primini, F. A., Lewin, W. H. G., Baity, W. A., Knight, F. K., Rothschild, R. E. & Petterson, J. A., *Ap. J. (Lett.)*, 1981, **246**, L21.
Liang, E. P. T. 1977, *Ap. J.*, **218**, 243.
Liang, E. P. T. & Price, R. H. 1977, *Ap. J.*, **218**, 247.
Liang, E. P. T. & Thompson, K. A. 1979, *MNRAS*, **189**, 421.
Lightman, A. P. 1974, *Ap. J.*, **194**, 419.
Lightman, A. P. & Eardley, D. M. 1974, *Ap. J.*, **187**, L1.
Lin, D. N. C. & Pringle, J. E. 1976, *In:* 'Structure and evolution of close binary systems', *IAU Symposium*, 73, eds. P. Eggleton, S. Mitton & J. Whelan, p. 327.
Livio, M. & Shaviv, G. 1977, *A. and A.*, **55**, 95.
Livio, M. & Shaviv, G. 1981, *Ap. J.*, **244**, 290.
Merritt, D. M. & Petterson, J. A. 1980, *Ap. J.*, **236**, 255.
Middleditch, J. & Nelson, J. 1976, *Ap. J.*, **208**, 567.
Novikov, I. D. & Thorne, K. S. 1973, *In: Les Astres Occlus, Les Houches*, eds. C. Dewitt & B. Dewitt, Gordon and Breach, New York, p. 344.
Paczynski, B. 1978, *Acta Astron.*, **28**, 91.
Papaloizou, J. & Pringle, J. E. 1977, *MNRAS*, **181**, 441.
Pedersen, H. *et al.* 1982, *Ap. J.*, **263**, 340.
Petterson, J. A. 1977*a*, *Ap. J.*, **214**, 550.
Petterson, J. A., 1977*b*, *Ap. J.*, **216**, 827.
Petterson, J. A. 1977*c*, *Ap. J.*, **218**, 783.
Petterson, J. A. 1978*a*, *Ap. J.*, **224**, 625.
Petterson, J. A. 1978*b*, *Ap. J.*, **226**, 253.
Petterson, J. A. 1980, *Ap. J.*, **241**, 247.
Piran, T. 1977, *MNRAS*, **180**, 45.
Piran, T. 1978, *Ap. J.*, **221**, 652.
Prendergast, K. H. & Burbidge, G. R. 1968, *Ap. J.*, **151**, 183.
Prendergast, K. H. & Taam, R. E. 1974, *Ap. J.*, **189**, 125.
Pringle, J. E. 1976, *MNRAS*, **177**, 65.
Pringle, J. E. 1981, *In: Ann. Rev. Astron. Ap.*, **19**, 137.
Pringle, J. E. & Rees, M. J. 1972, *Astron. Ap.*, **21**, 1.
Pringle, J. E., Rees, M. J. & Paczolchyk, A. G. 1973, *Astron. Ap.*, **29**, 179.
Pringle, J. E. & Savonije, G. J. 1979, *MNRAS*, **187**, 777.
Robinson, E. L. 1976, *Ann. Rev. Astron. and Ap.*, **14**, 119.
Rossi, B. & Olbert, S. 1970, *Introduction to the Physics of Space*, McGraw-Hill, New York, p. 374.
Shakura, N. I. 1972, *Astron. Zh.*, **49**, 291.
Shakura, N. I. & Sunyaev, R. A. 1973, *Astron. Ap.*, **24**, 337.
Shakura, N. I. & Sunyaev, R. A. 1976, *MNRAS*, **175**, 613.
Shakura, N. I., Sunyaev, R. A. & Zilitinkevich, S. S. 1978, *Astron. Ap.*, **62**, 179.
Shapiro, S. L., Eardley, D. M. & Lightman, A. P. 1976, *Ap. J.*, **204**, 187.
Shu, F. H. & Lubow, S. H. 1981, *In: Ann. Rev. Astron. Ap.*, **19**, 277.
Taylor, R. J. 1980, *MNRAS*, **191**, 135.
Thorne, K. S. & Price, R. H. 1975, *Ap. J.*, **195**, 101.
van Paradijs, J. & Zuiderwijk, E. 1977, *Astron. Ap.*, **61**, 119.
van den Heuvel, E. P. J. 1975, *Ap. J.*, **198**, L109.
van den Heuvel, E. P. J., Ostriker, J. P. & Petterson, J. A. 1980, *Astron. Ap.*, **81**, L7.
Warner, B. 1976, *Proc. IAU Symp.*, No. 73, 'Structure and evolution of close binary systems', eds, P. Eggleton *et al.*, p. 85.

11

SPINUP AND SPINDOWN OF ACCRETING NEUTRON STARS

Huib F. Henrichs
Astronomical Institute, University of Amsterdam

11.1 Introduction and overview

About 20 galactic X-ray sources are known to show strictly periodic variations in X-ray intensity with periods ranging from about 0.069 to 835 s. These objects are called 'X-ray pulsars' (see Chapter 1, this book for a more precise description of objects belonging to this class). The pulse periods of the majority of these 20 sources were measured many times, and it has become clear that in several aspects they behave differently from what is observed in the case of radio pulsars, although both kinds of objects are accepted to be rotating, magnetized neutron stars.

The pulse period is in both cases identified with the rotation period of the star (although the precise mechanisms for producing X-ray pulses as well as radio pulses are still not well understood). Major differences are: (a) for radio pulsars the observed range of periods is 0.001 56-~5 s, i.e. there are no 'long-period' radio pulsars, and (b) all radio pulsars are spinning down on timescales (defined as $P/\dot{P}$) greater than 10^3 yr. In the case of X-ray pulsars five sources are continuously spinning up, while the others show hardly any change in period, or display an erratic behavior of rapid spindown and spinup.

The explanation for this different behavior is obviously to be found in a difference in the nature of these two classes of objects. The vast majority - over 99 per cent - of the radio pulsars are single objects and radiate away their rotational energy and angular momentum in the form of magnetic dipole radiation and relativistic particles, whereas X-ray pulsars are members of interacting binary systems in which matter with angular momentum flows from the companion star towards the rotating neutron star. The response of this rotation rate to the supplied mass and angular momentum leads to a change in the rotation period. The analysis of this interaction is the subject of this article. The goal in this study of period changes of X-ray pulsars is to understand:

(1) the cause of the observed spinup and spindown behavior, and
(2) the origin of the observed range in pulse periods.

Helpful information might come from the study of observable phenomena that are correlated with pulse-period changes, such as X-ray luminosity, variations in X-ray absorption, etc. and from the study of binary evolution.

The outline of this article is guided by the above-given considerations. In Section 11.2 the relevant observational data are briefly reviewed. In Section 11.3 characteristic parameters of the theory are introduced and a comparison is made between the calculated accretion torque and the torque as can be derived from the available data. Attempts to explain the observed range of X-ray pulse periods are reviewed in Section 11.4.

11.2 Observed characteristics of X-ray pulsars

Table 11.1 lists characteristic properties of the 20 presently-known X-ray pulsars. We may summarize the data as follows.

(1) The pulse periods range from 0.069 to 835 s.

(2) Four objects are transient X-ray sources.

(3) Optical counterparts are known for 18 sources. In 14 cases (including the transients) the companion is a massive star with early spectral type (O, B or Be). Two sources (Her X-1 and 4U 1626−67) do have a low-mass companion star. One source (GX 1+4) is identified with an M6 giant.

(4) More than two pulse-period determinations during the last 10 yr are available for 13 pulsars, out of which only five show a more-or-less steady spinup behavior (SMC X-1, Her X-1, Cen X-3, 4U 1626−67 and GX 1+4). See Figures 11.1 and 1.5 for pulse-period histories of many of the sources.

(5) Direct evidence for the presence of a disk exists for five systems. Out of these five, four belong to the class of sources which are steadily spinning up, suggesting a one-to-one correspondence between the members of the two classes. Exceptions are LMC X-4 and GX 1+4. We notice, however, that in LMC X-4 a disk is found (Chevalier *et al.* 1981, van der Klis *et al.* 1982) but that only one pulse-period measurement of this source is known (Kelley *et al.* 1983). On the other hand GX 1+4 does steadily spin up but there exists no information about the presence of a disk in its optically faint counterpart.

(6) All pulsating X-ray sources in which direct evidence for a disk is found do have a short pulse period.

(7) An empirical timescale for the lifetime of the binary system X Per/4U 0352+30 is derived by Guetter (1977), who gives the value of 1.5×10^6 yr as an upper limit for the age of the Per OB 2 association of which X Per is likely to be a member.

395

Table 11.1. *The binary X-ray pulsars*

X-ray source name	Pulse period	Orbital period	Companion star name	Companion star spectrum	$-P/\dot{P}$ (yr)	L_X (10^{37} erg s^{-1})	Disk ?
A 0538−66	0.069 s	16.7 d	—	B7 IIe–B2 IVe	—	80 tr	—
SMC X-1	0.714	3.89	Sk 160	B0 I	1.7×10^3	50	D
Her X-1	1.24	1.70	HZ Her	A9-B	3.0×10^5	1	D
4U 0115+63	3.61	24.3	Johns	Be	±	<3 tr	—
Cen X-3	4.84	2.09	Krzeminski	O6.5 IIIe	3.4×10^3	5	D
4U 1627−67	7.68	41 min	—	Accr. disk	5.0×10^3	3 at 10 kpc	D
LMC X-4	13.5	1.41 d	Sk/Ph	O7 III-Ve	—	35	D
2S 1417−62	17.6	—	—	Hα	—	—	—
OAO 1653−40	38.2	—	—	—	186	—	—
A 0535+26	104	>40	HD 245770	O9.7 IIIe	±	<2 tr	—
GX 1+4	122	—	—	M6 giant	50	4	—
4U 1230−61	191	—	—	—	—	—	—
GX 304−1	272	>13	MMV	B2 V ne	—	0.2	—
4U 0900−40	283	8.96	HD 77581	B0.5 Ib	±	0.15	—
4U 1145−61	292	>45	HD 102567	B0−1 Ve	±	0.025	—
1E 1145.1−6141	297	>25	—	~B1 I	—	0.3	—
A 1118−61	405	—	Hen 3-640 ?	O9.5 III-Ve	—	<0.5 tr	—
4U 1538−52	529	3.73	Cowley	B0 Ib	±	0.4	—
GX 301−2	696	41.4	WRA 977	B1.5 Ia	±	0.1	—
4U 0352+30	835	580 ?	X Per	O9.5 III-Ve	±	4×10^{-4}	—

Columns 1 & 4: The nomenclature is arbitrary. Many sources are known under different names. For instance 4U 0900−40 = Vela X-1 and GX 301−2 = 4U 1223−62 (see Bradt *et al.* 1979 and Chapter 5, this book).

Column 5: A ± sign means that $P/\dot{P}$ may have either sign and/or over the last decade no regular trend is apparent. No value is listed when only one pulse-period determination exists. For OAO 1653−40 only two measurements are available.

Column 6: Typical 2−11 keV X-ray luminosities are quoted (Bradt *et al.* 1979). These values are in most cases uncertain by at least a factor of 2 due to uncertainties in the distance to the source and/or time variability. A transient source is denoted by tr. No value is listed when the distance to the source is unknown.

Column 7: Direct (optical) evidence for the existence of a disk is denoted by D.

11.3 Accretion torques on a neutron star in a binary system
11.3.1 Introduction

Consider a neutron star in a binary system with the following configuration. Plasma flows from the companion star towards the neutron star. The neutron star rotates and has a magnetic field. The magnetic field is so strong that a magnetosphere exists. In our context a magnetosphere is defined as the region of space surrounding the star within which the stellar magnetic field determines the motion of the plasma (see Section 11.3.7 for a more precise account of the concept of magnetosphere). The distance to which the magnetosphere extends from the stellar surface depends on the mechanical pressure (or stress in other form) exerted by the ambient plasma. Inside the magnetosphere, plasma ultimately corotates with the magnetic field which is thought to be rigidly anchored in the star. There must be a transition region between the magnetosphere and the outside region. In this transition region torques will arise which are transmitted to the neutron star by the magnetic field, leading to a change in the rotation period. The net torque is a function of the supplied mass and angular momentum. First, we consider the mass-accretion rate onto the neutron star. We distinguish capture from a stellar wind from the companion star and capture in the case of Roche-lobe overflow. Second, we discuss the capture of angular momentum with respect to the neutron star and formulate criteria for the formation of a disk. Then we define a magnetospheric radius in order to calculate the torque. Finally

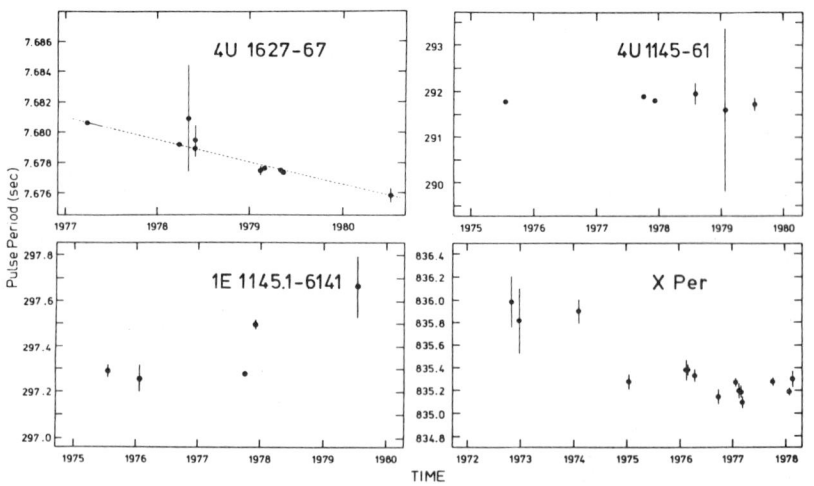

Fig. 11.1. Pulse-period histories for four X-ray pulsars (for the histories of other sources see Chapter 1, Figure 1.5). In the case of 4U 1627−67, a straight-line fit to the data is indicated. The remaining three sources do not show such a regular trend.

we confront the outlined theory with the observational data on X-ray pulsars and discuss proposed refined models.

11.3.2 The mass-accretion rate as inferred from the X-ray luminosity

If accretion takes place with a rate $\dot{M}$ onto a neutron star with mass M_x and radius R the resulting steady-state release of gravitation potential energy ('accretion luminosity') will be:

$$L_{acc} = GM_x \dot{M}/R, \tag{11.1}$$

where G is the gravitational constant, or

$$L_{37} = 1.33 \dot{M}_{17} m R_6^{-1}, \tag{11.2}$$

where we used the notation $L_{37} = L_{acc}/(10^{37} \text{ erg s}^{-1})$, $\dot{M}_{17} = \dot{M}/(10^{17} \text{ g s}^{-1}) \approx \dot{M}/(1.6 \times 10^{-9} \text{ M}_\odot \text{ yr}^{-1})$, $R_6 = R/(10^6 \text{ cm})$ and $m = M_x/M_\odot$.†

Most of this energy release appears as photons in the X-ray region (0.1–30 keV). In order to derive $\dot{M}$ from the observed mean X-ray luminosity L_x, one should know a bolometric correction (i.e. the difference between L_x and L_{acc}) which is largely determined by the transfer processes of the outcoming X-ray radiation through the accreting matter.

11.3.3 The mass-accretion rate in a stellar wind

The majority of the observed X-ray pulsars do have companions of early spectral type. Such stars are known to be more massive than about 15 M$_\odot$ and to lose mass in the form of a stellar wind. Suppose that the primary star with mass M_p loses mass at a rate $\dot{M}_p$. What fraction of $\dot{M}_p$ will be captured by the neutron star?

A neutron star with mass M_x moving with velocity v_{rel} through a medium with soundspeed c_s will gravitationally capture matter from a roughly cylindrical volume with radius r_a given by (e.g. Hoyle & Lyttleton 1939, Bondi & Hoyle 1944, Davidson & Ostriker 1973):

$$r_{acc} = \frac{2GM_x}{v_{rel}^2 + c_s^2}, \tag{11.3}$$

which is called the accretion radius or gravitational capture radius. (For details of the flow within the accretion radius in the case of accretion from a homogeneous medium the reader is referred to Hunt (1971) and Hunt (1979). Accretion from an inhomogeneous medium, which is likely the case, is considered by Davies & Pringle (1980) (see also Section 11.3.5). We notice, however, that a precise and realistic hydrodynamic description of the flow still does not exist.)

† This notation will be used throughout this article.

In our applications we will neglect the c_s^2 term in Equation (11.3) because in most cases $c_s/v_{rel} \ll 1$ (X-ray heating may, however, cause c_s to be comparable to v_{rel}). The relative velocity is given by

$$v_{rel}^2 = v_{orb}^2 + v_w^2, \qquad (11.4)$$

where v_{orb} is the orbital velocity and v_w is the velocity of the wind at the orbital radius. The orbital velocity follows from Kepler's third law:

$$v_{orb}^2 = \frac{G(M_x + M_p)}{a}. \qquad (11.5)$$

Notice that the accretion radius is usually smaller than the orbital radius a:

$$\frac{r_{acc}}{a} = \frac{2}{1 + M_p/M_x} \frac{1}{1 + (v_w/v_{orb})^2} \ll 1. \qquad (11.6)$$

For the typical values $M_x = M_\odot$ and $v_{rel} = 1000$ km s^{-1} we have $r_{acc} \approx 0.4$ R$_\odot$.

The accretion rate onto the neutron star will be

$$\dot{M} = \pi r^2 \rho v_{rel},$$

where ρ is the mass density of the stellar wind near the accretion radius. The density can be estimated using the continuity equation (assuming spherically symmetric and steady mass loss):

$$\dot{M} = 4\pi a^2 \rho v_w. \qquad (11.7)$$

Combination of these equations gives finally:

$$\frac{\dot{M}}{\dot{M}_p} = \frac{1}{(1 + M_p/M_x)^2} \left(\frac{v_{orb}}{v_w}\right)^4 \frac{1}{(1 + \alpha(v_{orb}/v_w)^2)^{3/2}}. \qquad (11.8)$$

We included for completeness the correction factor α which is intended to take into account the effect of rotation on the relative velocity and which is unity for a non-rotating primary star. In the case of a rotating primary with radius R_p and a stationary stellar wind that conserves its angular momentum we have $\alpha = [1 - \lambda(R_p/a)^2]^2$, where λ denotes the ratio of the angular velocities of the rotating primary and the orbital motion of the neutron star respectively.

We emphasize that the assumption of spherical symmetry is crucial in the derivation of Equation (11.8). Especially when the binary system contains a (rapidly rotating) Be star with a Keplerian envelope, the condition of spherical symmetry is not expected to be fulfilled.

We notice here that in the case of X-ray binaries with known distance, orbital radius and masses, the value of v_w can be solved from Equations (11.1) and 11.8 by elimination of $\dot{M}$. This has been done by Conti (1978) and updated by Henrichs (1980).

We conclude this section by mentioning that when a massive star evolves, its mass-loss rate $\dot{M}_p$ increases and its wind velocity v_w decreases as a function of time, while the other parameters in Equation (11.8) are approximately constant. Hence, before Roche-lobe overflow takes over, the mass-accretion rate will evolve as:

$$\dot{M}(t) \propto \dot{M}_p(t) v_w^{-4}(t), \tag{11.9}$$

which is a strongly increasing function of time.

11.3.4 The mass-accretion rate during Roche-lobe overflow

Roche-lobe overflow will occur when during the evolutionary expansion of the companion star its radius (including the atmosphere) exceeds the critical Roche radius which is defined by the critical equipotential surface surrounding the stars. In such a case the mass-accretion rate is determined by the expansion rate and structure of the companion star. For a detailed and critical review of calculations of the precise mass-transfer rate by Roche-lobe overflow under various circumstances the reader is referred to Chapter 9.

11.3.5 Angular momentum capture from a stellar wind

In the case of capture of stellar-wind matter, one calculates the net angular momentum with respect to the neutron star of all particles entering the cylinder-like volume with radius r_{acc} (Equation (11.3)) and one assumes that all angular momentum is accreted. The basic approach is given by Davidson & Ostriker (1973) (see also Illarionov & Sunyaev 1975, and Shapiro & Lightman 1976) and is refined and extended by Wang (1981), while an important criticism is found in Davies & Pringle (1980).

A non-orbiting neutron star in a spherically symmetric stellar wind will capture no angular momentum at all. For an orbiting neutron star the non-zero result can be explained as follows. The target face for incoming wind particles is perpendicular to the relative velocity vector of the neutron star according to Equation (11.4) and thus makes a non-zero angle with the orbital velocity vector. (Figure 11.2). This means that if there exists a density gradient in the stellar wind across the accretion cross-section, the target face will capture different amounts of particles at the far-side and at the near-side and, analogously, if there exists a velocity gradient, the particle velocities at the far- and the near-side will differ. Of course, azimuthal density and/or velocity gradients may also contribute to the asymmetry.

Taking into account only the radial density gradient in the assumedly-spherical steady expansion of the stellar wind (Equation 11.7), Shapiro & Lightman (1976) obtain for the captured specific angular momentum with respect to the neutron star:

$$l = \tfrac{1}{2} \Omega_{orb} r_{acc}^2 \eta, \tag{11.10}$$

where Ω_{orb} is the orbital angular velocity and $\eta = 1$. Expressions for the numerical factor η in cases where radial and/or azimuthal density and velocity gradients exist are given by Wang (1981). Wang showed that such gradients may lead to large deviations from $\eta = 1$ or even to a change of sign. Rotation of the companion star in the same sense as the orbital motion requires an additional factor α in the right-hand side of Equation (11.10), the same factor as was used in Equation (11.8).

Actually, Equation (11.10) gives an estimate of the amount of specific angular momentum that is captured in the accretion cylinder. What fraction of this amount is transferred to the neutron star is still an unanswered question. Davies & Pringle (1980) showed that, to first-order approximation in r_{acc}/a, the net accreted angular momentum should be zero in order to achieve any accretion at all, irrespective of the existence of a density and/or velocity gradient across the target face of the accretion cylinder. This result follows from the assumptions made by these authors that:

(1) The accreting star is point like. This is equivalent to the condition that the magnetospheric radius is much smaller than the accretion radius (see Section 11.3.7), which is usually fulfilled.
(2) The accretion line (where the downstream particle-orbits intersect) joins the point mass.
(3) No dissipation occurs as particles stream along the accretion line into the point mass.

Fig. 11.2. A neutron star that accretes from a stellar wind may capture angular momentum, depending on existing gradients across the target face r_{acc}.

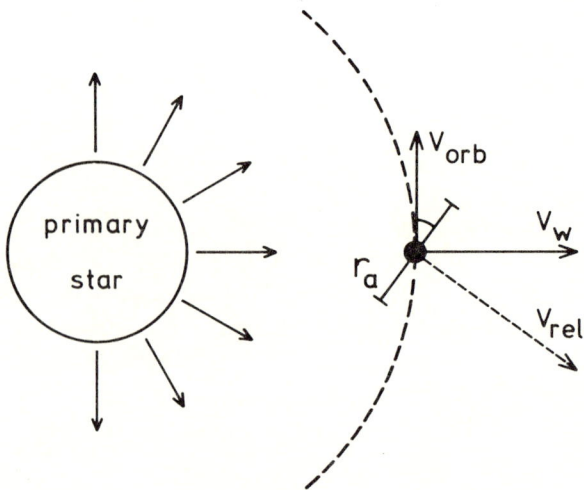

Whether these assumptions are justified is still open to question but, anyhow, it is important that Davies & Pringle (1980) and Wang (1981) at least showed that deviations from the expected angular momentum capture according to Equation (11.10) with $\eta = 1$ might be, in fact, very large. That deviations from $\eta = 1$ indeed must occur is supported by observations of the presumably wind-powered sources such as A 0535+26, GX 301−2 (= 4U 1223−62) and notably 4U 0900−40 (= Vela X-1) (Nagase et al. 1981, Nagase 1981) where temporarily spindown episodes alternate spinup episodes. Such an occasional spin-reversal would be impossible with Equation 11.10 and $\eta = 1$.

In conclusion, the presently-existing description of accretion of angular momentum is far from complete and can only be improved if progress is made in both (1) a proper treatment of the flow inside the accretion cylinder, and (2) knowledge of stellar-wind properties, in particular the (time-dependent) velocity and/or density gradients.

11.3.6 Angular momentum capture during Roche-lobe overflow

A usual assumption in the case of Roche-lobe overflow is that all matter from the companion star leaves the critical lobe near the inner Lagrangian point L_1. The specific angular momentum with respect to the orbiting neutron star will then be (Petterson, 1978):

$$l = \Omega_{\text{orb}}(a - r_L)(a - \lambda r_L), \tag{11.11}$$

where r_L is the location of the inner Lagrangian point and λ is, again, the ratio of the angular velocities of the rotational motion of the companion and of the orbital motion of the neutron star respectively. (We notice that, strictly, the Roche potential can only be calculated if $\lambda = 1$.) We see that this gives a specific angular momentum which is roughly a factor of $(a/r_{\text{acc}})^2$ higher than in the case of accretion from a wind (Equation 11.10)). In Section 11.3.8 it is shown that if the companion star overflows its Roche lobe, a Keplerian disk will likely form around the neutron star. The amount of accreted angular momentum is then determined by the location of the inner edge of the disk.

11.3.7 The magnetospheric radius

Although there exist conceptual differences in the various definitions for the radius of the magnetosphere (sometimes referred to as Alfvén radius), for the sake of arguments it is sufficient to define a magnetospheric radius r_m in a rather simplified way (as has been used by numerous authors). In our context r_m is located such that the motion of the plasma within r_m is determined by the magnetic field of the neutron star. We notice that in other contexts (notably in the field of geo- and planetary magnetospheres) this radius confines the region in which the magnetic field is completely screened by currents. For an excellent review on existing theories of magnetospheres around compact objects, the reader is referred to Vasyliunas (1979) (see also Lamb 1979) in which a critical discussion is given of the concept of magnetosphere, the location of the magnetospheric boundary, magnetospheric models and plasma flows inside magnetospheres, both for spherical and disk accretion.

We consider *radial* infall of plasma with velocity $v_{in}(r)$ and density $\rho(r)$ onto a neutron star with mass M_x, radius R, dipole magnetic field with strength B_0 at the stellar surface and magnetic moment $\mu \equiv B_0 R^3$. Balance of magnetic pressure and ram pressure of the infalling material at the magnetospheric boundary requires (Davidson & Ostriker 1973, Lamb, Pethick & Pines 1973):

$$\frac{B^2(r)}{8\pi} = \rho(r) v_{ff}^2(r), \tag{11.12}$$

where gaussian units are used. For a dipole field we have $B(r) = \mu r^{-3}$. The ambient density is related to the (assumedly steady) mass-accretion rate $\dot{M}$ by

$$\dot{M} = 4\pi \rho(r) v_{in}(r) r^2. \tag{11.13}$$

We write the infall velocity as a fraction ζ ($\leqslant 1$) of the free-fall velocity:

$$v_{in}(r) = \zeta v_{ff}(r) = \zeta (2GM_x/r)^{1/2}. \tag{11.14}$$

Solving for $r = r_m$ yields:

$$r_m = (\tfrac{1}{2}\zeta)^{2/7} \mu^{4/7} (2GM_x)^{-1/7} \dot{M}^{-2/7}. \tag{11.15}$$

Because of the weak power dependence of ζ we will use, throughout this, article $\zeta = 1$. An order of magnitude estimate for the value of r_m is given by the expression

$$r_m = (2.7 \times 10^8 \text{ cm}) \mu_{30}^{4/7} m^{-1/7} \dot{M}_{17}^{-2/7} \tag{11.16}$$

or

$$r_m = (2.9 \times 10^8 \text{ cm}) \mu_{30}^{4/7} m^{1/7} R_6^{-2/7} L_{37}^{-2/7}, \tag{11.17}$$

where we substituted $\mu_{30} = \mu/10^{30}$ gauss cm^3. The latter expression is only valid if all matter is accreted and converted into observable X-rays. We notice that the typical value for r_m is a few hundred times larger than the radius of a neutron star.

The limiting assumptions underlying the derivation of Equation (11.16) are: (1) spherical infall, (2) infall velocity equals free-fall velocity and outgoing velocity is neglected, (3) thermal pressure of the infalling gas is negligible with respect to the ram pressure, and (4) plasma pressure inside the magnetosphere is negligible with respect to the magnetic pressure (Lamb & Pethick 1974). Assumptions (1) and (2) imply that $r_m \ll r_{acc}$.

In the case of *disk* accretion, the magnetospheric radius is obtained by replacing the pressure balance expressed in Equation (11.12) by a balance between the pressure gradient against the gravitational force density in the direction normal to the plane of the disk (Pringle & Rees 1972). The steep radial dependence of the magnetic pressure, however, results in the magnetospheric radius in the case of disk accretion not differing very much from the estimate for r_m as given above. Therefore we will use Equation (11.16) and notice that for a given neutron star – with given mass and magnetic moment – the magnetospheric radius is determined only by the mass-accretion rate.

11.3.8 The formation of a disk around a neutron star

The radius r_K at which a fluid element with specific angular momentum l would go into a circular Keplerian orbit around a neutron star with mass M_x is given by:

$$r_K = l^2/GM_x. \tag{11.18}$$

A disk is expected to form if r_K is larger than the magnetospheric radius.

In the case of a steady stellar wind, we obtain with the aid of Equations (11.10) and (11.3), for the condition of disk formation (Shapiro & Lightman 1976):

$$r_K = \frac{\eta^2 \Omega_{orb}^2 (2GM_x)^3}{2v_{rel}^8} > r_m. \tag{11.19}$$

Solving for v_{rel} yields the criterion (Wang, 1981):

$$v_{rel} < (340 \text{ km s}^{-1}) \eta^{1/4} \mu_{30}^{-1/14} m^{11/28} \dot{M}_{17}^{1/28} P_{10}^{-1/4}, \tag{11.20}$$

where P_{10} denotes the orbital period in units of 10 d.

In spite of the large uncertainty in the parameter η, the conclusion is that a disk is not likely to form because in wide systems the stellar-wind velocity will exceed the escape velocity at the primary's surface (>600 km s^{-1}) and in close systems the orbital velocity will exceed the value 340 km s^{-1}). A direct measurement of the wind velocity at the orbital radius is available in the system HD 77581/4U 0900–40 (= Vela X-1, P_{orb} = 9.6 d) where the value $v_w \approx 850$ km s^{-1} is obtained (Dupree *et al.* 1980, see also Henrichs 1980). We stress, however, that the above-given conclusion only holds for a *steady* stellar wind. In contrast, it is well known that stellar winds of massive stars are strongly variable phenomena (see Snow, 1979, for a review) and hence it is quite conceivable that, *temporarily*, a disk can be formed. At present, however, the available observations of stellar winds do certainly not permit a numerical estimate of the parameters entering Equation (11.20).

When Roche-lobe overflow occurs, Equation (11.11), together with Kepler's third law, yields the criterion:

$$r_K = a(1 + M_p/M_x)(1 - r_L/a)^2(1 - \lambda r_L/a)^2 > r_m, \qquad (11.21)$$

where r_L, the location of the inner Langrangian point, is only a function of the ratio M_p/M_x (see, for instance, Pratt & Strittmatter 1976). Again, strictly, the Roche potential can only be calculated when the primary star corotates with the orbit, i.e. $\lambda = 1$. In that case, for mass ratios $M_p/M_x = 1/5$ and $M_p/M_x = 40$ (likely to be extreme values), the condition for disk formation becomes $r_m \lesssim a/4$ and $r_m \lesssim a/20$ respectively. In view of the usually very small value of the magnetospheric radius (a few thousand km), this criterion is usually satisfied and a disk is expected to form (see Chapter 10, Section 10.1.2, this book).

11.3.9 Accretion torques

Consider a rotating neutron star with moment of inertia I (which should include the contribution of its magnetosphere) and angular velocity Ω. Its response to accretion of matter at rate $\dot{M}$ with specific angular momentum $\mathbf{l}$ is governed by conservation of angular momentum for the system (Lamb, Pethick & Pines, 1973):

$$\frac{d}{dt}(I\Omega) = \dot{M}l - \alpha, \qquad (11.22)$$

where α represents the non-material torque including torques arising from magnetic and/or viscous forces and torques arising from matter that leaves the system. Rewriting of this equation yields the timescale τ for the change of angular frequency of the rotating neutron star:

$$\frac{1}{\tau} \equiv \frac{\dot{\Omega}}{\Omega} = \frac{\dot{M}l - \dot{I}\Omega - \alpha}{I\Omega}, \qquad (11.23)$$

where l and α are the components of $\mathbf{l}$ and $\boldsymbol{\alpha}$ parallel to $\boldsymbol{\Omega}$ respectively. To understand the different contributions to the right-hand side of Equation (11.23) we consider first the case of purely spherical accretion (where l is equal to zero) and in the absence of non-material torques. Then Equation (11.23) reduces to $\dot{\Omega}/\Omega = -\dot{I}/I$ where $\dot{I} = \dot{M}(dI/dM)$ has to be calculated from the internal structure of the neutron star (see, for instance, Canuto 1977, and Lamb 1977). It appears that the timescale in this idealized case is in general very long, i.e. $\sim 10^9$ yr for a typical neutron-star model, as constructed by Pandharipande et al. (1976) and assuming the highest observed accretion rates. This is the timescale for the neutron star to change its structure in response to an increase of mass, and, because for neutron stars a larger mass corresponds to a smaller radius, this timescale is negative which means spindown, or an increase of the rotation period. In view of the very long timescale involved we will ignore the $\dot{I}/I$ term throughout this chapter.

Consider now accretion from a *disk* were $l \neq 0$ and ignore again the non-material torques. We obtain for the torque:

$$I\dot{\Omega} = \dot{M}l \equiv N_0, \tag{11.24}$$

which apparently provides, under these idealized conditions, the strongest possible torque (and thus the shortest timescale), resulting in a spinup or spindown depending on the relative sense of rotation of the star and the disk.

It is very instructive to calculate the value of $\dot{\Omega}$ under these approximations and compare the result with the observational data. For simplicity it is assumed that matter which arrives at the inner edge of the disk has a Keplerian azimuthal velocity v_ϕ. This assumption basically means that the magnetic field of the neutron star is completely screened from the (Keplerian) disk and hence no interaction between the disk and the magnetic field exists. In addition, the fundamental assumption is made that accretion indeed takes place. (In the next section we shall relax some of these restrictions.) The specific angular momentum of the accreting matter will then be:

$$l = r_m v_\phi(r_m) = (GM_x r_m)^{1/2}, \tag{11.25}$$

which, together with Equation (11.16) yields the torque:

$$N_0 = 2^{-9/14} \mu^{2/7} (2GM_x)^{3/7} \dot{M}^{6/7}. \tag{11.26}$$

Hence, for a given neutron star, $\dot{\Omega}$ is proportional to $\dot{M}^{6/7}$:

$$\dot{\Omega} = k\dot{M}^{6/7}, \tag{11.27}$$

where k is a constant for that neutron star. We notice that in this idealized description the value of $\dot{\Omega}$ is apparently independent of Ω, the rotation rate of the neutron star. According to Equation (11.1), $\dot{M}$ can be estimated from the observed X-ray flux if the distance to the source is known. If we explicitly assume that $\dot{M}$ is a constant, the timescale on which the stellar rotation changes is:

$$\tau = \frac{\Omega}{\dot{\Omega}} = -\frac{P}{\dot{P}} = (1.4 \times 10^4 \text{ yr}) I_{45} \mu_{30}^{-2/7} m^{3/7} R_6^{-6/7} L_{37}^{-6/7} P^{-1}, \tag{11.28}$$

where P is the rotational period in seconds and $I_{45} = I/(10^{45} \text{ g cm}^2)$, which is a likely scaling value for a neutron star (see for instance the review paper by Canuto & Bowers 1981). It thus becomes possible to test the proportionality in Equation (11.27) (and hence the validity of the simplifications made) for different sources by making a comparison between the observed value for $\dot{\Omega}$ against the observed L_x. If all neutron stars had the same mass, the same moment of inertia and the same magnetic moment, and if the maximum observed values for $\dot{\Omega}$ were indeed found to be proportional to $\dot{M}^{6/7}$, we would have a strong indication that, for instance, non-material torques play no role for these sources and that, apparently, the specific angular momentum of the inflowing matter is equal to, or above, the threshold value for movement in a Keplerian orbit, and in addition that the disk rotates in the same direction as the star (which produces l parallel to Ω). Conversely, and at least of similar importance, the proportionality factor k contains the mass, moment of inertia and magnetic moment of the central object and hence the observationally deduced numerical value of k would provide an indication about the nature of the object (viz. a neutron star or a white dwarf).

Pringle & Rees (1972) were the first to notice that for Cen X-3 the observed parameters were in agreement with those of a neutron star. Lamb, Pethick & Pines (1973) closely examined the domain of validity of the approximations made, and concluded that the timescales observed in Her X-1 and Cen X-3 are in 'excellent agreement' with the timescale predicted for a neutron star (Equation (11.28)) and that, apparently, the orbital motion is in the same sense as the stellar rotation. Rappaport & Joss (1977) and Mason (1977) published a diagram in which the collected data of many other sources have been plotted (see Figure 1.6 in Chapter 1, this book). For the majority of the pulsars the agreement between the predicted and observed relation is quite good. All these authors stressed that this agreement strongly confirmed the supposition that the compact object in these systems is a neutron star, rather than a white dwarf.

Inspection of Equation (11.28) shows that it is mainly the stellar radius (which enters this equation also via the moment of inertia) which will determine the spinup rate for a given accretion luminosity and hence enables us to discriminate between magnetized white dwarfs and neutron stars (under the assumption of equal magnetic moments).

We point out here that the coordinates in Figure 1.6 are $-\dot{P}/P$ against $PL^{6/7}$. The choice of these coordinates is, of course, convenient for comparing timescales, but it suggests a dependence of the timescale on the rotation period P, which is, as we have seen, not the case. Further, we show in Section 11.3.10.3 that the observed value of $-P/\dot{P}$ is not necessarily a characteristic timescale of the system. We shall also see that another combination of the parameters P, $\dot{P}$ and L is, in some applications, more useful. We emphasize, however, that the basic purpose of a diagram like Figure 1.6 (namely the discrimination between neutron stars and white dwarfs) remains valid (see Chapter 6, Section 6.2.1, this book).

It is instructive to notice that, in this prescription, for a given neutron star, a line parallel to the (best-fit) solid line in Figure 1.6 gives in fact a lower limit to the predicted timescale for that neutron star (i.e. the shortest timescale).

That some of the well-studied X-ray pulsars, especially Her X-1, are situated far below the expected line and that some sources (those with $\dot{P} \lesssim 0$) find no place at all in the diagram shows obviously the simplicity of the outlined model.

Before we discuss more refined models we consider briefly the assumption that accretion indeed takes place. A necessary condition for accretion is that the matter can cool (Arons & Lea 1976, Elsner & Lamb 1977, see also the review by Vasyliunas 1979). The details of the accretion process are, however, still subject to study.

Let us suppose that accretion continues to take place and hence (in the simple picture described above) the neutron star continues to spin up. What will be the ultimate rotation period? An often-quoted condition for a quenching of the mass accretion is that this will happen when the neutron star rotates so fast that the (Keplerian) angular velocity of the plasma that arrives at the magnetospheric boundary just corotates with the neutron star (e.g. Pringle & Rees 1972, Davidson & Ostriker 1973, Lamb, Pethick & Pines 1973, Illarionov & Sunyaev 1975, Wickramasinghe & Whelan 1975, Schreier 1977, Savonije 1978). For that plasma the centrifugal forces just balance the local gravity. The distance at which this balance is satisfied, is given by the corotation radius r_{co}

$$r_{co} = (GM_x/\Omega^2)^{1/3} \tag{11.29}$$

or

$$r_{co} = (1.5 \times 10^8 \text{ cm}) m^{1/3} P^{2/3}, \tag{11.30}$$

which plays an essential role in more refined disk models. Assuming that the matter cools fast enough, the condition for accretion becomes $r_m < r_{co}$. The so-called 'equilibrium period' P'_{eq} is derived from the condition $r_m = r_{co}$:

$$P'_{eq} = (2.7 \text{ s}) \mu_{30}^{6/7} m^{-2/7} R_6^{-3/7} L_{37}^{-3/7}. \tag{11.31}$$

We stress, however, that the existence of this 'equilibrium' is a direct consequence of the restrictions used in the derivation and application of Equation (11.24) and hence the numerical value of P'_{eq} is only a rough estimate. We shall see in Section 11.3.10 that an equilibrium period (in the sense that $\dot{P} = 0$) may depend on many other parameters. Some authors (e.g. Wickramsinghe 1975, van den Heuvel 1977) argued that the existence of long-period pulsars is indicative of a low accretion rate (at least in the past), with the neutron star spinning at its equilibrium periods according to Equation (11.31). See, however, Davies & Pringle (1981) and Section 11.4 for a discussion.

11.3.10 Refined models

The sometimes large differences between the observed changes in rotation period and those expected on the basis of a simple-minded accretion-torque model, as described in the foregoing section, may be explained in three ways (or a combination thereof): (1) non-material torques are present, (2) the mass-accretion rate is not constant during the timespan of the observations, or cannot correctly be calculated from the X-ray flux, (3) accretion takes place from a stellar wind, rather than from a disk.

First, we shall review results concerning (1). Second, we shall present, in the light of these results, a critical discussion of (2). Finally we focus on (3).

11.3.10.1 Magnetic coupling with a disk

We review a series of papers by Ghosh and Lamb *et al.* (1977, 1978, 1979a, 1979b), in which the authors have developed a disk model that includes magnetic and viscous effects. In these papers, as well as those of Vasyliunas (1979) and Lamb (1979), a useful and critical discussion of previous disk models is given, to which the reader is referred for further details.

Let us start from the simple model of the foregoing section. We assumed an infinitely thin transition layer between the disk and the magnetosphere where all of the angular momentum transfer takes place, such that a strong spinup torque is expected, resulting from the excess of specific angular momentum of the accreting matter with respect to the neutron star. In reality, however, this transition region must have a finite extent where the velocity drops from a Keplerian to the corotation velocity well inside the magnetosphere. This layer is called the boundary layer and has a width of δ. Furthermore, the magnetic field lines will, in the case of a disk, not be completely screened from the disk and hence will certainly extend beyond the magnetospheric radius (as in the sense of our definition) and will thread the matter in the disk in the so-called outer transition layer which has a thickness d and which extends to a radial distance $r_s = r_m + \delta + d$ (see Figure 11.3, after Ghosh & Lamb).

Figure 11.4 depicts a corresponding diagram for analyzing the magnetic coupling in this outer transition zone of the disk. In this diagram the azimuthal velocity v_ϕ in a plane through the stellar center and perpendicular to the rotation axis of the star is displayed as a function of the distance r measured from the center of the neutron star. In such a diagram the location of the fluid elements in the (undisturbed) Keplerian disk flow is given by

$$v_\phi = (GM_x)^{1/2} r^{-1/2}. \tag{11.32}$$

The location of the corotation radius is also indicated. We assume that $r_{co} > r_m$. The flow inside the radius r_m is symbolically represented by a straight line (we do not consider the flow inside the magnetosphere, see the review papers mentioned above and Ghosh, Lamb & Pethick 1977 for this particular model). This straight line makes, by definition, an angle arctan Ω (Ω being the angular velocity of the neutron star) with the r-axis. This angle is symbolically

Fig. 11.3. Schematic picture of a possible configuration of an accretion disk around a magnetic neutron star. In this picture the magnetic field is not completely screened by currents in the magnetospheric boundary and hence the field lines may thread the disk. This leads to an additional contribution to the accretion torque (see text and Figure 11.4). (Reprinted courtesy of P. Ghosh & F. K. Lamb and the *Astrophysical Journal*.)

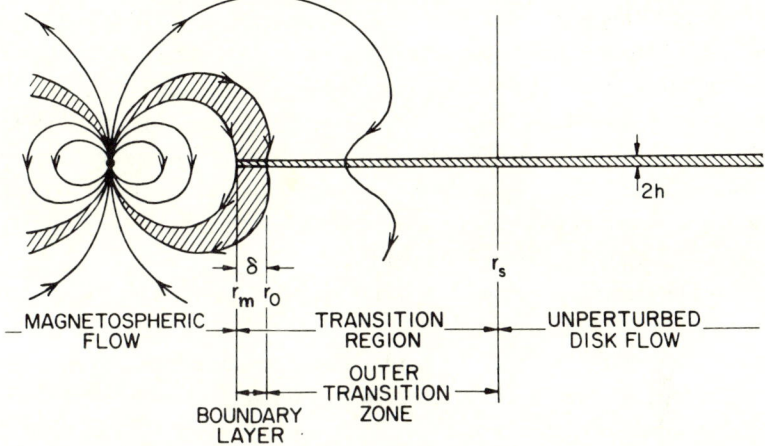

Fig. 11.4. A diagram showing the azimuthal velocity of the accretion flow in the plane of the disk corresponding to Figure 11.3. The symbol Ω denotes the quantity arctan Ω where Ω is the rotational frequency of the neutron star. For simplicity the flow inside the magnetospheric radius is drawn as a straight line. Because plasma in region A rotates faster than the star, its magnetic coupling with the star will give rise to a positive contribution to the total torque. Correspondingly, region B will give a negative contribution. The star will rotate at an equilibrium period if the two contributions cancel.

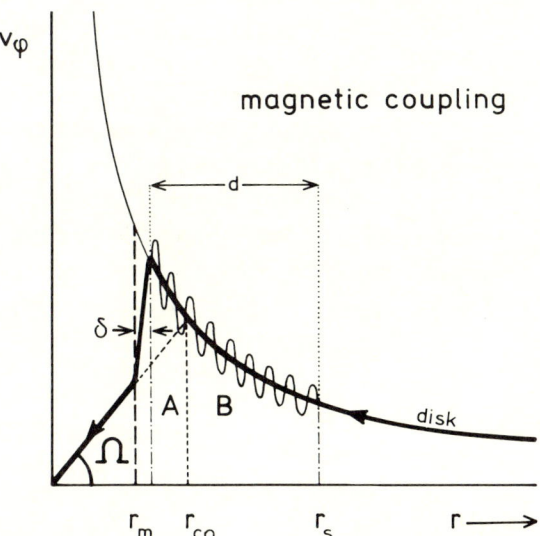

denoted by Ω. The magnetic field lines threading the disk are symbolically represented by the wavy line. Ghosh & Lamb have shown that the boundary layer is thin ($\delta \ll r_m$) and that, on a poloidal flux surface extending vertically through the disk at $r_0 (= r_m + \delta)$, the torque caused by the matter crossing this layer dominates completely over the magnetic and viscous stresses. Therefore the resulting torque is, to a very good approximation, the same as we calculated before in the case without magnetic field (Equation (11.26)). Although throughout the disk the viscosity determines the mass transport inwards by means of transport of angular momentum outwards, its contribution to the total torque appears to be unimportant throughout the whole transition layer $\delta + d$. The authors showed, further, that in the outer transition region an important phenomenon may occur, which is an essential property of any magnetically-coupled disk model.

Consider the matter in the disk inside the corotation radius r_{co} (region A in Figure 11.4). This matter rotates faster than the magnetosphere and the threading field lines will be dragged forwards, resulting in an additional contribution to the spinup torque. Thus if the region d of the disk does not extend beyond r_{co}, as would be the case in very slow rotators, (i.e. $r_m \ll r_{co}$), the total spinup torque will be greater than the torque exerted by matter stress alone. If, on the other hand, the star is a sufficiently 'fast' rotator, such that the field lines thread the disk also outside r_{co}, in this outer region (region B in Figure 11.4) a magnetic spin*down* contribution to the torque is expected. So the threading of the magnetic field may either increase or reduce the spinup (spindown) torques, depending on the relative values of r_{co}, r_m and r_s.

The question, of course, is whether there exists a certain fastness such that the negative contribution caused by the matter in region B just cancels the positive torque (spinup) exerted by the matter in region A. If this were the case, perhaps an equilibrium could be established such that accretion continued, the neutron star would emit X-rays, but no spinup or spindown would occur.

Ghosh & Lamb (1979) have constructed a numerical model of a disk, basically consisting of a Shakura-Sunyaev disk (see the review papers of Petterson in Chapter 10, this book and Verbunt, 1982) in which the magnetic coupling terms are included. The torque calculated in this model is most conveniently expressed as a function of the 'fastness parameter' (Elsner & Lamb 1977), which is defined as the ratio of the angular velocity of the star and the Keplerian angular velocity at r_m:

$$\omega_s \equiv \frac{\Omega}{\Omega_K(r_m)} = \left(\frac{r_m}{r_{co}}\right)^{3/2}, \tag{11.33}$$

which implies $\omega_s \ll 1$ for the present model. The torque is written as (compare

Equation (11.24)):

$$I\dot{\Omega} = N_0 n(\omega_s), \qquad (11.34)$$

where $n(\omega_s)$ is a dimensionless function.

Their numerical calculations show that in the case of a very slow rotator ($\omega_s \approx 0$), the torque is ~ 1.4 times the previously calculated torque N_0 as produced by matter stresses alone (Equation (11.26)) (see Figure 11.5 horizontal line). In addition, a critical fastness ω_c indeed exists, at which a rotating neutron star will undergo no spinup or spindown while accretion continues. In their model the value of ω_c is about 0.35, corresponding to $r_m = 0.5\, r_{co}$ (see Equation (11.33)). For $\omega_s > \omega_c$ the torque is negative, which means spindown. The change in period in this particular model is given by:

$$-\dot{P} = \left(\frac{1}{1.7 \times 10^4}\, \text{s yr}^{-1}\right) \mu_{30}^{2/7} m^{-3/7} R_6^{6/7} I_{45}^{-1} n(\omega_s) P^2 L_{37}^{6/7}, \qquad (11.35)$$

where for the magnetospheric radius the (numerically obtained) value of $0.63\, r_m$ (Equation (11.17)) has been substituted.

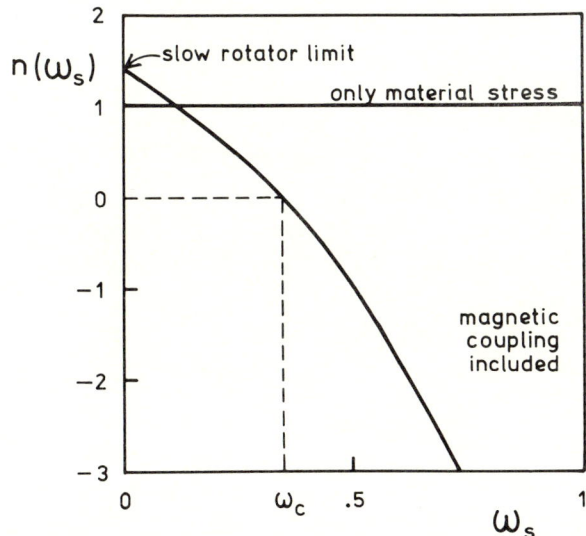

Fig. 11.5. The dimensionless torque $n(\omega_s)$ as a function of the fastness parameter ω_s is compared for accretion-torque models with and without magnetic coupling between the star and the disk. When magnetic coupling is taken into account the numbers have to follow from a numerical solution of the equations describing the disk. In the model of Ghosh & Lamb (1979b) the torque becomes zero at a certain critical fastness ω_c. At this critical value the contributions of the torque in the regions A and B of Figure 11.4 just cancel.

In analogy with Equation (11.27) we may write:

$$\dot{\Omega} = kn(\omega_s)\dot{M}^{6/7}, \tag{11.36}$$

where k is the same proportionality factor as used before. We see that in contrast to the case of Equation (11.27), now $\dot{\Omega}$ does depend on the rotation period of the star. Therefore a comparison between the data and the predictions from this description demands a diagram different from the $-P/\dot{P}$ vs. $PL^{6/7}$ diagram used before (Figure 1.6). Remembering that

$$\omega_s = \Omega(GM_x)^{-1/2} r_m^{3/2} \propto P^{-1}\dot{M}^{-3/7} \propto (PL^{3/7})^{-1}, \tag{11.37}$$

and that $\dot{\Omega} = -2\pi\dot{P}/P^2$, we obtain

$$-\dot{P} \propto n(PL^{3/7}) \cdot (PL^{3/7})^2, \tag{11.38}$$

which shows that $\dot{P}$ is only a function of the equivalent fastness parameter $PL^{3/7}$. Since the derivation is completely general (the function n may, in principle, have any form), it appears that a more natural spin-history-versus-luminosity diagram is that in which $-\dot{P}$ is plotted as a function of $PL^{3/7}$ (Ghosh & Lamb, 1979b) as depicted in Figure 11.6.

In this way of plotting the data, the points are arranged according to the fastness (the faster, the more to the left). Assuming for $n(\omega_s)$ the function of Figure 11.5, a collection of X-ray pulsars with the same mass and moment of inertia should be located (in Figure 11.6) on a (solid) line of which the location is determined by the value of the magnetic moment of the neutron star. In principle, from the many parameters entering Equation (11.35), the mass of the neutron star might be inferred from the orbital motion (e.g. Chapter 1, this book) while values for the radius and moments of inertia have mainly to follow from theoretical predictions. It appears that the magnetic moment is the most uncertain parameter which, therefore, can be used as a free adjustable parameter in the model. The improvement in the agreement between the predicted and observed relation with respect to the simple accretion-torque model of the foregoing section is, of course, only relevant for sufficiently fast rotators and, notably, the spinup rate of Her X-1 can be explained by the Ghosh & Lamb model. (See, however, the next paragraph for an alternative explanation in terms of the same model.) Other applications of this model are described in Section 11.4.1.

In the Ghosh & Lamb model the equilibrium period P_{eq} (defined by $\dot{P}=0$) is given by the relation $\omega_s = \omega_c \approx 0.35$ which yields, with Equation (11.37),

$$P_{eq} = (3.9 \text{ s})\mu_{30}^{6/7} m^{-2/7} R_6^{-3/7} L_{37}^{-3/7}. \tag{11.39}$$

(Notice the expected similarity to Equation (11.31).) This relation suggests (see Table 11.2) that, within the range of reasonable values for the magnetic moment, Her X-1, SMC X-1, Cen X-3 and possibly 4U 1627−67 might be spinning close to their equilibrium periods.

How fast will an X-ray pulsar evolve towards this equilibrium pulse period? This is the subject of the next section.

11.3.10.2 Spinup timescales due to changes in the mass-accretion rate

If all torques on the neutron star just cancel one another we will not observe changes in the rotation period, i.e. $\dot{P} = 0$. In the previous section we described that this situation can occur for a constant non-zero mass-accretion rate in any magnetically coupled disk model, in particular in the disk model of Ghosh & Lamb.

When the accretion rate is variable, two timescales are of importance for the resulting period changes (Henrichs 1983). The first is τ_Ω, which is the timescale on which the angular frequency evolves towards the equilibrium frequency Ω_{eq} if the mass-accretion rate is constant. Second, when the binary system evolves, the mass-accretion rate will be always an increasing function of time

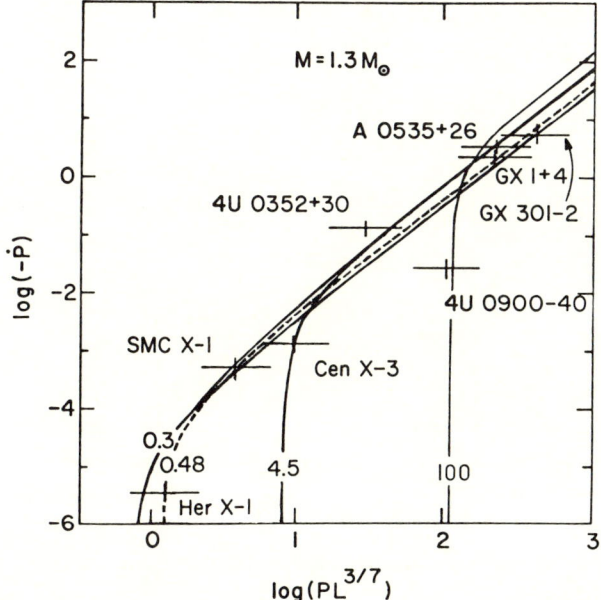

Fig. 11.6. The relation between the spinup rate and the quantity $PL_{37}^{3/7}$ as expected when magnetic coupling between the star and the disk is taken into account, is compared with the observational data (units of $\dot{P}$ are s yr^{-1}). The curves are labelled with the value of the magnetic moment in units of 10^{30} G cm^3, which is a free parameter in the model. As compared to Figure 1.6 (which is based on a model without magnetic coupling) the agreement for the fast rotator Her X-1 is much better. See, however, Section 11.3.10.4 for another interpretation of this source. (Reprinted courtesy of P. Ghosh & F. K. Lamb and the *Astrophysical Journal*.)

with a characteristic timescale:

$$\tau_{ev} \equiv \dot{M}/\ddot{M}, \qquad (11.40)$$

which we shall call the evolutionary timescale.

If $\dot{M}(t)$ is a continuous function of time we expect that if $\tau_\Omega \ll \tau_{ev}$, the long-term change in rotation period will reflect the change in the mass-accretion rate, and we expect to observe τ_{ev}. On the other hand, if $\tau_\Omega \gg \tau_{ev}$ we expect that $\dot{P}$ reflects the timescale required for the neutron star to adjust its rotation rate to the actual value of the mass-accretion rate, and we expect to observe τ_Ω. If $\tau_\Omega = \tau_{ev}$ we still expect to observe τ_{ev}, but the rotation rate may not be equal to the equilibrium value. The value of τ_Ω follows from the theory of accretion torques, whereas τ_{ev} must follow from detailed binary evolution calculations in the case of Roche-lobe overflow (Savonije 1978, 1979), or from stellar evolution with variable mass loss in the case of a stellar wind (van der Linden & Henrichs, 1983). (An independent check on the calculated value of τ_{ev} is that the calculated X-ray lifetime of X-ray binary systems ($\approx 6\tau_{ev}$, see Chapter 9, this book) has to be in agreement with the observed number of X-ray sources.)

Let us estimate τ_Ω in the Ghosh & Lamb model. For a *constant mass-accretion rate* ω_s is proportional to Ω and with the aid of the linear approximation for the (dimensionless) torque

$$n(\omega_s) \approx -4\omega_s + 1.4, \qquad (11.41)$$

which is a reasonable approximation in the domain $0 < \omega_s < \omega_c$ (see Figure 11.5), we may write:

$$\dot{\Omega} = a\Omega + b, \qquad (11.42)$$

where a and b are constants. Solving for $\Omega(t)$ we obtain:

$$\Omega(t) = \Omega_{eq} - (\Omega_{eq} - \Omega_0)e^{-t/\tau_\Omega}, \qquad (11.43)$$

where Ω_0 is the angular velocity at $t=0$, and $\tau_\Omega = -a^{-1} = \Omega_{eq}/b$ or, numerically,:

$$\tau_\Omega \approx (3 \times 10^3 \text{ yr})\mu_{30}^{-8/7} m^{5/7} R_6^{-3/7} I_{45} L_{37}^{-3/7}. \qquad (11.44)$$

The observable timescale for pulse-period changes becomes

$$\tau_{obs} = \frac{\Omega(t)}{\dot{\Omega}(t)} = \frac{\tau_\Omega}{\Omega_{eq}/\Omega(t) - 1}, \qquad (11.45)$$

which indeed approaches infinity for $t \to \infty$, (since $\Omega(t) \to \Omega_{eq}$). Notice that, in this case, the *observed* timescale is in general very different from the timescale τ_Ω which characterizes the duration of the spinup or spindown phase towards an equilibrium period.

On the other hand, if the *mass-accretion rate increases* on a timescale $\tau_{ev} \gg \tau_\Omega$ and the star rotates at its equilibrium value, we expect to observe, according to Equation (11.39) or Equation (11.31):

$$\tau_{\text{obs}} = \frac{\Omega_{\text{eq}}(t)}{\dot{\Omega}_{\text{eq}}(t)} = \frac{7}{3}\frac{\dot{M}}{\ddot{M}} = \frac{7}{3}\tau_{\text{ev}}, \qquad (11.46)$$

where we notice that this relation holds for any equilibrium spin rate provided that $\Omega_{\text{eq}} \propto \dot{M}^{3/7}$. Savonije (1978) also arrived at this equation but started from the assumption $r_m = r_{co}$, or, equivalently, $\omega_s = 1$ at $\Omega = \Omega_{\text{eq}}$, cf Equation (11.31).

Consider now, as an example, Her X-1 where Roche-lobe overflow is the mass-transfer mechanism. We have $P_{\text{obs}} = 1.24\,\text{s}$ and $\tau_{\text{obs}} = 3 \times 10^5$ yr. In the model of Ghosh & Lamb the magnetic moment is a free parameter that can be adjusted such that $P_{\text{eq}} \approx 1.19\,\text{s}$, and $\tau_\Omega \approx 1.2 \times 10^4$ yr (for a neutron star of $1.3\,M_\odot$ and radius and moment of inertia as calculated by Pandharipande et al. 1976). Hence, for the above-given equations of τ_{obs}, these values can only match Equation (11.35) for a source with a *constant* $\dot{M}$ that is spinning up towards its equilibrium value if it has almost reached equilibrium. On the other hand the calculation of Savonije (1979, improving on earlier results obtained in 1978) gives $\tau_{\text{ev}} \approx 1 \times 10^5$ yr which would perfectly match Equation (11.46) if τ_{obs} is explicitly due to the predicted increase of $\dot{M}$. Which of these two possibilities is right? Notice that in both applications we have used the same model of Ghosh & Lamb. Because in the case of Her X-1 we indeed have the situation $\tau_\Omega \ll \tau_{\text{ev}}$, the conclusion must be that Her X-1 is probably spinning at its equilibrium value (as predicted by a disk model including magnetic coupling) but that its observed long-term change in period is governed by the evolution of the companion star. We notice that the observed quasi-oscillations around the mean trend (Figure 1.5) suggests 'some' equilibrium value. As a consequence of this conclusion there would be no ground for plotting the observed value for $\dot{P}$ for Her X-1 in Figures 1.6 or 11.6 because the observed rate is simply no longer related to the mechanisms for which such diagrams were constructed. The long-term value of $\dot{P}$ depends now on $\dot{L}_x$ rather than on L_x. We should keep in mind, however, that it may be possible that, for some reason, $\dot{M}$ varies, as a fluctuation, on a timescale much shorter than τ_Ω, while τ_Ω is still small compared to the characteristic timescale τ_{ev} of the long-term trend. Of course, we cannot exclude this rather *ad hoc* hypothesis but observations of $\dot{P}(t)$ and $L_x(t)$ over a time interval which is long enough to cover such a fluctuation should enable us to discriminate between the two described modes.

Table 11.2 gives, for all known pulsars (i.e. five, including Her X-1) that are steadily spinning up the observed properties, together with the calculated value for τ_Ω from Ghosh & Lamb's model (assuming $\mu = 10^{30}\,\text{G cm}^3$), the observed value for $-(3/7)\,P/\dot{P} = \tau_{ev}$ and the calculated value τ_{ev} as follows from binary evolution (Savonije 1979), in so far as is known. We see from the table that only Her X-1 is a 'relaxed' system (i.e. $P \approx P_{eq}$, $\tau_\Omega \ll \tau_{ev}$ and $\tau_{obs} \approx (7/3)\,\tau_{ev}$), mainly because of its slow evolutionary timescale. Also in agreement with this interpretation is the large height above the galactic plane of the HZ Her/Her X-1, system which indicates an old age (7×10^7 yr) since the supernova explosion, giving ample time for the pulsar to evolve towards an equilibrium situation.

11.3.10.3 Accretion from a stellar wind

In the case of a stellar wind the specific angular momentum of the fluid elements of the plasma that arrives at the magnetospheric boundary is usually not above the threshold value for forming a disk (Section 11.3.8). The accreted angular momentum cannot be calculated, but should be smaller than in the case of disk accretion and may have either sign, depending on the unknown and, probably time-dependent, local wind velocity and/or density gradients. These gradients may even be a function of the (time-dependent) X-ray intensity of the source (see for instance McGregor & Vitello 1979), which might or might not be a major complication. In any case, it is this kind of complexity that prevents a detailed description of the spinup and spindown behavior as observed in wind-fed sources. Winds of early-type stars are known to vary irregularly on almost any timescale in the range from minutes to years and it is this type of variation that is probably reflected in the observed anomalous

Table 11.2. *Timescales of X-ray pulsars with 'steady' spinup*

Source	P (s)	L_{37}	$\tau_{obs} = -P/\dot{P}$ (yr)	$(7/3)\,\tau_{ev}$ (yr)	τ_{ev} (yr)	τ_Ω (yr)	P_{eq} (s)
SMC X-1	0.71	50	1.7×10^3	7×10^3	3×10^3	6×10^2	0.72
Her X-1	1.2	1	3.0×10^5	4×10^5	2×10^5	3×10^3	3.9
Cen X-3	4.8	5	3.4×10^3	2×10^4	8×10^3	2×10^3	1.9
4U 1627−67	7.7	3 at 10 kpc	5.0×10^3	–	–	2×10^3	2.4
GX 1+4	122	4	50	–	–	2×10^3	2.1

Columns 2, 3, 4: values taken from Table 11.1
Columns 5, 6: values based on Savonije (1978, 1979 and Chapter 10, this book).
Columns 7, 8: values from Equations (11.44) and (11.39) with $\mu_{30} = m = R_6 = I_{45} = 1$.

Using the model of Ghosh & Lamb (1979) we conclude from the table that Her X-1 is probably a 'relaxed' system (Henrichs 1983) because (1) its rotation period P is approximately equal to the equilibrium period P_{eq} (Equation (11.39)), where we arbitrarily used $\mu_{30} = m = R_6 = 1$, (2) the expected timescale to reach such an equilibrium (Equation (11.44)) is much smaller than the evolutionary timescale which also produces a spinup (Equation (11.44)) or $\tau_\Omega \ll \tau_{ev}$, and (3) the observed timescale τ_{obs} is in perfect agreement with the timescale predicted by evolutionary calculations (Equation (11.46)), or $\tau_{obs} \approx (7/3)\tau_{ev}$. For the other sources in the table no such conclusion can be drawn.

changes of the rotation period of wind-fed X-ray pulsars. Good examples are HD 77581/4U 900−40 (Nagase *et al.* 1981) and Wray 977/GX 301−2 where we sometimes see spinup and sometimes spindown.

It is important to point out that the maximum possible accreted angular momentum from a wind is comparable with that in the case of a disk. Therefore the maximum observed value for $\dot{P}$ (positive or negative) may be compared with the disk-fed sources in diagrams like those in Figures 1.6 and 11.6. This is the reason why sources like 4U 0115+63, A 0535+26, GX 301−2 and 4U 0900−40 can usually be included in these diagrams. Also for those sources, the observed high value of the spinup rate favors the interpretation of a spinning object with a small moment of inertia (i.e. a neutron star).

In transient X-ray sources (such as A 0535+26 and 4U 0115+63) the situation is different. Only during outburst the spinning neutron star is observable and is expected to adjust its rotation rate on a timescale given by Equation (11.35) with $\omega_s \approx 0$, or, approximately, by Equation (11.28). When little or no X-ray emission (i.e. little accretion) occurs, the star presumably spins down (see Section 11.4), in accordance with the observations.

Details of the accretion process and limits to the spinup rate in the case of (almost) spherical accretion from a stellar wind have been studied in particular by Arons & Lea (1980).

11.3.11 Short-term $\dot{P} - L_x$ correlations

In many disk-fed X-ray sources, episodes of temporary spindown occur. (See Figures 1.5 and 11.1.) We have seen that, in a model of accretion from a Keplerian disk without magnetic coupling, such a sequence of spinup and spindown can occur only if the disk alternates its sense of rotation. It is likely that such a reversal will be accompanied by considerable change in the accretion rate. On the other hand, in the case of a magnetically coupled disk, spindown only indicates that the rotation period is close to its equilibrium value and that the accretion rate has a predictable dependence on the spinup or spindown rate. Simultaneous measurements of $\dot{P}$ and L_x are clearly of extreme importance to test the different accretion torque models.

It is important to notice that, in fact, the theory predicts a relation between $\dot{\Omega}$ and $\dot{M}$. Usually $\dot{\Omega}$ can be measured directly, but $\dot{M}$ must be inferred from the X-ray luminosity (Equation (11.1)). This might be a problem because during a strong increase of $\dot{M}$ it is possible that absorption in the accreting matter may inhibit the X-rays escaping from the neutron star. In such a situation an *in*crease of $\dot{M}$ might be correlated with a *de*crease in L_x, and a change in $\dot{P}$ would still correspond to a change in $\dot{M}$ but no longer to a change in L_x.

Discrimination between a 'smothered' and a 'clean' X-ray source is possible in the case of eclipsing sources which show 'turn-on' behavior (Her X-1, Cen X-3, and SMC X-1). If during the turn-on the eclipse duration is much longer than during the 'on' state, the source was probably smothered during the 'off' state and cleaned up during the turn-on phase. This behavior has been observed for Cen X-3 (Giacconi 1974, Schreier et al. 1976, Fabbiano & Schreier 1977, Bonnet-Bidaud & van der Klis 1979) and SMC X-1 (Bonnet-Bidaud & van der Klis 1981). In the case of Cen X-3, changes in the X-ray spectrum were also reported. The (few) presently-existing simultaneous $\dot{P}$-L_x measurements for these fast rotators are, however, not sufficiently conclusive to discriminate between different accretion models. See, for instance, van der Klis, Bonnet-Bidaud & Robba (1980) and references therein for Cen X-3. For Her X-1, no published data on correlated $\dot{P}$-L_x measurements are available.

The best example of a 'clean' source is GX 1+4 for which a detailed comparison between $\dot{P}$ and L_x is made by Doty, Hoffman & Lewin (1981). The source is steadily spinning up, is far from its equilibrium (or a slow rotator, $\omega_s \ll 1$, see Table 11.2), and thus the accretion torque is similar in disk models with and without magnetic coupling. (Although there exists no direct observational evidence for a disk in this system, we infer from the observed pulse-period history ($-P/\dot{P} \approx 50$ yr) that this, probably low-mass, system is a disk-fed source powered by (beginning) Roche-lobe overflow.) Hence, we expect in both types of models:

$$\frac{\dot{\Omega}}{\Omega} = \frac{6}{7} \frac{\dot{L}_x}{L_x} \qquad (11.47)$$

or

$$\frac{\ddot{P}}{2\dot{P}} = \frac{\dot{P}}{P} + \frac{3}{7} \frac{\dot{L}_x}{L_x}. \qquad (11.48)$$

For GX 1+4 fluctuations in L_x and $\dot{P}$, observed over a 9 month period, appear to agree well with this predicted relation, qualitatively as well as quantitatively. For both the spinup and spindown episode the observed change in L_x which accompanied the change in $\dot{P}$, corresponds within a factor of 1.2 with that given by Equation (11.48). (This quantitative agreement is not mentioned explicitly by Doty et al., but can be derived from the data presented by these authors.) This agreement is an additional strong confirmation for the theory of spinup and spindown for slow rotators.

11.3.12 Summary and concluding remarks

With the presently available theory on accretion torques the observed spinup and spindown behavior of X-ray pulsars can be qualitatively understood in cases where an accretion disk is present around the magnetic neutron star, i.e. in X-ray binary systems with Roche-lobe overflow and in systems powered by a stellar wind in which, temporarily, a disk is formed. The systems with Roche-lobe overflow, the accreting matter transports (via a disk) orbital angular momentum towards the neutron star, which causes a long-term spinup trend. The direction of the angular momentum that is captured from a stellar wind is expected to be more erratically distributed. Disk models including magnetic effects contain, however, many simplifications (such as zero-obliquity of the rotation axis, stationary accretion flow, immediate response of the neutron star to the magnetic torque, see eg. Lamb, Pines & Shaham 1978, etc.) of which the relative importance and uncertainties deserve further study. Model tests have to come from simultaneous $\dot{P}$ and L_x measurements which should include spectral information.

In wind-fed sources the situation is much less clear. This is mainly due to lack of knowledge of stellar-wind properties of massive stars, such as deviation from spherical symmetry and mass-loss fluctuations, and lack of description of the accretion of angular momentum from a stellar wind within the accretion cylinder.

11.4 The rotational history of a neutron star in a binary system
11.4.1 Introduction

The subject of this section is: how can we understand the observed range of X-ray periods?

The initial period of a single, newborn star is of the order of 0.01 s. This has been estimated from ages and spindown rates of the two youngest-known radio pulsars (the Crab and the Vela pulsar, see e.g. Manchester & Taylor 1977). About one-half of the presently known X-ray pulsars have a period of more than 100 s. If we assume that the initial rotation period of a neutron star in a binary is comparable to that of a single one, this would imply that, before the X-ray phase, strong braking torques must have been acting on these stars.

All optical counterparts of slow X-ray pulsars ($P > 100$ s) are stars of early type. Therefore most attention has been given to a system in which a neutron star spins down in the wind of its massive companion.

Before the discovery of slow X-ray pulsars (in 1975), Illarionov & Sunyaev (1975) proposed that the newborn neutron star in a massive binary system behaves like a normal (single) radio pulsar and accordingly spins down by the standard radio-pulsar mechanism until the ambient stellar wind material overcomes the pulsar-wind pressure and inhibits further functioning as a radio pulsar. Subsequently, a 'propeller' mechanism operates, by which the accreting plasma is thrown out until the rotation has slowed down enough to make real accretion onto the stellar surface possible. In this theory only X-ray pulsars with periods of less than 10 s are expected because the timescale for producing a slow pulsar is much longer than the expected lifetime of a massive star. Consequently, the neutron star will be swallowed by the companion star long before it can have been slowed down to periods of hundreds of seconds.

Fabian (1975) applied this theory to the first-discovered slow X-ray pulsar (A 1118−61) and concluded that the spindown to the observed period of 405 s must have been taking place in a very weak stellar wind (or, in a widely-separated binary system) originating from a not too massive star ($M \lesssim 10 M_\odot$).

In contrast, in order to solve the 'long-period problem' Lea (1976) proposed that tidal forces in a binary system would be able to synchronize the stellar rotation of the pre-supernova companion with the orbital period, within the lifetime of the system. The implosion of the core might then produce a slowly rotating neutron star. This would largely remove the difficulty of producing long-period rotators. Savonije & van den Heuvel (1977), however, reconsidered tidal synchronization including the effects of radiative damping in the outermost layers of the pre-supernova helium star and concluded that no presently known tidal-dissipation mechanism is able to produce a torque strong enough to slow down the star before the supernova explosion. Thus, the 'long-period problem' remained to be solved.

Since then many authors have proposed mechanisms for producing strong torques in the boundary layer between the magnetosphere and the stellar-wind material, with the main aim of achieving a spindown timescale within the main-sequence time of a massive companion star.

Another point of view to explain the existence of slow X-ray pulsars is given by Elsner, Ghosh & Lamb (1980). They argue that spindown to long periods has not necessarily to occur in a stellar wind, but instead may take place after the onset of accretion. They advocate that very soon a disk is formed (as is described in Section 11.3.10.1) in which the neutron star spins faster than the critical fastness ω_s (see Figure 11.5). In that model, accretion continues and the star spins rapidly down on a timescale as given by Equation (11.44). Obviously, an additional essential requirement for this interpretation is the occurrence of a strongly variable mass-transfer rate. This is indeed observed in the Be–X-ray binaries, most of which are transients. At least one of them (A 0535+26, Li *et al.* 1979) has shown rapid spindown between two bright X-ray phases. Further observations of such behavior would be in strong favor of such a model in the case of transient sources.

Here we sketch the basic picture as it emerges from the different papers. We give a short account of some of the specific proposed spindown models (in particular that of Davies & Pringle, 1981) and outline the underlying physical assumptions. We emphasize here that the theory explaining the presently observed X-ray pulse-period distribution is still far from being settled.

11.4.2 The early spin history: the 'dark ages' of a neutron star

Why a newborn radio pulsar in a binary system has never been found is explained by Illarionov & Sunyaev (1975). The large optical depth of the stellar wind for free-free absorption hinders the detection of pulsars in the radio-wavelength region where radio-pulsar searches are carried out. The young pulsar initially emits a very strong relativistic wind mixed with high-energy, magnetic dipole radiation. Possibly the radio-gamma binary source LS 61°303 is presently in such a phase (Maraschi & Treves 1981). The pulsar wind blows a cavity in the stellar wind and no material can accrete. Because the source of the pulsar emission is rotational energy the neutron star slows down and the size of the cavity decreases.

Under which condition the radio pulsar is finally switched off (and hence the cavity structure is lost) is then decisive for the rotation period at which the neutron star starts to interact with the surrounding material. Davies & Pringle (1981) argued that the most commonly used condition for radio-pulsar cessation (Svartzmann 1971, Ostriker, Rees & Silk 1970, Lamb, Pethick & Pines 1973, strong-wind case of Illarionov & Sunyaev 1975, Kundt 1976, Maraschi & Treves 1979), namely that the cavity radius becomes smaller than the accretion radius (Equation (11.3)), causing a collapse, is not self-consistent because in such a picture the outward energy flow outside the cavity is not taken into account (see Section 11.4.4). Pulsar emission might be quenched when the cavity radius (determined by pressure balance) becomes smaller than the radius of the light cylinder:

$$r_{1c} = c/\Omega, \tag{11.49}$$

which plays an essential role in any radio-pulsar model. Other switch-off conditions, however, are conceivable. Before we give a more detailed account of this process, we describe, in short, the next stages of the rotational history.

When the radio-pulsar phase ends, the conditions for accretion down to the surface of the neutron star (and hence X-ray emission) are not expected to be immediately satisfied (see Section 11.3.9) and the neutron star spins down further without being observable. If this so-called sleeping phase (van den Heuvel 1977) takes a relatively long time (i.e. for not too massive stars, $M_p < 20\,M_\odot$, in wide systems), spindown to long periods is feasible; otherwise the newborn X-ray pulsar starts with a relatively short period (massive stars $M_p > 20\,M_\odot$, in narrow systems). Subsequent spinup by accretion torques will determine the ultimate distribution of X-ray pulse periods. In the case of transient X-ray sources, intermittent periods of spinup during X-ray emission will interrupt the sleeping phase during which the star spins down. This behavior has been observed in A 0535+26 (Li et al. 1979) and in 4U 0115+63 (Ricketts et al. 1981). The suggested mechanism that the pulsar spins rapidly down after the onset of accretion (Elsner, Ghosh & Lamb 1980) may operate as well.

Before we discuss the proposed relevant timescales and rotation periods we discuss some fundamental starting points.

11.4.3 The loss of energy and angular momentum

If a neutron star spins down on account of its rotational energy, the corresponding loss of angular momentum, N, is dictated by the rotation frequency Ω:

$$\frac{L}{N} = \left[\left(\frac{d}{dt}(\tfrac{1}{2}I\Omega^2)\right)\right]/I\dot{\Omega} = \Omega, \tag{11.50}$$

where L denotes the loss of rotation energy and we neglect the contribution from $\dot{I}$ in the last equality (cf. Section 11.3.9). This requirement opens, in principle, two different ways of calculating the characteristic timescale of the spindown process.

The first and most frequently encountered approach is that one calculates the torque N at the magnetospheric boundary (e.g. Kundt 1976, Holloway, Kundt & Wang 1978, Wang 1979). Essentially one evaluates $N = \dot{M}v_c r_m$ by specifying the characteristic velocity v_c of the particles that are flung away with the rate $\dot{M}$ (identified with the infall rate) at the magnetospheric radius r_m (Equation (11.16), which may, however, be modified. Afterwards it should be verified that the rate at which energy is given to the escaping particles is of order $N\Omega = L$. (This condition is not satisfied by the model of Kundt (1976) but was taken care of in Holloway, Kundt & Wang (1978).) If this is not the case, one has to invoke additional heat or radiation losses to ensure the energy balance according to Equation (11.50). The timescale is finally found from the equation

$$\tau \equiv \Omega/\dot{\Omega} = I\Omega/N. \tag{11.51}$$

The difficulty in this 'particle approach' is that one has to account for the detailed interaction of the particles with the (too) rapidly rotating magnetosphere. A fluid approximation would be more appropriate. One might argue that the maximum velocity that can be given to the outflowing matter is the velocity of the magnetospheric boundary. Several reasons why this is not necessarily true were pointed out by Illarionov & Sunyaev (1975) and further stressed by Davies, Fabian & Pringle (1979). These authors argue that the characteristic velocity v_c should be comparable to the sound speed in the external medium because a substantial, supersonic, hydrodynamic outflow of matter that carries all the energy is improbable. Furthermore, acceleration of matter in the magnetospheric boundary up to a velocity much greater than the local Alfvén velocity, which is comparable to the local sound speed, is also problematic. Finally, the outflowing high-velocity particles are not allowed to interact with the incoming stream, which makes it difficult to imagine how such a configuration could arise.

A useful classification of the different outcomes of the various proposed models is given by Davies, Fabian & Pringle (1979) and Wang (1979) has reviewed the physical assumptions made by the different authors. The shortest timescale produced in this type of models is 10^5-10^6 yr, comfortably within the evolutionary timescale of a massive companion star, but the relevance of this result is apparently subject to serious doubt.

A second and completely different treatment of the spindown problem has been put forward by Davies, Fabian & Pringle (1979). This approach is concerned with the energy transfer, taking into account the surroundings of the rotating neutron star. The energy is generated at rate L at the interaction radius r_i which is not necessarily equal to the previously value of r_m. The location of r_i (and hence the value of L) is rather determined by the requirement that the structure of the external medium is such that the outward energy transport can be properly accommodated. The timescale follows from

$$\tau = I\Omega^2/L. \tag{11.52}$$

A dynamical description of the behavior of the plasma at r_i is thus avoided.

This approach has been consistently applied by Davies & Pringle (1981) to all possible phases of spindown which occur between the birth of a rapidly rotating neutron star and the final long-period state.

11.4.4 Spindown in the case of no accretion

We summarize the model proposed by Davies & Pringle.

When no accretion occurs and the neutron star spins down, the liberated (rotational) energy is deposited in the surrounding stellar-wind matter. This will affect the flow up to a certain distance. Therefore, the authors envisage the neutron star to be embedded in a quasi-static atmosphere which is assumed to be spherically symmetric. The energy is transferred to the atmosphere at the inner radius r_i. In the radio-pulsar phase, the energy input is provided by a radio-pulsar mechanism, e.g. vacuum magnetic-dipole radiation. Davies & Pringle assumed that in subsequent phases the energy is transported at a rate:

$$L(r_i) = 4\pi r_i^2 \cdot \tfrac{1}{2} v_t^3(r_i)\rho(r_i), \tag{11.53}$$

where $v_t(r_i)$ is the turbulent velocity at r_i and $\rho(r_i)$ is the density. We notice that the justification for using this equation might be subject to more study. Because the rotating non-axisymmetric magnetosphere shocks the gas roughly once per revolution, the turbulent velocity $v_t(r_i)$ is identified with $r_i\Omega$, i.e. the rotational velocity of the magnetosphere at r_i, except when $r_i\Omega$ exceeds the local sound speed $c_s(r_i)$, in which case, as has been argued in the previous section, $v_t = c_s$ is used.

If the fundamental assumption is made that there are no radiative losses in the atmosphere, the only means of energy transport outwards is convection and the energy-transfer equation becomes:

$$L(r_i) = L(r) = \text{constant} = L, \tag{11.54}$$

where L is the quantity that is desired for the calculation of the timescale of the process (see Equation (11.52)). The energy and angular momentum are convected upwards and are lost from the system at the outer boundary of the atmosphere. The outer boundary is defined as the radius beyond which the density and pressure of the external medium are unaffected by the presence of the neutron star.

The location of the inner radius r_i, which enters Equation (11.53), is found, as usual, from pressure balance arguments. This requires knowledge of the pressure $p(r)$ throughout the atmosphere. Now, the structure of any static atmosphere in which the gas pressure is thermal is characterized by the ratio of the sound speed $c_s(r)$ and the free fall velocity $v_{ff}(r)$ according to the equation of hydrostatic equilibrium:

$$\frac{d \ln p}{d \ln r} = -\frac{\gamma}{2}\left(\frac{v_{ff}}{c_s}\right)^2, \tag{11.55}$$

where γ is the usual ratio of specific heats and $c_s^2 = \gamma p/\rho$. Because $v_{ff}(r)$ is fixed by gravity, the atmospheric structure is essentially determined by the parameter $c_s(r)$. The outer-boundary conditions, which are provided by the stellar-wind properties of the massive companion star, together with Equations (11.53) and (11.54), finally fix $p(r)$ (and hence the value for r_i) and $\rho(r_i)$.

According to Davies & Pringle there are four distinct spindown phases, each of which is characterized by a different combination of $v_t(r_i)$ and $c_s(r_i)$ with respect to $v_{ff}(r_i)$. The four cases are:

case a: the radio pulsar-phase with $c_s \gg v_{ff}$;
case b: the very rapid rotator with $r_b\Omega \approx c_s \gg v_{ff}$;
case c: the supersonic rotator with $r_c\Omega \gg c_s \approx v_{ff}$;
case d: the subsonic rotator with $r_d\Omega \ll c_s \approx v_{ff}$;

where r_b, r_c and r_d denote r_i in each case. Only in case d is the inner radius r_i given by the previously defined magnetospheric radius r_m (Section 11.3.7). It turns out that the neutron star spins down through the different phases in alphabetical order. For convenience we summarize in Table 11.3, for each of the

Table 11.3. *Characteristic properties of the subsequent spindown phases with no accretion in the model of Davies & Pringle (1981). The symbols are explained in Section 11.3. The relative velocity in units of 1000 km s^{-1} is denoted by v_8*

Case	Luminosity	Transition period P_{ij}	Timescale $\tau \equiv P/\dot{P}$
a radio pulsar	$\dfrac{2}{3c^3}\mu^2 \Omega^4$	$P_{ab} = (0.8\,\text{s})\, m^{1/3} (\mu_{30}^2/\dot{M}_{15})^{1/6} v_8^{-5/6}$	$\tau_a = (3.3 \times 10^7\,\text{yr})\, P_{ab}^2 I_{45}\, \mu_{30}^{-2}$
b very rapid rotator	$4B^2(r_b) r_b^3 \Omega$	$P_{bc} = (5.3\,\text{s})\, m^{-1} (\mu_{30}^2/\dot{M}_{15})^{1/2} v_8^{1/2}$	$\tau_b = (3.0 \times 10^6\,\text{yr})\, P_{ab}^{-1} I_{45}\, m^{-1}\, \mu_{30}^{-1} \dot{M}_{15}^{-1/2} v_8^{-5/2}$
c supersonic rotator	$8\dot{M} v_{\text{rel}}^2$	$P_{cd} = (51\,\text{s})\, m^{-1/3} (\mu_{30}^2/\dot{M}_{15})^{1/2} v_8^{-2/3}$	$\tau_c = (1.6 \times 10^7\,\text{yr})\, P_{bc}^{-2} I_{45} \dot{M}_{15}^{-1} v_8^{-2}$
d subsonic rotator	$\dfrac{2\mu^2 \Omega^3}{GM}$	$P_{dx}^\dagger = (175\,\text{s})\, m^{-4/21} \mu_{30}^{16/21} \dot{M}_{15}^{-5/7}$	$\tau_d = (3.3 \times 10^2\,\text{yr})\, P_{dx} I_{45}\, m\, \mu_{30}^{-2}$

† This particular value for P_{dx} is derived assuming that the bremsstrahlung's cooling rate determines the onset of accretion.

four cases, the thus-obtained expressions for the luminosity, the rotation period at which the transition to a next stage occurs, and the corresponding timescale. The transition from phase a to b has been described in Section 11.4.2. The transition to case c occurs when the rotation rate becomes too slow to support the top-heavy atmosphere (characterized by $p \approx$ constant and $\rho \propto r^4$), and the atmosphere has to change its structure drastically. Case d is attained once the centrifugal force at the magnetospheric boundary can be overcome (or $r_m < r_{co}$) but still no accretion can take place. We notice from Table 11.3 that, for a given neutron star, the transition periods are determined by the particular value of $\mu^2/\dot{M}$, which is not accurately known.

In this model the maximum period is reached when accretion starts. As has been stressed by Davies & Pringle, accretion can only occur if the material can cool (Arons & Lea 1976, Elsner & Lamb 1977) because the atmosphere collapses as soon as the cooling exceeds the rotational-energy input. This might occur in case c or case d, depending on the accretion rate and the cooling mechanism.

The total spindown time t_{sd} is given by the sum of the successive timescales and will be close to the maximum value of τ_i ($i =$ a, b, c, d). If the massive primary star evolves away from the main sequence in a time shorter than t_{sd}, not all spindown phases might be reached and, correspondingly, a shorter maximum period is reached.

One can show that in this theory τ_a is necessarily the longest timescale and that it will take a time duration of $\frac{1}{2}\tau_a$ to reach the end of the radio-pulsar phase. The numerical value turns out to be:

$$t_{sd} \approx (1.2 \times 10^7 \text{ yr}) I_{45} m^{2/3} \mu_{30}^{-4/3} \dot{m}_{15}^{-1/3} v_8^{-10/6} \qquad (11.56)$$

which is rather long, but comparable to the main-sequence lifetime of not too massive stars (notice, however, the strong dependence on the parameters).

The conclusion is that the long duration of the radio-pulsar phase is the bottle neck for a neutron star in a binary system in order to be spun down to a long rotation period within the main-sequence lifetime of the companion star.

We conclude by noting some limitations in the applicability of the model of Davies & Pringle. The validity of the existence of such an atmosphere is, of course, crucial. For instance, the outer boundary must fit well inside the dimensions of the binary system. This may not be true, especially in narrow systems that are in the beginning of the radio-pulsar phase. Typical parameters for a 20 $M_\odot$ zero-age main-sequence primary star are $\dot{M}_p \approx 1 \times 10^{-7} M_\odot \text{ yr}^{-1}$ (Lamers 1981) and $v_w \approx 3000 \text{ km s}^{-1}$ (Abbott 1978). For a binary period of 10 d and a neutron-star mass of 1 $M_\odot$, we find $\dot{M} \approx 1 \times 10^{12} \text{ g s}^{-1}$ (Equation (11.8)) and the radius of the outer boundary of the atmosphere will exceed the binary separation ($\sim 50 R_\odot$) as long as the pulse period is shorter than $P = 0.15$ s. Another obvious requirement is that the outer radius of the atmosphere should be larger than the inner radius. In the cases c and d this is equivalent to the condition that $r_m < r_{acc}$. For wide systems ($P > 10$ d) where $\dot{M} < 10^{12}$ g s^{-1} this might not be fulfilled and a neutron star in such a system might resemble the Earth's magnetosphere in the solar wind. What happens in such systems is still an open question.

11.4.5 Summary and concluding remarks

We close this chapter with a summary of the weakest points in the theory as given in these sections.

The assumption that all neutron stars are born with a short rotation period ($P \lesssim 0.1$ s) may not be established. Under which conditions the subsequent radio-pulsar phase (if it exists) ends is not well understood. Further spindown caused by the surrounding material that is not able to accrete should take place on a reasonably short timescale ($\sim 10^6$ yr), but which mechanism operates is still a matter of discussion. The conditions which determine the onset of accretion down to the surface of the neutron deserve more study. Finally, a possible decay of the magnetic field has not been considered.

The transient pulsating X-ray sources provide at present the most useful test for models of the spindown process in the latest stage.

Acknowledgements

After the completion of this work it is a pleasure to acknowledge the many stimulating and useful discussions with Leo van den Horn, Max Kuperus, Walter Lewin, Theo van der Linden, Saul Rappaport, Gertjan Savonije and Roelf Takens. The manuscript profited considerably from many inspiring discussions, comments and suggestions made by Ed van den Heuvel, Michiel van der Klis, Jan Kuijpers, Jim Pringle, Frank Verbunt, Yi-Ming Wang and, in particular, Fred Lamb, whose detailed and critical comments have lead to a major improvement of an earlier version of this paper. Drafts of this manuscript have been typed with great skill by Inge Hoonhout and Marianne van Oene. Permission for reproduction of Figures 11.3 and 11.6 is acknowledged. The other figures are carefully drawn by Maria Moesman.

Finally, this work would have never been completed without the continuous encouragement of Ed van den Heuvel who supported the author with his stimulating suggestions, inspiring questions, and helpful remarks.

References

Abbott, D. C., 1978, *Ap. J.*, **225**, 893.
Anzer, U. & Börner, G., 1980, *Astron. Ap.*, **83**, 133.
Anzer, U. & Börner, G., 1982, preprint.
Arons, J. & Lea, S. M., 1976, *Ap. J.*, **207**, 914.

Arons, J. & Lea, S. M., 1980, *Ap. J.*, **235**, 1016.
Bondi, H. & Hoyle, F., 1944, *MNRAS*, **104**, 273.
Bonnet-Bidaud, J. M. & van der Klis, M., 1979, *Astron. Ap.*, **73**, 90.
Bradt, H., Doxsey, R. & Jernigan, J., 1979, *COSPAR X-ray Astronomy* (eds. W. A. Baity & L. E. Peterson), Oxford, Pergamon, p. 3.
Canuto, V., 1977, *Proc. 8th Texas Symp. Rel. Astroph. (Ann. NY Ac. Sci.)*, **302**, 514.
Canuto, V. & Bowers, R. L., 1981, *Pulsars* (eds. W. Sieber & R. Wielebinski), Reidel, Dordrecht, p. 321.
Chevalier, C. et al., 1981, *Space Sci. Rev.*, **30**, 405.
Conti, P. S., 1978, *Astron. Ap.*, **63**, 225.
Davidson, K. & Ostriker, J. P., 1973, *Ap. J.*, **179**, 585.
Davies, R. E., Fabian, A. C. & Pringle, J. E., 1979, *MNRAS*, **186**, 779.
Davies, R. E. & Pringle, J. E., 1980, *MNRAS*, **191**, 599.
Davies, R. E. & Pringle, J. E., 1981, *MNRAS*, **196**, 209.
Doty, J. P., Hoffman, J. A. & Lewin, W. H. G., 1981, *Ap. J.*, **243**, 257.
Dupree, A. K. et al., 1980, *Ap. J.*, **238**, 969.
Elsner, R. F., Ghosh, P. & Lamb, F. K., 1980, *Ap. J.*, **241**, L155.
Elsner, R. F. & Lamb, F. K., 1977, *Ap. J.*, **215**, 897.
Fabbiano, G. & Schreier, E. J., 1977, *Ap. J.*, **214**, 235.
Fabian, A. C., 1975, *MNRAS*, **173**, 161.
Ghosh, P. & Lamb, F. K., 1978, *Ap. J.*, **223**, L83.
Ghosh, P. & Lamb, F. K., 1979a, *Ap. J.*, **232**, 259.
Ghosh, P. & Lamb, F. K., 1979b, *Ap. J.*, **234**, 296.
Ghosh, P., Lamb, F. K. & Pethick, C. J., 1977, *Ap. J.*, **217**, 578.
Giacconi, R., 1974, *Proc. 16th Solvay Conf. on Phys.* (Univ. of Brussels), p. 27.
Guetter, H. H., 1977, *Astron. J.*, **82**, 598.
Henrichs, H. F., 1980, *Highlights of Astronomy*, vol. 5 (ed. P. A. Wayman), Reidel, Dordrecht, p. 541.
Henrichs, H. F., 1983, in preparation.
Holloway, N., Kundt, W. & Wang, Y-M., 1978, *Astron. Ap.*, **70**, L23.
Hoyle, F. & Lyttleton, R. A., 1939, *Proc. Camb. Phil. Soc.*, **112**, 205.
Hunt, R., 1971, *MNRAS*, **154**, 141.
Hunt, R., 1979, *MNRAS*, **188**, 83.
Illarionov, A. F. & Sunyaev, R. A., 1975, *Astron. Ap.*, **39**, 185.
Kelley, R. L., Jernigan, J. G., Levine, A. & Petro, L. D. 1983, *Ap. J.*, **264**, 568.
Kelley, R. L., Rappaport, S., Brodheim, M. J., Cominsky, L. & Stothers, R., 1982, *Ap. J.*, **251**, 630.
Kundt, W., 1976, *Phys. Lett.* A**57**, 195.
Lamb, F. K., 1977, *Proc. 8th Texas Symp. Rel. Astroph. (Ann. NY Ac. Sci.)*, **302**, 482.
Lamb, F. K., 1979, *Proc. S. Chapman Conf. on Magn. Boundary Layers*, Alpbach, Austria (ed. B. Battrick) ESA SP 148 Addendum, 1.
Lamb, F. K. & Pethick, C. J., 1974, 'Astrophysics and gravitation', *Proc. 16th Solvay Conf. on Phys.* (Univ. of Brussels), p. 135.
Lamb, F. K., Pethick, C. J. & Pines, D., 1973, *Ap. J.*, **184**, 271.
Lamb, F. K., Pines, D. & Shaham, J., 1978, *Ap. J.*, **224**, 969.
Lamers, H. J. G. L. M., 1981, *Ap. J.*, **245**, 593.
Lea, S. M., 1976, *Ap. J.*, **209**, L69.
Li, F., Rappaport, S., Clark, G. W. & Jernigan, J. G., 1979, *Ap. J.*, **228**, 893.
Manchester, R. N. & Taylor, J. H., 1977, *Pulsars*, Freeman, San Francisco.
Maraschi, L. & Treves, A., 1979, *Nature*, **279**, 401.
Maraschi, L. & Treves, A., 1981, *MNRAS*, **194**, 1P.
Mason, K. O., 1977, *MNRAS*, **178**, 81P.
McGregor, K. & Vitello, P. A. J., 1979, *COSPAR X-ray Astronomy* (eds. W. A. Baity & L. E. Peterson), Oxford, Pergamon, p. 267.
Nagase, F., 1981, *Space Sci. Rev.*, **30**, 395.
Nagase, F. et al., 1981, *Nature*, **290**, 572.
Ostriker, J. P., Rees, M. J. & Silk, J., 1970, *Ap. Lett.*, **6**, 179.

Pandharipande, V. R., Pines, D. & Smith, R. A., 1976, *Ap. J.*, **208**, 550.
Petterson, J. A., 1978, *Ap. J.*, **224**, 625.
Pratt, J. P. & Strittmatter, P. A., 1976, *Ap. J.*, **204**, L29.
Pringle, J. E. & Rees, M. J., 1972, *Astron. Ap.*, **21**, 1.
Rappaport, S. & Joss, P. C., 1977, *Nature*, **266**, 683.
Ricketts, M. J., Hall, R., Page, C. G. & Pounds, K. A., 1981, *Space Sci. Rev.*, **30**, 399.
Savonije, G. J., 1978, *Astron. Ap.*, **62**, 317.
Savonije, G. J., 1979, *Astron. Ap.*, **71**, 352.
Savonije, G. J. & van den Heuvel, E. P. J., 1977, *Ap. J.*, **214**, L19.
Schreier, E. J., Swartz, K., Giacconi, R., Fabbiano, G. & Morin, J., 1976, *Ap. J.*, **204**, 539.
Schreier, E. J., 1977, *Proc. 8th Texas Symp. Rel. Astroph. (Ann. NY Ac. Sci.)*, **302**, p. 445.
Shapiro, S. L. & Lightman, A. P., 1976, *Ap. J.*, **204**, 555.
Shvartzman, V. F., 1971, *Soviet Astron. A. J.*, **15**, 342.
Snow, T. P., 1979, *Mass Loss and Evolution of O-type Stars* (eds. P. S. Conti & C. W. H. De Loore) Reidel, Dordrecht, p. 65.
van den Heuvel, E. P. J., 1977, *Proc. 8th Texas Symp. Rel. Astroph. (Ann. NY Ac. Sci.)*, **302**, p. 14.
van der Klis, M., Bonnet-Bidaud, J. M. & Robba, N. R., 1980, *Astron. Ap.*, **88**, 8.
van der Klis, M. *et al.*, 1982, *Astron. Ap.*, **106**, 339.
van der Linden, Th. J. & Henrichs, H. F., 1983, in preparation.
Vasyliunas, V. M., 1979, *Space Sci. Rev.*, **24**, 609.
Verbunt, F., 1982, *Space Sci. Rev.*, **32**, 159.
Wang, Y.-M., 1979, *Astron. Ap.*, **74**, 253.
Wang, Y.-M., 1981, *Astron. Ap.*, **102**, 36.
Wickramasinghe, D. T. & Whelan, J. A. J., 1975, *Nature*, **258**, 502.

Author index

Abell, 290-2
Abott, 351
Abramowicz, 330
Abt, 324, 325
Alme, 272
Andereck, 127, 129
Anderson, 299, 300
Anzer, 263-5
Apparao, 82
Arnett, 28, 312, 316, 332
Arons, 8, 12, 91, 261, 407, 417, 426
Arp, 124, 129
Avni, 2, 20, 21, 26, 352
Ayasli, 1, 94-9

Baade, 124, 129
Baan, 51
Bahcall, J., 20, 21, 24-6, 48, 117, 211, 308
Bahcall, N., 211
Bailey, 150
Bardeen, 379
Barton, 281
Basinska, 65, 75, 81
Basko, 6, 7
Bath, 178, 382, 384
Battistini, 124-5, 128, 129
Baym, 27, 28, 34
Bechtold, 122
Beck, 136
Becker, 2, 3, 6, 18, 19, 154
Begelman, 292, 298, 386
Belian, 50
Bell, 136
Bell-Burnell, 2, 3
Benson, 351
Benvenuti, 124
Berg, 169
Berkhuijsen, 124

Bernard, 127, 129
Bethe, 27
Bisnovatyi-Kogan, 7, 386
Blaauw, 303, 325
Blanford, 331
Blinnikov, 386
Blumenthal, 117
Bodenheimer, 325, 331
Boersma, 303
Bolton, 235
Bonazzola, 271, 272, 281
Bondi, 281, 351, 377
Bonnet-Bidaud, 418
Bouwman, 306
Bowers, 28
Bowyer, 1
Boynton, 212-14, 388
Bradt, 1, 53, 141, 192
Brecher, 82, 150
Burbidge, 367, 373
Burdyuzha, 265
Burnell, 50
Bussard, 8, 282
Bywater, 219

Calla, 53, 83, 84
Canal, 28, 336, 338
Canizares, 48, 66, 67, 201, 203
Cannizzo, 178
Canuto, 272, 404, 405
Cassinelli, 326
Castor, 8
Cavaliere, 51
Chanan, 371, 384
Chanmugam, 171
Chau, 357
Cherepashchuk, 244, 246, 293, 294
Chester, 24, 25

Author index

Chevalier, 19, 23, 212, 219, 229, 330, 394
Ciatti, 291
Clark, 47, 50, 57, 117, 125, 127, 129, 131, 141, 143, 229, 287, 294, 333
Colgate, 281
Collins, 299
Combes, 124
Cominsky, 2, 15, 32, 45, 57
Conner, 50
Conti, 21, 23, 311, 317, 324, 398
Cooper, 281
Corbet, 14, 18, 29
Córdova, 45, Chap. 4, 147, 154, 167, 176
Coroniti, 382
Counselman, 329
Cowie, 100
Cowley, 16, 19, 21, 23, 26, 30, 209, 293
Cowling, 349
Cox, 281, 363
Crampton, 16, 19, 21, 23, 26, 208, 209, 214, 293, 306, 308
Crawford, 147
Crosa, 214, 388
Cunningham, 378
Czerny, 98

Damahsek, 307, 308, 331
Darbro, 12
Darwin, 329
Davidsen, 1, 222
Davidson, 6, 7, 12, 271, 298, 351, 368, 369, 377, 399, 402, 407
Davies, 13, 397, 399, 401, 420–6
Davis, 28
Davison, 2, 18, 19
Day, 334, 363
DeCampli, 298
de Cuyper, 316, 323
Deeter, 13, 18, 19, 388
de Freitas Pacheco, 246
de Greve, 319, 327, 329, 353
de Jager, 326
Delgado, 312, 331
de Loore, 317, 319, 324, 329, 331, 353, 355
Dennis, 278
de Vaucouleurs, 137, 140
Dewitt, 281
D'Odorico, 124
Dopita, 124
Doty, 48, 51, 64, 97, 418
Downes, 295, 299
Doxsey, 1, 53, 141, 192
Dube, 143
Duerbeck, 219
Dulk, 171
Dupree, 403
Duthie, 169

Eardley, 382, 385, 386, 388
Eddington, 331, 351
Eggleton, 350, 358
Elsner, 9, 278, 407, 410, 420, 422, 426
Elvis, 135
Emerson, 122, 124
Engel, 268
Epstein, 18, 123
Ergma, 98, 281
Evans, A. J., 268
Evans, W. D., 50

Fabbiano, 8, 12, 18, 19, 143, 163, 166, 167
Fabian, 13, 48, 288, 298, 333, 420, 422, 423
Fahlman, 2, 306
Faulkner, 357
Feast, 1
Fechner, 27, 28, 35
Feigelson, 140, 143
Fenimore, 280
Ferguson, 165
Finzi, 336
Flannery, 322
Ford, 130
Foreman, 19
Fowler, 26
Fox, 248
Frank, 6, 7, 165, 271
Frechner, 18
Freedman, 27
Fryxell, 280, 316
Fujimoto, 95

Gabelle, 136
Galeev, 382
Gallagher, 132
Galloway, 6, 273
Garmany, 311, 324, 351
Garmire, 45
Geldzahler, 295
Gerend, 372, 384, 388
Ghosh, 8, 9, 12, 13, 160, 178, 378, 408–16, 420, 422
Giacconi, 2, 8, 12, 118, 418
Giannone, 89
Giess, 235
Gilmore, 296
Giora, 389
Giuli, 363
Glass, 1, 83
Gnedin, Y. N., 7, 271, 272
Gold, 263
Goldman, 98
Goldreich, 261, 263
Gottlieb, 207
Grabelsky, 119, 120
Graboske, 281
Grandi, 300

Author index

Gregory, 2, 306
Griffiths, 219
Grindlay, 45, 48-51, 66, 67, 82, 100, 117, 124, 125, 128, 190, 191
Gruber, 266, 268
Guetter, 394
Gunn, 261, 263, 295, 299
Gursky, 8, 46, 48, 51, 100, 333

Habets, 313
Hadjidemetriou, 327
Hammerschlag, 246
Hammerschlag-Hensberge, 16
Hameury, 281
Hanawa, 95
Hanes, 334
Hansen, 89
Härm, 89
Harris, 125
Harwitt, 281
Hatchett, 298, 368, 387, 389
Heggie, 334, 336
Heise, 50, 232, 317
Helfand, 119, 120, 121
Henrichs, 13, 246, Chap. 11, 398, 403, 413, 414, 416
Henriksen, 51
Herold, 265, 272
Hewish, 261
Heyvaerts, 271, 272, 281
Hills, 48, 334, 336, 363
Hjellming, 296, 297
Hodge, 120
Hoffman, 48, 50, 51, 52, 57, 64, 74, 75, 84-8, 97, 102, 418
Holloway, 422
Holt, 53, 57, 220
Hoshi, 91
Howard, 281, 314
Hoyle, 351, 397
Huang, 327
Huckle, 2
Hunt, 397
Hut, 48, 294, 298, 329, 330, 335
Hutchings, 16, 19, 21, 23, 26, 171, 209, 214

Iben, 28, 314
Ichimaru, 281, 382
Icke, 390
Illarionov, 399, 407, 419, 421, 422
Ilovaisky, 19, 23, 212, 219, 224, 229
Inoue, 6, 7, 75, 77
Isern, 336, 338
Itoh, 281
Ives, 2, 3

Jameson, 168
Jaroszynski, 98

Jedrezejec, 346
Jernigan, 1, 53, 141, 192
Jessen, 246
Johnson, 27, 83
Johnston, K. I., 296, 297
Johnston, M., 3
Jones, 83, 127, 129
Joss, Chap. 1, 1, 3, 9, 15, 17, 18, 19, 21, 22, 26, 27, 28, 29, 30, 31, 35, Chap. 2, 46, 51, 89-106, 117, 281, 303, 329, 355, 357, 406
Julian, 261, 263

Kaluzienski, 53, 57, 220
Kanno, 7
Katz, 46, 163, 292, 298, 299, 330, 333, 382
Kelley, 1, 2, 12, 14, 18, 23, 29, 32, 330, 394
Kemp, 249, 293
Khruzina, 246
King, 165, 168
Kiplinger, 162, 165
Kippenhahn, 309, 310, 317, 322, 323, 345, 351, 362
Kirk, J. G., Chap. 6, 272, 273, 282
Kirk, J. R., 6
Klare, 196
Klebesadel, 278
Klein, 8
Koo, 214
Kopal, 28, 345, 349
Kraft, 147
Kraicheva, 325
Kron, 214
Kruszewski, 172, 345, 349
Krzeminski, 172
Ku, 144
Kuiper, 323
Kulkarni, 53, 82, 83
Kundt, 422

Labay, 336, 338
Lacy, 136
Lahav, 293
Lamb, D. Q., 100
Lamb, F. K., 2, 6, 8, 9, 12, 13, 51, 100, 263, 264, 278, 282, 378, 402, 404, 406-16, 419, 420, 422, 426
Lamb, R. C., 2
Lamb, S. A., 314
Lamers, 22, 24, 326, 327, 351, 352, 426
Landau, 30, 345
Lang, 388
Langer, 7, 8, 271
Larson, 137, 140, 322
La Sala, 246
Lauterborn, 357

Lawrence, 83, 84
Lea, 8, 12, 91, 261, 407, 417, 420, 426
Lecar, 21, 303, 316
Leibowitz, 293
Levy, 324, 325
Lewin, 2, Chap. 2, 45, 48, 50, 51, 52, 56, 57, 64-6, 70, 72, 74, 75, 84, 97, 102, 117, 217, 229, 231, 281, 303, 418
Li, 13, 18, 23, 44, 45, 69, 91, 106, 420, 422
Liang, 51, 386, 390
Liebert, 148, 287, 294
Lifshitz, 3C, 345
Lightman, 12, 49, 334, 351, 382, 385, 386, 388, 399
Liller, 50, 67, 207
Limber, 349
Lloyd, 219
Lockhart, 296
Lodenquai, 272
Long, Chap. 3, 119-21, 131, 132, 137, 336, 337
Lubow, 319, 322, 325, 349, 368
Lucke, 2
Lucy, 324, 325, 351
Lyne, 333
Lyttleton, 397

Madore, 324
Maeder, 326, 332
Malina, 1
Malone, 27
Mammano, 290, 291
Manchester, 307
Mantegazza, 325
Maraschi, 51, 304, 421
Margon, 118, 205, Chap. 7, 287-92, 295, 297-300, 371, 384
Markert, 127, 129, 132
Marshall, F., 34, 154
Marshall, H., 52, 74, 75, 77-80, 84, 102
Marshall, N., 222
Martin, 294, 298
Masereel, 326, 331, 332
Mason, 2, 10, 14, 18, 28, 71, Chap. 4, 154, 355, 406
Massevitch, 311, 312, 314, 317, 324
Massey, 311, 327
Matese, 292, 293
Matilsky, 219, 246
Matsuoka, 195, 219
Mattei, 150
Mauder, 219
Mayo, 171
Mazeh, 293
Mazets, 34, 262
Mazurek, 312

McClintock, 2, 18, 30, 43, 66, 67, 71, 192, 201, 203
McCray, 271, 298, 352
McCulloch, 26
McGee, 119
McGregor, 416
McKee, 21, 272, 308, 316
McLean, 294
McLerran, 27
Melrose, 272
Merritt, 382
Mestel, 350
Mészáros, 7, 271, 272
Meyer, 178, 311, 332
Meyer-Hofmeister, 178, 311, 322, 332, 351
Michalsky, 294
Middelkoop, 350
Middleditch, 1, 19, 23, 24, 25, 176, 382
Milgrom, 46, 105, 190, 217, 243, 244, 288, 298, 364
Millit, 220
Milton, 119
Mitchell, 121
Miyaji, 95, 312, 336
Miyamoto, 195
Morrison, 82, 150
Morton, 129
Motch, 171
Murakami, 62, 65, 95
Murdin, 287, 291, 294

Nagase, 9, 13, 30, 262, 264, 265, 401
Nagel, 7, 271-3, 275
Nariai, 323
Nather, 147, 175
Neckel, 196
Nelson, 19, 23-5, 160, 382, 384
Newman, 281
Newsom, 299
Niell, 296
Noble, 219
Noels, 326, 331, 332
Nomoto, 28, 312, 315, 336, 338
Novikov, 367, 378, 383

Oda, 81, 83, 95
Odenwald, 49
Ögelman, 12
Ohashi, 65
Oke, 163, 214
Olbert, 380
Olson, 278
Oort, 136
Oppenheimer, 28
Ostriker, 12, 100, 117, 205, 261, 263, 282, 294, 308, 309, 330, 331, 351, 368, 369, 397, 399, 402, 407

Author index

Pacharintanakul, 163
Pacini, 261
Paczolczyk, 388, 390
Paczyński, 178, 308, 309, 311, 317, 332, 345, 346, 348, 357, 359, 361
Pakull, 120
Palumbo, 135, 142
Pandharipande, 26, 404, 415
Panek, 169
Panov, 272
Papaloizou, 30, 172, 327, 349
Parsignault, 45
Pauls, 295
Pavlov, 6, 7
Pavlov-Verevkin, 265
Payne-Gaposchkin, 148
Pedersen, 68, 70, 117, 221, 222, 372, 384
Peimbert, 135
Pentson, 219
Peterson, B. A., 224
Peterson, C. J., 47
Pethick, 6-9, 13, 27, 28, 34, 263, 402, 404, 406-8
Petre, 13, 14, 18, 29
Petro, 229-31
Petterson, 21, 214, 287, 294, 298, 330, 349, 352, Chap. 10, 369, 372, 379, 382, 387-9, 401, 410
Picklum, 95, 98
Pines, 6-9, 13, 263, 402, 404, 406, 407, 419
Piran, 330, 389
Plavec, 20, 319, 345
Pounds, 19, 84
Pratt, 349, 404
Pravdo, 2, 3, 6, 8, 18, 19, 23, 268, 270
Prendergast, 367, 373, 375
Press, 48, 49, 127, 334
Preston, 296
Prialnik, 177
Price, 390
Primini, 15, 18, 19
Pringle, 6, 8, 13, 30, 48, 159, 161, 172, 178, 349, 355, 367, 378, 382, 384, 388, 389, 397, 399, 400, 403, 406, 420, 421-6
Puget, 271, 272
Pye, 18, 19

Quinby, 268

Rabi, 269
Racine, 125
Rappaport, Chap. 1, 1-3, 7-9, 13-20, 23, 26, 29, 30-2, 45, 46, 51, 84, 106, 241, 271, 303, 305, 318, 329, 334 406

Raymond, 168
Rees, 6, 8, 48, 282, 288, 298, 351, 355, 367, 388, 403, 406
Refsdal, 363
Remie, 22
Ricco, 325
Ricker, 2
Ricketts, 32, 33, 422
Rieke, 136
Ritter, 309, 332
Robba, 418
Robertson, 219
Robinson, 148, 175, 371
Rose, 248
Rosenberg, 2
Rosner, 382
Rossi, Foreword, 380
Rubin, 130
Ruder, 265, 272
Ruderman, 272
Russell, 28
Ryle, 287, 295

Sadun, 82, 150
Salomon, 351
Salpeter, 7, 271-3, 281
Salter, 295
Sandage, 135, 140
Sanduleak, 287
Sanford, 14, 18, 23, 29
Sarazin, 298, 387
Sargent, 124-6, 128, 129
Sato, 83
Savonije, 159, 329, 330, 334, Chap. 9, 346, 349, 353, 355, 356, 360, 363, 378, 407, 414-16, 420
Schaefer, 282
Schatzman, 177, 336
Schiffer, 169
Schilizzi, 296
Schramm, 28
Schreier, 8, 12-14, 18, 19, 407, 418
Schvartsman, 281
Schwarzenberg-Czerny, 178
Schwarzschild, 29, 89
Seaquist, 136, 287
Seward, 121, 294
Shaham, 13, 246, 419
Shakura, 165, 330, 367, 386, 388
Shapiro, 12, 271, 334, 385, 386, 399
Shara, 177
Share, 280
Shaviv, 177, 389
Sherrington, 168
Shibanov, 7, 272
Shklovskii, 293
Shu, 319, 322, 345, 369
Shvartzman, 421

Sienkiewicz, 311, 357, 359
Silk, 282
Skinner, 2, 118, 141
Skumanich, 357
Slettebak, 32
Smak, 147, 179
Smarr, 331
Smith, L. F., 317
Smith, M. A., 357
Smith, R. A., 27
Smith, W. B., 229
Sorvari, 222
Spada, 45
Spencer, 296
Spitzer, 271
Srinivasan, 306, 331
Stark, 100
Steiner, 245, 246
Stephenson, 287
Sterken, 172
Sterne, 30
Strittmatter, 349, 403
Stoneham, 282
Stothers, 19, 22, 28, 30, 31
Strong, 278
Sugimoto, 312, 336, 338
Sutantyo, 306, 319, 323, 329, 332, 333, 335, 363
Sunyaev, 6, 7, 165, 271–3, 276, 330, 367, 368, 386, 388, 399, 407, 419, 421, 422
Svestka, 51
Swank, 50, 51, 53, 57, 64, 97, 154, 161, 219
Szkody, 168

Taam, 51, 95, 98, 331, 335, 337
Takagishi, 219
Tammann, 135, 140
Tananbaum, 2, 13, 18
Tapia, 169, 294
Tarter, 176
Taylor, J. H., 4, 23, 26, 34, 306, 307, 386
Teukolsky, 48, 127, 334
Thomas, 308, 309, 312, 323
Thompson, 390
Thorne, 351, 367, 378, 382
Thorstensen, 121
Tinsley, 137, 140
Torres-Peimbert, 135
Totsuji, 281
Treves, 234, 304, 421
Trimble, 248, 325
Trümper, 6, Chap. 6, 262, 265, 282
Truran, 167
Tsuruta, 7
Tucker, 117
Tueller, J., 268

Tutukov, 98, 281, 309, 311, 312, 314, 317
Tylenda, 159, 165

Ubertini, 268
Ulmer, 3, 77
Ulrich, 136, 322, 351
Underhill, 22

Vader, 338
Vaiana, 382
van Albada, 325
van Beveren, 311, 314, 317, 353
van den Bergh, 143
van den Heuvel, 13, 16, 21, 23, 234, 248, 294, 298, Chap. 8, 303–6, 308, 309, 311, 317, 318, 326, 330–3, 335, 337, 368, 388, 420
van der Klis, 330, 394, 418
van der Linden, 24, 319, 414
van der Woerd, 334
van Horn, 89
van Paradijs, 16, 51, 63, 65, 70, 143, Chap. 5, 217, 247, 372
van Speybroeck, Chap. 3, 122, 123, 125, 126, 128, 130, 131, 135, 336, 337
Vasyliunas, 402, 407, 409
Ventura, 7, 271–3, 281, 409
Verbunt, 48, 335, 350, 358, 410
Vitello, 416
Voges, 266, 268, 278
Vogt, 172
Volkoff, 28

Wade, 163
Walker, 132
Wallace, 103, 280, 281
Walter, 219
Wang, 6, 7, 12, 13, 271, 399–401, 403, 422, 423
Warner, 147, 148, 150, 161, 179, 371
Warren, 219
Warwick, 14, 18, 29
Wassermann, 271–3
Watson, 14, 18, 19, 29, 118
Weaver, 103, 281, 314
Webbink, 106, 177, 310, 311, 334, 335, 346, 357, 360, 363
Weber, 248
Weigert, 89, 310, 363
Weisberg, 306
Wheaton, 6, 268, 269
Wheeler, 21, 51, 178, 303, 312, 316
Whelan, 163, 165, 384, 407
White, 2, 3, 6, 14, 18, 23, 29, 34, 50, 80, 268
Whitmire, 293, 294
Whyte, 151
Wickes, 143

Wickramasinghe, 163, 165, 407
Wilkinson, 143
Williams, 165, 177
Wilson, 248, 272, 281
Wolf, 48
Woltjer, 261
Woosley, 51, 103, 280, 281, 314
Wright, 207
Wunner, 265, 272

Yahel, 7, 8, 271
Yakovlev, 6, 273

Yungelson, 311, 312, 314, 317

Zahn, 21, 30, 329, 349, 350
Zamorani, 140, 143
Zel'dovich, 271
Zheleznyakov, 272
Zilitinkevitch, 328
Ziolkowski, 13, 328, 348, 353, 355
Zuiderwijk, 16, 247
Zwaan, 350, 358
Żytkow, 348

Object index

Sources are listed in order of increasing right ascension and declination. For a listing of special source names, *see* separate list.

4U 0015+63, 9, 13, 16, 18
1E 002151−7221.5 (in 47 TUC), 42
4U 0026−73, 42
0042+327, 198, 208
0050−727, 232
0053−739, 232
0114+634, 232
0114+65, 43
4U 0115+63, 2, 3, 15, 17, 32, 43, 236, 395, 417, 422
0115−737, 232
E 0139−681, 152, 169
0142+61, 43, 198
0236+61, 232
2A 0311−227, 152, 157, 174
4U 0352+30, 2, 3, 232, 394, 395, 413
2A/2S/4U/MX/MXB 0512−40 (in NGC 1851), 42, 53, 54, 56, 65, 66
0521−720, 198
2A 0526−328, 152, 156, 171
0532−664, 232
A 0535+26, 2, 3, 12, 13, 43, 232, 264, 395, 401, 417, 420, 422
0538−641, 232
A 0538−66, 1, 141, 232
0540−697, 232
0544−665, 232
XB 06??+??, 54
4U 0614+09, 43, 54, 198, 208
A 0620−00, 43, 67, 68, 198
4U 0621+11, 54
MXB 0???+??, 54
1E 0643−1648, 152, 157
0726−260, 232
4U 0900−40 (*see* Vela X−1)
0918−549, 198
4U/2S 0921−63, 68, 69, 193, 197, 198, 208

E 1013−477, 152, 157, 169
A 1118−61, 2, 3, 43, 395, 419
4U 1145−61, 2, 43, 395, 396
4U 1223−62 (*see* GX 301−2)
4U 1230−61, 2, 395
A 1239−59, 2
2S 1254−690, 43, 71, 198
MXB 14??−6?, 54, 71
E 1405−451, 152, 157, 169, 174
2S 1417−62, 54, 394, 395
1455−31 (*see* Cen X−4)
A 1524−61, 43, 198, 225
4U 1535−29?, 54
4U/A 1538−52, 2, 3, 17-19, 21, 23, 25, 43, 249, 323, 328-30, 355, 395
1543−63, 43, 198
PG 1550+191, 152, 169
E 1551+718, 157
2S 1553−54, 1
1556−60, 43, 198
XB/4U 1608−52, 43, 52, 54, 56, 62, 63, 65, 68, 95, 198
4U 1627−67, 1, 2, 3, 9, 43, 45, 67-71, 105, 190, 192, 193, 198, 203, 208, 222, 224, 237, 270, 306, 394-6, 412, 416
4U/MXB 1636−53, 43, 45, 53, 56, 57, 59, 62-70, 190, 198, 200, 203-5, 207, 208, 217, 220-3
OAO 1653−40, 395
1656+354 (*see* Her X−1)
4U 1658−48 (*see* GX 339−4)
MXB 1659−29, 43, 45, 54, 56, 62, 63, 66-8, 199, 200, 204, 217
1659−487 (*see* GX 339−4)
4U 1700−37, 16, 26, 43, 242, 245, 246, 249, 328-30, 355
4U/2S/XB 1702−42 (*see* GX 349+2)
1705−25, 43, 199

Object index *439*

4U 1708−23, 87
NXB 1715−32, 54
XB 1716−31, 88
4U 1722−30, 42, 64
1E/XB 1724−30 (in Terzan 2), 42, 54, 64, 66
4U 1728−16 (*see* GX 9+9)
4U 1728−24 (*see* GX 1+4)
MXB 1728−34, 42, 45, 50, 54, 56, 57, 58, 64, 65, 73, 77, 81, 86, 87
MXB 1730−335 (*see* Rapid Burster)
XB 1732−30 (in Terzan 1 ?), 54
4U/MXB 1735−44, 43, 45, 52−4, 56, 62, 63, 65−9, 72, 190, 199, 200, 204−8, 217, 220, 226, 384
A 1742−28, 67
MXB 1742−29, 54, 56
MXB 1743−28, 55, 63
MXB 1743−29, 55, 56
4U/XB/2S 1744−26, 55
XB 1745−24, 55, 56, 63
MX 1746−20, 42
4U/MXB/2S 1746−37 (in NGC 6441), 42, 54
1755−33, 43, 199
XB 1802−22, 54
3U 1809+50, 169
4U/XB 1813−14 (*see* GX 17+2)
1820−000, 199
4U/MXB/2S 1820−30 (in NGC 6624), 42, 45, 48, 50, 54, 56, 57, 100
2A/4U 1822−37, 68, 152, 157, 190, 193, 199, 208
XB 1832−22, 54
MXB 1837+05 (=4U 1837+04 =Ser X−1) (*see* Ser X−1)
4U 1849−31 (=V1223Sgr), 171, 174
A/4U/MXB 1850−08 (in NGC 6712), 42, 49, 55
4U 1857+01, 55
4U/2S 1905+00, 55
MXB 1906+00, 55, 56
1907+09, 43
1908−00 (*see* Aql X−1)
4U/MXB 1916−05, 45, 55, 56, 68, 104, 190, 193, 199, 208, 217, 219
1921−63, 43
1957+11, 43, 199
XB 2???+3?, 55
2030+401, 208
4U 2058+32, 55
4U 2104+31, 55
4U 2120+32, 55
4U 2129−05, 68
4U 2129+47, 43, 121, 144, 152, 157, 190, 193, 199, 201, 208, 214−17
4U 2131+11, 42

2142+380 (*see* Cyg X−2)
1E 2259+586, 1, 306, 330

List of sources by name

AB Dra, 156
AC Cnc, 152, 156
AE Aqr, 150, 152, 156, 168, 171, 173, 174
AH Eri, 156
AH Her, 156, 175, 176
AM CVn, 151, 152, 156, 174
AM Her, 149, 152, 154, 155, 156, 158, 160, 166, 167, 168, 169, 170, 171, 173, 174
AN UMa, 152, 155, 157, 169, 170
Aql X−1 (=MXB 1908+00), 43, 54, 55, 62, 67, 68, 71, 199, 200, 208, 220, 221, 226, 228, 356
AY Lyr, 152, 157

Beta Lyrae, 319
BD Pav, 152
BP Cru, 233
Br Cir, 233
BT Mon, 152
BV Cen, 150, 152, 156

Cen X−1, 117
Cen X−3, 2, 3, 8, 13, 17−19, 23−5, 43, 117, 233−5, 240, 241, 245, 247, 248, 304, 322, 323, 329, 330, 352, 355, 394, 395, 405, 412, 413, 416, 418
Cen X−4 (=1455−31), 43, 53, 54, 56, 57, 62, 66, 67, 68, 198, 200, 208, 220, 221, 226, 228, 356
Cir X−1, 43, 141, 307, 331
CK Vul, 157
CN Ori, 157, 174
CP Lac, 157
CP Pup, 157
CQ Cep, 317, 323
Crab, 34, 62, 87, 419
CV Ser, 317
CX Cephei, 323
Cyg X−1, 43, 84, 117, 229, 233, 235, 240, 241, 242, 248, 249, 308, 385, 386
Cyg X−2 (=2142+38), 16, 26, 43, 68, 69, 192, 193, 199, 205, 208-11, 217, 306, 332
Cyg X−3 (=2030+47), 141, 199, 208, 306, 307, 331

30 Doradus, 119, 121
DQ Her, 150, 152, 154, 156, 168, 171, 173, 174, 175, 177, 371, 383

EF Eri, 152, 154, 155, 168, 169, 170

EH TrA, 152
EM Cyg, 152, 156, 172, 174
Eta Carinae, 121
EX Hya, 152, 154, 155, 156, 384, 164, 165, 172

FH Ser, 177

γ^2 Vel, 317
Gamma Cas, 43, 232
GK Per, 52, 149, 151, 152, 157, 174
GP Com (=G 61−29), 151, 152, 156
GP Vel, 233
Grindlay-Hertz 1, 42, 54
GX 1+4 (=4U 1728−24), 1, 2, 3, 43, 190, 192, 394, 395, 413, 416, 418
GX 3+1 (=4U 1744−26), 56, 190
GX 9+9 (=4U 1728−16), 43, 199
GX 17+2 (=4U 1813−14), 43, 54, 190, 208
GX 301−2 (=4U 1223−62), 2, 3, 12, 13, 17, 18, 29, 30, 43, 84, 233, 236, 249, 319, 328, 354, 355, 395, 401, 413, 417
GX 304−1 (=4U 1258−61), 2, 3, 43, 84, 233, 395
GX 339−4 (=4U 1658−48), 43, 199, 229
GX 349+2 (=4U/2S/XB 1702−36), 54, 208

H 1705−25, 219, 228
H 2215−086, 152, 157, 174
H 2252−035, 34, 152, 157, 171, 174
HD 63099, 317
HD 77581) 22, 30, 31, 233, 244, 245, 403, 417 (see also Vela X−1)
HD 90657, 317
HD 94546, 317
HD 97152, 317
HD 152270, 317
HD 153919, 233, 246, 317
HD 186943, 317
HD 190198, 317
HD 226868, 233 (see also Cyg X−1)
HD 311884, 317
HDE 245770, 232
HDE 320102, 317
He 3−640, 233
Her X−1 (=1656+354), 2, 3, 8, 13, 17-19, 23-5, 43, 198, 201, 203, 208, 210−15, 217, 220, 222-5, 243, 249, 261, 265-9, 276, 306, 329, 332, 333, 371, 382, 384, 294, 395, 406, 418
HR Del, 152, 156
HT Cas, 152, 154, 156, 174
HZ Her, 23, 198, 210-16, 224, 243, 244, 416 (see also Her X−1)

Kron 3, 42
Krz-star, 22, 233

KT Per, 157, 174
KY Tra, 198
KZ Tra, 198

Lanning 10, 152, 157
Liller I, 42, 44, 50, 54
LL Lyr, 157
LMC, 118-121, 138
LMC trans, 43
LMC X−1, 43, 119, 232
LMC X−2, 43, 119, 198
LMC X−3, 43, 119, 232
LMC X−4, 2, 16-19, 23, 25, 43, 84, 119, 141, 220, 232, 240, 241, 243, 245, 249, 322, 323, 329, 330, 355, 388, 395
LS 61 303, 421
LSI+65 010, 232
LX Ser, 150, 152, 157
M 31, 118, 121, 333, 336-8, 364
M 32, 124
M 33, 131-3, 141
M 81, 135
M 82, 141
M 86, 140
M 87, 142
M 100, 135
M 101, 132-5
MM Ser, 199
MMV, 233
MV Lyr, 149, 152, 157

N 253, 135
NP 0532 (Crab pulsar), 34
NGC 205, 134
NGC 224, 138
NGC 247, 138
NGC 253, 138
NGC 520, 138
NGC 524, 138
NGC 598, 138
NGC 628, 138
NGC 720, 138
NGC 936, 138
NGC 1073, 138
NGC 1097, 138
NGC 1300, 138
NGC 1313, 138
NGC 1316, 138
NGC 1332, 138
NGC 1350, 138
NGC 1380, 138
NGC 1398, 138
NGC 1533, 138
NGC 1559, 138
NGC 1574, 138
NGC 1851, 42, 44, 54, 56

Object index

NGC 2366, 138
NGC 2403, 134, 138
NGC 2763, 138
NGC 2775, 138
NGC 2835, 138
NGC 2841, 138
NGC 2848, 138
NGC 2859, 138
NGC 2903, 138
NGC 3031 (=M 81), 134, 138
NGC 3077, 138
NGC 3125, 140
NGC 3368, 138
NGC 3377, 138
NGC 3489, 139
NGC 3585, 139
NGC 3593, 139
NGC 3628, 139
NGC 3818, 139
NGC 3923, 139
NGC 3991, 141
NGC 3994, 141
NGC 3995, 141
NGC 4236, 139
NGC 4244, 139
NGC 4251, 139
NGC 4382, 139
NGC 4449, 139
NGC 4459, 139
NGC 4472, 140
NGC 4579, 139
NGC 4594, 139
NGC 4636, 139
NGC 4638, 139
NGC 4643, 139
NGC 4649, 139
NGC 4697, 139
NGC 4753, 139
NGC 4826, 139
NGC 5068, 139
NGC 5078, 139
NGC 5101, 139
NGC 5102, 139
NGC 5236 (=M 83), 134, 139
NGC 5253, 139
NGC 5457 (=M 101), 134, 139
NGC 5532, 139
NGC 5566, 139
NGC 5866, 139
NGC 5898, 139
NGC 5907, 139
NGC 6440, 42, 44
NGC 6441, 42, 44, 54
NGC 6624, 42, 44, 48, 50, 54, 56, 100
NGC 6712, 42, 44, 49, 54
NGC 6744, 139, 141
NGC 7078, 42, 44

NGC 7793, 139
Nova Cygni 1975, 177
Nova Cygni 1978, 177
Nova Vul 1976, 177

Orion nebula, 121
OY Car, 152

Per OB 2, 394
Phi Per, 319
PSR 0656+64) 307
PSR 0820+02, 307
PSR 1913+16, 23, 26, 306, 307, 331

QV Nor, 233
QX Nor, 199

Rapid Burster (=MXB 1730−335), 45, 49, 50, 53, 54, 56, 72−84
RR Pic, 152, 157, 174, 175
PS Oph, 157
RS Oph, 177, 178
RU Peg, 152, 157, 174, 175
RW Sex, 152, 157, 174
RW Tri, 152, 164, 165, 169, 172
RX And, 152, 156, 174, 179

SMC, 118, 121
SMC X−1, 2, 3, 15, 19, 21, 22, 25, 43, 121, 141, 232, 235, 240-5, 247, 248, 303, 323, 329, 330, 352, 355, 356, 372, 394, 395, 412, 413, 416, 418
SMC X−2, 43, 232
SMC X−3, 43, 232
Ser X−1 (=MXB 1837+05 =4U 1837+04), 43, 45, 54, 56, 62, 63, 69, 70, 199, 200, 220
Sco X−1 (=1617−155), 43, 52, 68, 190, 193, 195, 198, 204-9, 217, 220, 226, 229-31
Sk 160, 22, 232, 243, 247
Sk Ph, 232
SS 433, 233, Chap. 7, 306, 308, 330, 331, 382, 388
SS Aur, 152
SS Cyg, 150, 152, 155, 156, 158-60, 162, 164-8, 174, 175, 176
SU UMa, 149, 151, 157, 172
SV CMi, 156
SY Cnc, 156, 174, 176

T Aur, 152, 154, 156
T CrB, 156, 177, 178
Terzan 1, 42, 54
Terzan 2, 42, 54, 64
Terzan 5, 42, 54, 56, 63

Object index

Theta Mus, 317
T Pyx, 157
TT Ari, 149, 152, 156, 173, 174
47 TUC, 42
TU Men, 152
TW Vir, 157

U 1343 Aql, 233
U Cephei, 319
U Gem, 149, 150, 152, 154–6, 158, 163, 166, 167, 172, 174–6, 178, 179
U Sco, 157, 177
UX UMa, 152, 154, 157, 162, 164, 165, 169, 174, 175, 383

V436 Cen, 152, 156, 174, 175
V444 Cyg, 317
V446 Her, 156
V533 Her, 156, 171, 173, 174
V603 Aql, 152, 156
V616, Mon, 198
V635 Cas, 232
V691 CrA, 199
V725 Tau, 232
V779 Cen, 233
V794 Aql, 156
V801 Ara, 198
V818 Sco, 198
V821 Ara, 199
V822 Cen, 198
V841 Oph, 157
V845 Hyi, 157
V884 Sco, 233
V926 Sco, 199
V1017 Sgr, 157
V1055 Orl, 198
V1059 Sgr, 157
V1223 Sgr, 157
V1333 Aql, 199
V1341 Cyg, 199
V1343 Aql, 293
V1357 Cyg, 233
V1500 Cyg, 152, 156
V1521 Cyg, 199
V1668 Cyg, 152
V1727 Cyg, 199
V2051 Oph, 152
V2107 Oph, 199
V2116 Oph, 199
V3885 Sgr, 152, 157, 174, 175
Vela pulsar, 419
Vela X−1 (=4U 0900−40), 2, 3, 15, 17, 32, 43, 233, 236, 240-5, 262, 264, 278, 354, 355, 395, 417, 422 (*see also* HD 77581)
V Sge, 152
VV Pup, 152, 155, 157, 169, 170, 172
VW Hya, 384
VW Hyi, 152, 156, 164, 174, 175
VZ Scl, 152

W50, 287, 295, 296
WRA 977, 233, 249, 417
WW Cet, 152, 156
WX Hyi, 152, 157
WZ Sge, 149, 152, 154, 157, 174, 175, 178, 294

X Leo, 157, 174
X Per, 43, 232, 394, 396

YZ Cnc, 152, 156, 172, 174

Z Cam, 149, 152, 156, 174
Z Cha, 152, 154, 156, 174, 175, 179

Subject index

absorption dips
  periodic, 104
absorption lines, 43, 66, 67, 147, 197, 201, 209, 214, 243, (*see also* cyclotron)
accretion, Chaps. 9 and 10
  angular momentum, 98, 100, 369, 370, 416, 417
  columns, 160, 169, 170, 173
  deceleration mechanisms, 6, 7, 271, 273
  disk (*see* accretion disk)
  geometry, 159, 169
  instabilities, 45, 51, 52, 75, 84, 88, 89, 100, 107, 220, 326, 328, 331; and pulsations, 176
  luminosity, 397
  mass transfer (*see* mass transfer)
  models (*see also* accretion disk, models): degenerate dwarfs, 159–61; disk (*see* accretion disk); dwarf novae, 178, 179; gamma-ray bursts, 280–2; globular cluster binaries, 335; strongly magnetised accreting neutron star, 271–8; type II X-ray bursts, 100
  radius, 369
  rate, 90, 91, 94, 101, 103, 104, 106, 159–62, 165, 178, 192–3, 214, 246, 280, 281, 328, 334, 338, 346–64, 389, 396, 397, 413–27, 422, 427; during Roche lobe overflow, 399; in stellar wind, 397
  torque, 394, 396, 404–419, 422 (*see also* spinup and spindown)
accretion disk, Chap. 10
  angular momentum, 370–2, 416
  bright spot, 166, 172, 173, 178, 202
  bulge, 217, 219
  eclipses, 172, 190
  emptying time, 150
  formation, 403, 404
  instabilities, 388
  magnetic coupling, 408–13, 417
  models: cataclysmic variables, 159–69, 178, 179; reprocessing, 222, 231; Shakura–Sunyaev, 383; stationary axisymmetric, 380–3; structure, 373–80; time dependencies, 388, 389; twists, 386–8
  observational evidence, 201–3, 371, 372
  optical emission, 66, 70, 71, 161–5, 190
  precession of, 214, 220, 249, 370, 372, 385
  properties, 203–7
  radiation, 384, 385
  tilting, 214, 217, 249
  twists (*see* models, above)
  viscosity (*see* viscosity)
  X-ray heating, 69–71, 202–207
AM Her-type objects, 155, 158, 169–71 (*see also* cataclysmic variables)
angular momentum, 323, 329, 344, 349, 368, 378, 398–400, 404, 409, 419, 424
  capture: Roche lobe overflow, 401; stellar wind, 399, 417
  loss, 263, 264, 309, 323, 327, 345, 349, 351, 355, 357, 378, 422
ANS, 81
apsidal motion
  test in X-ray binaries, 28–32
Ariel 5, 33, 50, 81, 83, 84

B emission stars (*see* massive X-ray binaries)

Subject index

blackbody
  cooling, 51, 52
  model for soft X-rays, 158, 160, 161
  radii, 64–6, 82, 200; variation of, 101
  spectra, 65, 97, 166, 167; deviations from, 98, 99
  temperature: disk, 221, 225, 226; neutron star, 90, 92, 93, 97, 98, 205
black holes, 48, 51, 100, 117, 313, 333, 343, 372, 378
  absence of, 331
  Cyg X–1, 248, 303, 308
  formation, 332
  galactic bulge sources, 48, 51, 106
  globular cluster sources, 51, 117
  M 81, 135
bulge sources (*see* low-mass X-ray binaries)
bursts (*see* burst sources)
burst sources (*see* gamma-ray bursts and X-ray burst sources)

Cataclysmic variables, Chap. 4
  accretion disks (*see* accretion disks)
  AM Her-type objects, 155, 158, 169–71
  binary systems, 105, 147, 371
  bright spot, 148, 165, 202
  characteristics, 147
  classical novae, 176, 338
  classification, 147
  disk emission, 161–65
  dwarf novae, 178, 179
  dynamical properties, 150–3
  ellipsoidal variations, 172
  evolution, 151
  flickering, 173
  hydrogen burning, 167, 168, 176, 177 (*see also* nuclear burning and flashes)
  magnetic fields, 168–71
  orbital modulations, 171, 172
  orbital parameters, 152, 153
  orbital periods, 151-3
  pulsations, 172-5
  radial velocity curves (*see* radial velocity curves)
  rapid oscillations, 173, 174
  recurrent novae, 177, 178
  short time scale variability, 171
  standard model, 147, 148
  sub classes, 151
  thermonuclear runaway, 176, 177
  UV emission lines, 168
  X-ray emission: models, 147, 159–161; spectra, 154–61
Chandrasekhar limit, 105, 312-315, 332, 336
class I sources (*see* massive X-ray binaries)

class II sources (*see* low-mass X-ray binaries)
close-binary evolution (*see* evolution)
CNO cycle, 90, 95, 168, 177, 280, 317
color index (*see* luminosity)
companion stars, 1, 190, 210, 224, 225, 298, 304, 334, 345, 367, 368, 372, 396, 400, 419, 423
  Be type, 234, 304, 305
  capture, 48, 106, 333, 335, 336, 368, 369
  close binaries, 309–14
  co-rotation, 20, 235
  galactic bulge sources, 67, 69, 71, 104–6, 192, 306
  low-mass X-ray binaries, 67, 69, 71, 104–6, 192, 306
  mass and radii, 23, 214, 219, 220, 222, 241
  massive X-ray binaries, 21-5, 67, 189, 234, 304, 317
  model for mass transfer in binaries, 147, 148, 189, 192
  need for, 71, 104
  O or B type, 43, 324
  properties, 21–5
  Roche lobe (*see* Roche lobe)
  spectral types and masses, 67, 68, 210, 227, 229, 234, 246, 248, 304, 405, 325
  SS 433, 293
  thermonuclear flash model, 88
  transients, 105, 195, 356, 357
  unipolar inductor, 171
Copernicus, 81, 83
co-rotation (*see* companion stars)
critical surface (*see* Roche lobe)
cyclotron, Chap. 6
  characteristic line energy, 7, 269
  emission from inner disk, 229
  frequency, 276, 277
  gamma-ray bursts, 278–82
  resonances, 261, 266, 268, 278, 279, 282
  theory, 268–78

Doppler delays, 13–17, 32, 33, 236
dwarf novae, 178, 179 (*see also* cataclysmic variables)

eclipses
  optical, 196, 201, 217
  X-ray (*see* X-ray eclipses)
Eddington limit, 64, 65, 271, 272, 278, 330, 331
Einstein Observatory, Foreword, Intr., 1, 33, 47, 49, 81, 82, 118–21, 154, 205, 294, 295, 360, 364
  imaging instruments, 118–21

Subject index

electromagnetic braking, 261
ellipsoidal variations (*see* cataclysmic variables, low-mass X-ray binaries and massive X-ray binaries)
emission lines
  cyclotron (*see* cyclotron)
  optical, 41, 202–4, 209, 214, 215, 226, 228, 241, 242, 287–93, 319
  X-ray, 5, 261, 265 (*see also* cyclotron)
equations of state, 27, 98, 379, 383, 385, 386
evolution
  capture, 333–6
  close binaries, 308–16
  conservative, 309, 316–22
  formation rates of X-ray binaries, 308
  galactic bulge sources, 335–8
  globular cluster sources, 333–8
  gravitational radiation, 357, 358, 360–2
  helium stars, 312
  low-mass X-ray binaries, 335–8, 356–64
  magnetic braking, 358
  massive X-ray binaries, 316–333, 351–6
  non conservative, 326, 348–51
  pulsars, 330–3, 419–26
  spiral-in, 332
  stellar wind effects, 326, 327
  Wolf-Rayet stars, 317, 318, 323, 326, 327
  X-ray binaries, Chap. 8
EXOSAT, 34
extragalactic sources (*see* X-ray sources)

galactic bulge sources (*see* low-mass X-ray binaries)
galaxies
  survey, 136–41
  X-ray emission, Chap. 3 (*see also* X-ray sources)
gamma-ray bursts, 262, 278–82
  cyclotron lines (*see* cyclotron)
  gravitational red-shift, 262, 279
  observations, 278–80
  theory, 280–2
general relativistic corrections, 65, 97, 99, 378
globular cluster X-ray sources, 42, 43, 46–53, 68, 71, 103, 117, 125–130, 193, 194, 306, 356
  ages, 306
  black holes, 48, 51, 117
  capture, 47, 48, 333–5
  companion stars, 334 (*see also* companion stars, low-mass X-ray binaries)
  core density, 47
  formation, 47, 48, 106, 333–5
  low-mass X-ray binaries, 49, 51–3, 190
  luminosity, 124, 205, 356

M 31, 124–30
  mass determinations, 48, 49, 194
  table of galactic sources, 42
gravitational radiation, 357, 358, 360–2
gravitational redshift, 97, 99, 262, 279

Hakucho, 13, 33, 70, 75, 77, 81–3, 95, 220
HEAO-1, 33, 81, 82, 167, 268
heating (*see* X-ray heating)
helium flash (*see* nuclear burning and flashes)
Hertzsprung–Russell diagram, 21, 23, 24, 310, 355, 362
hydrogen flash (*see* nuclear burning and flashes)

infrared emission, 307 (*see also* Rapid Burster)
instabilities (*see* accretion, mass transfer, thermonuclear flash, tidal instabilities and radiation pressure)
interstellar extinction, 70, 196

jets (*see* SS 433)

Kolmogorov test, 125–7
Konus, 278–80

Lagrangian point, 147, 170, 214, 242, 323, 345, 351, 355, 368, 401, 404
life time X-ray, 353, 354 (*see also* time-scales)
low-energy cut-off (*see* X-ray spectra)
low-mass X-ray binaries
  absolute visual magnitude, 197–201
  accretion (*see* accretion)
  accretion disks (*see also* accretion disks); evidence for, 69–71, 201–3; properties, 203–7
  ages, 103, 306, 356
  angular momentum (*see* angular momentum)
  blackbody (*see* blackbody)
  black holes, 48, 106
  bursters, 44, 66–71, 192–4 (*see also* X-ray burst sources)
  catalogue, 198, 199: X-ray bursters, 54, 55
  characteristics, 41, 106, 343, 356
  companions, 67–9, 71, 72, 192 (*see also* comapanion stars)
  disks, evidence for, 69–71, 201
  eclipses, absence of, 45, 190
  ellipsoidal variations, 210, 212
  Eddington limit (*see* Eddington limit)
  formation and evolution, 106, 308, 332–8, 356–64

Subject index

evolution (*see* evolution)
galactic distribution, 53
globular clusters (*see* globular cluster X-ray sources)
gravitational radiation, 360-2
luminosity, 41-3, 356 (*see also* luminosity)
L_X/L_{opt} (*see* luminosity)
magnetic fields, 91, 103
mass determinations, 48, 49, 51, 67, 207-20
mass transfer, 356-64
neutron stars (*see* neutron stars)
optical: bursts, 69-71; emission lines (*see* emission lines); identifications, 43, 54, 55, 66, 191, 192, 198, 199; light curves, 207-20; properties, 66, 67, 190, 195-201; reprocessing (*see* optical reprocessing); variability, 56, 67, 69-71, 207-31
orbital periods, 67-9, 208-20
population (*see* populations)
properties, 41, 207
pulsations, absence of, 41, 45, 190
radial velocity curves (*see* radial velocity curves)
Roche lobe geometry (*see* Roche lobe)
soft transients, 56, 57, 190, 193, 194, 225-8
source types, 190
spectra (*see* X-ray spectra, below)
tidal interaction, 350
transients (*see* X-ray transients)
X-ray bursts (*see* X-ray burst sources)
X-ray heating, 69-71, 201-7 (*see also* X-ray heating)
X-ray spectra, 189, 190
luminosity
absolute visual magnitudes, 41, 197-201
alpha values for burst sources, 63
accretion, 104, 397
cataclysmic variables, Chap. 4 (*see also* cataclysmic variables)
color index, 138, 139, 196-9, 232, 233
disk, 161, 163, 167
distribution of X-ray sources, 131
extra-galactic sources, 118, 131
flash models, 90, 92, 94, 281
galactic bulge sources, 41, 42, 352
galaxies, 117, 132-9, 143
gamma-ray bursts (*see* gamma-ray bursts)
globular clusters, 54, 55, 124-30, 205, 337, 356 (*see also* globular cluster X-ray sources)
low-mass X-ray binaries, 41, 42, 352
L_X/L_{opt}, 42, 43, 69-71, 137, 143, 196, 197, 203, 205, 232, 233

massive X-ray binaries, 8-13, 189; companions, 21-5, 241 (*see also* companion stars)
maximum burst luminosity, 200
novae, 154 (*see also* Chap. 4)
period derivative, 263
persistent emission, 63, 92, 99-104, 192, 193; relation to burst intervals, 102

Magellanic clouds (*see* X-ray sources)
magnetic braking, 350, 358
magnetic field
cataclysmic variables, 169-71
cyclotron lines, 7, 269
low-mass X-ray binaries (*see* low-mass X-ray binaries)
neutron star (*see* neutron star)
thermonuclear flash (*see* thermonuclear flash)
X-ray burst sources (*see* X-ray burst sources)
magnetosphere, 396, 401-3, 408
magnitudes (*see* luminosity)
mass determinations (*see* neutron stars, low-mass X-ray binaries and massive X-ray binaries)
mass function, 17, 236, 239, 308
massive X-ray binaries
accretion (*see* accretion)
angular momentum (*see* angular momentum)
apsidal motion, 28-32
Be stars, 234, 236, 304, 315, 319
catalogue, 232, 233
characteristics, 343, 355
companion star, 21-25, 67, 189, 234, 304, 317, (*see also* companion star)
co-rotation (*see* co-rotation)
cyclotron lines (*see* cyclotron)
Doppler delays, 13-17, 32, 33, 236
eccentricities (*see* orbital parameters, below)
eclipses (*see* eclipses)
eclipse duration, 237
Eddington limit (*see* Eddington limit)
ellipsoidal variations, 238, 239, 244-9, 304
evolution (*see* evolution)
formation and evolution, 316-22, 351-6 (*see also* evolution)
iron emission lines, 5
L_{opt}/L_X, 236
mass determinations, 21-8, 236-9, 241
mass function, 17, 236, 239, 308
mass transfer, 351-6
neutron stars (*see* neutron stars)

Subject index

optical identifications, 232, 233
optical properties, 21-5, 232, 233, 237-49; light curves, 244-9; polarization, 237, 238; radial velocity curves, distortions, 241-4; reprocessing (*see* optical reprocessing); stellar wind (*see* stellar wind)
orbital parameters, 13-21, 235, 304
orbital periods, 13-20, 235, 304
populations (*see* populations)
pulsars, Chap. 1, 189, 355
pulsations (*see* pulsations)
radial velocity curves (*see* radial velocity curves)
radio pulsars, 306
Roche lobe geometry (*see* Roche lobe)
spinup-spindown of neutron stars (*see* spinup and spindown)
tidal forces (*see* tidal forces)
transients (*see* X-ray transients)
X-ray heating, 243 (*see also* X-ray heating)
X-ray spectra, 4-8, 265-8
mass ratio, 18, 206, 293, 310, 311, 322, 323, 324, 325-7, 333
mass transfer, Chaps. 8 and 9 (*see also* accretion, angular momentum and Roche lobe)
instabilities, 45, 51, 52, 75, 84, 88, 89, 100, 107, 220, 326, 328, 331
low-mass X-ray binaries, 356-64
massive X-ray binaries, 351-6
modes, 349
orbital evolution, 344, 345
rates, 346-351 (*see also* accretion rates)
Monte Carlo techniques, 8, 21, 22, 24

neutrino flux, 28
neutron stars
blackbody radii, 65, 66
burst sources, 51, 66, 97
core temperature, 91
equations of state (*see* equations of state)
formation from degenerate dwarfs, 105
magnetic properties: accreting star, Chap. 6; axes, 276, 277; dipole radiation, 263, 393, 421; field, 46, 91, 98, 103, 105, 265, 266, 268, 273, 276, 279; moment, 265, 412, 413
mass determinations, 25-8
period, initial, 419
photosphere, 66, 90, 91, 97
polar cap, 6, 7, 91, 103, 105, 217, 265, 281
rotational history in binaries, 419-27

spinup-spindown (*see* spinup and spindown)
spinup timescales, 416
surface layers, 89, 90, 92, 95, 98
thermonuclear flash model, 88-100
novae, 147, 148 (*see also* cataclysmic variables)
burst energy vs. waiting time to next burst, 150
classical, 176, 338
classification, 148
dwarf, 178
nova-like objects, 149
recurrent novae, 177
thermonuclear runaway, 176
X-ray luminosity, 154
NRAO, 83
nuclear burning and flashes (*see also* thermonuclear flash)
carbon burning, 89, 90
helium burning: core, 313, 317, 332, 362; fast transients, 102, 103; flahes, 90, 95 (*see also* thermonuclear flashes)
hydrogen: burning, 89-91, 95, 97, 102, 167, 168, 280, 310, 314, 326-329, 331, 334, 353, 354; classical novae, 176, 177; flashes, 102, 176, 177, 281
shells, 89, 90

Ooty radio telescope, 84
optical bursts (*see* burst sources)
optical emission lines (*see* emission lines)
optical light curves, 207-20, 244-9
optical identifications (*see* low-mass X-ray binaries and massive X-ray binaries)
optical pulsations (*see* pulsations)
optical reprocessing, 24, 25, 44, 68, 69, 104, 203-5, 222, 223, 228, 231
orbital periods (*see* cataclysmic variables, low-mass X-ray binaries and massive X-ray binaries)
OSO-8, 50

photodisintegration, 312, 314-16, 318
populations, 42, 103, 191, 306, 307, 338
p-p chain, 90
precursors, 85-8, 102
pulsations
optical, 44, 68, 69, 222-5, 248, 249 (*see also* optical reprocessing)
radio, 261, 262, 306, 307, 330-4, 393, 407, 419-26
X-ray (*see* X-ray pulsars)

quark star, 28, 34

Subject index

radial velocity curves, 152, 153, 208, 209, 211, 215, 240-4
radiation pressure, 118, 326, 331 (*see also* Eddington limit)
radio emission, 307, 330 (*see also* low-mass X-ray binaries, massive X-ray binaries and SS 433)
radio pulsars, 306, 419-21
Rapid Burster (MXB 1730-335) (*see also* X-ray burst sources, type II bursts)
  anomalous bursts, 77-9
  blackbody radii, 82
  burst energy vs. waiting time to next burst, 77, 78
  infrared bursts, 53, 82
  observations, complete record, 81
  persistent emission, 82
  radio bursts, 53, 82
  recurrent transient behavior, 79, 80
  type I bursts, 72, 74, 75
recurrent novae, 177, 178 (*see also* cataclysmic variables)
risetimes
  anamolous bursts, 77
  radio bursts, 83
  thermonuclear flash model, 90
  type I X-ray bursts, 57, 90
  type II X-ray bursts, 72
Roche lobe, Chaps. 9 and 10
  critical radius, 345, 346, 352, 356, 357
  filling, 234, 235
  mass transfer rate, 346-8
  overflow: from a main sequence dwarf, 356-8; from a low-mass degenerate contact star, 359-61; from a non-degenerate He star, 361-2; from a red giant, 362-4

SAS-3, 3, 15, 31, 32, 47, 49, 50, 52, 56, 73, 75, 76, 77, 81, 82, 84, 85, 118
shielding (*see* X-ray shielding)
spinup and spindown, 8-12, 261, 262-4, 350, 355, 356, Chap. 11
spiral-in (*see* evolution)
SS 433, Chap. 7, 306, 308, 330, 331, 382
  jets, 295-7
  kinematic model, 291-3
  observations, radio and X-ray, 294-7
  photometry and polarimetry, 293
  spectroscopy, 287-91
standard candles, 51, 64-6, 194, 200, 201
stellar wind, 12, 13, 104, 189, 192, 234, 235, 244, 326, 331, 344, 349, 351, 353-5, 368, 396-9, 403, 408, 416, 417, 419-23
  accretion, 416, 417

supernova explosion, 303, 306, 310, 312, 314, 315, 317, 318, 322, 326, 328, 330-3, 420
synchronisation (*see* tidal forces and tidal instabilities)

thermonuclear flash (*see also* nuclear burning and flashes)
  classical novae, 176, 177
  gamma-ray bursts, 280, 281
  general relativistic corrections (*see* general relativistic corrections)
  helium flashes, 90, 95
  hydrogen flashes, 102, 176, 177, 281
  magnetic field, 91, 103
  models for X-ray bursts, 51, 52, 62-4, 88-106, 117
  radiative and conductive processes, 91, 103
  suppression of, 46
  surface layers, 51
  theory versus X-ray burst observations, 101, 102
  thin shell instability, 88-91, 102
  transients, 88, 102
  type I X-ray bursts, 51, 52, 65, 88-102
Thomson scattering (*see* Eddington limit)
tidal forces, 21, 48, 348-50, 370, 386, 420
tidal instabilities, 328, 329
time scales
  evolutionary, 414-16, 423
  gravitational radiation evolution, 357, 358, 361
  spinup, 413-16
  thermal relaxation, 347, 351, 352, 357, 359, 361-3
  X-ray life time, 353, 354
transients
  gamma-ray, 34
  X-ray (*see* X-ray transients)
type I sources (*see* massive X-ray binaries)
type II sources (*see* low-mass X-ray binaries)
type I X-ray bursts (*see* X-ray bursts)
type II X-ray bursts (*see* X-ray bursts)

Uhuru, 13, 32, 81, 118, 143
UK-6, 32

viscosity
  forces, 376, 379, 380, 404
  law, 379, 383
  parameter, 382

white dwarfs, Chap. 4, 11, 88, 104, 105, 192, 264, 315, 316, 330-2, 335, 336, 338, 372, 406 (*see also* cataclysmic variables)

Subject index

Wolf-Rayet stars (*see* evolution)

X-ray burst sources, Chap. 2 (*see also* low-mass X-ray binaries)
Ages, 103
anamolous bursts, 77
brief history, highlights, 50–3
burst classification, 52, 72, 102
eclipses, lack of, 46, 104, 105, 190
Eddington limit, 64, 65
fast transients, 84–8, 102
galactic distribution, 53, 56
infrared emission, 53, 72, 82
luminosity: alpha values, 63; burst maximum, 200; persistent, 63, 102, 192, 193; standard candles, 51, 64–6, 194, 200, 201
magnetic fields, 45, 46, 103
optical identifications, 43, 66, 198, 199
optical properties, 53, 192, 195–201
orbital periods, 67, 68 (*see also* low-mass X-ray binaries)
precursors, 85–8, 102
pulsations, lack of, 42, 45, 190
radio emission (*see* type I bursts and type II bursts, below)
Rapid Burster (*see* Rapid Burster and type II bursts, below)
risetimes (*see* risetimes)
Roche lobe overflow (*see* Roche lobe)
spectra (*see* type I bursts and type II burst, below)
standard candles, 51, 64–6, 194, 200, 201
theory, 88–106
type I bursts: alpha values, 63; double peaked, 57, 61; galactic distribution, 53; intervals, 62, 63; optical bursts, 44, 52, 53, 66, 67, 69–71, 220–2, 305; optical properties, 66–71; persistent emission, periodicities, 45; profiles, 57–62; radio emission, 72; spectra, 64–6; theory versus experiment, 101, 102; thermonuclear flash model, 88–99 (*see also* thermonuclear flash); transients, 84–8
type II bursts (*see also* Rapid Burster): accretion instability models, 100; blackbody radii, 82; burst energy vs. waiting time to next burst, 77; double peaked, 77; interburst emission, 80, 82; light curve, 79; other sources, 84; patterns (modes), 77; profiles, 72–77; quasi-periods, 72; radio emission, 53, 82–4; recurrence intervals, 77, 102; saturation, 72–7; spectra, 82

X-ray eclipses, 17, 20, 46, 131, 189, 224, 237, 246, 249, 304, 419
absence in galactic bulge sources, 41, 46, 104, 105, 190
duration, 17, 20, 224, 237
X-ray heating, 44, 66, 68, 71, 217, 221, 243, 246, 247, 249
X-ray pulsars, Chap. 1, 103, 107, 171, 189, 261–5, 268, 306, 393, 397, 406, 412, 416–26
characteristics, 4, 6, 394, 395
companion star spectrum, 232, 233, 395
cyclotron lines (*see* cyclotron)
low-mass X-ray binaries, 68, 190, 224, 306, 333
luminosity, X-ray, 395
massive X-ray binaries, Chap. 1, 189
optical companions, 232, 233, 394 (*see also* companion stars)
pulse: period, 3, 4, 393, 395 – changes, 8–13, 262–5, 394, 417–27; histories, 10, 396; range, 394, 419–27 – phases, 3, 4, 270, 277; spectra, 2–8, 265–8
radio pulsars, difference with, 393
runaway pulsars, 333
spectra, 2–8, 265–8
spin-up time scales, 416
theory, 268–78, 419–26
timing measurements, 262–5
white dwarfs, 264
X-ray shielding, 46, 105, 190, 206, 214, 217, 243
X-ray sources
absorption dips (periodic), 104
Be stars, 234, 236, 304, 305
bright galactic, 343
bursters (*see* X-ray burst sources)
class I (*see* massive X-ray binaries)
class II (*see* low-mass X-ray binaries)
extra-galactic: M 31, 121–32, 333, 336–8; M 33, 132; M 81, 135; M 100, 135; M 101, 132–5; Magellanic Clouds, 119–21, 144, 247; N 253, 135; NGC, 247, 135; non-nuclear spiral galaxies, 144
galactic: bright, 343; mass, 143; optical observations, Chap. 5
galactic bulge (*see* galactic bulge sources)
globular clusters (*see* globular clusters)
low mass X-ray binaries (*see* low-mass X-ray binaries)
luminosity (*see* luminosity)
massive X-ray binaries (*see* massive X-ray binaries)
number of sources and galactic mass, 143

optical identifications, 43, 66, 198, 199, 232, 233
soft X-ray transients (*see* X-ray transients)
transients (*see* X-ray transients)
type I (*see* massive X-ray binaries)
type II (*see* low-mass X-ray binaries)
X-ray spectra (*see* cataclysmic variables, low-mass X-ray binaries, massive X-ray binaries and X-ray pulsars)

X-ray transients, 56, 417, 426
fast, 84–8, 102, 103, 107
hard, 234; recurrent, 236
soft, 194, 224–8: accretion flow, 220; low-mass X-ray binaries, 56, 67–9, 190, 194, 201
optical: observations, 67–9, 225–8; variability, 219, 220